Clinical Anatomy
A Case Study Approach

Clinical Anatomy
A Case Study Approach

AUTHORS

Mark H. Hankin, PhD
Professor
Department of Biomedical Sciences
Director of Anatomy Programs
Oakland University
William Beaumont School of Medicine
Rochester, Michigan

Dennis E. Morse, PhD
Emeritus Professor of Anatomy
Department of Neurosciences
The University of Toledo
College of Medicine and Life Sciences
Toledo, Ohio

Carol A. Bennett-Clarke, PhD
Professor of Anatomy
Department of Neurosciences
Associate Dean for Pre-Clinical Medical Education
The University of Toledo
College of Medicine and Life Sciences
Toledo, Ohio

Mc Graw Hill Education | Medical

New York   Chicago   San Francisco   Lisbon   London   Madrid   Mexico City
Milan   New Delhi   San Juan   Seoul   Singapore   Sydney   Toronto

Clinical Anatomy: A Case Study Approach

1 2 3 4 5 6 7 8 9 0   CTP/CTP   18 17 16 15 14 13

ISBN 978-0-07-162842-6
MHID 0-07-162842-8

This book was set in Times LT Std by Aptara, Inc.
The editors were Michael Weitz and Karen G. Edmonson.
The production supervisor was Sherri Souffrance.
Project management was provided by Indu Jawwad, by Aptara, Inc.
The cover designer was Brad Foltz Design.
China Translation and Printing Services, Ltd. was printer and binder.

This book is printed on acid-free paper.

Library of Congress Cataloging-in-Publication Data

Clinical anatomy : a case study approach / editors, Mark H. Hankin,
Dennis E. Morse, Carol A. Bennett-Clarke.
     p. ; cm.
   ISBN 978-0-07-162842-6 (alk. paper) – ISBN 0-07-162842-8 (alk. paper)
   I. Hankin, Mark Harris. II. Morse, Dennis E. III. Bennett-Clarke, Carol A.
   [DNLM: 1. Anatomy–Case Reports. 2. Anatomy–Examination Questions.
3. Outlines–Case Reports. 4. Outlines–Examination Questions. QS 18.2]

   616.07–dc23

                                                        2012037643

McGraw-Hill books are available at special quantity discounts to use as premiums
and sales promotions, or for use in corporate training programs. To contact a
representative, please e-mail us at bulksales@mcgraw-hill.com

Contents

Contributors

AUTHORS

Mark H. Hankin, PhD

Professor
Department of Biomedical Sciences
Director of Anatomy Programs
Oakland University
William Beaumont School of Medicine
Rochester, Michigan

Dennis E. Morse, PhD

Emeritus Professor of Anatomy
Department of Neurosciences
The University of Toledo
College of Medicine and Life Sciences
Toledo, Ohio

Carol A. Bennett-Clarke, PhD

Professor of Anatomy
Department of Neurosciences
Associate Dean for Pre-Clinical Medical Education
The University of Toledo
College of Medicine and Life Sciences
Toledo, Ohio

ILLUSTRATORS

Tonya L. Floyd-Bradstock, MFA

Medical Illustrator
Center for Creative Instruction
The University of Toledo
Toledo, Ohio

Roy E. Schneider, MFA

Manager, Medical Illustration
Center for Creative Instruction
The University of Toledo
Toledo, Ohio

Joshua M. Klein

Intern
Center for Creative Instruction
The University of Toledo
Toledo, Ohio

Preface

Traditional medical curricula have considered preclinical and clinical undergraduate education as pedagogically distinct. Historically, instruction in the anatomical sciences followed this model. However, changes in licensure examinations and accreditation standards have necessitated the adoption of different pedagogies for human anatomy courses at many institutions. In many contemporary courses, students apply their knowledge of anatomy to clinical problems at the earliest stages of their training.

Anatomy: A Clinical Case Approach is written to promote the integration of basic anatomy with clinical findings in the context of specific patient conditions. **Chapter 1** presents an overview of the nervous system. **Chapters 2–10** use the following format to present clinical cases relevant to each body region:

- **Patient Presentation.** This section outlines the presenting complaint(s) as might be described by a patient.

- **Relevant Clinical Findings.** A standard sequence for clinical evaluation is presented: *History, Physical Examination, Laboratory Tests, Imaging Studies,* and *Procedures*. It is understood that vital signs are collected and assessed for every patient, but only those that have significance for the particular case are noted. Reference values for laboratory tests are derived primarily from Harrison's Online (http://www.accessmedicine.com).

- **Clinical Problems to Consider.** The relevant clinical findings are used to generate a list of diseases or conditions that correlate with the anatomical region involved in the case. The list is not intended for use in making a differential diagnosis.

- **Relevant Anatomy.** The essential anatomy is reviewed for each disease or condition presented in the list of *Clinical Problems to Consider*. This section assumes students have knowledge of basic anatomical information and concepts.

- **Clinical Reasoning.** The *Clinical Problems* for each case are defined and *Signs and Symptoms*, and *Predisposing Factors* outlined.

- **Diagnosis.** A likely diagnosis is presented and the basis for the signs and symptoms associated with each *Clinical Problem* is discussed.

Review Questions. USMLE-type (clinical-vignette) questions that emphasize important aspects of regional anatomy are presented at the end of each chapter. Explanations for the correct answers are given in an appendix.

Anatomical terminology established by the Federative Committee on Anatomical Terminology (*Terminologica Anatomica*, 1998) is used throughout. *Anatomy: A Clinical Case Approach* assumes that students acquire **anatomical terminology** during a basic anatomy course. **Clinical terms**, including eponyms, familiarize students with terminology used commonly in medical practice.

Acknowledgments

Anatomy: A Clinical Case Approach would not have been written without the inspiration provided by decades of interaction with our students and fellow anatomists. We are also grateful for the suggestions and feedback provided by colleagues and friends, including Joshua Barden, MD, Carol Cheney, MD, James Kleshinski, MD, Carl Sievert, PhD, and Martin Skie, MD.

We are grateful to Michael Weitz, Acquisitions Editor for McGraw-Hill Professional Publishing, for his confidence in our abilities, and his patience, flexibility, and encouragement at every stage of development of this book. Karen G. Edmonson, Managing Editor, Development, Editing, and Production, Medical Publications Division, was invaluable for guidance in formatting and organizing the book. Her knowledgeable and timely responses to our queries kept the process moving. Her skills in layout for the text and illustrations provided the capstone for our efforts. The final stage of the book's journey was overseen by Indu Jawwad (Aptara, Inc.); her patience with final corrections in proof was remarkable.

Special thanks are due Sherry Andrews, Director of the Center for Creative Instruction at The University of Toledo. Her skill in keeping the authors and illustrators synchronized and her willingness to allocate resources when needed are greatly appreciated.

Success at all levels depends on a solid personal support system. Sharyl, Jayne, and David, our spouses, listened, counseled, and encouraged us at every stage as we rejoiced, complained, and worked extended days, evenings and weekends.

CHAPTER 1

Nervous System

FUNCTIONAL CLASSIFICATION OF NEURONS

The **neuron**, or nerve cell, is the basic functional unit of the nervous system. A neuron includes its **cell body** and **processes** (axons and dendrites). Long neuronal processes are frequently referred to as **fibers**. Neurons are generally classified as either afferent or efferent:

- **Afferent**, or sensory, neurons receive input from peripheral structures and transmit it to the spinal cord and/or brain.
- **Efferent**, or motor, neurons transmit impulses from the brain and/or spinal cord to effectors (skeletal muscle, cardiac muscle, smooth muscle, glands) throughout the body.

ANATOMICAL DIVISIONS OF THE NERVOUS SYSTEM

The nervous system has two anatomical divisions:

1. The **central nervous system (CNS)** includes the brain and spinal cord.
2. The **peripheral nervous system (PNS)** consists of spinal nerves, their roots, and branches; cranial nerves (CN) and their branches; and components of the autonomic nervous system (ANS).

Collections of nerve cell bodies in the CNS form **nuclei**, whereas those in the PNS form **ganglia**. Ganglia and nuclei contain either motor or sensory neurons. Bundles of axons in the CNS are called **tracts**. Similar neuronal processes collected in the PNS form **nerves**. Nerves are categorized based on their CNS origin:

- **Spinal nerves** are attached to the spinal cord. They transmit both motor and sensory impulses and are, thus, considered **mixed nerves**.
- Most **CN** are attached to the brain. Some CN are either motor or sensory only, while others are mixed.

Spinal Nerves

The spinal cord is composed of segments, as indicated by the 31 pairs of spinal nerves. Each segment has numerous **dorsal** (posterior) and **ventral** (anterior) **rootlets** that arise from the respective surfaces of the spinal cord (**Fig. 1.1**). Dorsal rootlets contain neuronal processes that conduct afferent impulses to the spinal cord, whereas the ventral rootlets conduct efferent impulses from the spinal cord. Respective rootlets from each segment unite to form **dorsal** and **ventral roots**:

- The **dorsal root** contains the **central processes** of sensory neuronal cell bodies that are located in the **dorsal root ganglion (DRG)**. The DRG is also called a spinal ganglion. The **peripheral processes** of these neurons are located in the spinal nerve, its rami, and their branches. These processes end at or form receptors.
- The **ventral root** contains motor fibers. Their neuronal cell bodies are found in the gray matter of the spinal cord: **ventral horn** if the axons innervate skeletal muscle; **lateral horn** if the axons supply smooth muscle, cardiac muscle, or glands.

The dorsal and ventral roots join to form a short, mixed **spinal nerve**. Almost immediately after its formation, the spinal nerve divides into mixed **dorsal** and **ventral rami**:

- **Dorsal rami** supply intrinsic (deep) muscles of the back and neck, joints of the vertebral column, and skin on the dorsal surface of the trunk, neck, and head.
- **Ventral rami** innervate all other muscles of the neck and trunk (including the diaphragm), skin of the anterior and lateral body walls, and all muscles, and skin of the limbs. In general, ventral rami are larger than dorsal rami.

The 31 pairs of spinal nerves arise from the five regions of the spinal cord (**Fig. 1.2**). Most of the spinal nerves (C2-L5) are formed in an intervertebral foramen. The C1 spinal nerve emerges between the skull and the first cervical vertebra. Sacral spinal

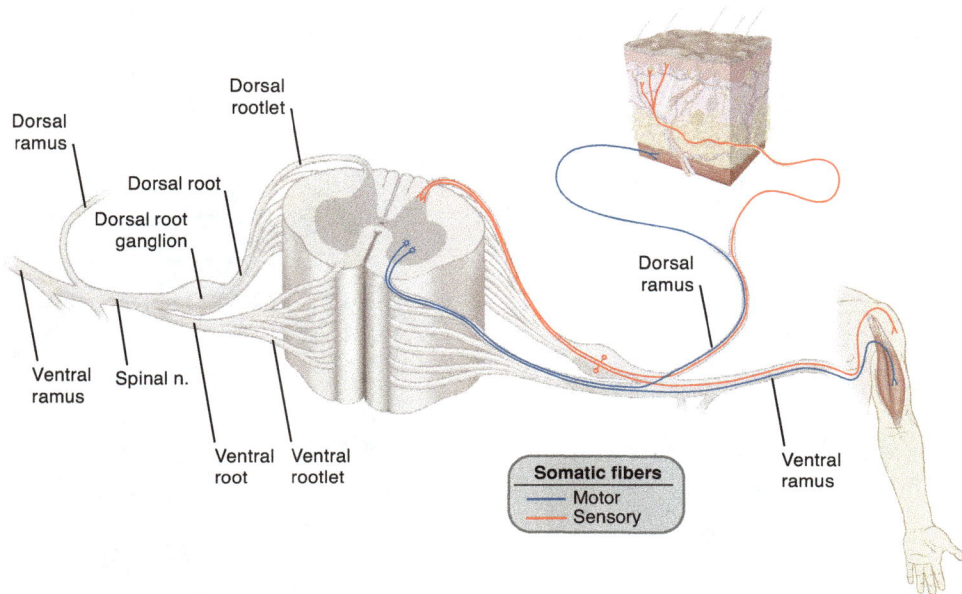

FIGURE 1.1 Somatic components of a spinal nerve.

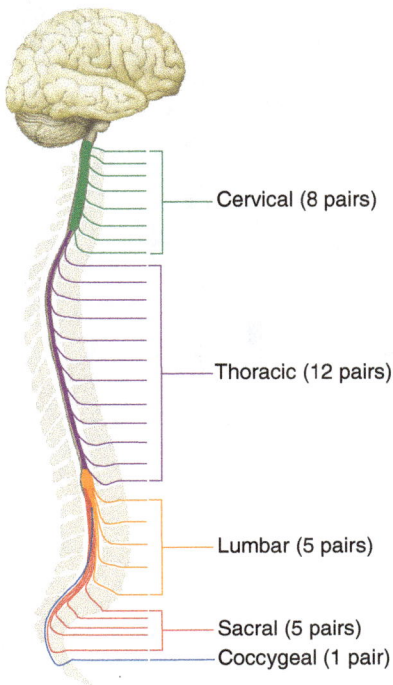

FIGURE 1.2 Regional organization of the spinal cord and spinal nerves.

nerves are formed in the vertebral canal and their rami exit through respective dorsal and ventral sacral foramina. The Co1 spinal nerve exits inferior to the rudimentary transverse process of the first segment of the coccyx.

Dermatomes

A **dermatome** is the region of skin innervated by a single-spinal nerve (**Fig. 1.3**). All spinal nerves, with the exception of C1, transmit sensory information from the skin. Dermatomes on the trunk, the neck, and the posterior head form consecutive bands that vary in width. Dermatomes on the limbs have more complex shapes and arrangements. Most of the skin of the face and scalp is supplied by the trigeminal nerve (CN V).

Spinal Nerve Plexuses

A **nerve plexus** is a network of nerves. The four **spinal nerve plexuses** are formed from the ventral rami of spinal nerves: cervical, brachial, lumbar, and

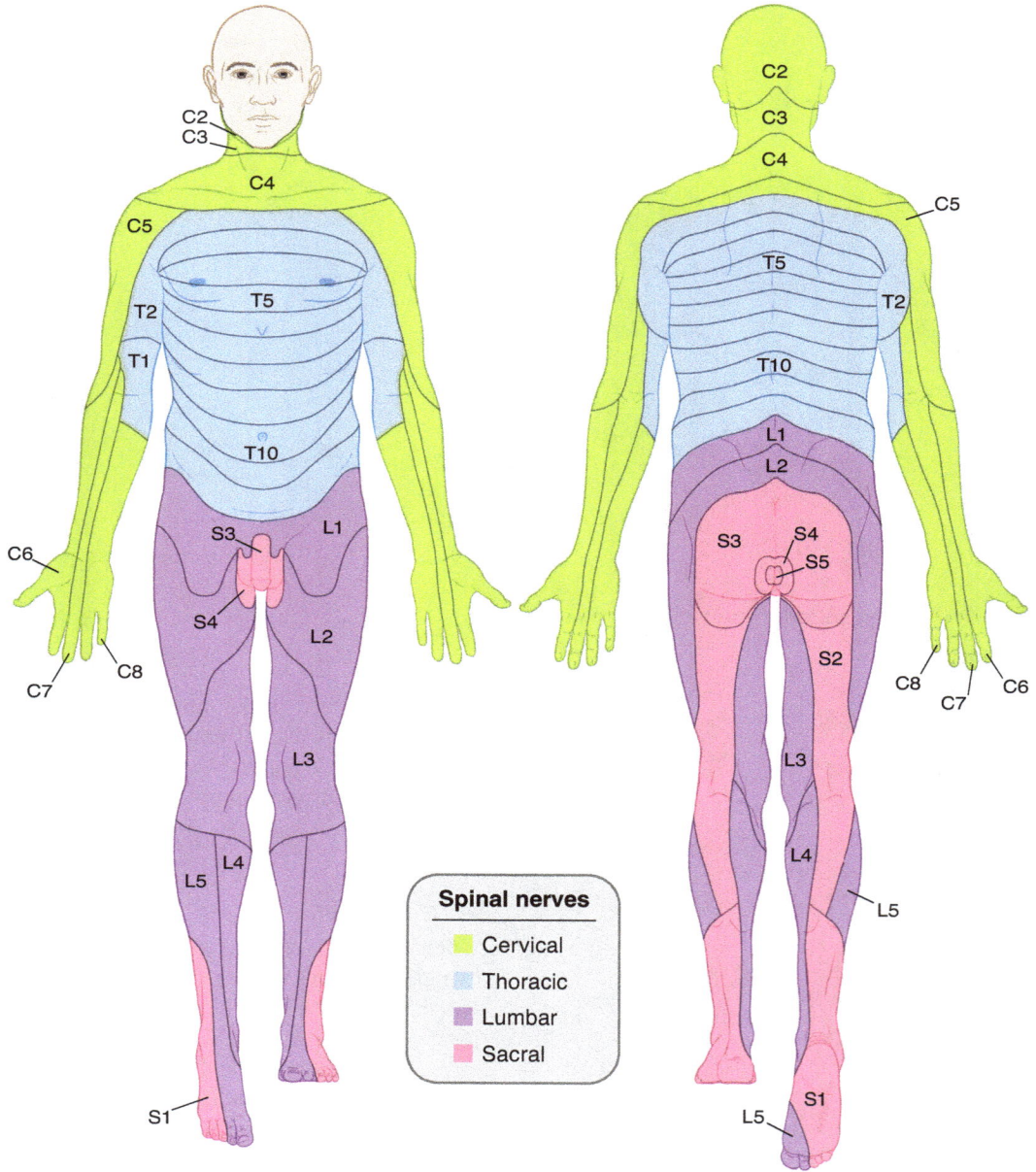

FIGURE 1.3 Dermatomes.

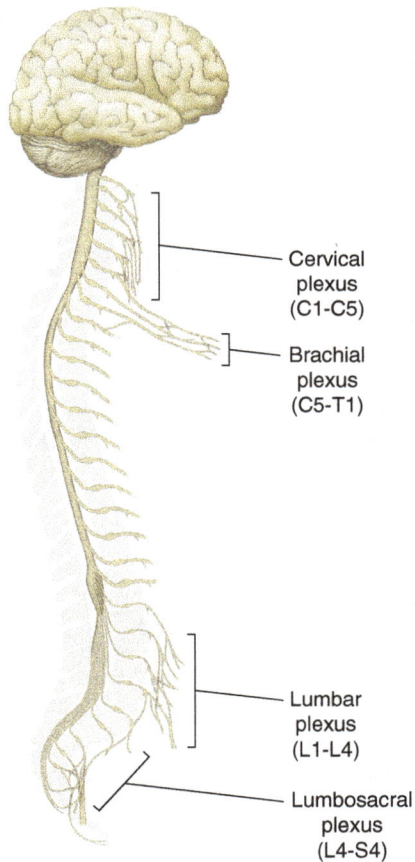

FIGURE 1.4 Spinal nerve plexuses.

Cervical
plexus
(C1-C5)

Brachial
plexus
(C5-T1)

Lumbar
plexus
(L1-L4)

Lumbosacral
plexus
(L4-S4)

lumbosacral (**Fig. 1.4**). Each plexus has a characteristic sensory and motor distribution (**Table 1.1**).

Cranial Nerves

There are 12 pairs of **cranial nerves (CN I-CN XII)**. The spinal part of CN XI (accessory) takes origin from upper cervical spinal cord, enters the cranial cavity through foramen magnum and, in the jugular foramen, joins a cranial root from the brain stem. CN II-XII are attached to the brain stem. CN I (olfactory) is located in the nasal cavity and anterior cranial fossa. Cranial nerves are sensory, motor, or mixed (**Table 1.2**). They supply structures in the head and neck; CN X (vagus) also supplies structures in the thorax and abdomen.

FUNCTIONAL DIVISIONS OF THE NERVOUS SYSTEM

Functionally, the nervous system can be divided into somatic and visceral parts.

- The **somatic** portion innervates structures derived from somites of the embryo (skin and skeletal muscles).
- The **visceral** portion is commonly known as the **ANS**.

Both functional divisions convey impulses through afferent and efferent fibers (**Table 1.3**).

TABLE 1.1	Spinal nerve plexuses and their distribution.		
Plexus	**Spinal levels**	**Sensory distribution**	**Motor distribution**
Cervical	C1-C4	▪ Skin of anterior and lateral neck, and shoulder	▪ Anterior neck muscles ▪ Thoracic diaphragm
Brachial	C5-T1	▪ Skin of upper limb	▪ Upper limb muscles
Lumbar	L1-L4	▪ Skin of anterior and inferior abdominal wall ▪ Anterior scrotum/labium majus ▪ Anterior and medial thigh ▪ Medial leg and foot	▪ Muscles of anterior and inferior abdominal wall, including cremaster ▪ Anterior and medial thigh
Lumbosacral	L4-S4	▪ Skin on posterior lower limb ▪ Anterior and lateral leg, and foot	▪ Muscles of pelvic floor ▪ Gluteal region ▪ Posterior thigh ▪ Leg and foot

TABLE 1.2 Cranial nerves and their distribution.

CN	Name	Sensory	Motor	Parasympathetic
I	Olfactory	Olfaction		
II	Optic	Vision		
III	Oculomotor		▪ Superior rectus ▪ Inferior rectus ▪ Medial rectus ▪ Inferior oblique ▪ Levator palpebrae superioris	Sphincter pupillae and ciliary muscles
IV	Trochlear		Superior oblique	
V	Trigeminal	▪ Skin of anterior scalp and face ▪ Mucous membranes of nasal and oral cavities ▪ Teeth ▪ Anterior ⅔ of tongue ▪ Anterior auricle	▪ Muscles of mastication ▪ Anterior belly of digastric ▪ Mylohyoid ▪ Tensor tympani ▪ Tensor veli palatini	
VI	Abducens		Lateral rectus	
VII	Facial	Taste: anterior ⅔ of tongue	▪ Muscles of facial expression ▪ Posterior belly of digastric ▪ Stylohyoid ▪ Stapedius	▪ Lacrimal glands ▪ Mucous glands of nasal and oral cavities ▪ Sublingual and submandibular salivary glands
VIII	Vestibulocochlear	Hearing and balance		
IX	Glossopharyngeal	▪ Taste: posterior ⅓ of tongue ▪ Posterior ⅓ of tongue ▪ Mucous membranes of pharynx and middle ear ▪ Carotid body and sinus	Stylopharyngeus	Parotid glands
X	Vagus	▪ External acoustic meatus and tympanic membrane ▪ Mucous membranes of larynx, thoracic, and abdominal organs through the transverse colon	▪ Muscles of soft palate (except tensor veli palatini) ▪ Muscles of pharynx (except stylopharyngeus) ▪ Muscles of larynx	▪ Mucous glands of pharynx and larynx ▪ Smooth muscle of cervical, thoracic, and abdominal viscera through transverse colon
XI	Accessory		▪ Trapezius ▪ Sternocleidomastoid	
XII	Hypoglossal		Muscles of tongue (except palatoglossus)	

TABLE 1.3	Distribution of functional divisions of the nervous system.	
Division	**Afferent**	**Efferent**
Somatic	Sensory information primarily from receptors in skin	Skeletal muscles
Visceral	Sensory information from receptors in body cavity organs	Smooth muscle, cardiac muscle, and glands

Autonomic Nervous System

The **ANS** has **sympathetic, parasympathetic**, and **enteric** divisions. The **enteric division** is a network of neurons located within the wall of the gastrointestinal tract. This part of the ANS can function autonomously, although it interacts with the sympathetic and parasympathetic divisions. Somatic efferent impulses utilize a single neuron to transmit information from the CNS to skeletal muscle (**Fig. 1.5A**). In contrast, *sympathetic and parasympathetic efferent information is transmitted through two neurons* (**Fig. 1.5B**):

1. A **preganglionic neuron,** whose cell body is located in the CNS, is the first neuron in this pathway.

2. A **postganglionic neuron,** whose soma is located in an autonomic ganglion, is the second neuron in the pathway.

Preganglionic axons synapse on postganglionic cell bodies in autonomic ganglia. Axons of postganglionic neurons exit these ganglia for distribution to smooth muscle, cardiac muscle, or glands. Anatomically, the **sympathetic** and **parasympathetic** divisions are distinguished both by the location of their pre- and postganglionic cell bodies (**Table 1.4**) and by the nerves that transmit impulses to the effectors.

Sympathetic Division

All sympathetic preganglionic cell bodies are located in the **lateral horn of the spinal cord at levels T1-L2** (Table 1.4). **All preganglionic axons** exit the spinal cord through ventral roots at these levels, enter respective spinal nerves, and follow the ventral rami. These axons branch from the ventral ramus and enter a **white ramus communicans**. This short nerve carries the

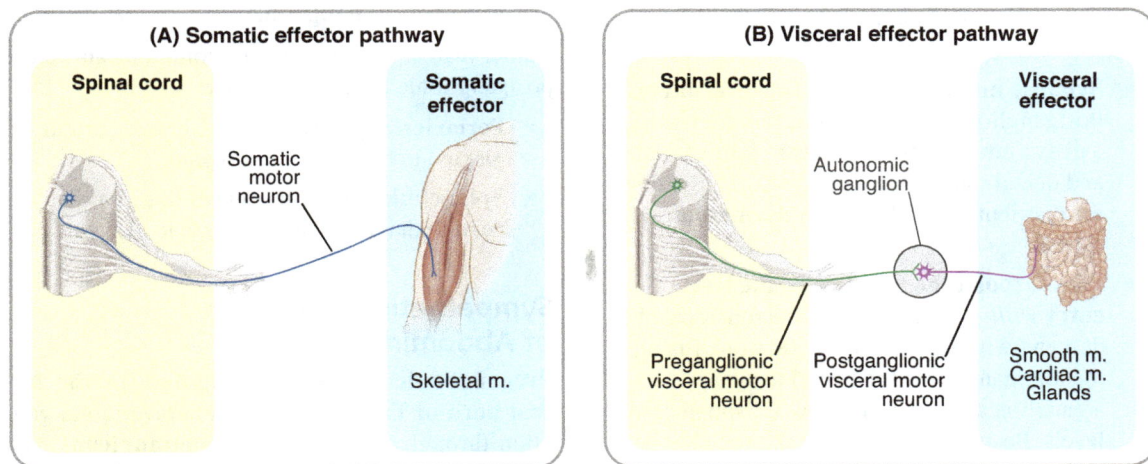

(A) Somatic effector pathway

Spinal cord   Somatic effector

Somatic motor neuron

Skeletal m.

(B) Visceral effector pathway

Spinal cord   Visceral effector

Autonomic ganglion

Preganglionic visceral motor neuron   Postganglionic visceral motor neuron   Smooth m. Cardiac m. Glands

FIGURE 1.5 Somatic and visceral efferent (motor) pathways.

TABLE 1.4	Location of sympathetic and parasympathetic nerve cell bodies.	
ANS division	**Preganglionic cell body**	**Postganglionic cell body**
Sympathetic	▪ Spinal cord (T1-L2)—lateral horn	▪ Paravertebral ganglia ▪ Prevertebral ganglia
Parasympathetic	▪ Brain stem nuclei of CN III, VII, IX, X ▪ Spinal cord (S2-S4)—lateral horn	▪ Ciliary ganglion ▪ Pterygopalatine ganglion ▪ Submandibular ganglion ▪ Otic ganglion ▪ Terminal ganglia

preganglionic fibers to a **paravertebral ganglion**. There are **22 pairs of paravertebral ganglia** adjacent to the vertebral column from C1 to the coccyx: 3 cervical, 11 thoracic, 4 lumbar, 3 sacral, and 1 coccygeal (unpaired). These ganglia contain postganglionic sympathetic cell bodies. Together with interganglionic segments, these ganglia form the **sympathetic trunk (sympathetic "chain")**. *Once in the sympathetic trunk, the pathway of a preganglionic axon depends on the effector (skin, regional viscera, adrenal medulla) that will ultimately be innervated by a postganglionic axon.*

Sympathetic Innervation of Peripheral Effectors in Skin (Sweat Glands and Arrector Pili)

Preganglionic axons that enter a paravertebral ganglion through a white ramus communicans can either (**Fig. 1.6A**):

- **Synapse in the ganglion at the level of entry.** Postganglionic axons exit the ganglion through a **gray ramus communicans** to join ventral and dorsal rami at the same level. These fibers are distributed to effectors in **dermatomes T1-L2**.
- **Pass through the ganglion at the level of entry** *without synapse.* These axons ascend or descend within the sympathetic trunk (using the interganglionic segments) to synapse in a ganglion above T1 or below L2 spinal cord levels. Postganglionic axons exit the ganglion at the level of synapse via a **gray ramus communicans** to join branches of spinal nerves that supply the skin of the C2-C8 and L3-Co1 dermatomes.

White rami communicantes are associated with only T1-L2 spinal nerves. In contrast, **gray rami communicantes** are associated with every spinal nerve.

Sympathetic Innervation of Cranial, Cervical, and Thoracic Viscera

Preganglionic axons that originate in the lateral horn of **T1-T5** and enter a paravertebral ganglion through a **white ramus communicans** can either (**Fig. 1.6B**):

- **Synapse in the ganglion at the level of entry.**
- **Ascend in the sympathetic chain to synapse in a cervical ganglion.**

Rather than enter a gray ramus communicans, these postganglionic fibers form either:

- **Periarterial plexuses** (e.g., internal carotid plexus) to head and neck organs.
- **Splanchnic (visceral) nerves** (e.g., cardiac nerves) that distribute to thoracic viscera.

Sympathetic Innervation of Abdominal Viscera

Preganglionic axons that originate in the lateral horn of **T5-L2** and enter a paravertebral ganglion through a **white ramus communicans** pass through the ganglion *without synapse* (**Fig. 1.6C**).

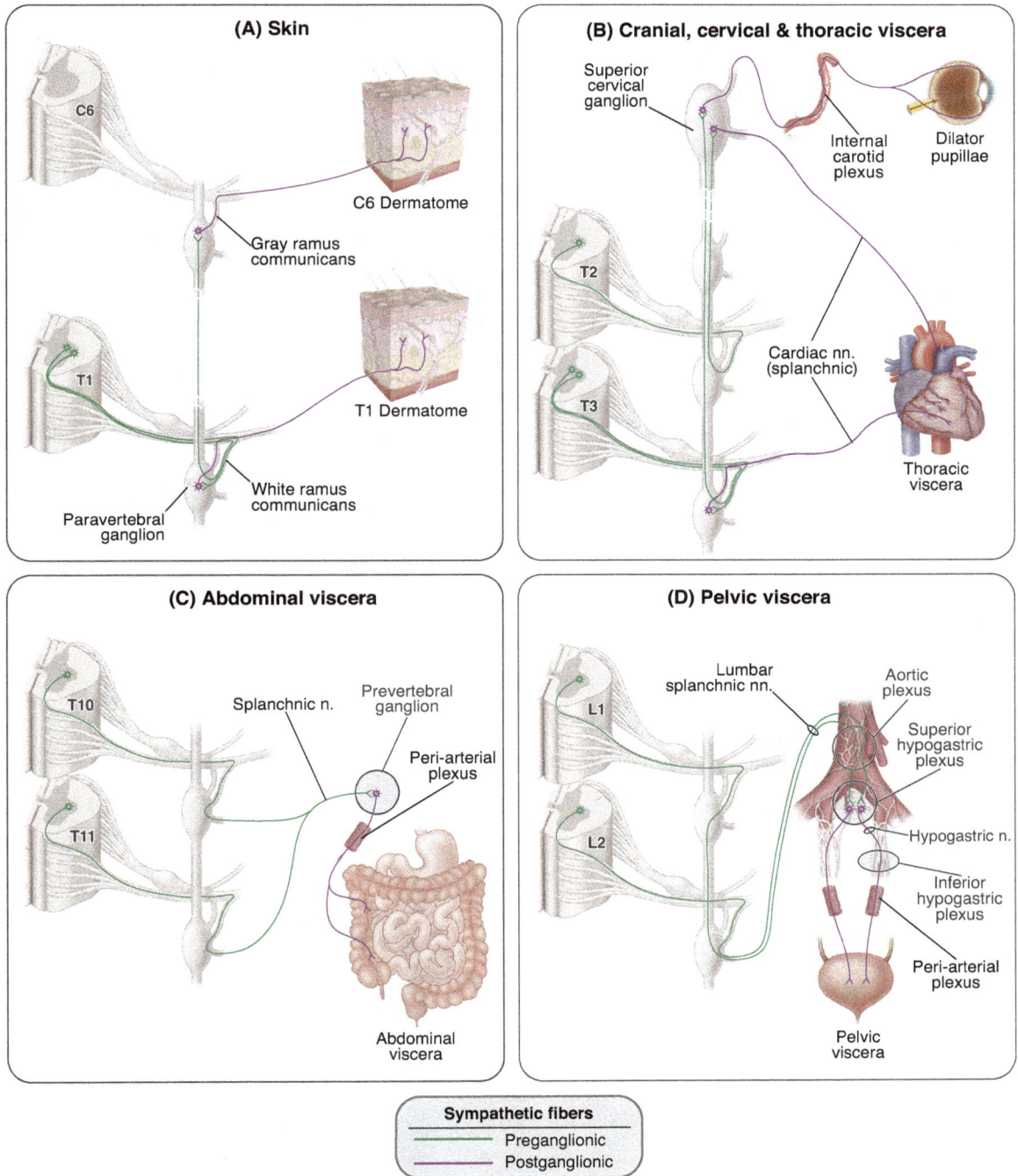

(A) Skin

C6

C6 Dermatome

Gray ramus communicans

T1

T1 Dermatome

White ramus communicans

Paravertebral ganglion

(B) Cranial, cervical & thoracic viscera

Superior cervical ganglion

Internal carotid plexus

Dilator pupillae

T2

T3

Cardiac nn. (splanchnic)

Thoracic viscera

(C) Abdominal viscera

T10

Splanchnic n.

Prevertebral ganglion

Peri-arterial plexus

T11

Abdominal viscera

(D) Pelvic viscera

Lumbar splanchnic nn.

Aortic plexus

Superior hypogastric plexus

L1

Hypogastric n.

L2

Inferior hypogastric plexus

Peri-arterial plexus

Pelvic viscera

Sympathetic fibers	
———	Preganglionic
———	Postganglionic

FIGURE 1.6 Sympathetic pathways.

These axons exit the sympathetic chain and form **splanchnic nerves** (e.g., **greater, lesser, and least thoracic,** or **lumbar**) that end in **prevertebral ganglia**. There are four collections of **prevertebral ganglia**, which are located adjacent to main branches of the abdominal aorta:

1. Celiac
2. Superior mesenteric
3. Aorticorenal
4. Inferior mesenteric

After synapsing in a prevertebral ganglion, postganglionic fibers follow the branches of abdominal aorta to abdominal viscera, pelvic viscera, and the testes (**Table 1.5**).

Adrenal Medulla

The adrenal (suprarenal) medulla is unique in that it "acts" as the ganglion. The **chromaffin** cells of the adrenal medulla are equivalent to postganglionic sympathetic neurons. Upon stimulation by preganglionic sympathetic axons, they release neurotransmitters (epinephrine and norepinephrine)

TABLE 1.5	Sympathetic innervation of abdominopelvic viscera and testes.			
	Preganglionic cell body in lateral horn	Preganglionic fibers	Prevertebral ganglion	Postganglionic fibers
Celiac plexus Esophagus (distal) Stomach Duodenum (proximal) Pancreas Liver Gall bladder Spleen	T5-T9	Greater splanchnic nerve	Celiac	Celiac plexus
Superior mesenteric plexus Duodenum (distal) Jejunum Ileum Cecum Appendix Ascending colon Transverse colon (proximal)	T10-T11	Lesser splanchnic nerve	Superior mesenteric	Superior mesenteric plexus
Inferior mesenteric plexus Transverse colon (distal) Descending colon Sigmoid colon Rectum Anal canal (upper)	L1-L2	Lumbar splanchnic nerves	Inferior mesenteric	Inferior mesenteric plexus
Aorticorenal plexus Kidneys, gonads	T12	Least splanchnic nerve	Aorticorenal	Renal or gonadal plexus
Adrenal medulla	T8-T12	Thoracic splanchnic nerves	None	Chromaffin cells of adrenal medulla

directly into the blood. **Preganglionic axons** that innervate the adrenal medulla originate in the lateral horn of T8-T12, enter a prevertebral ganglion via a white ramus communicans, pass through the paravertebral ganglia without synapsing, and join thoracic splanchnic nerves to distribute to the gland (**Table 1.5**).

Sympathetic Innervation of Pelvic Viscera

Preganglionic axons that originate in the lateral horn of T12-L2 and enter paravertebral ganglia via white rami communicantes, pass through the ganglion *without synapse*, and form **lumbar splanchnic nerves**. These nerves contribute to the inferior part of the **abdominal aortic plexus** and the **superior hypogastric plexus (Fig. 1.6D)**.

- The **abdominal aortic plexus** is located along the anterior and lateral surfaces of the abdominal aorta.
- The **superior hypogastric plexus** is formed by preganglionic fibers of the aortic plexus as they descend onto the S1 vertebral body. Some of these fibers will synapse on postganglionic sympathetic cell bodies located in small ganglia within the plexus. Other fibers synapse in small, inferior hypogastric and sacral paravertebral ganglia.
- Postganglionic fibers from the superior hypogastric plexus unite to form the **right or left hypogastric nerves**.

- The **hypogastric nerves** descend into the pelvis and distribute into the right and left **inferior hypogastric plexuses.** These postganglionic fibers follow periarterial plexuses along branches of the internal iliac artery to innervate pelvic organs (**Table 1.6**).

The pathway described above is characteristic for sympathetic innervation of pelvic viscera.

Parasympathetic Division

There are two locations for preganglionic cell bodies of the parasympathetic division:

1. Brain stem nuclei of CN III, VII, IX, and X
2. Lateral horn of S2-S4 spinal cord segments

Preganglionic fibers that emerge as part of a CN form the **cranial outflow**, while those that exit the sacral spinal cord form the **sacral outflow**.

Cranial Outflow

Cranial outflow is associated with CN III, VII, IX, and X (**Fig. 1.7**). **Preganglionic fibers** in CN III, VII, and IX synapse in named parasympathetic ganglia in the head (**Table 1.7**). **Postganglionic fibers** innervate smooth muscle in the eye and salivary and lacrimal glands. The **vagus nerve** (CN X) has 80% of the parasympathetic cranial outflow. In addition to providing branches to the head and neck, it contributes to several plexuses in the thorax (e.g., cardiac, pulmonary, and esophageal) and abdomen (e.g., celiac and superior mesenteric). **All preganglionic**

TABLE 1.6	Sympathetic innervation of pelvic viscera.			
	Preganglionic cell body in lateral horn	**Preganglionic fibers**	**Sympathetic ganglion**	**Postganglionic fibers**
Rectum Anal canal (upper) Urinary bladder Uterus, uterine tubes Vagina Prostate Seminal glands	T12-L2	Lumbar splanchnic nerves to aortic and superior hypogastric plexuses	Ganglia within superior hypogastric plexus	Hypogastric nerves to inferior hypogastric plexuses to periarterial plexuses for specific organs

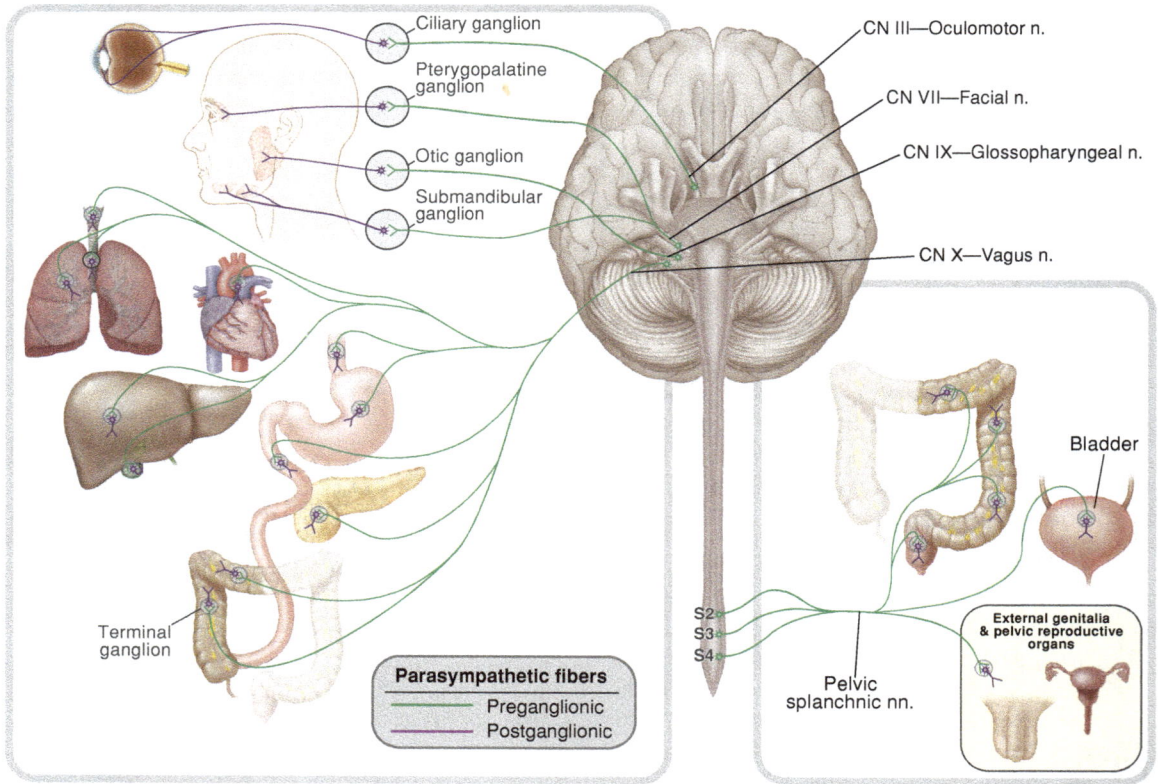

FIGURE 1.7 Parasympathetic pathways.

TABLE 1.7 Parasympathetic cranial outflow.

CN	Preganglionic fibers	Parasympathetic ganglion	Postganglionic fibers	Effectors
III	Oculomotor nerve	Ciliary	Short ciliary nn.	▪ Pupillary sphincter muscle ▪ Ciliary muscle
VII	Greater petrosal nerve	Pterygopalatine	Branches of maxillary nerve (CN V2)	▪ Lacrimal gland ▪ Mucous glands of nasal and oral cavities
VII	Chorda tympani	Submandibular	Lingual nerve	▪ Submandibular gland ▪ Sublingual gland
IX	Lesser petrosal nerve	Otic	Auriculotemporal nerve	▪ Parotid gland
X	Vagus nerve	Terminal	Short nerve fibers near/in wall of organ	▪ Mucous glands of pharynx and larynx ▪ Smooth muscle of cervical, thoracic, and abdominal viscera

TABLE 1.8 Parasympathetic innervation of pelvic viscera.

Organs	Preganglionic cell body in lateral horn	Preganglionic fibers	Parasympathetic ganglion	Postganglionic fibers
Transverse colon Sigmoid colon Rectum Anal canal (upper) Urinary bladder Uterus, uterine tubes Vagina Prostate Seminal glands Erectile tissues	S2-S4	Pelvic splanchnic nerves to inferior hypogastric plexuses From inferior hypogastric plexuses to adjacent periarterial plexuses (e.g., uterovaginal, rectal)	Terminal	Short nerve fibers near/in wall of organ

fibers in the vagus nerve end in small, **terminal ganglia** located near or in thoracic and abdominal organs (**Fig. 1.7**). Postganglionic fibers innervate the smooth muscle and glands of these organs. Vagal distribution to the gastrointestinal tract ends at the transverse colon, where it overlaps with the parasympathetic sacral outflow.

Sacral Outflow

Preganglionic parasympathetic cells in the lateral horn of S2-S4 distribute their axons in respective ventral roots, spinal nerves, and ventral rami. These preganglionic fibers branch from the S2 to S4 ventral rami as **pelvic splanchnic nerves** (**Table 1.8**). These nerves **pass through** the inferior hypogastric plexus *without synapse* and follow periarterial plexuses to pelvic organs and the distal gastrointestinal tract. These nerves synapse on postganglionic parasympathetic cells in **terminal ganglia**.

Visceral Afferent Pathways

Visceral afferent fibers conduct sensory information (pain or reflexive) from organs to the CNS. Their neuronal cell bodies are located either in dorsal root ganglia or a sensory ganglion associated with CN IX and X. They are considered part of the ANS, but not classified as sympathetic or parasympathetic.

Pain from Thoracic and Abdominal Organs

Pain receptors of visceral afferents are located in thoracic and abdominal organs. In general, these receptors are responsive to ischemia, chemical insult, or over-distension. Visceral afferent fibers enter the periarterial plexus for an organ. From there, they travel to and through the prevertebral ganglia (i.e., without synapse). These fibers continue along splanchnic nerves (specific for the visceral efferent of the organ) to the sympathetic chain. They leave the chain in a white rami communicans and enter the ventral rami and spinal nerves. Afferent fibers follow the dorsal root to their cell bodies in a dorsal root ganglia. The central process of the visceral afferent neuron enters the dorsal horn of the spinal cord. The spinal cord segments that receive pain *afferent* input from thoracic and abdominal viscera are, in general, the same segments as those that provide sympathetic visceral *efferent* innervation for an organ (**Fig. 1.8**).

Pain from Pelvic Organs

Pain information from pelvic organs is conducted along afferent fibers that travel sequentially through organ-specific periarterial plexuses, the inferior hypogastric plexus, and pelvic splanchnic nerves. From pelvic splanchnic nerves, afferent fibers course in order through ventral rami, spinal nerves, and the dorsal roots of S2-S4. The afferent neuronal cell bodies are located in the S2-S4 dorsal root ganglia

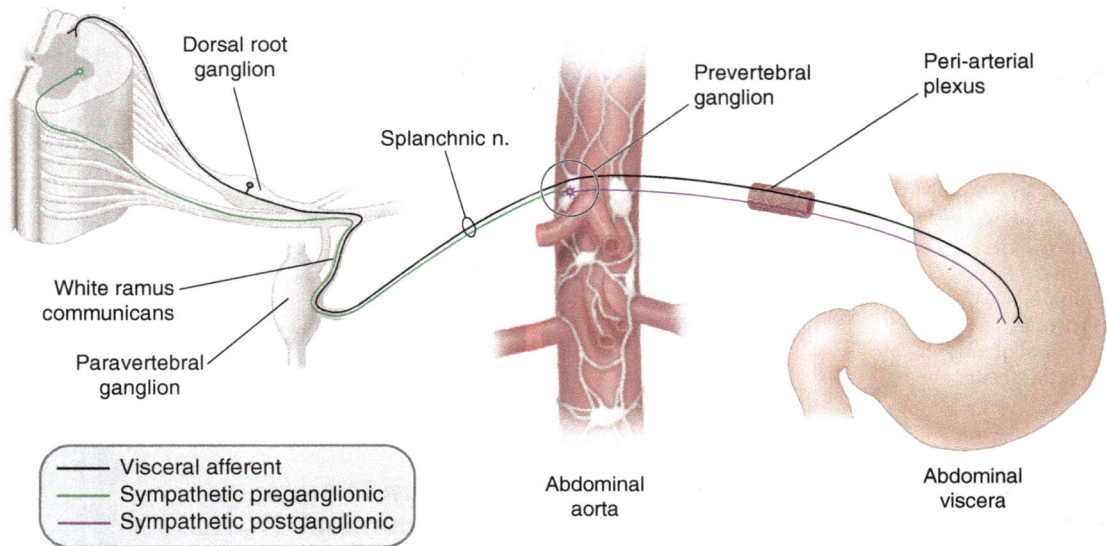

FIGURE 1.8 Visceral pain afferent pathway.

and their central processes enter the dorsal horn of spinal cord at these levels.

Referred Pain

Pain that originates in an organ but is perceived in a somatic structure, such as skin or underlying muscle, is considered **referred pain**. Visceral afferent fibers conduct impulses in response to painful stimuli from thoracic and abdominal organs. These fibers enter the spinal cord through dorsal roots of levels T1-L2. In these dorsal roots, visceral afferents intermingle with somatic afferents from the body wall. In the

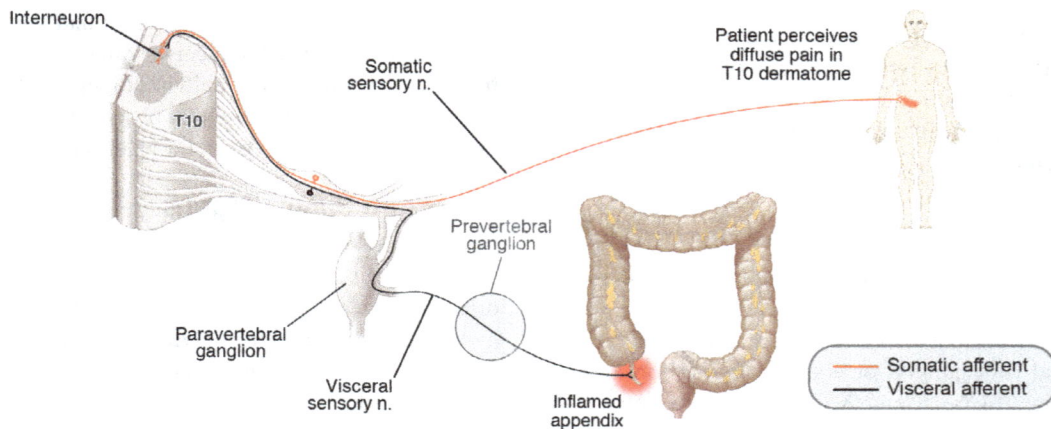

FIGURE 1.9 Referred pain pathway.

dorsal horn of the spinal cord, visceral afferents may synapse on the same interneurons as somatic afferents. This "sharing" of a common interneuron pool is one explanation for referred pain (**Fig. 1.9**). Most pelvic viscera lie inferior to the parietal peritoneum and therefore are located below the "pelvic pain line." The majority of pain afferent fibers from pelvic organs join, sequentially, the pelvic splanchnic nerves, ventral rami, and spinal nerves to reach their cell bodies in the S2-S4 dorsal root ganglia. Therefore, pain from these organs is normally referred to the S2-S4 dermatomes. Although the distal sigmoid colon and superior rectum do not lie inferior to the parietal peritoneum, pain afferents from these regions of the intestinal tract are also associated with the S2-S4 dorsal root ganglia.

Visceral Reflexes

Afferent impulses from organs can evoke reflexes that influence visceral functions such as gastric motility, urination, and defecation. These visceral afferents primarily detect stretch and/or distention of organs. The afferent fibers travel from these receptors along either the vagus nerve (abdominal and thoracic viscera) or pelvic splanchnic nerves (pelvic viscera). Afferent neuronal cell bodies are therefore located in vagal ganglia or the S2-S4 dorsal root ganglia, respectively.

REVIEW QUESTIONS

1. A male patient is admitted to the hospital complaining of severe headaches. During the neurological examination, his tongue deviates to the left upon protrusion. A CT scan reveals a tumor affecting a cranial nerve. Which nerve if affected?

   A. Glossopharyngeal (CN IX)
   B. Hypoglossal (CN XII)
   C. Mandibular (CN V3)
   D. Maxillary (CN V2)
   E. Vagus (CN X)

2. A female patient is unable to close her left eye tightly. Which nerve is most likely affected?

   A. Facial (CN VII)
   B. Maxillary (CN V2)
   C. Oculomotor (CN III)
   D. Ophthalmic (CN V1)
   E. Trochlear (CN IV)

3. During surgery to repair the gastro-esophageal junction, the surgeon damaged the vagal trunks. Which part of the large bowel would be affected in this patient?

   A. Ascending colon
   B. Descending colon
   C. Left colic flexure
   D. Sigmoid colon
   E. All of the above

4. Amyotrophic lateral sclerosis (ALS) is a neurodegenerative disease that specifically targets motor neurons that control skeletal muscles. The cell bodies for these neurons are located in the:

   A. Dorsal (posterior) root ganglia
   B. Dorsal (posterior) horn of the spinal cord
   C. Lateral horn of the spinal cord
   D. Sympathetic chain ganglia
   E. Ventral (anterior) horn of the spinal cord

5. A patient is suffering from intractable abdominal pain due to a large, malignant tumor. The pain persists despite removal of the tumor. A pain management physician suggests a rhizotomy (nerve transection) be performed as a treatment for pain relief. What structures would be cut during this procedure?

   A. Dorsal (posterior) rami
   B. Dorsal (posterior) roots
   C. Spinal nerves
   D. Ventral (anterior) rami
   E. Ventral (anterior) roots

6. An 85-year-old woman visits a neurology clinic complaining of intermittent pains "shooting" down her legs. An MRI of her lumbar spine reveals severe arthritic changes, including stenosis (narrowing) of the lumbar intervertebral foramina. The neurologist explains that her pain is related to the pressure on the nerves exiting these constricted openings. What nerve is located in the intervertebral foramen?

   A. Dorsal (posterior) ramus
   B. Dorsal (posterior) root
   C. Spinal nerve
   D. Ventral (anterior) ramus
   E. Ventral (anterior) root

7. A 25-year-old female patient is seen in the family medicine clinic with a complaint of palmar hyperhidrosis (excessive sweating on her hands). After numerous tests, her physician suggests endoscopic thoracic sympathectomy to disconnect the sympathetic chain ganglia from adjacent nerves. Sympathetic chain ganglia contain:

   A. Postganglionic neuronal cell bodies
   B. Preganglionic neuronal cell bodies
   C. Somatic afferent neuronal cell bodies
   D. Visceral afferent neuronal cell bodies

8. A dermatome is an area of skin supplied by the afferent component of one spinal nerve. The neuronal cell bodies for the sensory axons that supply the skin of the T4 dermatome would be located in the T4:

   A. Dorsal (posterior) horns of spinal cord
   B. Dorsal (posterior) root ganglia
   C. Intercostal nerves
   D. Sympathetic ganglia
   E. Ventral (anterior) horns of spinal cord

9. Which structure would *not* be in the pathway for the sympathetic innervation of the heart?

   A. Cardiac splanchnic nerves
   B. Gray rami communicantes
   C. Neuronal cell bodies in T2 lateral horn
   D. Sympathetic paravertebral ganglia
   E. White rami communicantes

10. Visceral afferents associated with pain receptors in the wall of the appendix are located in:

   A. Greater splanchnic nerve
   B. Least splanchnic nerve
   C. Lumbar sympathetic ganglia
   D. Superior cervical ganglion
   E. Superior mesenteric arterial plexus

CHAPTER 2

Thorax

Patient Presentation

During a scheduled prenatal visit, a 34-year-old in the 24th week of her first pregnancy complains of shortness of breath and fatigue. During a couple of recent exercise sessions she noticed that her heart was "racing." During one of these episodes, she noticed a small amount of blood in her saliva.

Relevant Clinical Findings

History

This patient emigrated from India to the United States 10 years ago. She indicates that there is no family history of cardiovascular or respiratory disease. She does not have records for childhood immunizations and illnesses. To date, her pregnancy has been unremarkable.

Physical Examination

Noteworthy vital signs:

- Pulse: 122 bpm
  Adult resting rate: 60–100 bpm

- Fetal pulse: 143 bpm
  Baseline fetal rate: 110–160 bpm

Results of physical examination:

- Auscultation over the heart apex with the patient in left decubitus position reveals a mid-diastolic murmur and a systolic "snap."
- Crackles are detected in all lung lobes.

Imaging Studies

- Echocardiography reveals calcification in both leaflets of the mitral valve, doming of the anterior leaflet, and thickened and shortened chordae tendineae. The left atrium is enlarged.

Electrocardiogram

- Atrial fibrillation

Clinical Problems to Consider

- Atrial myxoma
- Mitral valve prolapse
- Mitral valve stenosis

LEARNING OBJECTIVES

1. Describe the anatomy of the left atrioventricular (mitral) valve.
2. Correlate the normal electrocardiogram (ECG) with events of the cardiac cycle.
3. Explain the anatomical basis for the signs and symptoms associated with this case.

RELEVANT ANATOMY

Left Atrioventricular (Mitral) Valve

A system of one-way valves guards the entry and exit channels of the ventricles of the heart (**Fig. 2.1.1**). These valves are divided into those that lie at the junction of each atrium with its respective ventricle (**atrioventricular valves**) and those that are located

---

Auscultation A diagnostic method, usually with a stethoscope, to listen to body sounds (e.g., heart, breath, and gastrointestinal sounds)
Decubitus Lying down

Myxoma Benign neoplasms derived from connective tissue
Prolapse To fall or slip out of place
Stenosis Narrowing a canal (e.g., blood vessel and vertebral canal)

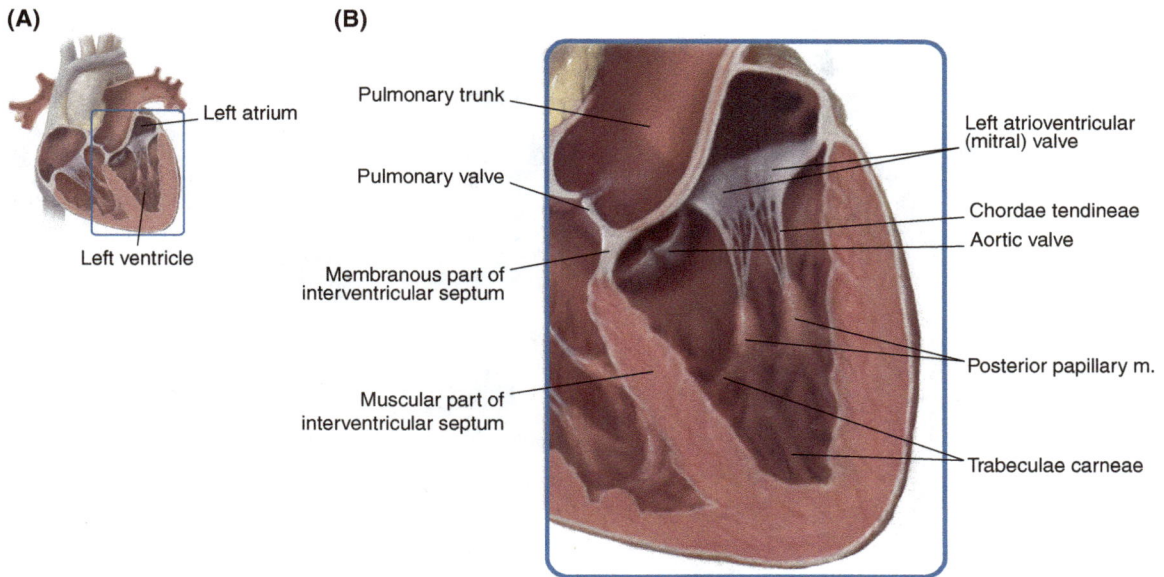

FIGURE 2.1.1 (**A**) Anterior view of the interior of the heart in a frontal section. The pulmonary trunk and arteries are sectioned in this view. (**B**) View of the left atrium and ventricle showing the left atrioventricular (mitral) valve.

in the root of the large vessel exiting each ventricle (**pulmonary and aortic valves**). The cusps for each heart valve are composed of a fibrous core that is coated with endocardium. All cusps are anchored to the **cardiac skeleton**, a fibrous ring that insulates atrial from that of the ventricular myocardium.

The **left atrioventricular (mitral or bicuspid) valve** lies at the junction of the left atrium and ventricle (**Fig. 2.1.1**). Its two cusps are termed anterior and posterior and the two papillary muscles of the left ventricle have the same distinctions. The area of the anterior cusp is approximately twice that of the posterior cusp. Chordae tendineae from each papillary muscle extend to both valve cusps. This arrangement is most important during ventricular systole since it prevents prolapse of the valve cusps into the atrium and separation of the valve cusps. Thus, as blood is forced from the ventricle during systole, it cannot regurgitate into the left atrium.

In the healthy heart, the left atrioventricular valve opens wide during ventricular diastole so that the left atrium and ventricle become a common chamber.

Surface Projections of Heart Valves

The position of the heart in the mediastinum places the four heart valves deep to the body of the sternum (**Fig. 2.1.2**).

- The **right atrioventricular (tricuspic) valve** lies slightly right of midline, deep to the 4th and 5th intercostal spaces.

- The **left atrioventricular (mitral) valve** lies just left of midline, deep to the 4th costal cartilage.

- The **pulmonary valve** lies just left of the midline at the level of the 3rd costal cartilage.

- The **aortic valve** lies near the midline at the level of the 3rd intercostal space.

Auscultation Points for Heart Valves

The opening and closing of the heart valves produces "heart sounds" that can be assessed with the aid of a stethoscope. Because of the orientation of the heart in the mediastinum, blood flow through the heart valves is closer to horizontal than vertical. The sound produced by each valve and the contracting ventricular wall is projected along vectors

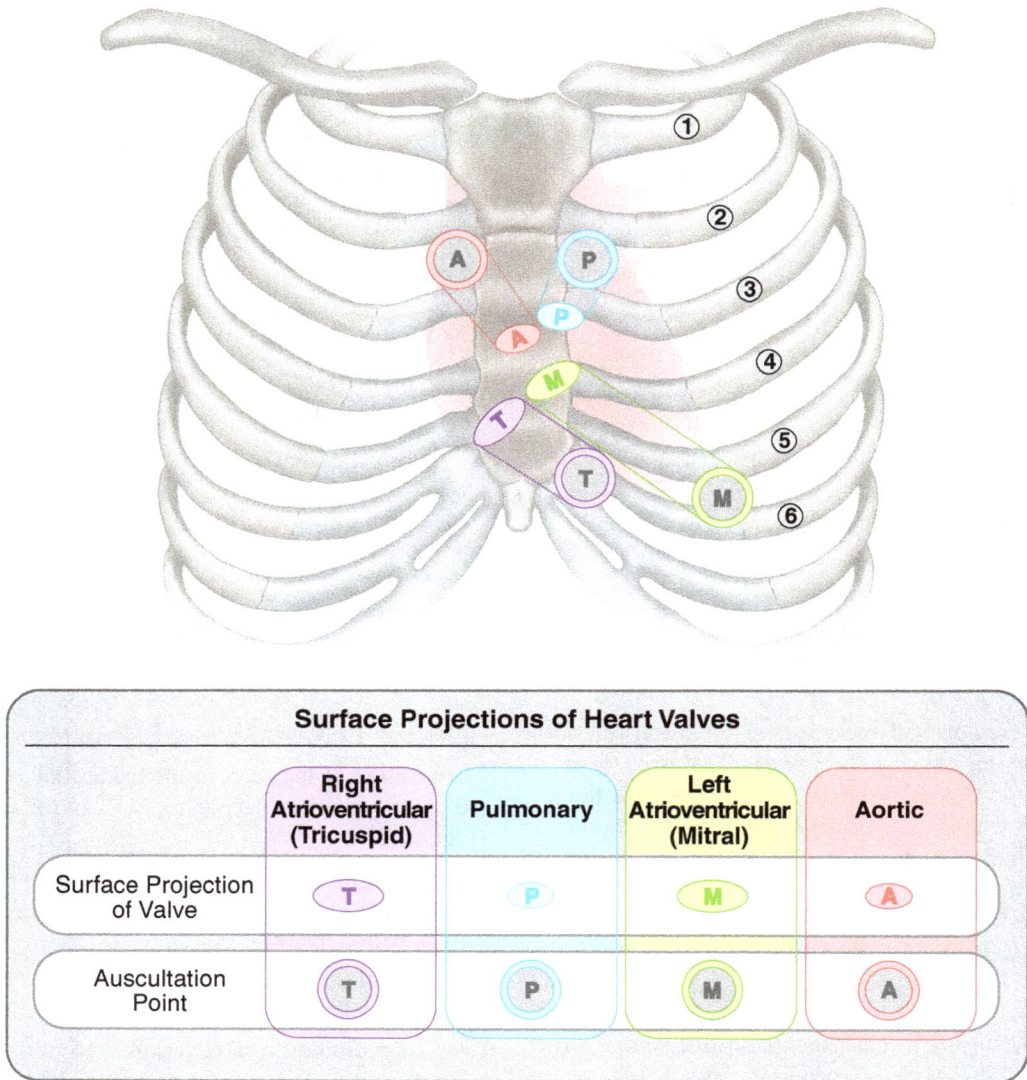

Surface Projections of Heart Valves

	Right Atrioventricular (Tricuspid)	Pulmonary	Left Atrioventricular (Mitral)	Aortic
Surface Projection of Valve	T	P	M	A
Auscultation Point	T	P	M	A

FIGURE 2.1.2 Surface projections and auscultation points for the heart valves.

aligned with the orientation of the blood flow. Thus, the valve sounds are projected to the sides of the sternum where the resonance through the thoracic wall is significantly better. In other words, the valve sounds are best heard at some distance from actual valve location (**Fig. 2.1.2**).

The aortic and pulmonary valve sounds are assessed with the bell of the stethoscope over, respectively, the right and left 2nd intercostal spaces, adjacent to the sternum. Both atrioventricular valve sounds can be heard in the 5th left intercostal space: the right atrio-

ventricular valve adjacent to the sternum and the left atrioventricular valve on the midclavicular line. The position in the left 5th intercostal space on the midclavicular line represents the location of the apex of the heart. This region is often referred to as the "Point of Maximal Impulse" (PMI) since heart sounds are generally loudest at this location.

Normal Electrocardiogram

The electrocardiogram (EKG or ECG) is a record of the electrical activity in the heart during each cardiac cycle. The **cardiac cycle** takes <1 second and includes one episode each of **diastole** and **systole**.

- **Diastole** is the portion of the cardiac cycle when the myocardium relaxes and the heart chambers fill.

- **Systole** is the portion of the cardiac cycle when the myocardium contracts and the heart chambers empty.

The electrical activity in the heart can be monitored by an electrocardiograph, an instrument that records and amplifies all action potentials in the conduction system and myocardium. This electrical activity is detected by a series of electrodes attached to the thoracic wall and to the extremities. The EKG is visualized on a paper tracing or on a video monitor.

The **EKG for a normal cardiac cycle** consists of a series of deflections above and below a reference or baseline. These deflections represent depolarization and repolarization activities in the heart. The EKG for one cardiac cycle is divided into three major parts (**Fig. 2.1.3**):

1. The **P wave** represents **depolarization** of atrial myocytes as impulses from the **sinu-atrial (SA) node** spread.

2. The **QRS complex** includes a prominent **R spike** above baseline and two smaller deflections, **Q and S**, below baseline (before and after the R wave, respectively). The QRS complex represents **ventricular depolarization**. The shape of this part of the EKG is created because of the difference in size and depolarization rate for the ventricles.

3. The **T wave** represents the time of **ventricular repolarization** and immediately precedes **ventricular diastole**.

During the **PQ segment**, the atria undergo systole. At the end of the PQ segment, impulses have reached the **atrioventricular (AV) node**.

Ventricular systole begins at the end of the QRS complex and extends into the T wave. It is during this period that the ventricles force blood into the pulmonary trunk and ascending aorta.

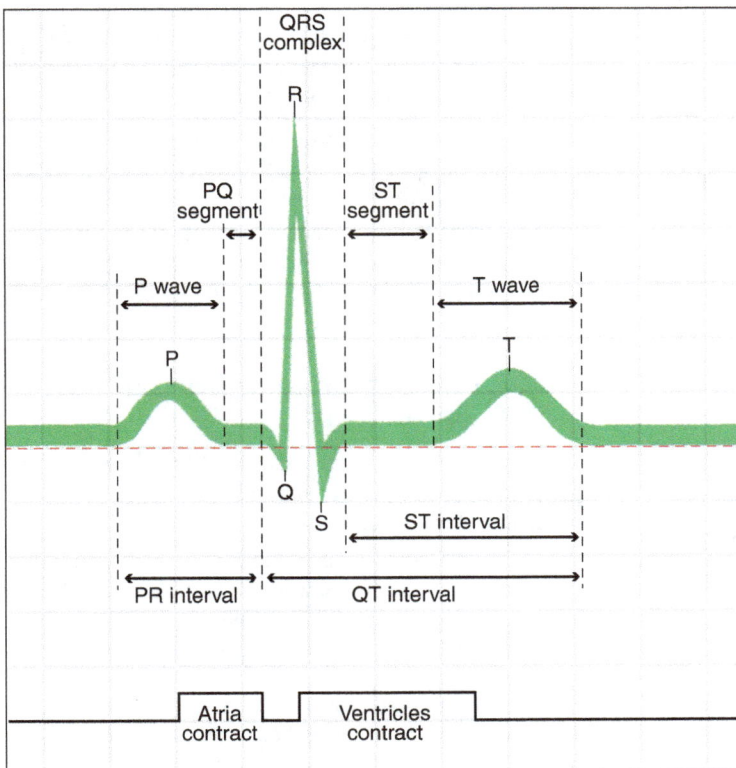

FIGURE 2.1.3 Normal electrocardiogram (EKG).

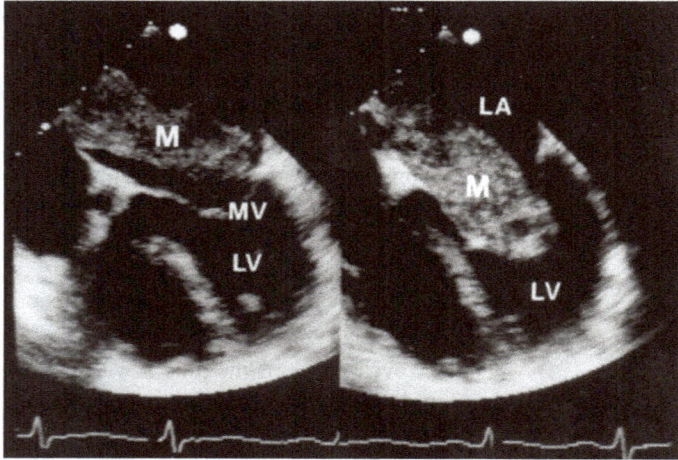

FIGURE 2.1.4 Transesophageal echocardiogram from a four-chamber view from a 50-year-old man who presented with exertional dyspnea and syncope. A large, left atrial myxoma (M) attached to the interatrial septum is seen prolapsing across the mitral valve (MV) into the left ventricle (LV) in diastole (*right panel*). *Source:* Fig. 90-6 in *Hurst's the Heart.*

Alterations to the heart conduction system, valvular function, and/or myocardium may alter the electrical activity in the heart. These alterations will manifest as changes in the different segments of the EKG.

CLINICAL REASONING

This patient presents signs and symptoms of **heart disease**.

Atrial Myxoma

An intracardiac myxoma, the most common primary tumor of the heart, is a benign neoplasm. It occurs most frequently (75%) in the left atrium. Typically, it is pedunculated with attachment to the interatrial septum near fossa ovalis. On average, it is 5–6 cm in diameter.

Signs and Symptoms

- Pulmonary hypertension
- Dyspnea on exertion
- Orthopnea
- Acute pulmonary edema
- Mitral valve murmur
- Heart failure

- Palpitations
- Fatigue

Predisposing Factors

Atrial myxomas are classified as familial or sporadic:

- Sex: sporadic myxomas are most common in women (75%).
- Age: mean age at presentation for sporadic myxoma is 56 years; mean age at presentation for familial myxoma is 25.

A left atrial myxoma may obstruct the mitral orifice (**Fig. 2.1.4**) or pulmonary venous orifices leading to pulmonary edema and hypertension. This may result in leading to decreased left ventricular output. Right ventricular hypertrophy frequently develops as a result of obstruction of pulmonary blood flow.

Left Atrioventricular (Mitral) Valve Prolapse

Mitral valve prolapse is a condition in which one or both cusps project partially or fully into the left atrium during left ventricular systole.

Dyspnea Difficulty breathing and shortness of breath
Syncope Loss of consciousness (fainting)
Pedunculate Having a stalk
Edema Swelling of skin due to abnormal accumulation of fluid in subcutaneous tissue
Murmur Variable vibrations produced by turbulence of blood flow

Palpitation Forcible or irregular pulsation of the heart that is perceptible to the patient
Hypertension Abnormal increase in arterial and/or venous pressure
Hypertrophy Increase in size of a tissue or organ due to increased cell size, i.e., without increased cell number (antonym: hyperplasia)

Signs and Symptoms

- Dyspnea
- Fatigue
- Palpitations
- One or more midsystolic clicks heard on auscultation
- Midsystolic clicks accentuated when patient standing

Predisposing Factors

- Sex: female (3:1)
- Thin stature
- Thoracic skeletal deformities (e.g., scoliosis)
- Heritable connective tissue diseases

Mitral valve prolapse is nearly always asymptomatic. The prolapse is always accompanied by some regurgitation of blood into the left atrium during systole. This regurgitation places a volume overload on the left ventricle.

The underlying cause of mitral valve prolapse is not well understood, but excessive (redundant) valve may play a role in many patients with inherited connective tissue disorders. The prolapse of one or both leaflets places additional stress on the associated papillary muscles and chordae tendineae. This may impair the function of the valve complex. Papillary muscle damage and/or rupture of chordae tendineae exacerbate regurgitation through the valve, establishing a vicious cycle of events that may culminate in **heart failure**, a failure of the ventricles to eject the volume of blood necessary to adequately supply oxygen to the organs and tissues of the body.

Mitral Valve Stenosis

Mitral valve stenosis is a progressive condition in which the valve cusps become rigid and prevent the valve from opening properly during ventricular diastole.

Signs and Symptoms

- Exertional and/or paroxysmal nocturnal dyspnea
- Orthopnea
- Atrial fibrillation
- Lung crackles
- Hemoptysis

Predisposing Factors

- History of rheumatic fever
- Systemic lupus erythematosus
- Sex: female (3:1)
- Congenital valve defect

Clinical Notes

Rheumatic Fever

- Forty percent of all patients with documented history of rheumatic fever (*Streptococcus pyogenes*) develop mitral valve stenosis. Most patients have lived in tropical and subtropical areas where rheumatic fever is still a common childhood infection. Rheumatic fever has not been common in the United States since the development of effective therapeutic medications.
- The latent period between initial infection with *S. pyogenes* and the onset of symptoms of mitral valve stenosis is about two decades.
- Cardiac involvement from rheumatic fever is thought to result from the body's immunological response to attack the "M" protein antigens found in *S. pyogenes* and in the heart tissue of some patients. In these patients, inflammation of the endocardium occurs.
- It is thought that the hemodynamic stress on valve cusps increases their vulnerability. Why the mitral valve cusps are more susceptible is unknown.

Paroxysmal Sudden onset of a symptom or disease
Orthopnea Difficulty breathing and shortness of breath when lying down
Fibrillation Rapid contraction or twitching of muscle fibrils but not of the muscle as a whole

Crackle Crackling noise heard with lung disease (also known as a rale)
Hemoptysis Blood in sputum from airway hemorrhage

DIAGNOSIS

The patient presentation, medical history, physical examination, imaging studies, and procedure results support a diagnosis of **mitral valve stenosis**.

Mitral Valve Stenosis

Mitral valve stenosis is a condition in which the canal between the left atrium and ventricle is narrowed due to disease of the cusps of the left atrioventricular (mitral) valve. Mitral valve stenosis may remain asymptomatic for years. When clinical symptoms develop, they may be similar to those of other heart diseases.

- The symptoms in this patient developed with pregnancy due to the elevated cardiovascular demands and associated increased transmitral pressure gradient.

- The systolic "snap", detected during auscultation, is associated with the stenotic mitral valve. Normally, the mitral valve cusps "float" into the closed position. In stenosis, the valve remains open longer for ventricular filling and closes from the full-open position with a "snap."

- Conclusive diagnosis is made from echocardiographic evidence of rigid cusps with calcifications.

Mitral valve stenosis is said to "protect" the left ventricle since there is no overload placed on this chamber. Output from the ventricle is decreased, but the chamber is not stressed. The stenotic valve is not allowing the left atrium to empty and it eventually enlarges. This change is reflected in the P wave of the EKG (**Fig. 2.1.5**).

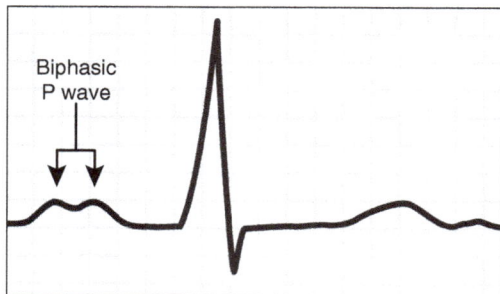

FIGURE 2.1.5 EKG with a biphasic ("notched") P wave from a patient with mitral stenosis. *Source:* Fig. 15-22 in *Hurst's the Heart.*

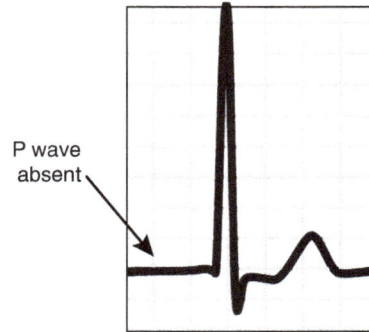

FIGURE 2.1.6 EKG with absent P wave from a patient with atrial fibrillation.

In 50–80% of cases, the stress on the left atrium results in atrial fibrillation (**Fig. 2.1.6**). This will, in turn, produce increased pulmonary vasculature pressure leading to:

- Hypertension
- Edema
- Lung crackles
- Hemoptysis

Clinical Notes

- Hemoptysis due to rupture of hypertensive lung capillaries is not common in other heart diseases.

- Patients may present with hoarseness due to pressure on the left recurrent laryngeal nerve by the enlarged left atrium.

Atrial Myxoma

This benign neoplasm is typically mobile due to the presence of a stalk of variable length. The pedunculated mass may enter the mitral orifice during diastole and return to the left atrium during systole. Symptoms arise due to the mechanical interference with the valve.

- Echocardiography in this patient did not reveal any masses in the left side of the heart.

Clinical Note

Cardiac myxomas account for most cases of tumor embolism.

Left Atrioventricular (Mitral) Valve Prolapse

Mitral valve prolapse is a condition in which one or both cusps project into the left atrium during systole. The underlying cause is often a larger than normal valve cusp associated with familial connective disorders. The posterior cusp is more often affected and may have chordae tendineae that are elongated or ruptured. Increased stress placed on the papillary muscles in mitral valve prolapse may lead to ventricular and atrial arrhythmias.

- Midsystolic clicks are postulated to be the result of sudden tension on elongated, slackened chordae tendineae, or by the prolapsing valve cusp reaching its maximal excursion.

- With arrhythmias, presenting symptoms may include palpitations, presyncope, and dyspnea.

- Echocardiography is useful in establishing the degree of prolapse. In **mitral valve prolapse**, the cusps are mobile. In contrast, the cusps in **mitral stenosis** are more rigid due to fusion along the commissures, the accumulation of foreign material on the cusp surfaces, and calcification.

Clinical Note

Mitral valve prolapse is present in approximately 10% of the general female population. For this reason, the clinical significance of this condition is disputed.

Presyncope Feeling faint or lightheaded (syncope is actually fainting)
Arrhythmia Irregular heart beat

Congestive Heart Failure

Patient Presentation

A 58-year-old male visits his cardiologist for biannual monitoring of cardiovascular disease. His current complaints are shortness of breath, fatigue, and swelling of his feet and ankles.

Relevant Clinical Findings

History

At age 49, this patient was diagnosed with advanced coronary artery disease. This diagnosis was made subsequent to a mild heart attack. At that time, he had a triple bypass on the anterior interventricular artery. For 30 years prior to the heart attack, he smoked 20–30 cigarettes daily, although he has not smoked for the last 9 years. He admits to regular use of illicit drugs "during my 20s and 30s."

He relates that he has trouble breathing in bed and is able to sleep best in his recliner chair. He also indicates that he commonly needs to urinate 3–4 times each night.

Physical Examination

Noteworthy vital signs:

- Pulse: 102 bpm
    Adult resting rate: 60–100 bpm
Results of physical examination:

- Mild edema of feet and hands
- Auscultation reveals wheezing and crackles in both lungs
- Orthopnea
- Distension of neck veins
- S3 heart sound over apex of heart

Clinical Note

The S3 heart sound is produced during ventricular diastole. It may be present in children and in adults up to age 40. The presence of S3 after age 40 suggests left-side heart failure.

Electrocardiogram

- Consistently wide and deep Q wave
- Deep S and tall R waves

Imaging Studies

- Plain film radiography shows left ventricular hypertrophy.
- Echocardiography reveals a left ventricular ejection fraction (EF) of 38%
    Normal: 55–70%.

Clinical Problems to Consider

- Chronic obstructive pulmonary disease (COPD)
- Congestive heart failure (CHF)
- Valvular disease

Auscultation A diagnostic method, usually with a stethoscope, to listen to body sounds (e.g., heart, breath, and gastrointestinal sounds)

Crackle Crackling noise heard with lung disease (also known as a rale)

Edema Swelling of skin due to abnormal accumulation of fluid in subcutaneous tissue

Orthopnea Difficulty breathing and shortness of breath when lying down

Wheezing Labored breathing that produces a hoarse, whistling sound

1. Describe the anatomy of the heart valves.
2. Describe the surface projections and auscultation points for the heart valves.
3. Describe the anatomy of the heart ventricles.
4. Describe the intrapulmonary airway.
5. Correlate the normal electrocardiogram (ECG) with the cardiac cycle.
6. Explain the anatomical basis for the signs and symptoms associated with this case.

RELEVANT ANATOMY

Heart Valves

A system of one-way valves guards the entry and exit channels of the ventricles of the heart (**Fig. 2.2.1**). These valves are divided into those that lie at the junction of each atrium with its respective ventricle (**atrioventricular valves**) and those that are located in the root of the large vessel exiting each ventricle (**pulmonary and aortic valves**). The cusps for each heart valve are composed of a fibrous core that is coated with endocardium. All cusps are anchored to the **cardiac skeleton**, a fibrous ring that insulates the atrial myocardium from that of the ventricles.

The **right atrioventricular (tricuspid) valve** lies at the junction of the right atrium and ventricle (**Fig. 2.2.1**). It has three cusps: anterior, posterior, and septal. **Chordae tendineae** are attached near the ventricular margin of each cusp. These fibrous tethers, in turn, attach to **papillary muscles** of the ventricular wall. Each papillary muscle (anterior, posterior, septal) has chordae tendineae that attach to two valve cusps.

- Chordae tendineae from the anterior papillary muscle attach to anterior and posterior cusps
- Chordae tendineae from the posterior papillary muscle attach to posterior and septal cusps

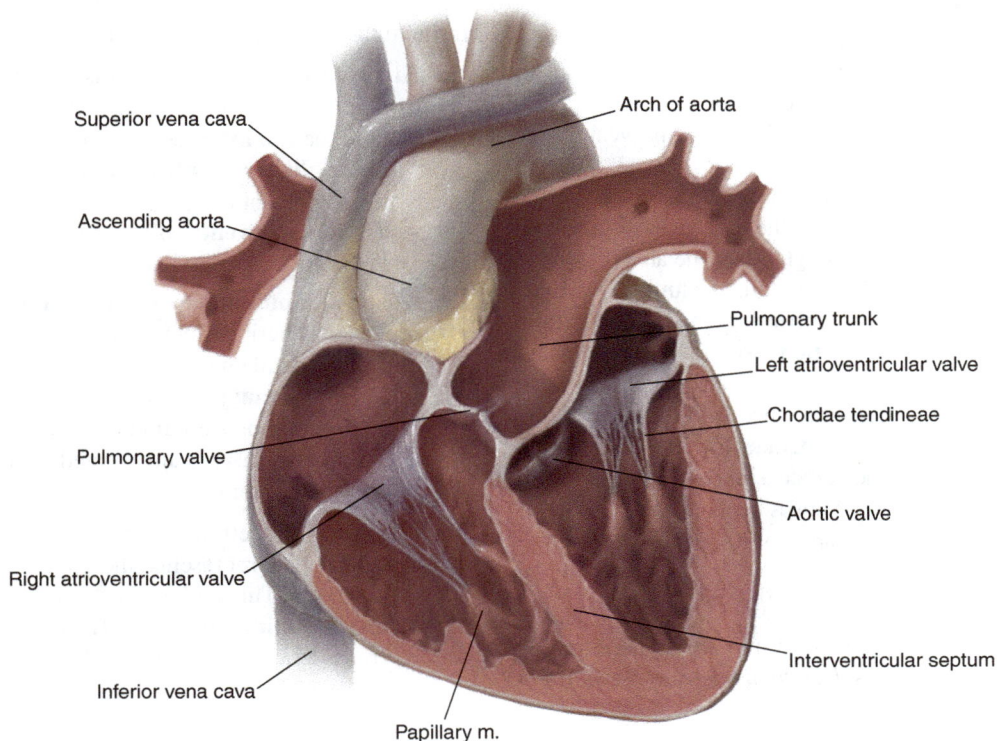

FIGURE 2.2.1 Anatomy of the heart valves.

■ Chordae tendineae from the septal papillary muscle attach to septal and anterior cusps

This arrangement is most important during ventricular systole since it prevents (a) prolapse of the valve cusps into the atrium and (b) separation of the valve cusps. Thus, as blood is forced from the ventricle during systole, it cannot regurgitate into the left atrium.

The **left atrioventricular (bicuspid or mitral) valve** lies at the junction of the left atrium and ventricle (**Fig. 2.2.1**). Its two cusps are termed anterior and posterior. The two papillary muscles of the left ventricle have the same names. Chordae tendineae from each papillary muscle extend to both valve cusps and function similarly to those on the right side of the heart.

The **pulmonary valve** is located at the junction of the **pulmonary trunk** with the right ventricle (**Fig. 2.2.1**). It consists of three semilunar cusps (anterior, right, and left) that lack chordae tendineae. Each cusp is concave on its distal surface. These concavities form the **pulmonary trunk sinuses**. During right ventricular systole, blood is forced from the chamber into the pulmonary trunk, causing the valve cusps to be displaced toward the wall of the vessel. Conversely, during right ventricular diastole, the elastic recoil in the pulmonary trunk forces blood back toward the right ventricle. Retrograde flow into the ventricle is prevented by closure of the pulmonary valve due to blood pooling in the pulmonary trunk sinuses. This hemodynamic force brings the free edges of the cusps back together to form a tight seal. The area of contact for each cusp is thickened to form the **lunule**. The midpoint of each lunule is further thickened as the **nodule**. The anatomy of the **aortic valve** (posterior, right, and left cusps) is very similar to that of the pulmonary valve (**Fig. 2.2.1**). Likewise, the function of the valves is the same—prevention of retrograde flow into the ventricle during diastole. The **aortic sinuses** associated with the right and left valve cusps contain the ostia for the right and left coronary arteries, respectively.

Surface Projections of Heart Valves

The position of the heart in the mediastinum places the four heart valves *deep to the body of the sternum* (**Fig. 2.2.2**).

Ostium   Opening (plural=ostia)

■ The right atrioventricular (tricuspic) valve lies slightly right of midline, deep to the 4th and 5th intercostal spaces.

■ The left atrioventricular (mitral) valve lies just left of midline, deep to the 4th costal cartilage.

■ The pulmonary valve lies just left of the midline at the level of the 3rd costal cartilage.

■ The aortic valve lies near the midline at the level of the 3rd intercostal space.

Auscultation Points for Heart Valves

The opening and closing of the heart valves produces "heart sounds" that can be auscultated with a stethoscope. Because of the orientation of the heart in the mediastinum, blood flow through the heart valves is closer to horizontal than vertical. The sound produced by the valve and the contracting ventricular wall is projected along vectors that align with the flow of blood. In other words, valve sounds are best heard at some distance from their actual location, in intercostal spaces where resonance through the thoracic wall is better (**Fig. 2.2.2**).

■ **Atrioventricular valve sounds** are projected toward the left and inferiorly and are auscultated in the **5th intercostal space**.
  ■ The **right atrioventricular valve** is auscultated adjacent to the sternum.
  ■ The **left atrioventricular valve** is auscultated in the midclavicular line.

■ **Aortic and pulmonary valve sounds** are projected superiorly and are auscultated in the **2nd intercostal space**.
  ■ The **pulmonary valve** is auscultated adjacent to the sternum on the left side.
  ■ The **aortic valve** is auscultated adjacent to the sternum on the right side.

The position in the left 5th intercostal space on the midclavicular line represents the location of the apex of the heart. This region is often referred to as the **"Point of Maximal Impulse" (PMI)** since heart sounds are generally loudest at this location.

Intrapulmonary Airway

In the **hilum**, each main bronchus divides into **lobar (secondary) bronchi** that distribute to lung lobes:

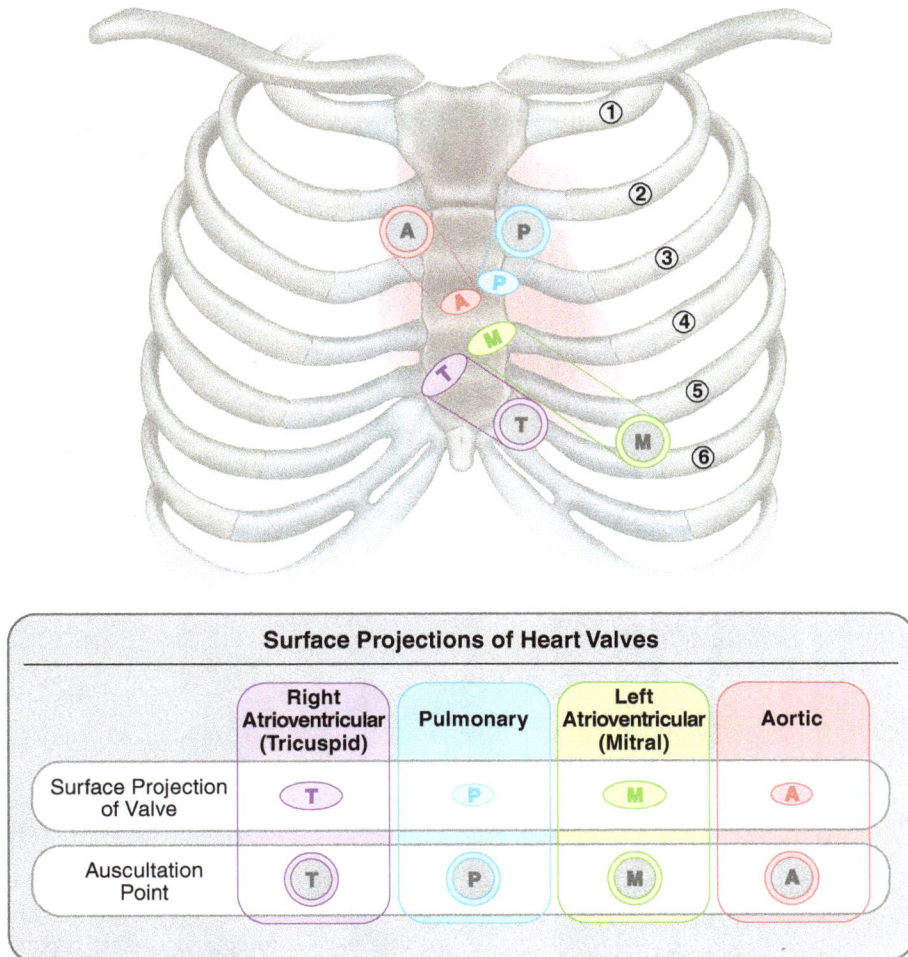

FIGURE 2.2.2 Surface projections and auscultation points for the heart valves.

three to the right lung and two to the left (see **Fig. 2.4.2**). Lobar bronchi further divide in a prescribed pattern to distribute as **segmental (tertiary) bronchi** to bronchopulmonary segments. The basic **functional unit of the lung** is the **bronchopulmonary segment**. There are 10 bronchopulmonary segments in the right lung and 8–10 in the left. Each bronchopulmonary segment is pyramidal: its base directed toward the surface of the lung and its apex toward the hilum.

Beyond the segmental bronchi, there is another 10 generations of branching until the bronchi are approximately 1 mm in diameter. At this point, the airways are termed **bronchioles** (**Fig. 2.2.3**), which lack carti-

lage. Bronchioles undergo two to three generations of branching, and end as **terminal bronchioles**.

Terminal bronchioles form the stem for the **respiratory unit of the lung**, the **acinus**. Each acinus gives rise to several **respiratory bronchioles**. The walls of the respiratory bronchioles contain a few **alveoli** and are the most proximal portion of the airway in which gas exchange can occur. Thus, the airway within each bronchopulmonary segment undergoes 20–25 generations of branching before they acquire alveoli. The respiratory bronchioles give rise to 2–11 **alveolar ducts**. Each of these, in turn, gives rise to five or six **alveolar**

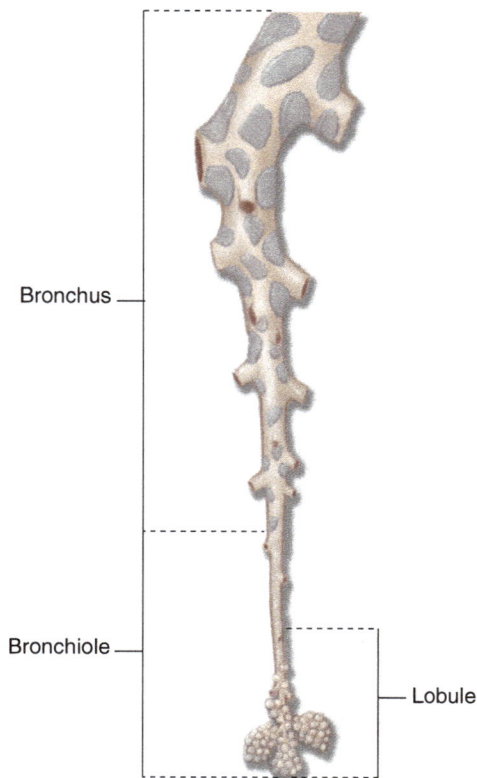

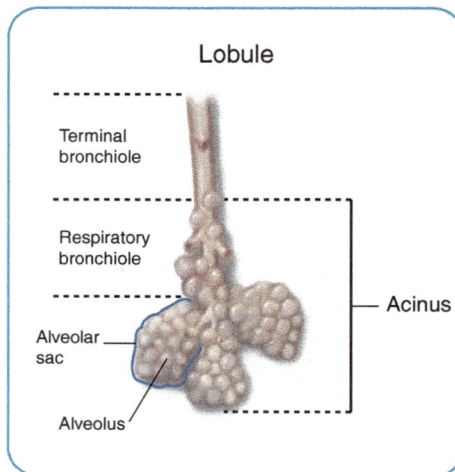

FIGURE 2.2.3 Branching of a segmental bronchus.

sacs. The number of acini exceeds 100,000, each of which contains approximately 3,000 alveoli.

Normal Electrocardiogram

The normal EKG is presented in **Figure 2.2.4**. Alterations to the heart conduction system, valvular function, and/or myocardium may affect electrical activity in the heart that will manifest as changes in the specific segments of the EKG. For further discussion of the normal EKG see pages 20–21.

CLINICAL REASONING

This patient presents signs and symptoms of **heart disease**.

Chronic Obstructive Pulmonary Disease

COPD is a lung disease that diminishes the flow of air through the lungs. It is often a combination of two diseases: chronic bronchitis and emphysema. The inflammation of the airways in **chronic bronchitis** causes increased mucus production, which leads to partial airway obstruction. The inflammation also causes bronchospasms, which further narrows the airways. In **emphysema**, there is permanent damage to elastic tissue in the lung. This results in the loss of the recoil function of the lungs and compromises the ability to move air out of the lungs.

Signs and Symptoms

- Chronic, productive (mucous) cough
- Dyspnea
- "Barrel chest" deformity in advanced stages (produced by chronic over-inflation of lungs)

Dyspnea Difficulty breathing and shortness of breath

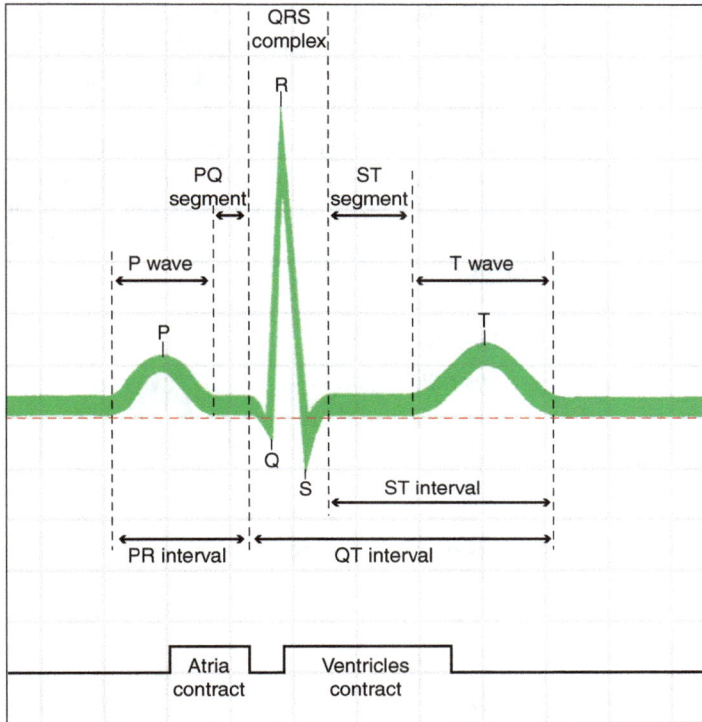

FIGURE 2.2.4 Normal electrocardiogram (EKG).

Predisposing Factors

- Cigarette smoking (active or former)
- Smoking cigarettes **and** marijuana have risk higher than either one alone
- Environmental air pollutants
- Occupation: exposure to dust and chemical fumes

The major cause of COPD is smoking. This is a progressive disease in which the rate of progression is determined by the number of cigarettes smoked and how long the patient has smoked.

Clinical Notes

- Nonsmoking patients with asthma rarely develop COPD.
- Spirometry is the most effective test for staging COPD.

Congestive Heart Failure

This disease of the heart is almost always progressive and the result of prior damage to the myocardium by coronary artery disease, myocardial infarction, valve disease, chronic hypertension, congenital heart defects, and/or substance abuse. CHF often occurs because of the combined effect of several diseases. "**Failure**" in this sense refers to the inability of the ventricles to eject the volume of blood necessary to adequately supply oxygen to the organs and tissues of the body. CHF may involve either or both ventricles. The failure may be either systolic or diastolic in nature.

- In **systolic heart failure**, the ventricular myocardium cannot contract with enough force.
- In **diastolic heart failure**, the myocardium does not relax adequately to allow proper filling of the ventricle(s).

Spirometry Pulmonary function test that measures volume and rate of airflow

Hypertension Abnormal increase in arterial and/or venous pressure

Signs and Symptoms

- Lung congestion with dyspnea
- Edema in extremities
- Tachycardia

Predisposing Factors

- Cardiovascular disease
- Prior heart attack
- Hypertension
- History of tobacco, alcohol, and/or illicit drug use
- Arrhythmias

Clinical Note

Some patients with CHF are asymptomatic.

Valvular Disease

Valvular heart disease is a condition in which one or more of heart valves (right atrioventricular, pulmonary, left atrioventricular, aortic) are not functioning properly. This disease is divided into two main categories:

1. **Valvular stenosis** occurs when the leaflets of a heart valve are rigid or fused (**Fig. 2.2.5**), causing the opening guarded by the valve to be more narrow than normal.
2. **Valvular insufficiency** is caused by failure of the valve to close properly. This allows blood to "leak" across the valve when it should form a seal.

Signs and Symptoms

- Dyspnea
- Dizziness
- Overall weakness
- Peripheral edema
- Murmurs
- Palpitations

Predisposing Factors

- Rheumatic fever
- Endocarditis
- Heart attack
- Hypertension

Arrhythmia Irregular heart beat
Tachycardia Increased heart rate: >100 bpm (normal adult heart rate: 55–100 bpm)
Murmur Variable vibrations produced by turbulence of blood

Palpitation Forcible or irregular pulsation of the heart that is perceptible to the patient
Stenosis Narrowing a canal (e.g., blood vessel and vertebral canal)

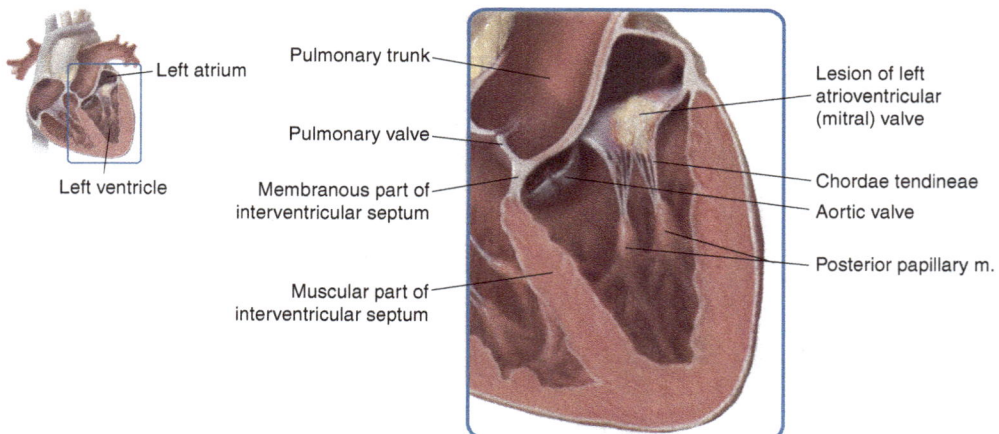

Left atrium
Left ventricle
Pulmonary trunk
Pulmonary valve
Membranous part of interventricular septum
Muscular part of interventricular septum
Lesion of left atrioventricular (mitral) valve
Chordae tendineae
Aortic valve
Posterior papillary m.

FIGURE 2.2.5 Lesion of the left atrioventricular (mitral) valve, secondary to streptococcal infection.

Clinical Notes

- Although rheumatic fever is usually a childhood streptococcal infection, the effect on heart valves may not manifest for several decades.

- Stenosis and insufficiency of the aortic and pulmonary valves are frequently due to congenital malformation of these valves.

DIAGNOSIS

The patient presentation, history, physical examination, EKG, and imaging studies support a diagnosis of **left side CHF**.

Congestive Heart Failure

A major contributing factor to CHF is a previous heart attack that causes the death of myocardial cells. The area of cell death will be replaced by fibrous tissue, compromising the ability of the ventricle to eject the normal volume of blood during systole (low EF). The decreased cardiac output will cause:

- Peripheral edema because less blood is filtered through the kidneys.

- Pulmonary hypertension and pulmonary edema with associated crackles.

- Increased pressure in the pulmonary circulation results from a sequence of cardiac events that cause blood to "back-up" into the lungs:

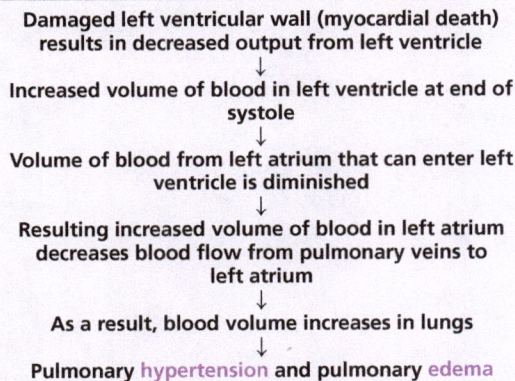

Damaged left ventricular wall (myocardial death) results in decreased output from left ventricle

↓

Increased volume of blood in left ventricle at end of systole

↓

Volume of blood from left atrium that can enter left ventricle is diminished

↓

Resulting increased volume of blood in left atrium decreases blood flow from pulmonary veins to left atrium

↓

As a result, blood volume increases in lungs

↓

Pulmonary hypertension and pulmonary edema

- Dyspnea and overall weakness. The pulmonary edema increases resistance to airflow and gas exchange.

It is common for a viscous cycle to evolve whereby pulmonary congestion will result in increased resistance to blood flow through the lung, leading to increased pressure in the right ventricle. In an effort to compensate for the decreased output from both ventricles, the heart force and rate will increase and, over time, cardiomegaly develops. Cardiomegaly will cause disruption of normal atrioventricular valve function, putting further load demands on the ventricles.

Ventricular failure is accompanied by EKG changes.

- Wide and deep Q wave

- Deep S wave

- Tall R wave

The S3 (third) heart sound is considered a cardinal sign for CHF in patients over 40 years of age. This sound is produced by overexpansion of the ventricular walls during diastole. The S3 heart sound is difficult to discern and may require a cardiologist for its discovery.

In this patient, symptoms of heart failure developed as a result of long-standing coronary artery disease associated with smoking and illicit drug use.

Chronic Obstructive Pulmonary Disease

COPD is a progressive disease of the respiratory system, in most cases caused by long-term smoking. Many of the presenting respiratory symptoms for COPD resemble those of CHF and valvular disease. A chronic cough will often be present with each of these diseases. However, in COPD, the cough typically produces thick mucous. In this patient, imaging and procedures revealed:

- Left ventricular hypertrophy. In COPD, the resistance to blood flow to the lungs will typically result in *right* ventricular hypertrophy.

Cardiomegaly Enlarged heart

- A low left ventricular EF would not be expected in COPD.
- Electrocardiography showed deviations that are characteristic of heart failure.

Valvular Disease

Valvular disease encompasses a spectrum of pathologies involving one or more of the heart valves. The valve is considered **stenotic** if cannot open fully due to narrowing or hardening. Alternatively, the valve is considered **incompetent** if it is unable to close completely.

Valvular disease is not indicated as the underlying problem in this patient because:

- No murmurs were noted during auscultation of the thorax.
- No opacities were observed in the radiographic images of the heart

The low left ventricular EF in the patient is due to weakness of the myocardium rather than stenosis/insufficiency of the aortic valve.

Patient Presentation

A 22-year-old male is admitted to the emergency department complaining of chest pain.

Relevant Clinical Findings

History

The patient relates that the initial pain began 2 days ago and has progressively worsened. He describes the pain to be beneath his sternum and it "comes and goes." He is having trouble sleeping because the pain increases when he lies down. Coughing or sneezing will initiate a bout of pain. A week ago, he "came down with the flu," which consisted of fever, sore throat, and sores in his mouth. He relates that he still has a mild sore throat. There is no history of heart disease.

Physical Examination

Noteworthy vital signs:

- Temperature: 37.2°C (99°F)
  Normal: 36.0–37.5°C = 96.5–99.5°F
- Pulse: 106 bpm
  Adult resting rate: 60–100 bpm

Results of physical examination of the thorax:

- Auscultation reveals a friction rub at the left sternal margin
- Inflammation of the oropharynx

Laboratory Tests

Test	Value	Reference value
Troponin I	0	0–0.04 ng/mL
Troponin T	0	0–0.01 ng/mL

Clinical Note

The troponin test is a blood test for levels of troponin T and I. These compounds will be elevated in serum when there is heart muscle damage. Troponin is usually detectable in a blood sample within 6 hours of a myocardial infarction and may remain detectable for up to 2 weeks. The levels of troponin T and I are indicative of the degree of myocardial injury.

Procedures

- Electrocardiogram (EKG) results show depression of the PR wave and elevation of the ST segment.
- Echocardiography shows a normal size heart and pericardial cavity (**Fig. 2.3.1**).

Clinical Problems to Consider

- Acute pericarditis
- Myocardial infarction
- Pleuritis

LEARNING OBJECTIVES

1. Describe the anatomy of the pericardium and pleura.
2. Describe the anatomy of the coronary arteries.
3. Describe the nerve supply of the pleura, pericardium, and myocardium.
4. Explain the anatomical basis for the signs and symptoms associated with this case.

Auscultation A diagnostic method, usually with a stethoscope, to listen to body sounds (e.g., heart, breath, and gastrointestinal sounds)

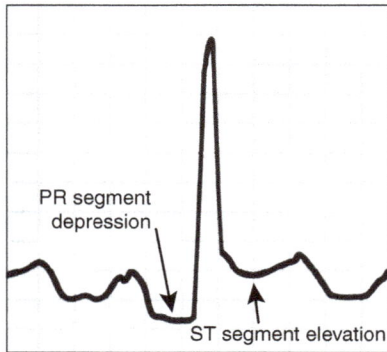

FIGURE 2.3.1 EKG of a patient with pericarditis showing ST segment elevation with PR segment depression.

RELEVANT ANATOMY

Pericardium

The **pericardium** forms a sac that envelops the heart (**Fig. 2.3.2**). The pericardium is divided into:

- **Fibrous pericardium**
- **Serous pericardium**

The **fibrous pericardium** is outermost, thick, and opaque. It encases the heart as well as the great vessels as they exit the heart. Thus, the ascending aorta and the pulmonary trunk are within the **pericardial sac**. The fibrous pericardium loses its identity as it blends with the adventitia of the great vessels. Inferiorly, it fuses with the central tendon of the diaphragm.

The **serous pericardium** is composed of parietal and visceral parts. The **parietal layer of serous pericardium** is adherent to, and inseparable from, the inner surface of the fibrous pericardium. At the great vessels, the parietal layer separates from the fibrous pericardium and reflects onto the surface of the heart as the **visceral layer of serous pericardium**. Coronary vasculature courses between the myocardium and the visceral layer of serous pericardium.

The **pericardial cavity** is the potential space between the parietal and visceral layers of serous pericardium. A few milliliters of **pericardial fluid** limit surface tension between these layers during the cardiac cycle.

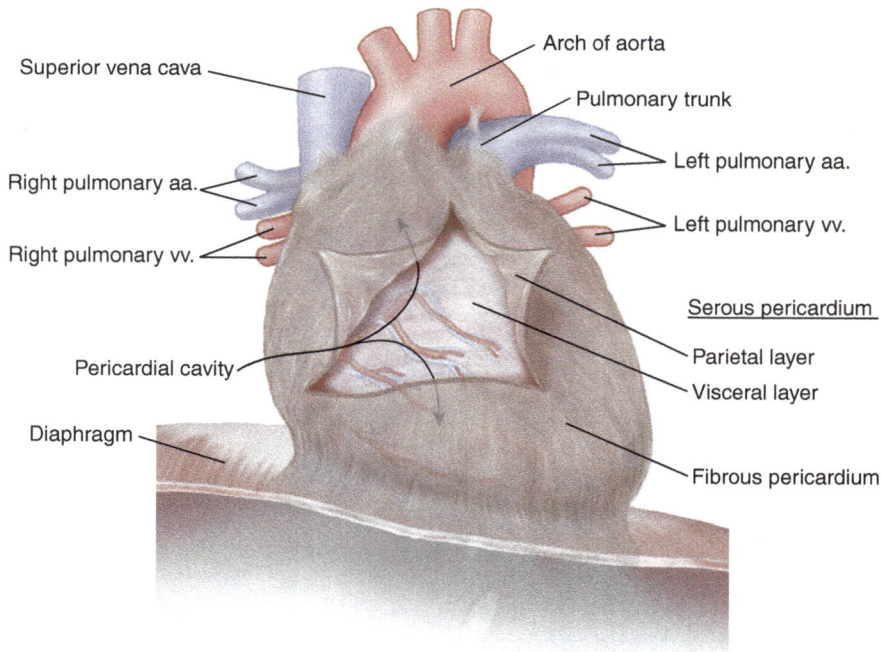

FIGURE 2.3.2 Anterior view of the heart in the pericardial sac (opened).

(A)

(B)

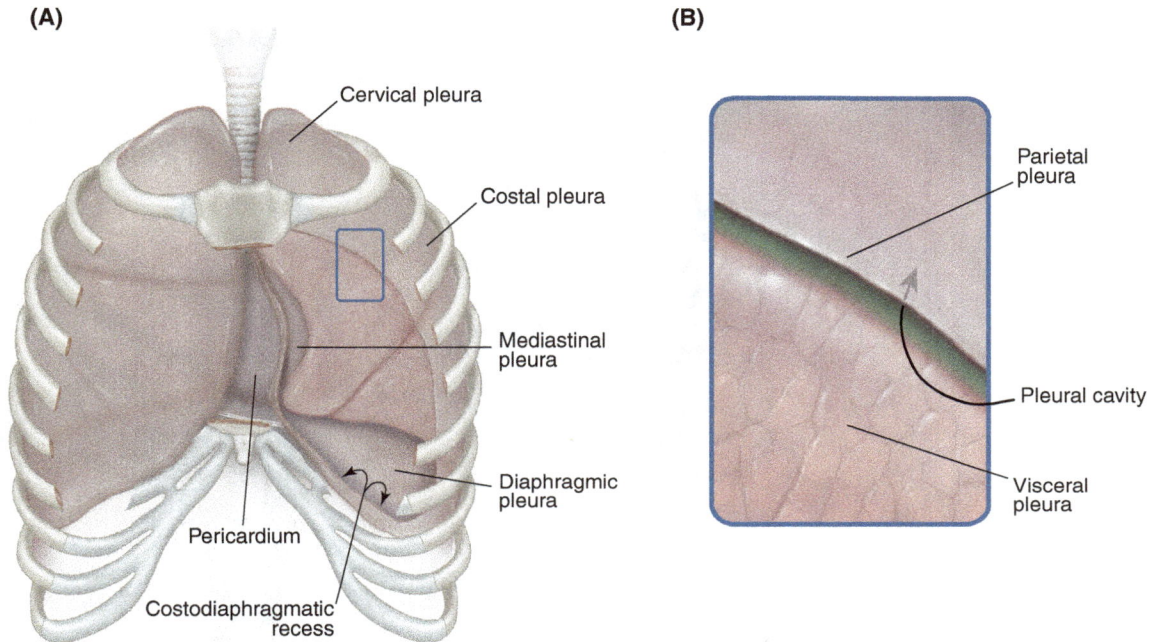

FIGURE 2.3.3 (**A**) Anterior view of the thoracic cavity (anterior wall removed). The parietal pleura is intact on the right side. On the left, the anterior part of costal pleura has been removed to show the pleural cavity and left lung. (**B**) High power view of the inset (in A) showing the relationship between visceral pleura, lung, pleural cavity, and parietal pleura.

Pleura

The **thoracic cavity** is divided into **right and left pulmonary cavities** and a central **mediastinum**. Each pulmonary cavity contains the respective lung and a sac of **pleura**. The pleural sacs do not communicate across the midline (**Fig. 2.3.3**). Pleura is a membrane that consists of two parts:

1. **Parietal**
2. **Visceral**

Parietal pleura is adherent to the walls of each pulmonary cavity and is divided into parts based on the structures with which it is associated:

- **Costal pleura**
- **Diaphragmatic pleura**
- **Mediastinal pleura**
- **Cervical pleura**

Points of reflection of parietal pleura create two recesses:

1. **Costodiaphragmatic**
2. **Costomediastinal**

A pleural sleeve (mesopneumonium) surrounds the pulmonary vessels and main bronchus as they extend from the mediastinum to the hilum of the lung. Collectively, these structures form the root of the lung. At the hilum, the pleural sleeve reflects onto the surface of the lung to become visceral pleura. **Visceral pleura** coats each lung lobe and lines the fissures between the lung lobes.

The potential space between parietal and visceral pleurae is the **pleural cavity**. This cavity contains a few milliliters of fluid to reduce friction when the parietal and visceral layers come in contact during a respiratory cycle.

(A) Anterior

Sinu-atrial
nodal a.

Right coronary a.

Left coronary a.

Circumflex a.

Right marginal a.

Anterior interventricular a.

(B) Posterior

Circumflex a.

Left marginal a.

Right coronary a.

Posterior interventricular a.

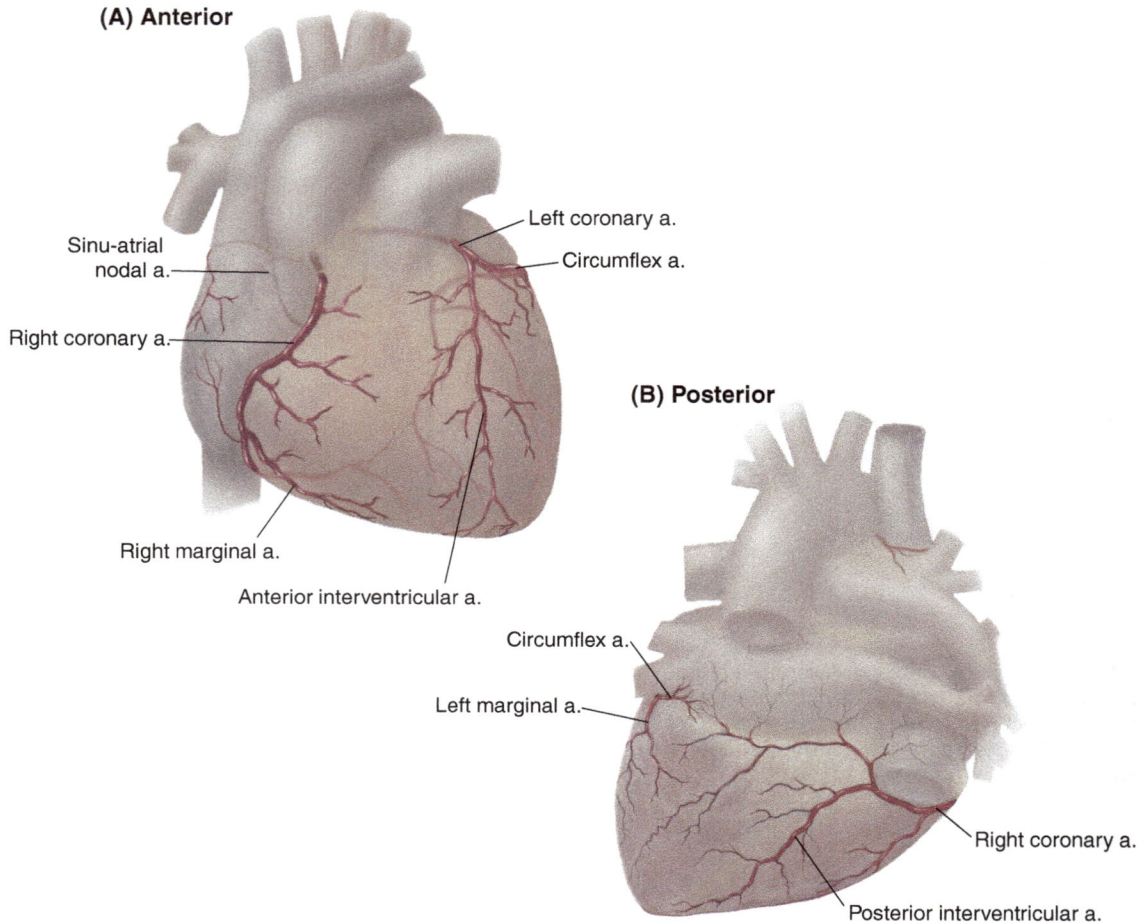

FIGURE 2.3.4 (**A**) Anterior/sternocostal and (**B**) posterior/diaphragmatic views of the coronary circulation.

Nerve Supply of the Pleura, Pericardium, and Myocardium

Nerve supply to the pleura, pericardium, and myocardium (**Table 2.3.2**) utilizes the phrenic (C3-C5) and intercostal nerves (T1-T11), as well as sympathetic (T1-T5) and parasympathetic nerves (vagus—CN X). A cardiac plexus, associated with the great vessels, extends along tracheobronchial tree as the pulmonary plexus. These plexuses have sympathetic and parasympathetic components:

- Cardiac (cardiopulmonary splanchnic) nerves carry **postganglionic sympathetic fibers** from the cervical and upper four thoracic paraverte-

bral ganglia to the cardiac plexus. Stimulation of these **visceral efferents** increases heart rate and contraction force, and dilates coronary arteries. These sympathetic fibers also conduct visceral afferent impulses (reflexive and nociceptive) from thoracic viscera.

- **Preganglionic parasympathetic fibers** originate from the vagus nerve (cardiac branches). Parasympathetic stimulation decreases heart rate and contraction force, and constricts coronary arteries.

Nociception Nerve modality related to pain

TABLE 2.3.1 | **Major branches of coronary arteries and their distribution.**

Artery	Branch	Distribution
Right	Sinu-atrial nodal (60%)	▪ Right atrium ▪ Sinoatrial node
	Right marginal	▪ Right ventricle and apex
	Posterior interventricular (67%)	▪ Right and left ventricles ▪ Interventricular septum
	Atrioventricular nodal	▪ Atrioventricular node
Left	Anterior interventricular	▪ Right and left ventricles ▪ Interventricular septum
	Circumflex	▪ Left ventricle ▪ Left atrium
	Sinu-atrial nodal (40%)	▪ Left atrium ▪ Sinoatrial node
	Left marginal	▪ Left ventricle
	Posterior interventricular (33%)	▪ Right and left ventricles ▪ Interventricular septum

Pain from pleura, pericardium, and myocardium may be referred to dermatomes C2-T4, depending on which organ, tissue, or nerve is involved. For additional information about the autonomic nervous system and referred pain, see Chapter 1: Visceral Afferent Pathways.

Normal Electrocardiogram

The normal EKG is presented in **Figure 2.3.5**. For further discussion of the normal EKG see pages 20–21. Alterations to the heart conduction system, valvular function, and/or myocardium may affect electrical activity in the heart that will manifest as changes in the specific segments of the EKG.

CLINICAL REASONING

This patient presents with **intermittent, substernal chest pain of recent onset**.

Acute Pericarditis

Acute pericarditis is an inflammation of the serous pericardium. It may occur with or without pericardial effusion, that is, an abnormal collection of fluid

---

Effusion Abnormal collection of fluid (e.g., blood, lymph, synovial, pleural, or pericardial)

TABLE 2.3.2 | **Innervation of pleura, pericardium, and myocardium.**

Tissue	Nerve(s)	Modality
Parietal pleura Costal	Intercostal	General sensation
Diaphragmatic Central Peripheral	Phrenic Intercostal	
Mediastinal	Phrenic	
Cervical	Phrenic	
Visceral pleura	Sympathetic	Anoxia and distention
Fibrous pericardium	Phrenic	General sensation
Serous pericardium Parietal Visceral	Phrenic Sympathetic	General sensation Anoxia and distention
Myocardium	Sympathetic	Increases heart rate and force; dilates coronary arteries
	Parasympathetic	Decreases heart rate and force; constricts coronary arteries

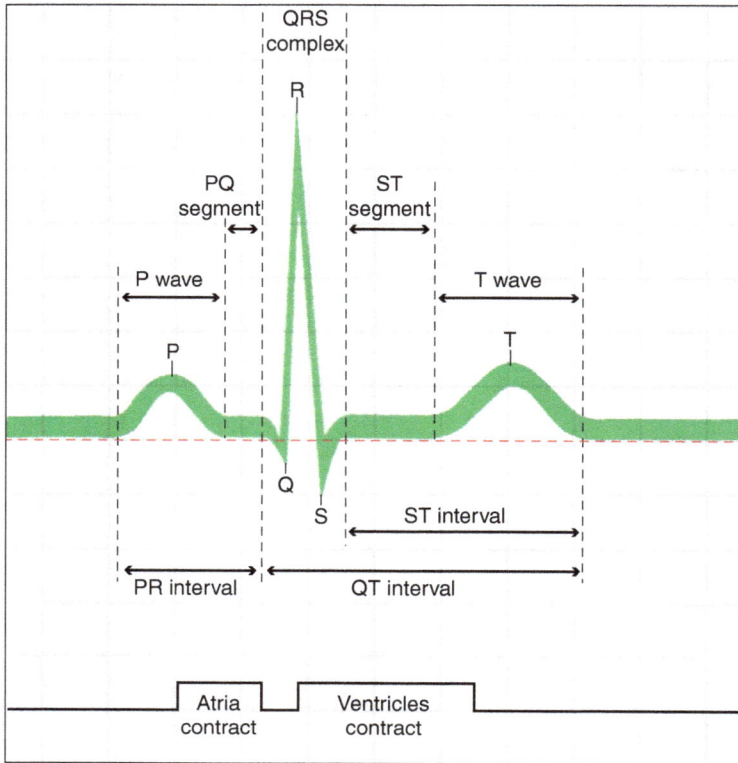

FIGURE 2.3.5 Normal electrocardiogram (EKG).

in the pericardial cavity. The most common cause is a viral infection and the usual agent is coxsackievirus B. Idiopathic acute pericarditis is also most likely viral.

Signs and Symptoms
- Febrile
- Substernal pain (dull or sharp), typically increased when recumbent, coughing, or sneezing
- Tachycardia
- Tachypnea
- Pericardial rub heard on auscultation

Predisposing Factors
- Recent history of viral infection, especially coxsackievirus B
- Recent (2–3 days) of myocardial infarction
- Open coronary artery surgery

Clinical Note

Dull pain is suggestive of primary involvement of the visceral serous pericardium (visceral afferent fibers). In contrast, sharp pain would indicate involvement of the parietal serous pericardium (somatic afferent fibers). Rapid, shallow breathing (tachypnea) reduces the pain.

Febrile Elevated body temperature, that is, a fever (normal body temperature: 36.0–37.5°C (96.5–99.5°F)
Tachycardia Increased heart rate: >100 bpm (normal adult heart rate: 55–100 bpm)

Tachypnea Increased respiration rate (normal adult respiration rate: 14–18 cycles/min)

Myocardial Infarction

A myocardial infarction (commonly called a "heart attack") occurs when a region of the heart wall is deprived of oxygen for a prolonged period and muscle cells die. Most myocardial infarctions are caused by blockage of one or more coronary arteries. The blockage frequently results from thrombus formation.

Signs and Symptoms

- Chest discomfort (female: aching, tightness, or pressure) or pain (male: often described as "crushing")
- Pain may be referred to arms (most commonly left), shoulders, neck, teeth, or back
- Dyspnea
- Hyperhidrosis
- Nausea and vomiting
- Malaise or fatigue
- Anxiety
- Elevated serum troponin T and I

Predisposing Factors

Nonmodifiable

- Sex: 2:1 male to female ratio
- Age: incidence increases with age
- Family history of heart disease
- Race: African American, Mexican American, Native American

Modifiable

- Smoking
- Diet high in cholesterol and triglycerides
- Obesity
- Stress

Thrombus A fixed mass of platelets and/or fibrin (clot) that partially or totally occludes a blood vessel or heart chamber. An embolism is a mobile clot in the cardiovascular system

Dyspnea Difficulty breathing, shortness of breath

Hyperhidrosis Excessive sweating

Malaise Feeling of general body weakness or discomfort, often marking the onset of an illness

Clinical Note

Women <65 years of age are 25–30% more likely than men of similar age to present without chest pain.

Pleuritis

Pleuritis (pleurisy) is an inflammation of pleura. Pleuritis may occur with our without pleural effusion. In young, healthy adults, the cause is usually a respiratory viral infection (especially coxsackievirus B).

Signs and Symptoms

- Sharp thoracic wall pain, aggravated by deep breathing, coughing, or sneezing
- Auscultation reveals "pleural friction rub" (may be inconstant)
- Febrile
- Tachypnea

Predisposing Factors

- Recent viral infection
- Chest injury (simple rib fracture)
- Pneumonia
- Tuberculosis

The chest pain is caused by inflamed pleural membranes rubbing against each other (friction rub). Both pleural sacs are often involved, so the symptoms may be bilateral. The breathing is often shallow and rapid since this limits the movement of lungs and thoracic wall and, thereby, reduces the pain.

DIAGNOSIS

The patient presentation, medical history, physical examination, laboratory tests, and procedures support a diagnosis of **acute pericarditis**.

Acute Pericarditis

Signs and symptoms related to pericarditis involve inflammation of the membranes around the heart. Normally, these membranes should glide smoothly on each other during the cardiac cycle. With irritation, increase in friction produces auscultatory sounds and stimulates pain receptors.

- The friction rub at the left border of the sternum, determined by auscultation, is pericardial. A friction rub associated with pleuritis is heard lateral to the mediastinum, and frequently is bilateral.

- Substernal pain (dull or sharp), typically increased when recumbent, coughing, or sneezing. Pain may be referred to dermatomes of the neck because pericardium is innervated by phrenic nerve (C3-C5).

- Electrocardiography is consistent with early pericarditis.

- Normal serum troponin levels make a myocardial infarction unlikely.

- The patient is at risk because he has symptoms of a respiratory viral infection, probably coxsackievirus B.

Pleuritis

Like pericardium, pleural membranes should glide smoothly on each other during the respiratory cycle. Pleural inflammation, likewise, produces auscultatory sounds and stimulates pain receptors.

- A friction rub associated with pleuritis is heard lateral to the mediastinum, and is frequently bilateral.

- Referred pain may indicate the involved portion(s) of pleura:
  - Involvement of costal or peripheral diaphragmatic pleura, supplied by intercostal nerves, may refer pain to the thoracic wall.
  - Involvement of cervical, mediastinal, or central diaphragmatic pleura, innervated by the phrenic nerve (C3-C5), may refer pain to dermatomes in the neck.

- Respiratory viral infection places the patient at risk. This is also a risk factor for pericarditis.

Myocardial Infarction

Heart attacks affect cardiac muscle. The associated pain is due to prolonged ischemia and necrosis of myocardium. Friction rubs are not present because the pleural and pericardial membranes are not inflamed.

- Characterized by "crushing" substernal pain (male) or chest tightness (female) as opposed to sharp stabbing pain associated with inflammation of pleural and pericardial membranes.

- Pain is not affected by body position or respiration, but may be referred to upper limb, neck, teeth, and back.

- Serum troponin is elevated, usually within 6 hours, with heart muscle damage, and is indicative of the degree of myocardial injury.

- Elevated body temperature is not a symptom because pathophysiology does not involve infection or inflammation.

- Hyperhidrosis, nausea, vomiting, malaise, fatigue, and anxiety are not typically associated with pericarditis or pleuritis.

Ischemia Local anemia due to vascular obstruction   **Necrosis** Pathologic death of cells, tissues, or organs

Patient Presentation

A 26-year-old male is admitted to the emergency department complaining of chest pain.

Relevant Clinical Findings

History

The patient relates that over the past 3 hours he has been experiencing sharp pain over the right side of his chest. The pain started suddenly when he was at home watching television. He relates that the pain worsens when he inhales. He also feels that his breathing is becoming progressively more difficult. The patient admits to smoking as many as 15 cigarettes each day. There is no history of heart or respiratory disease.

Physical Examination

Noteworthy vital signs include:

- Height: 6'4"
- Weight: 168 lbs
- Pulse: 112 bpm
  Adult resting rate: 60–100 bpm
- Respiratory rate: 24 cycles/min
  Normal adult: 14–18 cycles/min; women slightly higher

Results of physical examination of the thorax:

- Right, upper thorax has mild hyperresonance on percussion.
- Respiratory sounds on the right are absent in the upper lobe and are weak in the middle and lower lobes.

Imaging Studies

- Anteroposterior chest radiography reveals a visceral pleural line in the right pulmonary cavity.
- No patchy infiltrates or increased bronchovascular markings.

Procedures

- Electrocardiography (EKG) was normal.

Clinical Problems to Consider

- Myocardial infarction
- Pneumonia
- Primary spontaneous pneumothorax

LEARNING OBJECTIVES

1. Describe the anatomy of the pleura.
2. Describe the anatomy of the airways in the thorax.
3. Describe the coronary arteries.
4. Explain the anatomical basis for the signs and symptoms associated with this case.

Hyperresonance Exaggeration of sound produced by transmission of vibrations generated by an organ within a cavity; often elicited by percussion

Percuss Diagnostic procedure in which a body part is tapped gently with a finger or instrument; used to assess organ density or to stimulate a peripheral nerve

RELEVANT ANATOMY

Pleura

The **thoracic cavity** is divided into **right and left pulmonary cavities** and a central **mediastinum**. Each pulmonary cavity contains the respective lung and a sac of **pleura** (**Fig. 2.4.1**). The pleural sacs do not communicate across the midline. Pleura is a membrane that consists of two parts:

1. **Parietal**
2. **Visceral**

Pneumothorax Air or gas in pleural cavity

(A)

(B)

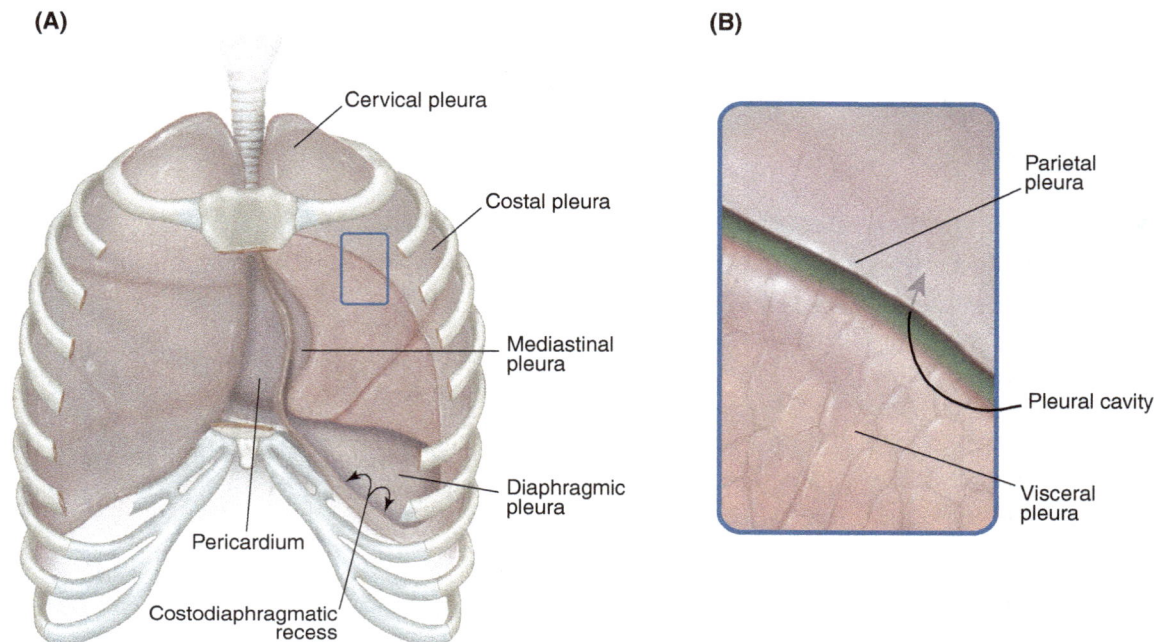

FIGURE 2.4.1 (**A**) Anterior view of the thoracic cavity (anterior wall removed). The parietal pleura is intact on the right side. On the left, the anterior part of costal pleura has been removed to show the pleural cavity and left lung. (**B**) High power view of the inset (in A) showing the relationship between visceral pleura, lung, pleural cavity, and parietal pleura.

Parietal pleura is adherent to the walls of each pulmonary cavity and is divided into parts based on the structures with which it is associated:

- **Costal pleura**
- **Diaphragmatic pleura**
- **Mediastinal pleura**
- **Cervical pleura**

Points of reflection of parietal pleura create two recesses:

1. **Costodiaphragmatic**
2. **Costomediastinal**

A pleural sleeve (mesopneumonium) surrounds the pulmonary vessels and main bronchus as they extend from the mediastinum to the hilum of the lung. Collectively, these structures form the root of the lung. At the hilum, the pleural sleeve reflects onto the surface of the lung to become visceral pleura. **Visceral pleura** coats each lung lobe and lines the fissures between the lung lobes.

The potential space between parietal and visceral pleurae is the **pleural cavity**. This cavity contains a few milliliters of fluid to reduce friction when the parietal and visceral layers come in contact during a respiratory cycle.

Nerve Supply

Nerve supply to the pleura (**Table 2.4.1**) utilizes the phrenic (C3-C5) and intercostal nerves (T1-T11), as well as sympathetic (T1-T5) and parasympathetic (vagus—CN X) nerves.

Innervation of the lungs is via the cardiac plexus. It is associated with the great vessels and extends along tracheobronchial tree as the pulmonary plexus. The pulmonary plexus has sympathetic and parasympathetic components:

- Cardiac (cardiopulmonary splanchnic) nerves carry **postganglionic sympathetic fibers** from the cervical and upper four thoracic paravertebral ganglia to the cardiac plexus. Stimulation of these **visceral efferents** increases respiratory rate

TABLE 2.4.1 | **Innervation of pleura.**

Pleura	Nerve(s)	Modality
Parietal		General sensation
Costal	Intercostal	
Diaphragmatic		
Central	Phrenic	
Peripheral	Intercostal	
Mediastinal	Phrenic	
Cervical	Phrenic	
Visceral	Sympathetic	Anoxia and distention

and leads to bronchodilation and decreased glandular secretion. These fibers also conduct visceral afferent impulses (reflexive and nociceptive).

- **Preganglionic parasympathetic fibers** (cardiac branches) originate from the vagus nerve. Parasympathetic stimulation decreases respiratory rate and leads to bronchoconstriction and increased glandular secretion.

Nociception Nerve modality related to pain

Pain from pleura and lungs may be referred to dermatomes C2-T4, depending on which nerve is involved. For additional information about the autonomic nervous system and referred pain, see Chapter 1: Visceral Afferent Pathways.

Tracheobronchial Tree
Extrapulmonary Airway
Extrapulmonary airways consist of the **trachea and main bronchi**. The trachea in the adult is a hollow tube approximately 2.5 cm in diameter. It is kept patent by a series of C-shaped, cartilaginous "rings" that are incomplete posteriorly. The **trachialis** muscle occupies the gap in each cartilage. The **trachea** passes from the neck to the thorax near the midline. It enters the superior mediastinum and descends to the level of the **manubriosternal angle** (Louis). Here, it divides into **right and left main bronchi** (**Fig. 2.4.2**). The **right main bronchus** is

- shorter,
- more vertical, and
- of greater diameter.

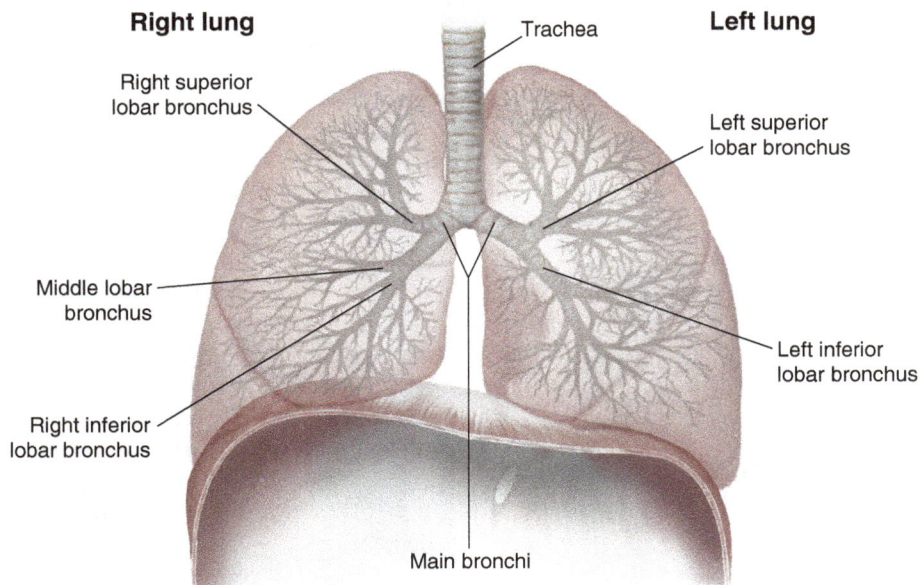

FIGURE 2.4.2 Anterior view of the tracheobronchial tree.

(A) Anterior

Sinoatrial
nodal a.

Left coronary a.

Circumflex a.

Right coronary a.

(B) Posterior

Right marginal a.

Anterior interventricular a.

Circumflex a.

Left marginal a.

Right coronary a.

Posterior interventricular a.

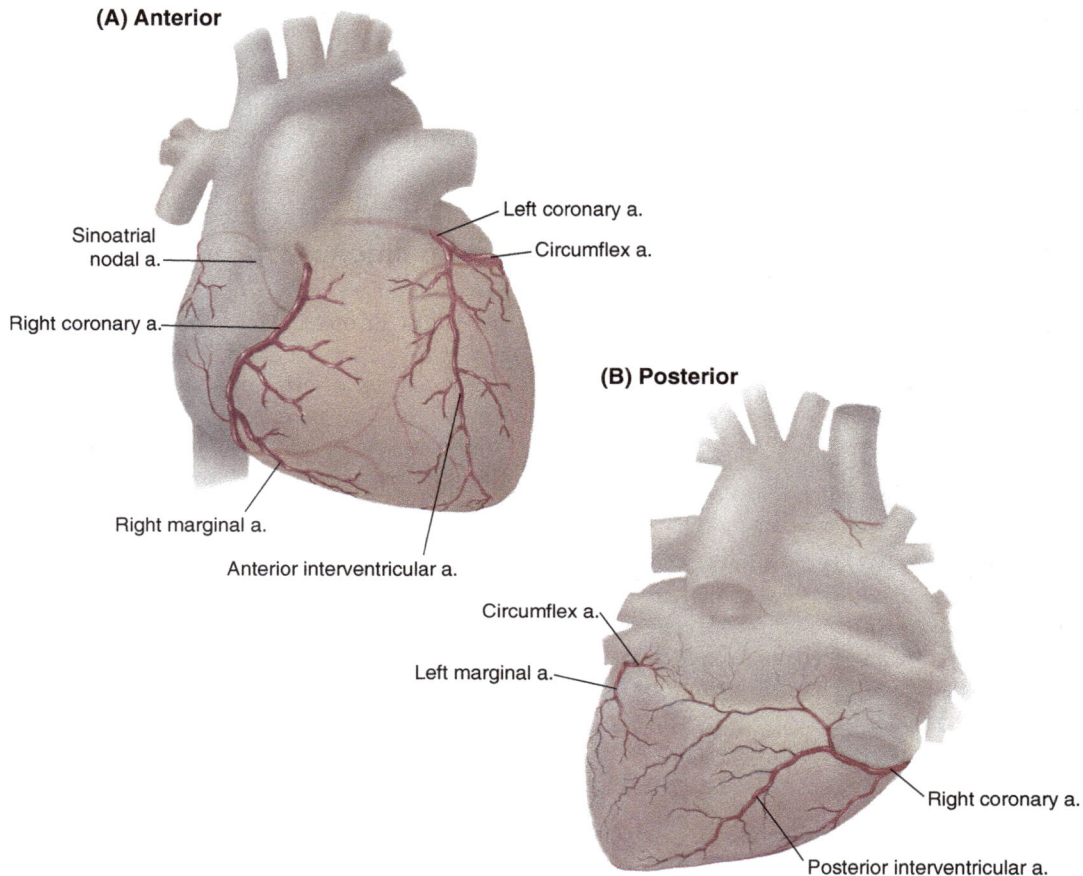

FIGURE 2.4.3 (**A**) Anterior/sternocostal and (**B**) posterior/diaphragmatic views of the coronary circulation.

Each **main bronchus** and its respective pulmonary artery and veins, enveloped by pleura (mesopneumonium), forms the **root of the lung**.

Coronary Arteries

The right and left coronary arteries branch from the respective sinuses of the aortic valve (**Fig. 2.4.3**). Each vessel immediately enters the coronary sulcus.

- The **right coronary artery** arises from the right aortic sinus and follows the **coronary sulcus** to the posterior aspect of the heart.

- The **left coronary artery** arises from the left aortic sinus and is shorter. It gives rise to its terminal branches (**anterior interventricular**

and **circumflex**) soon after passing between the pulmonary trunk and auricle of the left atrium.

Collateral circulation and anastomoses between branches of the coronary arteries do exist; however, they are not well established in the healthy heart and are highly variable. The branches of the coronary arteries and their **distribution** are outlined in **Table 2.4.2**. Heart circulation is classified as "**right dominant**" or "**left dominant**," based on which coronary artery gives rise to the **posterior interventricular artery**. In 67% of cases, this artery is derived from the right coronary artery. The sinu-atrial artery is a branch of the right coronary artery in 60% of cases. In the remainder, this artery is a branch of the circumflex artery from the left coronary artery.

TABLE 2.4.2	Major branches of coronary arteries and their distribution.	
Artery	**Branch**	**Distribution**
Right	Sinu-atrial nodal (60%)	• Right atrium • Sinoatrial node
	Right marginal	• Right ventricle and apex
	Posterior interventricular (67%)	• Right and left ventricles • Interventricular septum
	Atrioventricular nodal	• Atrioventricular node
Left	Anterior interventricular	• Right and left ventricles • Interventricular septum
	Circumflex	• Left ventricle • Left atrium
	Sinu-atrial nodal (40%)	• Left atrium • Sinoatrial node
	Left marginal	• Left ventricle
	Posterior interventricular (33%)	• Right and left ventricles • Interventricular septum

CLINICAL REASONING

This patient presents with **sudden-onset pain on the right side of his chest**.

Myocardial Infarction

A myocardial infarction (commonly called a "heart attack") occurs when a region of the heart wall is deprived of oxygen for a prolonged period and muscle cells die. Most myocardial infarctions are caused by the blockage of one or more coronary arteries. The blockage frequently results from thrombus formation.

Signs and Symptoms

- Chest discomfort (female: aching, tightness, or pressure) or pain (male: often described as "crushing")
- Pain may be referred to the arms (most commonly left), shoulders, neck, teeth, or back
- Dyspnea
- Hyperhidrosis
- Nausea and vomiting
- Malaise or fatigue
- Anxiety
- Elevated serum troponin T and I

Predisposing Factors

Uncontrollable
- Sex: 2:1 male
- Age: incidence increases with age
- Family history of heart disease
- Race: African American, Mexican American, Native American

Controllable
- Smoking
- Diet high in cholesterol and triglycerides
- Obesity
- Stress

Clinical Notes

- Women <65 years of age are 25–30% more likely than men of similar age to present without chest pain.
- The troponin test is a blood test for levels of troponin T and I. These compounds will be elevated in serum when there is heart muscle damage. Troponin is usually detectable in a blood sample within 6 hours of a myocardial infarction and may remain detectable for up to 2 weeks. The levels of troponin T and I are indicative of the degree of myocardial injury.

Dyspnea Difficulty breathing and shortness of breath
Hyperhidrosis Excessive sweating
Malaise Feeling of general body weakness or discomfort, often marking the onset of an illness

Thrombus A fixed mass of platelets and/or fibrin (clot) that partially or totally occludes a blood vessel or heart chamber. An embolism is a mobile clot in the cardiovascular system

Pneumonia

Pneumonia is an infection of the lung. While over 100 infectious agents have been linked to pneumonia, the most common cause is the gram-positive bacterium, *Streptococcus pneumoniae*. The common etiology is an initial infection of the nasopharynx with subsequent spread to other areas of the respiratory tree. Pneumonia may develop if there is a failure to clear the bacterium from the lower respiratory system. Infections may last for weeks to months. Viral co-infection may lead to excess mucus production resulting from impaired ciliary action. Smoking may have a similar effect.

Signs and Symptoms

- Dyspnea
- Cough, with or without sputum
- Febrile
- Crackling and wheezing breath sounds
- Percussion dullness
- Patchy lung infiltrate and increased bronchovascular markings in radiographs (**Fig. 2.4.5**)

Predisposing Factors

- Recent history of nasopharyngeal infection
- Age: >65 years
- Smoking
- Chronic cardiovascular or respiratory illnesses
- Long-term use of inhaled corticosteroids
- Compromised immune system (chemotherapy or long-term treatment with immunosuppressant drugs)

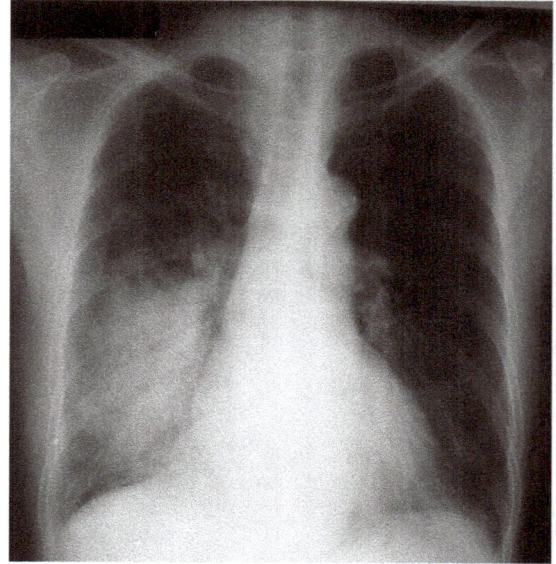

FIGURE 2.4.5 Anteroposterior chest radiograph showing lobar pneumonia in the lower lobe of the right lung. *Source:* Fig. 134-6, *Harrison's Online.*

Clinical Note

Most cases are considered to be community-acquired pneumonia (CAP). Areas of crowding (e.g., nursing homes and child-care centers) have a higher incidence.

Spontaneous Pneumothorax

The pleural cavity is normally under negative pressure compared to alveolar pressure in the lung. A pneumothorax is a condition in which air accumulates in the pleural cavity. This will compromise the expansion of the lung on the affected side. Excessive accumulation of air in the pleural cavity will cause the lung to collapse (atelectasis).

Atelectasis Reduction or absence of air in all or part of a lung (lung collapse)

Crackle Crackling noise heard with lung disease (also known as a rale)

Febrile Elevated body temperature, that is, a fever (normal body temperature: 36.0–37.5°C (96.5–99.5°F)

Wheezing Labored breathing that produces a hoarse, whistling sound

Spontaneous pneumothorax is divided into:

- **Primary:** spontaneous pneumothorax without underlying lung disease or trauma
- **Secondary:** spontaneous pneumothorax with a history of lung disease or recent trauma

Signs and Symptoms

- Sudden onset of unilateral, constant chest pain
- Tachypnea
- Tachycardia

Predisposing Factors

- Age: 20–30 years
- Sex: male (6:1)
- Stature: tall and lean individuals
- Smoking: 20-fold increase in risk for males; 10-fold for females
- Previous pneumothorax (15–40% of patients have a recurrence)

Tall, young males, especially those who smoke, have a higher incidence of blebs on the visceral surface of the apical regions of the lungs. Rupture of a bleb will allow alveolar air to escape into the pleural cavity, producing a pneumothorax.

Clinical Notes

- Any type of pneumothorax has the potential to become a **tension pneumothorax**. This life-threatening situation occurs when a piece of injured tissue (on the lung surface or in the thoracic wall) forms a one-way valve:
  - Air can enter the pleural cavity during inspiration, but cannot escape during expiration.

  The atmospheric air in the pleural cavity can cause complete collapse of the lung and a shift of mediastinal structures toward the unaffected side. This will compromise venous return to the heart.

DIAGNOSIS

The patient presentation, medical history, physical examination, imaging studies, and procedures support a diagnosis of **primary spontaneous pneumothorax**.

Primary Spontaneous Pneumothorax

A pneumothorax occurs when the partial vacuum in the pleural cavity is compromised. Primary spontaneous pneumothorax develops when a bleb on the lung surface ruptures. Alveolar (atmospheric) air that enters the pleural cavity may have several consequences: atelectasis, tachypnea, tachycardia, chest pain, and potentially compromised venous return.

- The patient is at risk because of his age, stature, and use of cigarettes.
- Visceral pleural line in radiographs indicates separation of the visceral and parietal pleura by a foreign gas or fluid. The lack of lung parenchymal markings eliminates pneumonia as a possibility.
- The absence of respiratory sounds is consistent with atelectasis. Tachypnea and tachycardia result from decreased lung capacity and decreased oxygen saturation.
- Unilateral chest pain is most likely the result of a small amount of blood being released into the pleural cavity with bleb rupture. Blood is an irritant to pleural membranes.
- The position of the pain (right lateral versus substernal) and a normal EKG is not consistent with myocardial infarction.

Pneumonia

Pneumonia is an infection of the lungs. Accumulation of mucus obstructs airways and impairs pulmonary function. Smoking exacerbates this condition.

- Dyspnea, cough, altered breath sounds, percussion dullness, and radiographic evidence

---

Bleb Small bulla (<1 cm diameter)
Tachycardia Increased heart rate: >100 bpm (normal adult heart rate: 55–100 bpm)

Tachypnea Increased respiration rate (normal adult respiration rate: 14–18 cycles/min)

of lung infiltrates and altered bronchovascular markings reflect obstructed airways and impaired pulmonary function.

- Increased body temperature is a sign of infection. A patient with pneumothorax or myocardial infarction would not be febrile.

Myocardial Infarction

Heart attacks affect cardiac muscle. The associated pain is due to prolonged ischemia and necrosis of myocardium. Friction rubs are not present because the pleural and pericardial membranes are not inflamed.

- Characterized by "crushing" substernal pain (male) or chest tightness (female) as opposed to sharp stabbing pain associated with inflammation of pleural and pericardial membranes.

- Pain is not affected by body position or respiration, but may be referred to upper limb, neck, teeth, and back.

- Serum troponin is elevated, usually within 6 hours, with heart muscle damage and is indicative of the degree of myocardial injury.

- Elevated body temperature is not a symptom because pathophysiology does not involve infection or inflammation.

- Hyperhidrosis, nausea, vomiting, malaise, fatigue, and anxiety are not typically associated with pericarditis or pleuritis.

Ischemia Local anemia due to vascular obstruction

Necrosis Pathologic death of cells, tissues, or organs

Patient Presentation

A 60-year-old female visits her primary care physician because she has discovered a "lump" in her left breast.

Relevant Clinical Findings

History

The patient relates that last week, while bathing, she discovered a lump in her left breast. While she has not noticed any pain in the breast, pushing against the lump causes discomfort. Her last mammogram was approximately 8 years ago. She has had three uncomplicated pregnancies and one first-trimester miscarriage during her 4th decade. She used birth control pills for approximately 15 years (age 35–50). She started menopause at age 56 and began combined hormone replacement medications soon after. She voluntarily stopped the hormone replacement regimen 2 years ago.

She has been a smoker (2–12 cigarettes/day) for most of her adult life. The patient was adopted at 3 months of age and does not know the health history of her biological parents.

Physical Examination

Results of physical examination:

- Left breast exhibits mild edema.
- Slight dimpling of the skin of left breast at the 11 o'clock position.
- Palpation of the left breast (patient either seated or supine) reveals a solid mass, approximately 2 cm in diameter, in its lateral half.
- Two 0.5 cm, movable masses are palpable in the left axilla.
- Both nipples are inverted and the patient indicates that they have been since puberty.
- Contraction of pectoralis major muscles with the patient's hands on her hips results in symmetrical movement of the breasts.

Imaging Studies

- Mammography reveals a 3 cm density in the superior lateral quadrant of the left breast.

Biopsy Results

- Core needle biopsy of the breast mass indicates the presence of adenocarcinoma cells.

Clinical Problems to Consider

- Breast carcinoma
- Fibroadenoma of the breast
- Fibrocystic condition of the breast

LEARNING OBJECTIVES

1. Describe the anatomy of the female breast.
2. Explain the anatomical basis for the signs and symptoms associated with this case.

RELEVANT ANATOMY

Female Breast

The female breast begins to grow and differentiate from the onset of puberty. An **intermammary cleft** separates the breasts. Rarely are the breasts bilaterally symmetrical.

Palpation Physical examination with the hand(s) to assess organs, masses, infiltration, heart beat, pulse, or body cavity vibrations

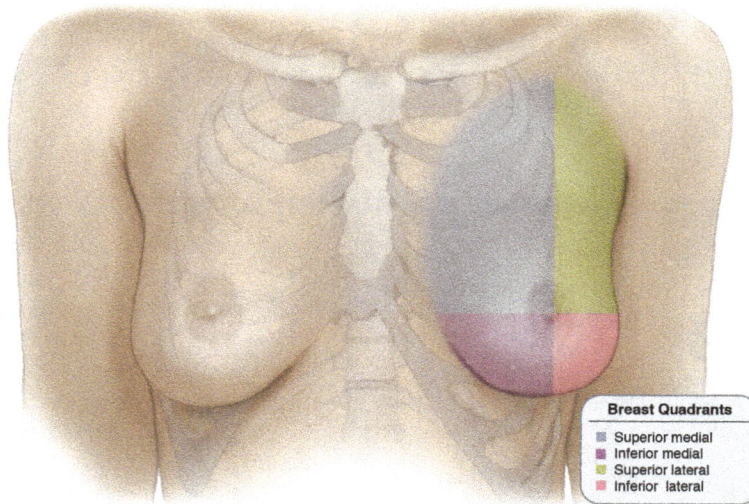

FIGURE 2.5.1 Position of female breast on thoracic wall; the quadrants of the female breast are shown on the left side.

The **base of the breast** is its deepest and broadest part. It overlies costal cartilages 2–6 and their associated ribs (**Fig. 2.5.1A**). On the transverse plane, the base extends from the parasternal region to near the midaxillary line. A superolateral extension of the base lies along the inferior border of the pectoralis major muscle and extends toward, or into, the axilla as the **axillary tail (Spence) of the breast**. Breast tissue lies within the subcutaneous layer of the upper thorax. The base of the breast is in contact with muscular (deep) fascia over:

- Pectoralis major
- Serratus anterior
- Aponeurosis of external abdominal oblique

The **retromammary space** is a layer of loose connective tissue between this muscular fascia and the base of the breast. This space allows the breast some movement independent of the deep fascia and musculature.

For descriptive purposes, the breast is divided into quadrants that intersect at the nipple (**Fig. 2.5.1B**):

1. Superior medial
2. Inferior medial
3. Superior lateral
4. Inferior lateral

Clinical Note

In advanced breast carcinoma, the tumor may invade the retromammary space and adhere the base of the breast to the deep fascia, "fixing" the breast to pectoralis major muscle.

The post-pubescent breast includes the **nipple**, **areola**, **mammary gland**, fat, and connective tissue (**Fig. 2.5.2**). The **nipple** is a conical projection of modified skin that usually lies near the center of the breast, just inferior to the equator. Fifteen to twenty lactiferous ducts open on its tip. Smooth muscle is arranged circumferentially in the nipple and contraction causes it to become erect.

The **areola** is an oval zone of skin that surrounds the nipple. The pigmentation of the areola is typically greater than that in other areas of skin for an individual. Areolar pigmentation increases during the second month of pregnancy. The width of the areola and its degree of pigmentation are variable. The areola usually has several small elevations. These indicate the position of sebaceous **areolar glands (Montgomery)** that are best developed during lactation.

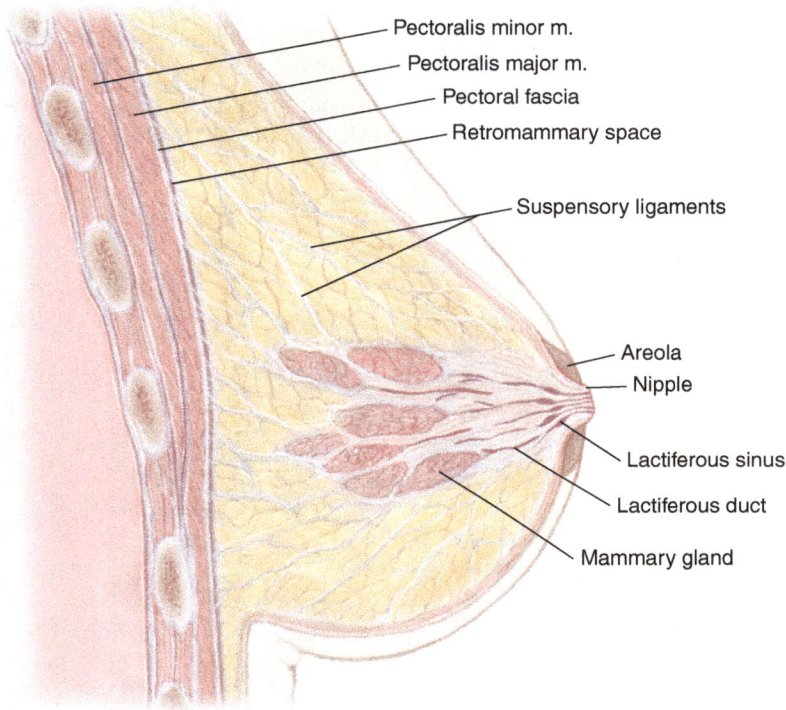

Pectoralis minor m.
Pectoralis major m.
Pectoral fascia
Retromammary space

Suspensory ligaments

Areola
Nipple

Lactiferous sinus

Lactiferous duct

Mammary gland

FIGURE 2.5.2 Sagittal section showing the anatomy of the nonlactating female breast.

Internal Structure

The size, shape, and consistency of the breast varies greatly among individuals and races. Likewise, breast anatomy changes with age, pregnancy, and lactation. In the premenopausal, nonpregnant, and nonlactating condition, the breast is comprised primarily of fat. It is divided into 16–20 lobes that are defined by fibrous septae. These septae extend in a spoke-like fashion from the nipple and areola through the thickness of the breast. Septae are most apparent superiorly (suspensory ligaments of Cooper) and have a role in breast support. Each lobe of the breast contains:

- A single **lactiferous duct**

Clinical Note

Malignant growths in the breast may invade the fibrous septae, placing traction on them. This may cause retraction of the nipple and/or dimpling of the breast skin.

- Clusters of **mammary glands** that produce milk during lactation
- Fat

As each lactiferous duct approaches the nipple, it dilates as a **lactiferous sinus**, which then narrows to open independently on the nipple. The suggested function of the lactiferous sinus is to act as a reservoir for a droplet of milk, which is released when the infant begins to nurse. The taste from this small amount of milk is thought to keep the infant suckling until the mother's neurohormonal response triggers "let down" of additional milk from the mammary gland. Some researchers maintain that the lactiferous sinus is an artifact.

The greatest glandular development occurs during the later months of pregnancy and during lactation. Under the influence of elevated estrogen and progesterone, glandular alveoli proliferate and become capable of producing milk.

Blood Supply

The breast is supplied by branches of vessels derived from the **axillary** and **internal thoracic arteries** and the **aorta** (Fig. 2.5.3A).

- The **lateral mammary arteries** are derived from the lateral thoracic artery, a branch of the axillary. They supply primarily the lateral breast quadrants.

- The **medial mammary arteries** arise from the internal thoracic and supply primarily the medial quadrants.

- Lateral cutaneous branches of **posterior intercostal arteries** from the aorta also have small branches that extend to the lateral breast.

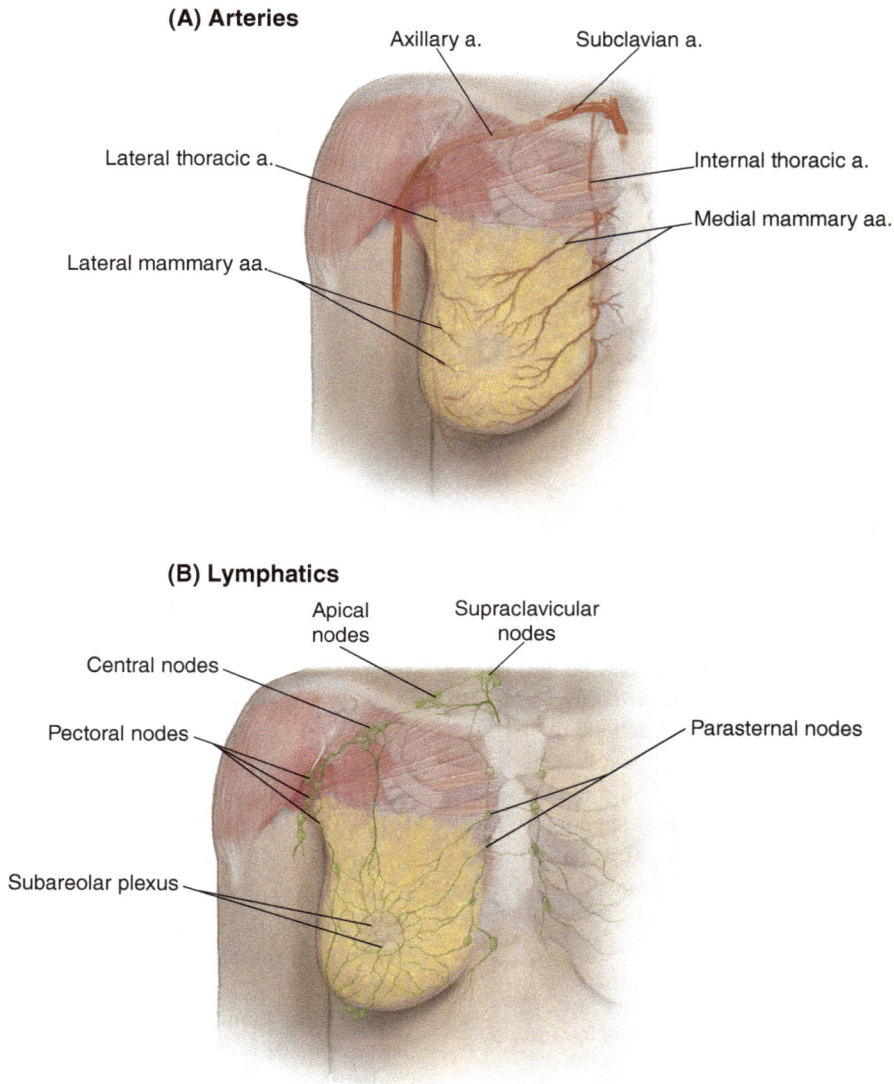

(A) Arteries

Axillary a.   Subclavian a.

Lateral thoracic a.   Internal thoracic a.

Medial mammary aa.

Lateral mammary aa.

(B) Lymphatics

Apical nodes   Supraclavicular nodes

Central nodes

Pectoral nodes   Parasternal nodes

Subareolar plexus

FIGURE 2.5.3 Anterior view of breast vasculature. (**A**) Arterial. (**B**) Lymphatic.

The **venous drainage** follows the arterial pattern, but the largest channels enter the **axillary vein**.

Lymphatics

Lymphatic drainage of the breast is predominately (75%) into **axillary lymph nodes** (**Fig. 2.5.3B**). Lymph from all breast quadrants enters these nodes. The medial quadrants also drain into **parasternal nodes** that lie along internal thoracic vessels. Breast lymphatics are divided into two plexuses.

1. A superficial, **subareolar plexus** drains the terminal portions of the lactiferous ducts and the subcutaneous tissues of the breast and surrounding skin.

2. A **deep plexus** receives lymph from the remaining tissues of each lobe.

There is little communication between these plexuses. However, there are anastomoses between the right and left subareolar lymphatic plexuses. Efferent lymphatic vessels tend to follow blood vessels in the subcutaneous tissues.

Clinical Notes

- Metastasis of breast tumor cells is preferentially via the lymphatic system.
- Anastomoses between the right and left subareolar lymphatic plexuses account for the fact that breast tumor cells can spread to the opposite side.

Axillary lymph nodes are typically divided into three levels.

1. **Level I nodes lie lateral to pectoralis minor muscle** and include the brachial, subscapular, and pectoral groups of axillary nodes.

2. **Level II nodes lie deep to pectoralis minor muscle** and are commonly termed the **central group**.

3. **Level III nodes lie between pectoralis minor muscle and rib 1** and are termed **apical nodes**.

Usually, lymph passes sequentially through levels I, II, and III.

CLINICAL REASONING

This patient presents with a recently self-discovered **mass in her left breast**.

Breast Carcinoma

Breast carcinoma is a malignancy in the breast and almost always originates from the ductal epithelial cells of the mammary gland.

Clinical Notes

- The name of a carcinoma refers to the organ of origin of the tumor cells. This is considered the *original* or *primary* cancer. For example, the term breast carcinoma indicates tumor cells that developed in the breast.
- Metastatic carcinoma has the name of the primary tumor. For example, breast carcinoma that spreads to the lungs and forms a tumor is considered *metastatic (secondary) breast carcinoma*, not lung cancer.
- Sites for metastasis of common primary carcinomas have been developed by the National Cancer Institute of the U.S. National Institutes of Health (**Table 2.5.1**).

Signs and Symptoms

- Painless, palpable breast mass
- Breast skin dimpling and/or nipple retraction
- Nipple discharge
- Edema in affected breast
- Changes in breast symmetry

Predisposing Factors

- Age: median age at diagnosis is 61 years
- Race: white
- Family history (breast or ovarian carcinoma; mutations on chromosomes 13 or 17)
- Previous documented breast carcinoma
- Nulliparous or first full-term pregnancy after age 30

Edema Swelling of skin due to abnormal accumulation of fluid in subcutaneous tissue
Nulliparous Having never borne children

TABLE 2.5.1	Metastatic sites for common primary carcinomas (National Cancer Institute[a]).
Primary cancer type	**Main sites of metastasis**
Breast	Lung, liver, bone
Colon	Liver, peritoneum, lung
Kidney	Lung, liver, bone
Lung	Adrenal gland, liver, lung
Melanoma	Lung, skin/muscle, liver
Ovary	Peritoneum, liver, lung
Pancreas	Liver, lung, peritoneum
Prostate	Bone, lung, liver
Rectum	Liver, lung, adrenal gland
Stomach	Liver, peritoneum, lung
Thyroid	Lung, liver, bone
Uterus	Liver, lung, peritoneum

[a]http://www.cancer.gov/cancertopics/factsheet/Sites-Types/metastatic

- Early menarche (<12 years of age) or late natural menopause (>56 years of age)
- Hormone replacement therapy (HRT) after menopause, especially with a combination of estrogen and progesterone

Clinical Notes

- Younger females have denser breast tissue, making mammography less reliable. The breasts of older females are more fatty, yielding much higher detection of breast lesions by imaging.
- Most women with breast carcinoma do not have identifiable risk factors.
- A woman with one first-degree relative (mother, daughter, sister) with breast carcinoma has double the chance of developing breast carcinoma. With two first-degree relatives with breast carcinoma, the likelihood triples.

Fibroadenoma of the Breast

Fibroadenoma of the breast is a benign neoplasm composed of encapsulated glandular and cystic epithelial structures.

Signs and Symptoms

- Painless, palpable breast mass
- Multiple, bilateral masses
- Progressive increase in size of masses

Predisposing Factors

- Premenopausal, within 20 years of menarche
- Race: more common in young African American women

Clinical Notes

- Fibroadenomas are considered by some to be focal, developmental anomalies. This argument for their etiology would not classify them as a neoplasm.
- They are common in adolescence and nearly always involute spontaneously at menopause.

Fibrocystic Condition of the Breast

This condition involves the presence of benign, solid, and/or cystic masses in the breast. Fibrocystic condition of the breast is the most common breast lesion. The causative factor is elevated estrogen levels.

Clinical Note

Fibrocystic condition is often termed "fibrocystic disease."

Signs and Symptoms

- Transient breast pain or tenderness
- Transient, multiple, palpable masses; often bilateral

Menarche Age of first menstrual cycle
Menopause Cessation of menstrual cycles
Neoplasm Abnormally increased tissue growth by cell proliferation

Etiology Underlying cause of a disease or condition
Cyst Abnormal membrane-bound sac that contains gas or fluid

- Nipple discharge
- Mass enlarges during premenstrual part of cycle

Predisposing Factors

- Age: 30–50 years
- Alcohol consumption between ages 18 and 22
- Postmenopausal women on HRT

Clinical Notes

- A cystic mass can be distinguished by ultrasound.
- Distinguishing fibrocystic condition and breast carcinoma by mammography can be difficult in young women because of the denser breast tissue. Mammography may be helpful, however, in establishing the extent of the masses.
- A lump should be considered malignant until proven otherwise through a biopsy analysis.

DIAGNOSIS

The patient presentation, medical history, physical examination, imaging studies, biopsy results support a diagnosis of **primary carcinoma of the breast**. Standard procedure in evaluation of breast masses is the "**triple test**":

1. Physical examination
2. Imaging
3. Needle biopsy

When all three tests indicate a benign mass, or when all three tests indicate malignancy, the triple test is said to be concordant. If the triple test is benign concordant, accuracy is >99%. If any of the three components of the triple test indicates malignancy, intervention is recommended.

Breast Carcinoma

In this patient, two components of the triple test (physical examination and needle biopsy) indicated malignancy. The strongest support for this diagnosis was confirmation of malignant cells in the biopsy. This information alone supports the diagnosis.

- The patient is at risk because of her age, the late onset of menopause, having her first full-term pregnancy after age 30, and the use of combined (estrogen/progesterone) hormone replacement after menopause.
- The position of the mass is relevant since 75% of carcinomas form in the lateral breast quadrants (**Fig. 2.5.4**).
- The symmetrical movement of the breasts with contraction of both pectoralis major muscles (hands on hip test) suggests that the carcinoma has not invaded the **retromammary space** and involved the pectoral fascia. Asymmetric movement of the affected breast would be expected had tumor cell invaded the deep (pectoral) fascia.
- The lack of nipple discharge and the failure to palpate enlarged axillary lymph nodes would indicate early detection.

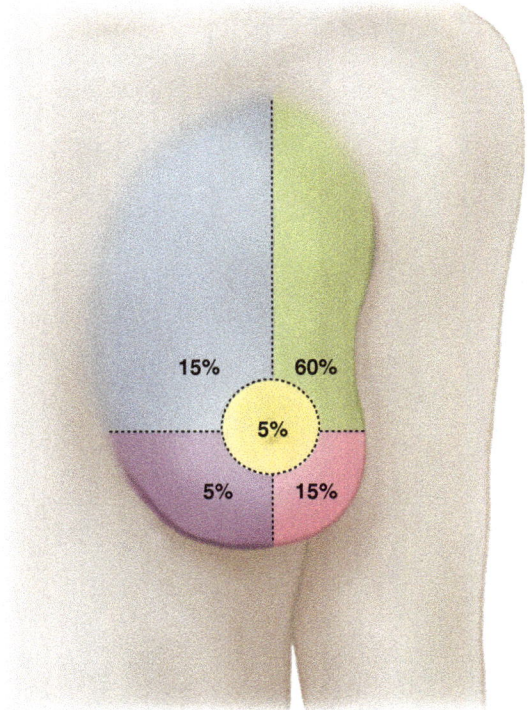

FIGURE 2.5.4 Quadrants of the female breast. Frequency distribution of carcinoma in different regions of the breast.

Clinical Notes

- **Palpation** of axillary lymph nodes of <1 cm diameter is considered a normal finding. Lymph nodes >1 cm should be monitored closely.

- When there is no evidence of regional spread beyond the breast, the clinical cure rate with most accepted methods of therapy is 75–80%. When the axillary lymph nodes are involved with the tumor, the survival rate drops to 50–60% at 5 years.

Fibrocystic Condition and Fibroadenoma of the Breast

These types of breast masses, while multiple, are restricted to breast lobes, and do not extend beyond the limits of the breast (i.e., they will not involve pectoral fascia). The connective tissue septae of the breast are not involved, and dimpling of the skin or nipple retraction is not seen.

- Fibrocystic condition and fibroadenoma are typically found in younger patients.

- These masses are commonly multiple and bilateral.

Patient Presentation

A 73-year-old male visits his primary care physician because he has recently noticed bloody streaks in his sputum.

Relevant Clinical Findings

History

The patient indicates that he has had a "smoker's cough" for several decades, but over the past 3 months the cough has become more productive and the sputum contains blood. He relates that he began smoking in his late teens and has smoked 20 or more cigarettes/day. Over the past several months, he has noticed that his clothes fit looser and he believes he has lost some weight (chart review indicated a 10 lb decrease since his last physical evaluation, 10 months ago). He has not altered his activity or diet. He often eats only one meal a day, as most foods do not appeal. The patient's son is with him in the examination room and indicates he has noticed that his father's voice has become hoarse.

Physical Examination

Noteworthy vital signs:

- Weight: 149 lb

Results of physical examination:

- Asthenic
- Dyspneic, with sibilant rhonchi in all lung lobes

Laboratory Tests

- Sputum analysis: hemoptysis and malignant squamous cells

Imaging Studies

- Chest radiographs indicate a mass along the left main bronchus, with extensions to the hilum of the lung and mediastinum. There is atelectasis of the superior lobe of the left lung.
- Abdominal radiographs show several 2–3 cm masses in the liver.

Clinical Problems to Consider

- Bronchogenic carcinoma
- Metastatic carcinoma to the lung
- Tuberculosis

LEARNING OBJECTIVES

1. Describe the anatomy of the lungs and tracheobronchial tree.
2. Describe the nerve and blood supply of the pleura, lungs, and tracheobronchial tree.
3. Describe the anatomical relationships and surface anatomy of the lungs and tracheobronchial tree.
4. Explain the anatomical basis for the signs and symptoms associated with this case.

Asthenia Overall weakness due to debility
Atelectasis Reduction or absence of air in all or part of a lung (lung collapse)
Dyspnea Difficulty breathing and shortness of breath

Hemoptysis Blood in sputum from airway hemorrhage
Sibilant rhonchus High-pitched whistling lung sound caused by airway narrowing or obstruction

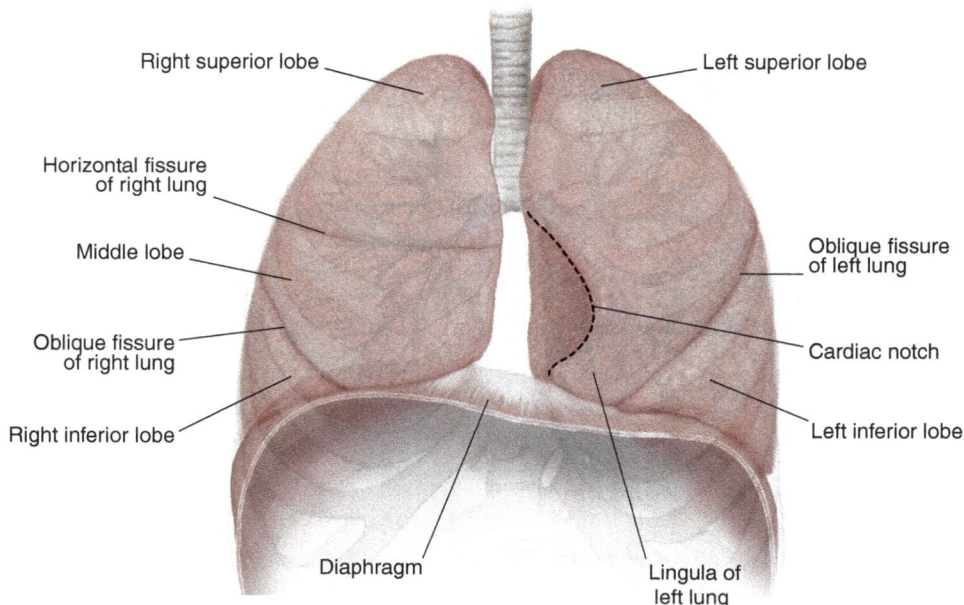

FIGURE 2.6.1 Anterior view of the lungs.

RELEVANT ANATOMY
Lungs and Tracheobronchial Tree
Lungs

Each **lung**, and its **pleura**, occupies its respective **pulmonary cavity** (**Fig. 2.6.1**). The lungs are conical, with the **apex** directed into the root of the neck, posterior to the medial one-third of the clavicle. The lung **base** assumes the contour of the diaphragm. Each lung has **costal**, **diaphragmatic**, and **mediastinal surfaces**, and **anterior** and **inferior borders**. Each lung has lobes that are separated by fissures.

Right Lung
Lobes

- **Superior**
- **Middle**
- **Inferior**

Fissures

- **Oblique fissure** separates superior from middle and inferior lobes.
- **Horizontal fissure** separates superior and middle lobes.

Left Lung
Lobes

- **Superior**
- **Inferior**

Fissures

- **Oblique fissure** separates superior from inferior lobe.

The anterior border of the superior lobe of the left lung is indented by the **cardiac notch**. A small projection of lung tissue inferior to the notch is the **lingula**.

Tracheobronchial Tree

The **trachea** passes through the superior thoracic aperture, anterior to the esophagus, to enter the mediastinum. This hollow tube is approximately 2.5 cm in diameter and is kept patent by a series of cartilaginous "rings." Tracheal rings are incomplete posteriorly; the gap in each cartilage is occupied by the **trachialis** muscle.

In the mediastinum, the trachea ends by dividing into a **right and left main (primary) bronchus** (**Fig. 2.6.2**). The right main bronchus is wider, shorter, and more vertical. Each unites with the

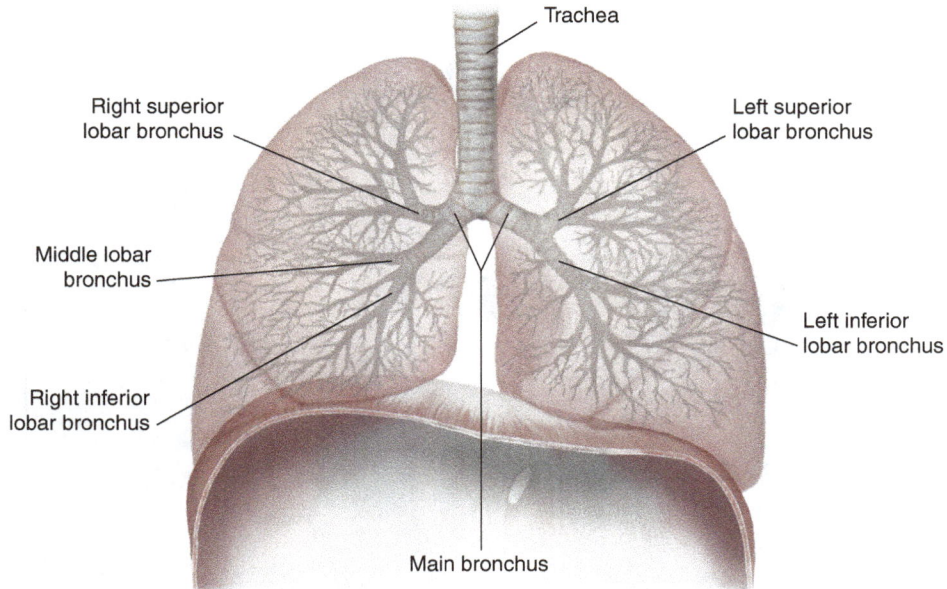

FIGURE 2.6.2 Tracheobronchial tree.

pulmonary artery, pulmonary veins, lymphatic vessels, and autonomic nerves to from the short **root of the lung**. This bundle of structures leaves the mediastinum and enters the **hilum of the lung** on its mediastinal surface. In the hilum, each main bronchus divides to provide an airway for each lung lobe (three on the right; two on the left). These are known as **lobar (secondary) bronchi**.

Each lobe is composed of **bronchopulmonary segments (Fig. 2.6.3)**, the basic **functional units of the lung**. Bronchopulmonary segments are separated by thin, connective tissue septae. Bronchopulmonary segments are pyramidal, with their base directed toward the lung surface and the apex pointed toward the hilum. Segments are named by their position within a lobe. The number of segments varies among lobes; however, each lung contains a total of 10.

Lobar bronchi divide into **segmental (tertiary) bronchi**, one to each bronchopulmonary segment. Beyond the segmental bronchi, there are another 10 generations of branching until the bronchi are approximately 1 mm in diameter (**Fig. 2.6.3**). At this diameter, the airways are termed **bronchioles**. This is the first part of the airway that lacks cartilage. No gas exchange occurs across the airway wall up to this point.

Bronchioles undergo two to three generations of branching to give rise to **terminal bronchioles**. Each terminal bronchiole forms the stem for an **acinus**, the **respiratory unit of the lung**. Each acinus gives rise to several **respiratory bronchioles**. The walls of the respiratory bronchioles contain a few **alveoli** and are the most proximal portion of the airway in which gas exchange can occur. The respiratory bronchioles give rise to 2–11 **alveolar ducts** that, in turn, give rise to five or six **alveolar sacs**. The two lungs, collectively, contain more than 100,000 acini. Each acinus contains approximately 3,000 alveoli.

Clinical Notes

- The full complement of alveoli is not formed until approximately 8 years of age.
- The tendency of foreign material to preferentially enter the right main bronchus results from its greater diameter and more vertical orientation.

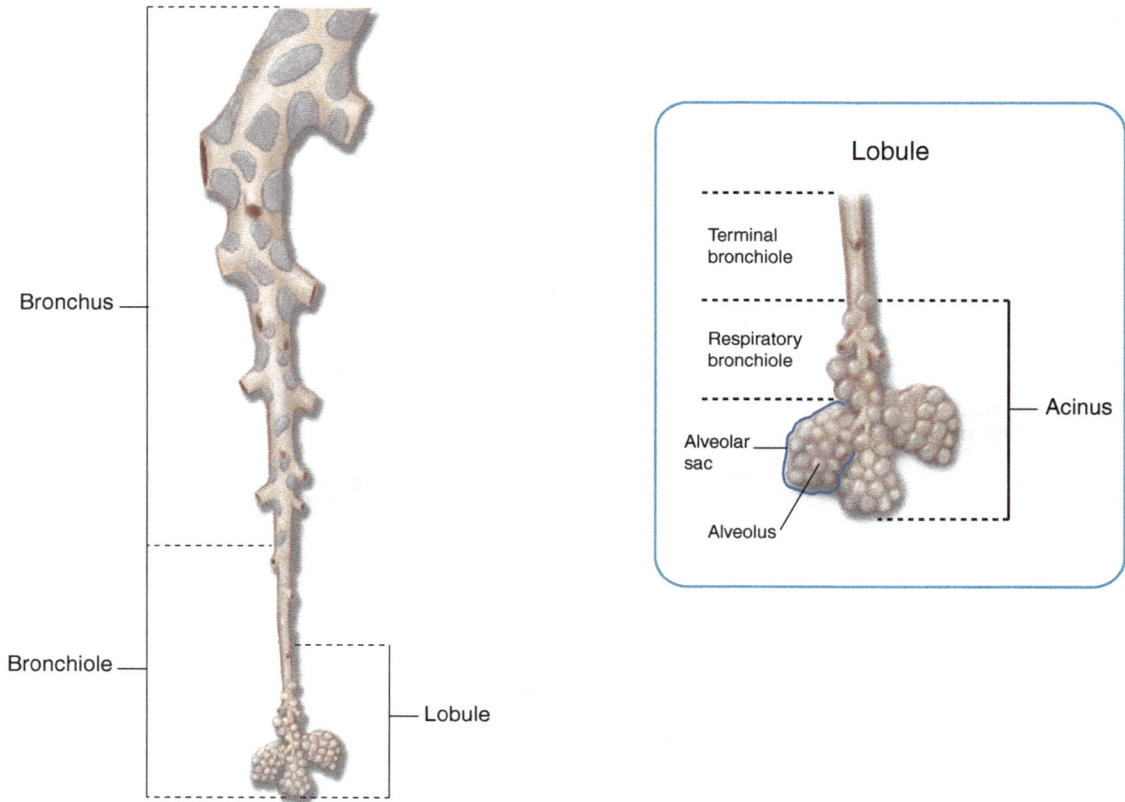

FIGURE 2.6.3 Branching of a segmental bronchus.

Blood Supply and Lymphatics

Two **pulmonary arteries** are derived from the **pulmonary trunk** (**Fig. 2.6.4A**). These carry blood to the lungs for oxygenation. From the hilum of the lung, pulmonary arteries follow the branching pattern of the airways, forming **lobar** and then **segmental arteries**. Lobar veins become **pulmonary veins** in the root of each lung (**Fig. 2.6.4B**). Four pulmonary veins (superior and inferior from each lung) enter the left atrium. On the right, the *middle lobe vein* unites with the *superior lobe vein* to form the **right superior pulmonary vein**. These vessels transport oxygen-enriched blood from the lung.

Unlike the arteries and airways, which are confined to a bronchopulmonary segment, **segmental veins** communicate with those in adjacent segments.

Segmental veins also receive blood from the visceral pleura and most of the bronchial venous system.

There are typically three small **bronchial arteries**.

- Two **left bronchial arteries** usually branch from the **thoracic aorta**.

- One **right bronchial artery** usually branches from the **right third posterior intercostal artery**.

Bronchial arteries course in the respective root of the lung and follow the airway branching pattern as far as the respiratory bronchioles. They supply:

- Structures in the root and hilum of the lung

- Supporting structures of the lung

- Visceral pleura

(A) Pulmonary and bronchial arteries

Right pulmonary a.

Right superior lobar a.

Pulmonary trunk

Left pulmonary a.

(B) Pulmonary and bronchial veins

Right pulmonary vv.

Left pulmonary vv.

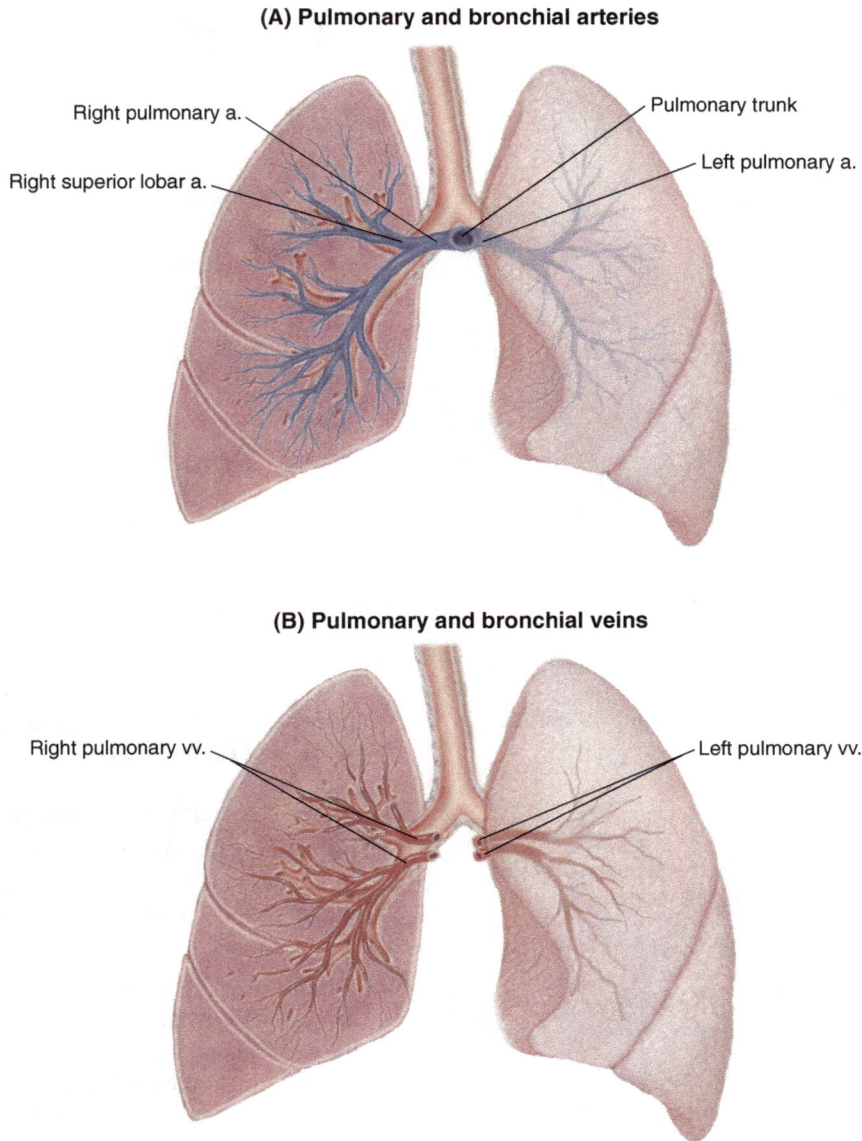

FIGURE 2.6.4 (**A**) Pulmonary and bronchial arteries (blue) supply to lungs and bronchi. (**B**) Pulmonary and bronchial veins (red) supply to lungs and bronchi.

Each lung has a single **bronchial vein**. They do not drain the same areas that are supplied by the bronchial arteries. These small veins drain tissues near the hilum of the lung. The remainder of the venous drainage of the lung, including the visceral pleura, enters the pulmonary veins.

- The **right bronchial vein** enters the **azygos vein**.
- The **left bronchial vein** enters the **hemiazygos vein**.

Two interconnected **lymphatic plexuses** (superficial and deep) drain the lungs and bronchi (**Fig. 2.6.5**).

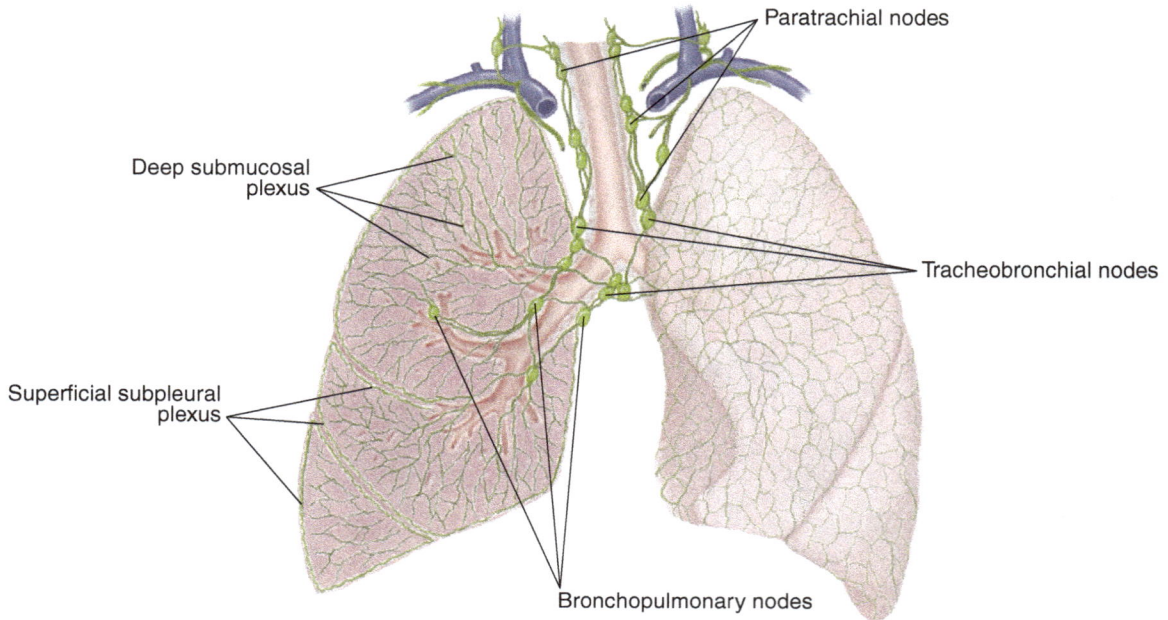

FIGURE 2.6.5 Lymphatics of the lungs and bronchi.

The lymphatic drainage of the lung is outlined in **Table 2.6.1**.

Clinical Note

Lymphatic vessels provide the primary route of spread for bronchogenic carcinoma.

Nerve Supply of the Pleura, Tracheobronchial Tree, and Lungs

Pleura

Innervation of the pleura utilizes the phrenic (C3-C5) and intercostal nerves (T1-T11), as well as sympathetic (T1-T5) and parasympathetic (vagus—CN X) nerves (**Table 2.6.2**).

Trachea, Bronchi, and Lungs

The **pulmonary plexus**, an extension of the cardiac plexus, supplies the trachea, bronchi, and lungs. The pulmonary plexus has sympathetic and parasympathetic components:

- Cardiac (cardiopulmonary splanchnic) nerves carry **postganglionic sympathetic fibers** from the cervical and upper four thoracic paravertebral ganglia to the cardiac plexus. Stimulation of these **visceral efferents** increases respiratory rate and leads to bronchodilation and

TABLE 2.6.1	Lymphatic drainage of the lungs and bronchi.		
Lymphatic plexus	**Location**	**Structures**	**Nodes**
Superficial	Subpleural: follows surface contour of lung	Visceral pleura Lung parenchyma	Bronchopulmonary (hilar) to tracheobronchial
Deep	Bronchial submucosal	Bronchi	Pulmonary to bronchopulmonary to tracheobronchial

TABLE 2.6.2	Innervation of pleura.	
Pleura	**Nerve(s)**	**Modality**
Parietal		General sensation
Costal	Intercostal	
Diaphragmatic		
Central	Phrenic	
Peripheral	Intercostal	
Mediastinal	Phrenic	
Cervical	Phrenic	
Visceral	Sympathetic	Anoxia and distention

decreased glandular secretion. These fibers also conduct visceral afferent impulses (reflexive and nociceptive).

- **Preganglionic parasympathetic fibers** (cardiac branches) originate from the vagus nerve. Parasympathetic stimulation decreases respiratory rate and leads to bronchoconstriction and increased glandular secretion.

Pain may be referred to dermatomes C2-T4, depending on which nerve is involved. For additional information about the autonomic nervous system and referred pain, see Chapter 1: Visceral Afferent Pathways.

Anatomical Relationships and Surface Anatomy

Anatomical Relationships

There are several important anatomical relationships of the airways and pulmonary vasculature:

- The **left main bronchus** lies inferior to the arch of the aorta.
- The **right main bronchus** lies inferior to the arch of the azygos vein.
- In the hila of the lungs:
  - The **superior pulmonary vein** lies anterior to other structures.
  - The **inferior pulmonary vein** lies inferior to other structures.
  - The **right pulmonary artery** lies anterior to the main bronchus.
  - The **left pulmonary artery** lies superior to the main bronchus.

- The **phrenic nerve** lies anterior to the root of each lung.
- The **vagus nerve** lies posterior to the root of each lung.
- The **left recurrent laryngeal nerve** loops around the ligamentum arteriosum and, at this point, is closely associated with the left main bronchus.

Surface Anatomy

The surface projections of components of the respiratory system are important landmarks used in physical examination.

The **manubriosternal angle** (Louis) lies at the junction of the manubrium and body of the sternum. It is subcutaneous and lies at the level of the T4-T5 intervertebral disc. The trachea bifurcates at this level.

The **cervical pleura** extends into the root of the neck, approximately 3 cm superior to the medial one-third of the clavicle. Thus, the **pleural cavity** extends through the superior thoracic aperture into the root of the neck.

The **oblique fissure** for each lung approximates a line connecting the spinous process of T3 or T4 with the costochondral junction of rib 6. In the midaxillary line, the oblique fissure lies deep to the 5th intercostal space. The **horizontal fissure** extends from the right oblique fissure in the midaxillary line (5th intercostal space) and parallels the right 4th costal cartilage.

Clinical Note

- Auscultation during physical examination requires evaluation of each lung lobe. The superior lobes are auscultated on the anterosuperior aspect of the thorax, the inferior lobes posteroinferiorly, and the middle lobe of the right lung on the right lateral chest wall near the midaxillary line.

- Auscultation can detect movement of air in the lungs ("breath sounds") during inspiration and early expiration. In the healthy lung, breath sounds should not be detected during most of expiration.

Auscultation A diagnostic method, usually with a stethoscope, to listen to body sounds (e.g., heart, breath, and gastrointestinal sounds)

(A)

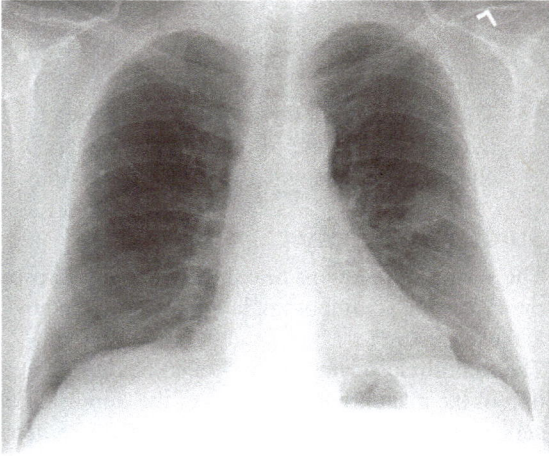

(B)

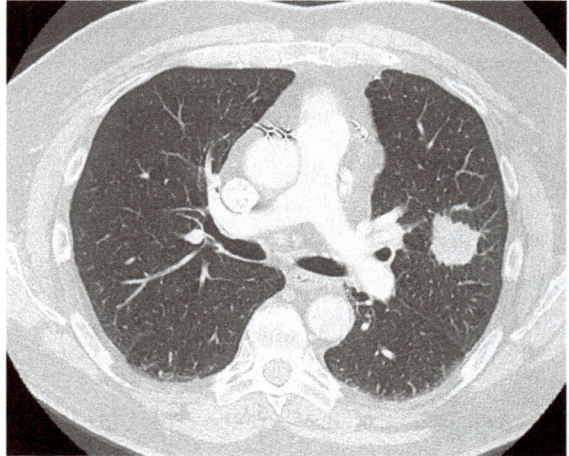

FIGURE 2.6.6 Bronchogenic carcinoma. (**A**) Postero-anterior radiograph showing a mass in the left lung. *Source:* Fig. 4-36A in *Basic Radiology,* 2e. www.accessmedicine.com. (**B**) Axial CT image of the same patient showing the mass in the left superior lobe. *Source*: Fig. 4-37A in *Basic Radiology,* 2e. www.accessmedicine.com.

CLINICAL REASONING

This patient presents with signs and symptoms of **respiratory disease**.

Bronchogenic Carcinoma

Bronchogenic carcinoma is a malignant disease of the lung (**Fig. 2.6.6**). It is one of the most preventable malignancies since >90% of cases are caused by cigarette smoking.

Bronchogenic carcinoma is divided into five histologically based categories (**Table 2.6.3**).

Clinical Note

Biopsy of the tumor or regional lymph nodes is required to confirm the histological type(s) of lung cancer present so that proper treatment can be administered. Two different types of bronchogenic carcinoma can exit simultaneously.

Signs and Symptoms

- Chronic cough
- Hemoptysis

TABLE 2.6.3	Types of lung cancers.	
Category	**Frequency (%)**	**Characteristics**
Adenocarcinoma	40	Arises from mucous glands in the airway; usually a peripheral mass or nodule
Squamous cell	25	Arises from bronchial epithelium; usually central and intraluminal
Small cell	15	Bronchial in origin; begins centrally and invades submucosa
Large cell	10	Heterogenous; may be central or peripheral
Other	10	Poorly differentiated lung cancers

- Hoarseness
- Atelectasis and dyspnea
- Dysphagia
- Asthenia
- Weight loss

Predisposing Factors

- Cigarette smoking (active or former)
- Environmental tobacco smoke ("second-hand" smoke)
- Prolonged exposure to radon
- Occupation: exposure to heavy metals, asbestos, or polycyclic aromatic hydrocarbons

If the tumor is restricted to the hilum and/or bronchi, there may be no pain and the presenting symptom is often a productive cough with hemoptysis. Somatic pain may be a symptom if the tumor is more peripheral in the lung, with involvement of parietal pleura. Regional tumor growth may compress bronchi, the trachea, and/or esophagus leading to atelectasis, dyspnea, and dysphagia, respectively.

Metastasis to the brain, bone, liver, and adrenal medulla is common. Liver involvement typically leads to weight loss, while metastasis to brain may present as headache or neurologic deficits. Bone metastases commonly present with pain and/or pathologic fractures.

Clinical Notes

- The typical patient with lung carcinoma is past 60 years of age, with a history of cigarette smoking. The median age at diagnosis in smokers (active or former) is 71.
- More than 50% of patients with lung cancer present with advanced disease.

Metastatic Cancer to the Lung

This condition involves lung malignancy due to metastasis of nonrespiratory primary carcinoma.

Dysphagia Difficulty swallowing
Lymphatogenous Spread via lymphatic vasculature

Signs and Symptoms

- Respiratory symptoms uncommon until advanced stages

Predisposing Factors

- Nonpulmonary malignancy

Nearly all cancers can metastasize to the lung (see **Table 2.5.1**). The pulmonary arteries are considered the primary route for transporting malignant cells *to the lung*. In contrast, the metastasis of a primary tumor *from the lung* is usually lymphatogenous. Imaging studies for metastatic involvement of the lung often reveal multiple, small bilateral tumors that tend to be located in the lower lung segments.

Tuberculosis

Tuberculosis is a bacterial infection (*Mycobacterium tuberculosis*) that commonly involves the respiratory system, but may involve other systems as well. The bacilli are spread most often in aerosolized droplets of saliva produced during coughing, sneezing, or speaking. Development of the disease following *M. tuberculosis* infection depends on immunologic status; immunocompromised patients are at higher risk. Tuberculosis is divided into two clinical categories: latent and active.

Signs and Symptoms

Latent Tuberculosis

- Patient has *M. tuberculosis* infection but bacterium is inactive and there are no symptoms
- Not contagious

Active Tuberculosis

- Cough
- Unexplained weight loss
- Fever and chills
- Fatigue
- Contagious

Predisposing Factors

- HIV infection
- Race: African American (U.S. born)
- Ethnicity: foreign-born U.S. resident
- Occupation: health-care worker

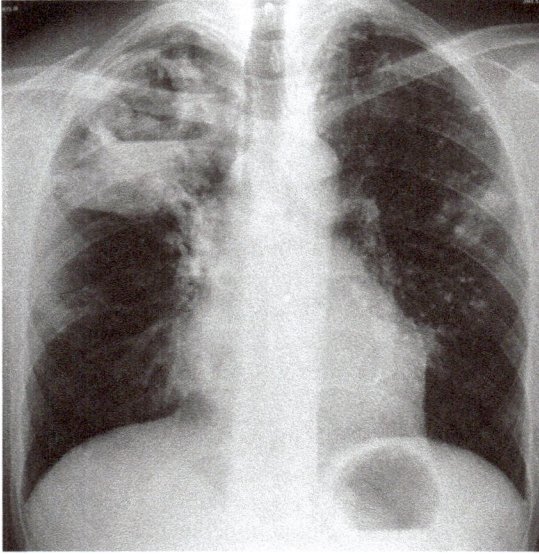

FIGURE 2.6.7 Postero-anterior chest radiograph showing a right upper lobe infiltrate in a patient with active tuberculosis. *Source*: Fig. 165-5 in *Harrison's Online*. www.accessmedicine.com.

Culture analysis (commonly of sputum) is considered the "gold standard" for the diagnosis of active tuberculosis. Radiographic findings of infiltrates and cavities in the superior lung lobes in patients with respiratory symptoms are consistent with tuberculosis (**Fig. 2.6.7**).

Clinical Notes

- In 2009, 5.8 million new cases of tuberculosis were reported to the World Health Organization.
- The **tuberculin skin test** (TST) is used commonly to assess *M. tuberculosis* in asymptomatic patients. Tuberculin is an extract of *M. tuberculosis* (and other species in the *Mycobacterium* genus). The TST, also known as the Mantoux or PPD (purified protein derivative) test, involves measuring induration (swelling) around an intradermal injection of 0.1 mL of PPD tuberculin after 48–72 hours.
- Most tuberculosis infections remain latent.
- Active tuberculosis may develop weeks to years after initial infection.

DIAGNOSIS

The patient presentation, medical history, physical examination, laboratory tests, and imaging studies support a diagnosis of **bronchogenic carcinoma** of the squamous cell type.

Bronchogenic Carcinoma

Bronchogenic carcinoma is primary to the lungs and bronchi.

- The patient's history of cigarette smoking and his age put him at high risk.
- Imaging studies show the mass along the main bronchus, extending into the hilum of the lung and the mediastinum. A mass at this position could constrict main airways and lead to atelectasis and dyspnea.
- Chronic cough, hemoptysis, and squamous cells in the sputum support the diagnosis.
- The voice change (hoarseness) in this patient is due to involvement of the left recurrent laryngeal nerve. The close relationship of this nerve with the left main bronchus may result in it becoming invested by an enlarging tumor. This compromises the innervation to most of the intrinsic muscles on the left side of the larynx. The unilateral paralysis of these muscles produces asymmetric tension on the vocal cords and presents as hoarseness.
- The weight loss and liver masses seen in films support the conclusion of metastatic spread to the liver.

Metastatic Cancer to the Lung

Metastasis of malignant cells to the lung from other nonpulmonary organs and tissues is common (see **Table 2.5.1**). These tumor cells are transported to the lungs via pulmonary arteries. Commonly, both lungs are involved and the tumors tend to be multiple and in lower bronchopulmonary segments. These tumors are also more peripheral in the lung tissue, that is, further from the hilum.

- This patient had a unilateral tumor located close to hilum of the lung with metastasis to

---

Paralysis Loss of muscle function, particularly related to voluntary movement

the liver. Sputum analysis suggested the tumor was of the squamous type.

Tuberculosis

Tuberculosis is a bacterial infection. In the active stage, it presents as a recent cough that has persisted for more than 3 weeks. The cough is usually accompanied by fever and chills.

- Hemoptysis may be present and sputum may contain *M. tuberculosis* (rather than tumor cells).
- A chest radiograph will show small white spots throughout the lung field. These represent the body's immune system isolating clusters of the bacillus.
- While active tuberculosis typically involves the lung, the kidneys, brain, and spinal column also may be involved.

REVIEW QUESTIONS

1. A 23-year-old male arrives at the outpatient clinic with complaints of fever, chills, and a "chest cold" of 3 days duration. Crackling is detected during inspiration with the stethoscope over the right triangle of auscultation. At this position on the posterior chest wall, air movement is being heard in which part of the respiratory system?
   A. Right inferior lung lobe
   B. Right main bronchus
   C. Right middle lung lobe
   D. Right superior lung lobe
   E. Trachea

2. A 68-year-old male presents with the primary complaint of shortness of breath during mild exertion. Auscultation of the thoracic wall detects a murmur adjacent to the sternum in the left 2nd intercostal space. This is suggestive of:
   A. Aortic valve disease
   B. Coronary artery disease
   C. Left ventricular failure
   D. Pulmonary valve disease
   E. Stenosis of the right atrioventricular (tricuspid) valve

3. A 35-year-old male driver was involved in a car accident with air-bag deployment. Radiography reveals costochondral joint separation for both the fourth and fifth ribs on the left. While still in the emergency department, he develops dyspnea (dif-

ficulty breathing), hypotension, cyanosis, and neck vein distension. These classic signs of cardiac tamponade result from accumulation of fluid in the:
   A. Costomediastinal recess
   B. Potential space between fibrous and parietal serous pericardium
   C. Potential space between parietal and visceral serous pericardium
   D. Potential space between visceral serous pericardium and myocardium
   E. Superior mediastinum

4. A 48-year-old obese man collapsed while playing basketball on his driveway. He could not be resuscitated. At autopsy, his left coronary artery and its two terminal branches were 95% occluded with sclerotic buildup. The terminal (end) branches of the left coronary artery are the:
   A. Anterior and posterior interventricular
   B. Anterior interventricular and circumflex
   C. Anterior interventricular and left marginal
   D. Circumflex and left atrial
   E. Left marginal and circumflex

5. Coronary angiography done in advance of coronary bypass surgery on a 54-year-old female reveals that she has a "right dominant heart." This condition:
   A. Indicates that the posterior interventricular artery is derived from the right coronary artery

---

Crackle Crackling noise heard with lung disease (also known as a rale)

Auscultation A diagnostic method, usually with a stethoscope, to listen to body sounds (e.g., heart, breath, and gastrointestinal sounds)

Murmur Variable vibrations produced by turbulence of blood flow

Stenosis Narrowing a canal (e.g., blood vessel and vertebral canal)

Hypotension Abnormal decrease in arterial and/or venous pressure

Cyanosis Bluish color of skin and mucous membranes from insufficient blood oxygen

B. Indicates that the right and left coronary arteries are derived from a common trunk

C. Is a major correlate with coronary artery disease

D. Is frequently associated with atrial fibrillation

E. Is present in <10% of female patients

6. A premenopausal 51-year-old female describes the recent discovery of a lump in her right breast. Physical examination reveals a mass of approximately 2.5 cm in the upper lateral breast quadrant. There is some dimpling of the skin overlying the mass. This dimpling is due to:

A. Decreased fat in the area of the mass

B. Displacement of glandular tissue by the mass

C. Increased vasculature to the mass

D. Influence of cyclical hormonal changes

E. Traction on connective tissue septae by the expanding mass

7. A transverse CAT scan at the level of the intervertebral disc between T4 and T5 in a 29-year-old male reveals normal anatomy. All of the following might be seen in this scan *except*:

A. Arch of aorta

B. Arch of azygos vein

C. Pulmonary valve

D. Superior lobe of right lung

E. Superior vena cava

8. Imaging of the thorax in an 18-month-old female reveals a diffuse structure in the superior mediastinum, representing the thymus gland. All of the following also would be found in the superior mediastinum *except*:

A. Ascending aorta

B. Brachiocephalic artery

C. Left brachiocephalic vein

D. Phrenic nerve

E. Vagus nerve

9. Thoracentesis is performed in the midaxillary line through the right 8th intercostal space of a 73-year-old female to drain excess pleural fluid. All of the following can be found in this intercostal space *except*:

A. Axons of parasympathetic neurons

B. Axons of sympathetic neurons

C. Somatic pain receptors

D. Tributaries of the azygos vein

E. Ventral ramus of T8 spinal nerve

10. During a domestic quarrel, a 38-year-old male is stabbed in the chest with a paring knife. The knife pierced the left 5th intercostal space beside the sternum. Emergency department workup reveals a hemopericardium, but there is no indication of a hemo- or pneumothorax. A senior intern explains during rounds that there is no pneumothorax because:

A. The left lung has only two lobes.

B. The parietal pleura reflects away from the midline beginning at the 4th costal cartilage on the left side.

C. The patient was exhaling at the time of the injury.

D. The patient was standing at the time of the injury.

E. There is a cardiac notch in the inferior lobe of the left lung.

11. A 61-year-old patient complains of a nagging cough that, over the past 2 months, produces blood-tinged sputum. Her husband indicates that he has noticed a progressive hoarseness in her voice. Imaging reveals a mass in the root and hilum of the left lung. A likely explanation for the hoarse voice is:

A. A compromised blood flow to and from the lung.

B. Displacement of the trachea by the mass.

C. Metastasis to the larynx.

D. Partial occlusion of the main bronchus by the mass.

E. The relationship of the left recurrent laryngeal nerve to the left lung root.

Hemopericardium Blood in pericardial fluid within the pericardial cavity

Pneumothorax Air or gas in pleural cavity

CHAPTER 3

Abdomen

Patient Presentation

During a routine physical examination, a small, round swelling was noted in the left groin of a 12-year-old male.

Relevant Clinical Findings

History

The patient reported that he had noted the swelling some time ago, but that it was not painful. The bulge seemed to change in size depending on his body position and time of day. Early in the morning it was barely noticeable, but later in the day or after exercise the enlargement was more obvious. The patient has had no prior major illnesses or surgeries.

Physical Examination

The following findings were noted on physical examination:

- Testes of normal size and position for patient's age
- Small palpable mass in left groin, just lateral and superior to pubic tubercle
- Mass is easily reducible

Laboratory Tests

Test	Value	Reference value
Erythrocytes (count)	5.3	$4.3–5.6 \times 10^6/mm^3$
Leukocytes (count)	8.2	$3.54–9.06 \times 10^3/mm^3$

Imaging Studies

- Ultrasonography of the left groin showed an abnormal mass.

Clinical Problems to Consider

- Enlarged superficial inguinal lymph nodes
- Hydrocele
- Direct inguinal hernia
- Indirect inguinal hernia

LEARNING OBJECTIVES

1. Describe the anatomy of the inguinal region.
2. Describe the anatomy of the spermatic cord.
3. Describe the development of the male inguinal canal.
4. Define the anatomy of the inguinal triangle.
5. Explain the anatomical basis for the signs and symptoms associated with this case.

RELEVANT ANATOMY

Inguinal Region

The inguinal (groin) region is the antero-inferior area of the anterior abdominal wall. Three flat muscles of the abdominal wall (external oblique, internal oblique, and transversus abdominis) contribute to the anatomy of this region. Each of these muscles has an aponeurosis (flat tendon). The inferior margin of the external oblique aponeurosis forms the **inguinal ligament**, a thickened band that stretches from the anterior superior iliac spine to the pubic tubercle.

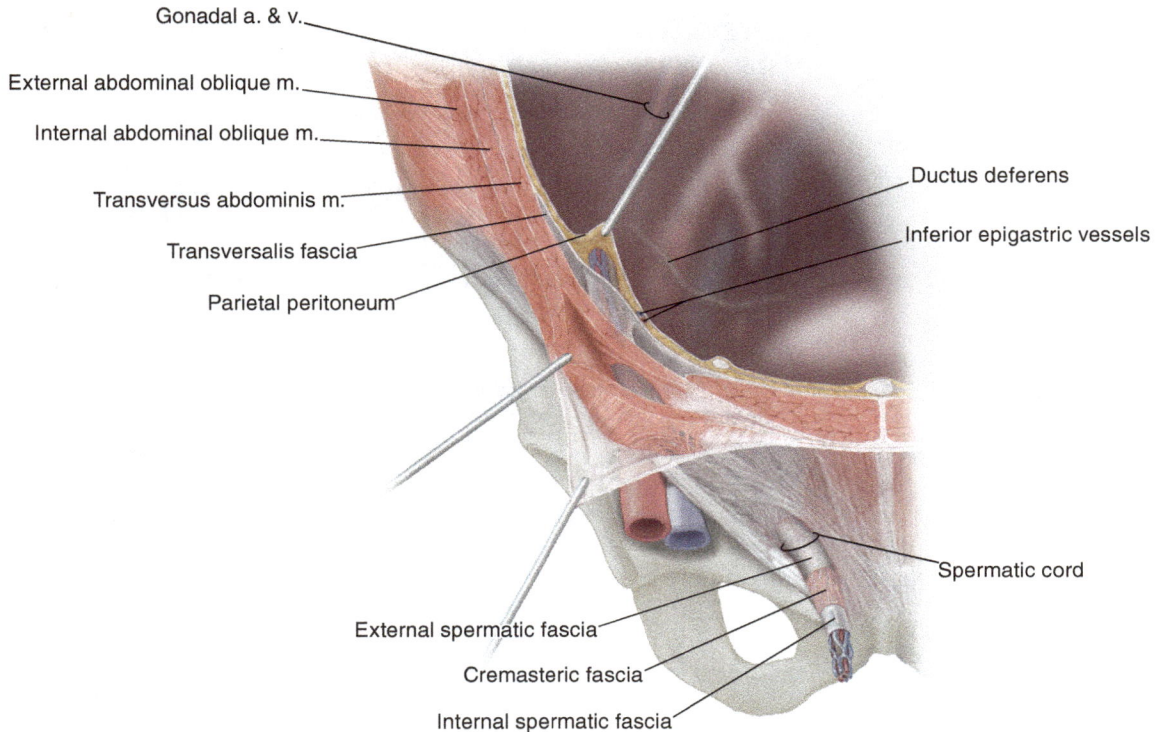

FIGURE 3.1.1 Antero-inferior abdominal wall (anterior view) showing anatomy of the inguinal region.

Inguinal Canal

The inguinal canal is an oblique passageway through the inferior part of the anterior abdominal wall (**Fig. 3.1.1**). The canal is approximately 5 cm in length and is directed inferomedially. In males, it serves as a passageway for the spermatic cord, which contains structures coursing to and from the testis. In females, the round ligament of the uterus passes through the canal.

The boundaries of the inguinal canal are formed by muscles and fascia of the abdominal wall (**Table 3.1.1**).

Spermatic Cord

The spermatic cord contains structures that support the testis (**Fig. 3.1.2**). The contents of spermatic cord are given in **Table 3.1.2**.

The spermatic cord begins at the deep inguinal ring, passes through the inguinal canal, exits the superficial inguinal ring, and ends in the scrotum at the root of the testis.

Development of the Male Inguinal Canal

Prior to descent of the testis, the **gubernaculum testis** (a mesenchymal condensation) extends from the inferior pole of each testis to the scrotal swellings in the inguinal region. The **processus vaginalis** (a hollow outgrowth of peritoneum) evaginates the fascial and muscular layers of the anterior abdominal wall as it extends into the scrotal swellings, following the course of the gubernaculum testis. This process creates the inguinal canal. The testis follows the course of the gubernaculum from its retroperitoneal position, through the inguinal canal, and into the scrotum. As a result, layers of the anterior abdominal wall contribute to the fasciae that surround the spermatic cord (**Table 3.1.3**).

Before birth, the proximal part of processus vaginalis becomes obliterated, closing off the entrance to the inguinal canal from the peritoneal cavity. The distal portion of the processus vaginalis forms

TABLE 3.1.1	Boundaries of the inguinal canal.
Boundary	**Structures**
Anterior	Aponeurosis of external abdominal oblique muscle
Posterior	Transversalis fascia
	Conjoint tendon (fused aponeuroses of internal abdominal oblique and transversus abdominis muscles that attach to pubic crest) reinforces medial part of posterior wall
Superior (roof)	Arching fibers of internal abdominal oblique and transversus abdominis muscles
Inferior (floor)	Inguinal ligament
Deep inguinal ring	Evagination of transversalis fascia, superior to inguinal ligament and lateral to inferior epigastric vessels
Superficial inguinal ring	Slit-like opening in external abdominal oblique aponeurosis, superior and medial to pubic tubercle
	Lateral crus is lateral margin of superficial ring
	Medial crus is medial margin of superficial ring

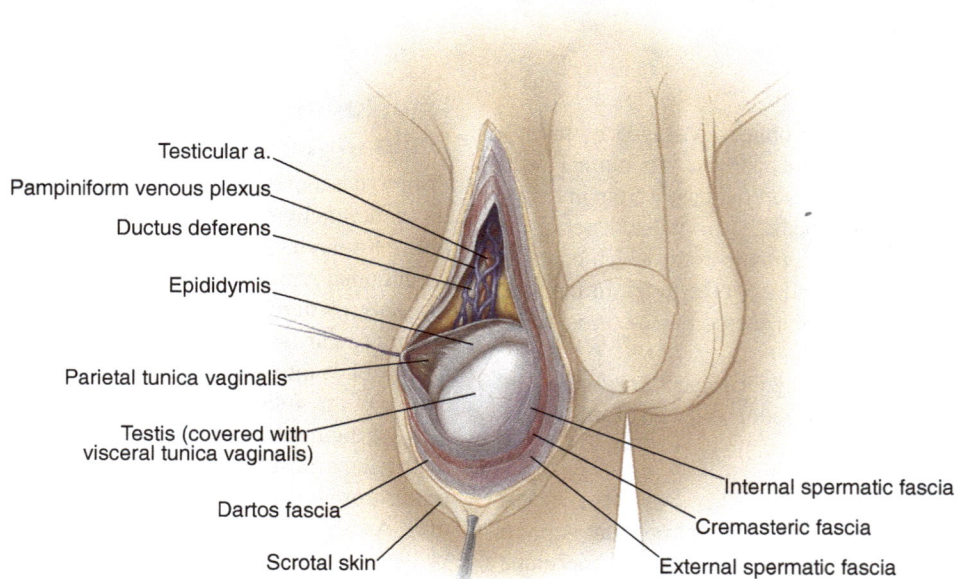

FIGURE 3.1.2 Anterior view of scrotum showing the spermatic fasciae, spermatic cord, and testis.

TABLE 3.1.2 Structures within the spermatic cord.

Structure	Description
Ductus deferens (vas deferens)	Muscular tube that conveys sperm from epididymis to ejaculatory duct
Testicular artery	Branch of abdominal aorta; supplies testis and epididymis
Pampiniform venous plexus	Vascular network that converges to form testicular vein(s)
Artery to ductus deferens	Branch of inferior vesical artery; supplies ductus deferens
Genital branch of genitofemoral nerve	Innervates cremaster muscle
Autonomic nerve fibers	Innervate smooth muscle in blood vessels, testis, ductus deferens, and epididymis
Lymphatic vessels	Drain lymph from testis and associated structures to lumbar and pre-aortic nodes

a reflected fold that partially covers the testis and epididymis. This remnant of processus vaginalis creates the visceral and parietal layers of the tunica vaginalis testis. In some individuals, the processus vaginalis remains patent through infancy, into childhood, and possibly into adulthood.

Clinical Note

A persistent processus vaginalis creates a potential pathway for herniation of abdominal contents through the inguinal canal and into the scrotum (indirect inguinal hernia).

Inguinal Triangle

The inguinal triangle (Hesselbach) is a region of the inferomedial anterior abdominal wall (**Fig. 3.1.3** and **Table 3.1.4**). The triangle is within the medial umbilical fossa. The inferior portion of the triangle is related to the medial end of the inguinal canal and superficial inguinal ring. Deep to the superficial ring, the posterior wall of the canal is composed of transversalis fascia and parietal peritoneum, which is reinforced medially by the conjoint tendon.

Clinical Note

Because the inferomedial aspect of the inguinal triangle is not well reinforced, it is an area that is susceptible to herniation, particularly in older males.

TABLE 3.1.3 Layers of scrotum and spermatic fascia.

Layers of scrotum and spermatic fascia	Corresponding layer from anterior abdominal wall	Notes
Skin	Skin	
Dartos muscle and fascia	Superficial fascia	Smooth muscle
External spermatic fascia	External abdominal oblique aponeurosis	
Cremaster muscle (fascia)	Internal abdominal oblique	Skeletal muscle
Internal spermatic fascia	Transversalis fascia	

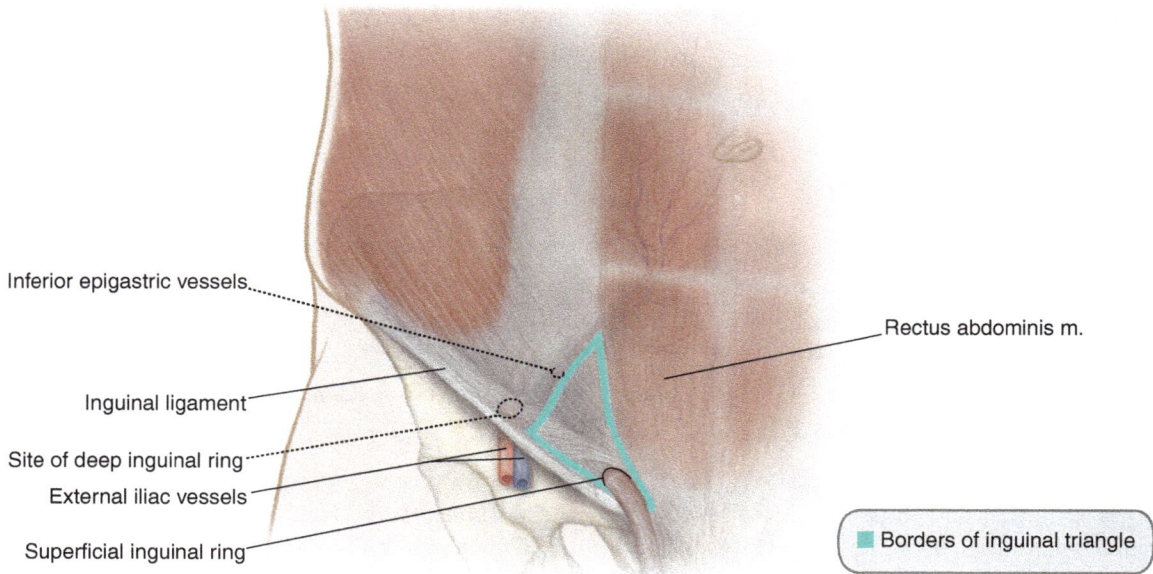

FIGURE 3.1.3 Anterior view of the male inguinal region, showing the borders of the inguinal triangle (Hesselbach) and the position of the deep inguinal ring.

CLINICAL REASONING

This patient presents with signs and symptoms related to a **swelling in the inguinal region**.

Enlarged Superficial Inguinal Lymph Nodes

The superficial inguinal lymph nodes are a collection of 12–20 nodes parallel to the inferior border of the inguinal ligament.

- **Medially placed nodes** receive lymph from the external genitalia (except the testis, epididymis and spermatic cord), inferior anal canal, and perianal region, as well as the uterus in the female (lymph drains along the round ligament). Also included in this group are four or five nodes that are located around the saphenous hiatus (where great saphenous vein passes through the fascia lata to join the femoral vein); these nodes receive lymph from all superficial lymphatics of the lower limb (except those that drain the posterolateral calf).

- **Laterally placed nodes** drain the lateral gluteal region and the lower part of the anterior abdominal wall.

Most superficial inguinal nodes drain into the external iliac nodes. Superficial inguinal nodes are vulnerable to enlargement or inflammation known as lymphadenitis. Enlargement of the superficial inguinal nodes may block drainage from the perineum, inferior abdominal wall, and superficial lower limb.

Signs and Symptoms

- Nontender, nonreducible swelling(s) along the inguinal ligament
- Fever

TABLE 3.1.4	Boundaries of the inguinal triangle.
Boundary	**Structure(s)**
Medial	Linea semilunaris (lateral margin of rectus sheath)
Lateral	Inferior epigastric vessels
Inferior	Medial portion of inguinal ligament

Lymphadenitis Inflammation of a lymph node or nodes

Predisposing Factors

- Regional bacterial or viral infection
- Regional cancer and resulting infiltration of malignant cells

Hydrocele

A hydrocele is formed when there is peritoneal fluid in a persistent (i.e., patent) processus vaginalis. It presents as an oblong, nontender mass that is either adjacent to the testis or along the course of the spermatic cord. There are two types of hydrocele:

1. A **communicating hydrocele** is most common in infants and children. It results when the proximal portion of the processus vaginalis remains patent, allowing peritoneal fluid from the peritoneal cavity to enter the scrotal sac.

2. A **noncommunicating hydrocele** occurs when the proximal part of the processus vaginalis is closed, but a distal segment remains patent. In this case, fluid is trapped within the patent part of the processus vaginalis and may be confined to the scrotum.

Signs and Symptoms

- Range from asymptomatic to intense scrotal pain

Predisposing Factors

- Persistent processus vaginalis (notably in children)
- Testicular tumor
- Tuberculous epididymitis
- Trauma
- Iatrogenic causes

Clinical Notes

- Transillumination is used to diagnose a hydrocele. During this procedure, a bright light is applied to the side of the scrotum. The fluid in a hydrocele usually is clear and, therefore, the light will outline the testicle indicating that the fluid surrounds it.
- Ultrasound imaging may be used to rule out a hernia, testicular tumor, or other causes of scrotal swelling (e.g., varicocele).

Iatrogenic Caused by medical or surgical treatment

Inguinal Hernia

An inguinal hernia is a protrusion of abdominal cavity contents, usually a portion of the small bowel and associated peritoneum, through a portion of the inguinal canal. A history of pain, swelling, or the presence of a mass in the groin is suggestive of an inguinal hernia. These hernias may be detected in the region of the superficial inguinal ring or in the scrotum.

Signs and Symptoms

- Swelling in groin at site of hernia; may be more conspicuous with straining
- A range of pain and tenderness: none to dull, aching regional pain, to acute, localized pain

There are two types of inguinal hernia, be defined by their relationship to the inferior epigastric vessels and inguinal canal.

Direct Inguinal Hernia

Direct inguinal hernias occur medial to the inferior epigastric vessels (**Fig. 3.1.4A**).

Distinguishing Characteristics

- Swelling in region of inguinal triangle
- May present at the superficial inguinal ring (may be palpable)
- Does not traverse the entire length of the inguinal canal
- Almost never enters the scrotum

Predisposing Factors

- Age: more common with increasing age
- Sex: majority in males
- Family history
- Chronic cough
- Chronic constipation
- Excess weight
- Pregnancy

(A)

(B)

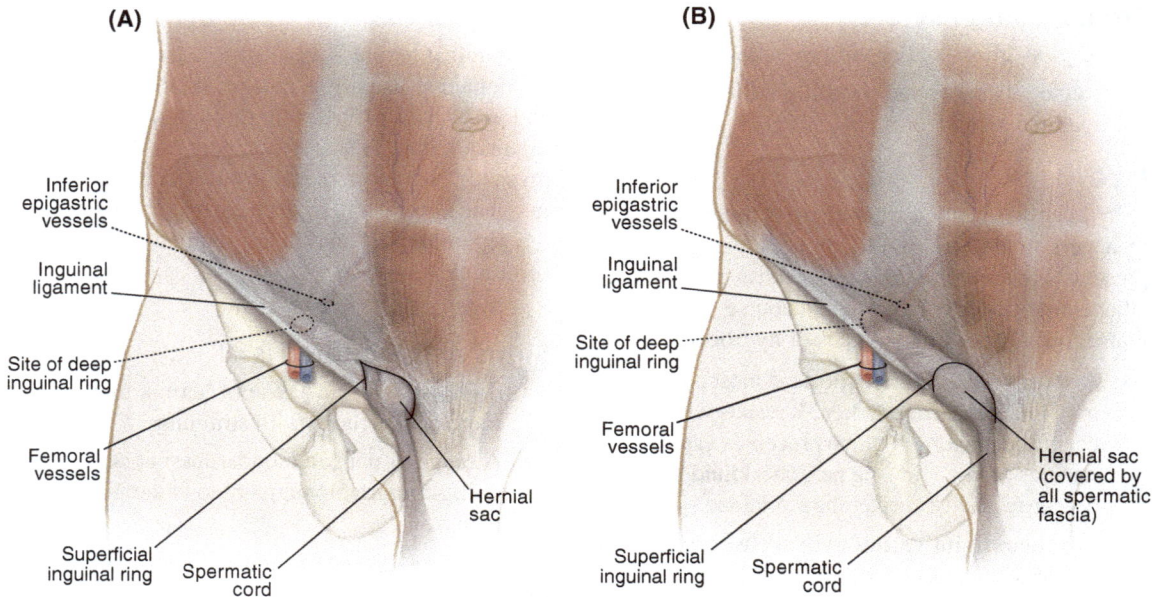

FIGURE 3.1.4 Anterior view of inferior abdominal wall showing (**A**) direct and (**B**) indirect inguinal hernias.

- Occupation: jobs that require standing for long periods or heavy physical labor have increased risk

Indirect Inguinal Hernia

The underlying cause for this type of hernia is a failure of the **processus vaginalis** to close at the deep inguinal ring during the fetal period. Therefore, an indirect inguinal hernia is considered **congenital**, even though it may not develop until the second or third decade. Intra-abdominal pressure may eventually cause a loop of small bowel to enter the patent processus vaginalis, forcing the hernial sac into the inguinal canal.

An indirect inguinal hernia enters the deep inguinal ring, lateral to the inferior epigastric vessels, and passes along the entire length of the inguinal canal (**Fig. 3.1.4B**). It may protrude through the superficial inguinal ring and extend into the scrotum.

Distinguishing Characteristics

- Passes along a patent processus vaginalis
- May traverse entire length of inguinal canal
- May emerge through the superficial inguinal ring
- May present at the superficial inguinal ring or in the scrotum

Predisposing Factors

- Sex: male (9:1)
- Age: <25 years
- Family history
- Past history of inguinal hernia
- Chronic cough
- Smoking
- Excess weight

Most inguinal hernias are reducible, meaning the protruding structure(s) can return to their normal position in the abdominal cavity. Inguinal hernias are very common (lifetime risk is 27% for men, 3% for women), and their repair is one of the most frequently performed surgical operations.

Clinical Notes

Examination of the inguinal region is best performed with the patient standing. The examiner places a finger at the site of the superficial inguinal ring and asks the patient to cough (coughing increases intra-abdominal pressure, which forces the hernia toward the superficial ring and may make it easier to detect during physical examination). The size of the superficial ring can be determined by palpating just lateral to the pubic tubercle.

- A direct hernia is suspected if a bulge is felt against the side of the examining finger.
- An indirect hernia is suspected if a bulge is felt at the tip of the finger, as the finger is directed toward the deep ring.

In most patients it is difficult on the basis of physical examination to distinguish between direct and indirect inguinal hernias; the type of inguinal hernia can be established accurately during surgery.

Any hernia mass that is tender to palpation, or associated with nausea and vomiting, should be considered possibly strangulated (compromised vascularity of entrapped bowel). This condition represents a surgical emergency.

DIAGNOSIS

The patient presentation, history, physical examination, and lab results support a diagnosis of an **indirect inguinal hernia**.

Indirect Inguinal Hernia

Inguinal hernias can be either **direct or indirect** based on the course they take through the anterior abdominal wall. **Indirect inguinal hernias** pass lateral to the inferior epigastric vessels and traverse the full length of the inguinal canal before they present at the superficial inguinal ring. This type of hernia is the result of a persistent processus vaginalis. A **direct inguinal hernia** appears as a swelling in the inguinal triangle. This type of hernia does not traverse the length of the inguinal canal but may present "directly" at the superficial inguinal ring.

- This patient had a small, reducible swelling at the site of the superficial inguinal ring indicating the presence of an inguinal hernia.
- Based on the patient's age, his inguinal hernia would most likely be indirect.
- This diagnosis and classification can be verified at the time of surgical repair.

Lymphadenitis

There is a collection of lymph nodes (superficial inguinal) located along the inguinal ligament. These nodes are prone to enlargement following infections in children. Lymphadenitis may affect a single node or a group of nodes, and may be unilateral or bilateral.

- If this patient had lymphadenitis, inguinal swellings would not be reducible and the patient would most likely have elevated white cell count and temperature.

Hydrocele

A hydrocele can form anywhere along the extent of a persistent processus vaginalis or within the tunica vaginalis of the scrotum.

- The ultrasound in this patient indicated an inguinal mass, which was not consistent with a fluid-filled hydrocele.
- Hydroceles typically present as a soft, nontender swelling anterior to the testis in the scrotum.
- Diagnosis of a hydrocele could be confirmed by transillumination of the scrotal sac, which was not performed on this patient.

Patient Presentation

A 58-year-old white male is admitted to the emergency department with severe upper abdominal pain.

Relevant Clinical Findings

History

The patient reports that he has a long-standing history of mild "stomach" pain, but today just after lunch he felt a sharp, stabbing pain in the region just inferior to his sternum. He described the pain as "piercing," and noted it spread quickly across the left upper part of his abdomen. The pain was very intense for hours, but then subsided. He was nauseous all afternoon (symptoms started after lunch) and vomited once. He described the vomitus as appearing like "coffee grounds." He now reports vague pain over his left shoulder.

Physical Examination

The following findings were noted on physical examination:

- Hypoactive bowel sounds
- Guarding of the superior and anterolateral abdominal wall
- Epigastric and left hypochondrial regions are tender during deep palpation

Clinical Note

The **urea breath test** is a noninvasive method for identifying *H. pylori* infection. It is based on the ability of *H. pylori* to convert urea to ammonia and carbon dioxide. The result of this test can be confounded by a patient's medications.

Laboratory Tests

Test	Value	Reference value
Erythrocytes (count)	4.9	$4.3–5.6 \times 10^6/mm^3$
Leukocytes (count)	13.2	$3.54–9.06 \times 10^3/mm^3$
Hemoglobin	11	14–17 gm/dL
Helicobacter pylori	Positive	Negative
Fecal occult blood (FOBT)	Negative	Negative

Imaging Studies

- Postero-anterior and lateral chest radiographs indicated a pneumoperitoneum under the left dome of the diaphragm. This was confirmed by abdominal computed tomography (CT).

Clinical Note

Pneumoperitoneum is often detected radiographically, but small amounts of air may be missed. CT is considered the standard of care in the assessment for pneumoperitoneum.

Procedures

- Esophagogastroduodenoscopy (EGD) revealed a discrete, well-circumscribed, 1.5 cm mucosal lesion with a "punched-out" base. Gastric biopsy indicated the presence of gram-negative rods consistent with *H. pylori*.

Clinical Problems to Consider

- Duodenal ulcer
- Gastric ulcer
- Gastritis

Guarding Muscle spasm (especially of anterior abdominal wall) to minimize movement at site, at or near injury or disease (e.g., inflammation associated with appendicitis or diverticulitis). May be detected with palpation during physical examination.

Pneumoperitoneum Air or gas in the peritoneal cavity

1. Describe the anatomy of the stomach.
2. Describe the anatomy of the peritoneum and peritoneal cavity.
3. Explain the anatomical basis for the signs and symptoms associated with this case.

RELEVANT ANATOMY

Stomach

The stomach is a hollow, muscular organ located predominantly in the left upper abdominal quadrant, just inferior to the left hemidiaphragm. Most often, it is shaped like the letter "J" (**Fig. 3.2.1**). It receives the food bolus from the esophagus. The stomach secretes hydrochloric acid and enzymes that aid in digestion. Smooth muscle in the wall of the stomach contracts in peristaltic waves, churning the gastric contents to enhance digestion. The pyloric sphincter is a muscular valve at the distal end of the stomach that regulates the passage of stomach contents (chyme) into the duodenum.

The regions of the stomach are outlined in **Table 3.2.1**.

Blood Supply

Arteries that supply the stomach are derived from the **celiac trunk**, the first unpaired branch of the abdominal aorta (**Fig. 3.2.2A**).

- The **left gastric** arises directly from the celiac trunk and supplies the proximal part of the lesser curvature. It anastomoses with the right gastric artery.
- The **right gastric** arises from the proper hepatic artery and supplies the distal part of the lesser curvature. It anastomoses with the left gastric artery.
- The **left gastro-omental** is a branch of the splenic. It supplies the proximal part of the greater curvature, and anastomoses with the right gastro-omental.
- The **right gastro-omental** is a branch of the gastroduodenal and supplies the distal part of the greater curvature. It anastomoses with the left gastro-omental.

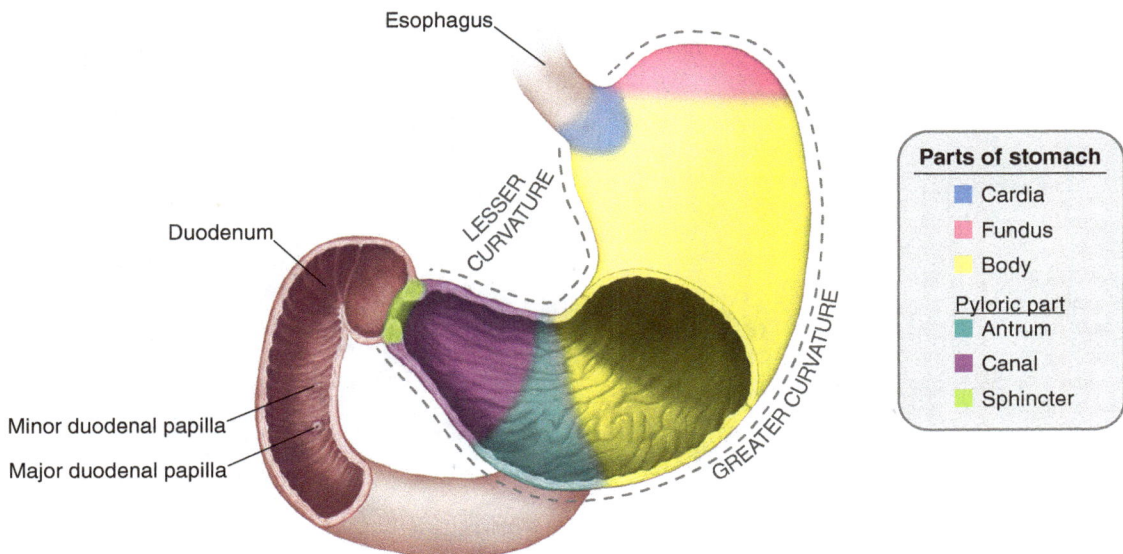

FIGURE 3.2.1 Anterior view of the stomach and duodenum showing anatomical regions of the stomach.

Peristalsis Waves of alternate contraction and relaxation along a muscular tube

TABLE 3.2.1	Regions and curvatures of the stomach.
Region	**Description**
Cardia	Area adjacent to esophageal opening
Fundus	Domed area superior to cardia
Cardiac notch	Between esophagus and fundus
Body	Largest area; between fundus and pyloric part
Pyloric part	Between body and duodenum
Antrum	Wide portion of pyloric part; leads to pyloric canal
Canal	Narrow portion of pyloric part; leads to pylorus
Pylorus	Thickened wall forms **pyloric sphincter** that regulates opening into duodenum
Greater curvature	Convex border
Lesser curvature	Concave border
Angular incisure	Indentation on lesser curvature that indicates transition from body to pyloric part

- **Short gastric arteries** arise from splenic and supply the fundus.
- The **posterior gastric** arises from the splenic artery. Its branches supply the posterior aspect of the body.

Veins that drain the stomach correspond to arteries listed above. All veins drain directly or indirectly into the **hepatic portal vein** (**Fig. 3.2.2B**).

Nerve Supply

The stomach is derived from the embryonic foregut and, thus, receives **sympathetic innervation** from the **thoracic splanchnic nerves** and **parasympathetic innervation** from the **vagus nerve** (CN X).

- **Sympathetic (visceral efferent)**. Sympathetic fibers are responsible for vasomotor control of gastric vessels. Preganglionic axons originate from cell bodies in the T5-T9 lateral horns of the spinal cord. These axons are carried by the **greater splanchnic nerves** to the **celiac ganglia** (prevertebral ganglia) in the celiac plexus. Postganglionic axons are distributed in periarterial plexuses along branches of the celiac trunk.
- **Parasympathetic (visceral efferent).** Vagal impulses mediate or facilitate digestion by increasing gastric secretions and motility. The

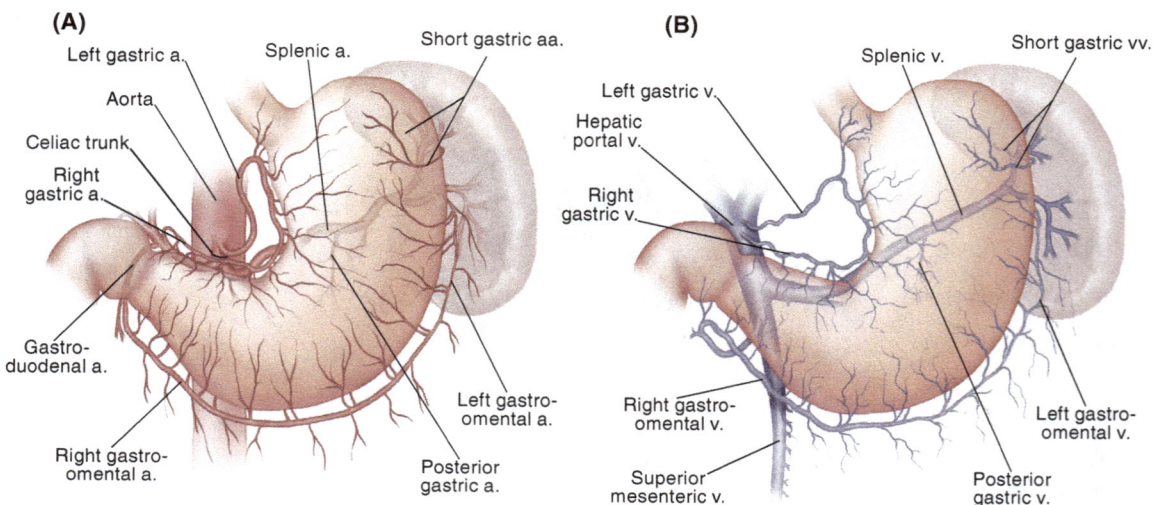

(A)
Left gastric a.
Splenic a.
Short gastric aa.
Aorta
Celiac trunk
Right gastric a.
Gastro-duodenal a.
Right gastro-omental a.
Posterior gastric a.
Left gastro-omental a.

(B)
Short gastric vv.
Splenic v.
Left gastric v.
Hepatic portal v.
Right gastric v.
Right gastro-omental v.
Superior mesenteric v.
Posterior gastric v.
Left gastro-omental v.

FIGURE 3.2.2 Anterior view showing the blood supply of the stomach. (**A**) Arteries and (**B**) veins.

CN X originates in the brainstem. In the thorax, the right and left CN X contribute to the esophageal plexus. The **anterior and posterior vagal trunks** coalesce from the esophageal plexus: the anterior trunk is derived primarily from the left CN X and the posterior trunk from the right CN X. These trunks enter the abdomen through the esophageal hiatus of the diaphragm. Vagal fibers reach the stomach as **gastric nerves** derived from the **vagal trunks**.

■ Visceral afferent impulses are carried in the **vagus** and **sympathetic nerves**. Receptors in the stomach that detect **stretch and distension** have their afferent fibers in the vagus. Their cell bodies are located in vagal ganglia. Receptors that detect **pain** have their afferent fibers in the sympathetic nerves. Their cell bodies are in the dorsal root ganglia of T5-T9. Therefore, stomach pain may be referred to the T5-T9 dermatomes.

For additional information about the autonomic nervous system and referred pain, see Chapter 1: Autonomic Nervous System.

Peritoneum

Peritoneum is a serous membrane that lines the abdominopelvic cavity. This membrane produces serous fluid that lubricates the peritoneal surfaces to minimize friction as organs move. The peritoneum is subdivided into two parts:

1. The **parietal peritoneum** lines the walls of the cavity (including the internal aspect of the anterior and posterior abdominal walls and the abdominal surface of the thoracic diaphragm). Parietal peritoneum continues into the pelvis as pelvic peritoneum, which reflects onto superior surface of the bladder, and the anterior and lateral surfaces of the proximal rectum. In the female, peritoneum coats the body of the uterus, uterine tubes, and ovaries as the broad ligament. A mesentery of peritoneum extends from each of these organs (mesometrium, mesosalpinx, and mesovarium, respectively) as the broad ligament. Parietal peritoneum is separated from the abdominal walls by variable amounts of extraperitoneal fat and connective tissue. This peritoneal layer is usually loosely attached to these walls which

allows for changes in size of certain organs (e.g., distensibility of the descending colon).

2. Visceral peritoneum is firmly attached to the external surface of most abdominal organs as it reflects over or invests them.

Mesenteries, Omenta, and Peritoneal Ligaments

Three general terms describe double layers of peritoneum that connect some organs with the abdominal wall or with other organs.

1. A **mesentery** is a peritoneal fold that suspends or supports an organ in the abdominopelvic cavity. Mesenteries have a connective tissue core with variable amounts of fat. They contain blood vessels, lymphatic vessels and nodes, and nerves.
   a. The mesentery of the small intestine refers to the peritoneal fold that supports the jejunum and the ileum. It anchors this part of the small intestine to the posterior abdominal wall.
   b. Other mesenteries include the name of the organ (e.g., transverse mesocolon and mesoappendix).

2. An **omentum** is a peritoneal fold that is associated with the stomach and proximal duodenum.
   a. The **lesser omentum** is attached to the lesser curvature of stomach and proximal duodenum and connects them to the liver. This omentum is divided into the **hepatogastric** (liver-to-stomach) and **hepatoduodenal** (liver-to-duodenum) ligaments. The hepatoduodenal ligament contains the hepatic portal vein, proper hepatic artery, and bile duct.
   b. The **greater omentum** is a prominent, apron-like, peritoneal fold that is suspended from the greater curvature of the stomach and proximal duodenum. This omentum is folded on itself and, therefore, is composed of four peritoneal layers. The greater omentum usually "hangs over" the small intestines. The two posterior lamellae of this omentum attach to the transverse colon and are continuous with the transverse mesocolon (which attaches the transverse colon to the posterior abdominal). The greater omentum also includes the **gastrosplenic**

(stomach-to-spleen) and **gastrophrenic** (stomach-to-diaphragm) ligaments.

3. A **peritoneal ligament** is a peritoneal fold that connects one organ to another, or to the abdominal wall. They are named after the structures involved (e.g., splenorenal ligament and phrenicocolic ligament).

Peritoneal Cavity

The **peritoneal cavity** is a potential space between the parietal and visceral layers of peritoneum. This cavity is closed in males. In females, however, the uterine tubes open into the cavity. Therefore, the female peritoneal cavity communicates, indirectly, with the external environment.

The peritoneal cavity is subdivided into two parts (**Fig. 3.2.3**):

1. The **greater sac** forms the majority of the peritoneal cavity. It extends from the thoracic diaphragm superiorly and continues into the pelvic cavity inferiorly. The transverse mesocolon divides the greater sac into two compartments.

   a. The **supracolic compartment** is superior to the transverse mesocolon. The stomach, liver, and spleen are located in this compartment.

   b. The **infracolic compartment** lies inferior to the transverse mesocolon and posterior to the greater omentum. This compartment includes the jejunum, ileum, and ascending and descending colon. The mesentery of the small intestine further divides this compartment into **right and left infracolic spaces**.

2. The **lesser sac (omental bursa)** is the smaller subdivision of the peritoneal cavity. It lies

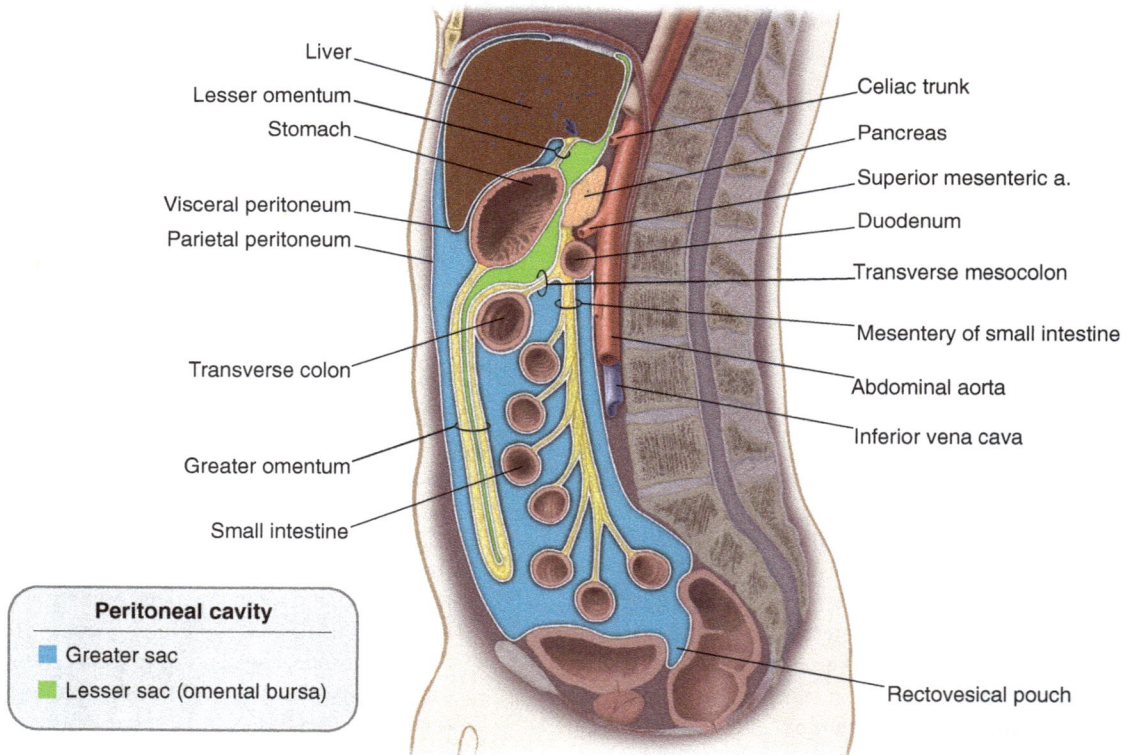

Peritoneal cavity

- Greater sac
- Lesser sac (omental bursa)

FIGURE 3.2.3 Midsagittal section through the abdominopelvic cavity showing the subdivisions of the peritoneal cavity. The greater sac is indicated in blue, the lesser sac in green.

posterior to the stomach and liver, and is continuous with the greater sac through the omental foramen (Winslow). This sac has superior and inferior extensions called recesses:

a. The **superior recess** is located posterior to the liver and extends superiorly to the diaphragm.

b. The **inferior recess** extends between the inner two layers of the greater omentum. The peritoneal layers that line the inferior recess are usually fused, obliterating most of the inferior recess.

Clinical Notes

- A loop of small intestine may pass through the omental foramen and into the lesser sac. Although rare, when this happens the intestine is vulnerable to strangulation in the foramen.

- The potential space of the peritoneal cavity has a large surface area that facilitates the spread of disease. Under certain pathological conditions, infectious exudate (e.g., pus) may spread throughout the peritoneal cavity.

- Ascites is excess peritoneal fluid. It is most commonly the result of liver disease. The excess peritoneal fluid may be removed by paracentesis.

- A perforated bowel may lead to the release of gas into the peritoneal cavity. This gas, known as a pneumoperitoneum, can usually be visualized on a chest radiograph with the patient erect.

CLINICAL REASONING

This patient presents with signs and symptoms of **epigastric pain** with guarding **of the anterior abdominal wall** and **referred pain to the shoulder**.

Duodenal Ulcer

Peptic ulcers usually occur when acid in the digestive tract erodes the mucosa of the **esophagus, stomach, or small intestine**. The mucosa of the gastrointestinal tract produces a layer of mucus that normally provides protection against acid. If the production of acid increases or mucus production decreases, an ulcer can develop. This can create a painful "open" sore in the luminal wall of the stomach or duodenum that may bleed.

Duodenal ulcers develop in the **superior part of the duodenum**. They are usually about a centimeter in diameter, but may be considerably larger. *H. pylori* bacterial infections are the cause of most duodenal ulcers. *H. pylori* colonizes the mucosa of the stomach and duodenum and causes local inflammation. Infection leads to a combination of increased gastric acid secretion and reduced duodenal bicarbonate secretion. This chemical imbalance makes the duodenal walls susceptible to erosion.

Signs and Symptoms

- Generalized abdominal pain
- Burning stomach or upper abdominal pain, typically 2–3 hours after eating
- Nocturnal abdominal pain (50–80% of patients)
- Bloating
- Feeling of fullness
- Loss of appetite
- Unexplained weight loss
- Nausea, with or without vomiting

More serious symptoms include:

- Severe abdominal pain
- Bloody stool
- Bloody or coffee ground-like vomitus

Predisposing Factors

- Alcohol abuse
- Tobacco use
- *H. pylori* infection
- History of radiation therapy

Paracentesis Puncturing a cavity, usually using a needle or other hollow tube, to aspirate fluid

Ulcer A lesion on skin surface or in a mucous membrane that extends through epidermis or endothelium, respectively.

- Long-term, regular use of aspirin or NSAIDs (e.g., ibuprofen and naproxen)
- Stress or severe illness

Gastric Ulcer

A **gastric ulcer** is a type of peptic ulcer that affects the stomach. As with duodenal ulcers, the majority of gastric ulcers involve *H. pylori* infection. The second most common causative factor is chronic or over-use of nonsteroidal anti-inflammatory drugs (NSAIDs). In this case, the hydrogen ion pump is affected, which changes the mucosal defense system and, subsequently, results in epithelial cell injury. Some gastric ulcers are associated with low gastric acids levels which impairs other mucosal defense factors.

Clinical Note

The classification of peptic ulcers is summarized in **Table 3.2.2**.

Sign and Symptoms

The signs and symptoms of a **gastric ulcer** are similar to those of a **duodenal ulcer;** the primary difference is in the timing and severity of the pain.

- Dull aching upper abdominal pain, often immediately after eating
- Eating does not relieve pain
- Meal preparation can cause increased pain
- Indigestion or acid reflux
- Nocturnal abdominal pain (30% of patients)

- Chronic pain in upper abdomen, inferior to sternum
- Episodes of nausea
- Loss of appetite
- Weight loss

Predisposing Factors

- The risk factors for gastric ulcers are similar to those for duodenal ulcers.

Gastritis

Gastritis is an inflammation of the mucosa of the stomach. This condition can last for weeks (acute gastritis) or it may linger for months to years (chronic gastritis).

Signs and Symptoms

- Upper abdominal pain
- Abdominal upset
- Loss of appetite
- Nausea and vomiting

If gastritis causes bleeding, symptoms may include:

- Black stools
- Bloody or coffee ground-like vomitus

Predisposing Factors

- Prolonged use of NSAIDs
- Caffeine abuse
- Alcohol abuse
- Cocaine abuse
- Eating or drinking caustic or corrosive substances

TABLE 3.2.2	Modified Johnson classification of peptic ulcers*.	
Type	**Location of ulcer**	**Notes**
Type I	In body of stomach, most often along lesser curve at angular incisure	Associated with low gastric acid
Type II	In body of stomach, in combination with duodenal ulcers	Associated with gastric acid oversecretion
Type III	In pyloric channel, within 3 cm of pylorus	Associated with gastric acid oversecretion
Type IV	Proximal gastroesophageal ulcer	Associated with low gastric acid
Type V	Throughout stomach	Associated with chronic NSAID use

*Peptic ulcers include gastric and duodenal ulcers.

- Bacterial infection (e.g., *H. pylori*)
- Viral infection (e.g., cytomegalovirus and herpes simplex virus)
- Autoimmune disorders (e.g., pernicious anemia)
- Bile reflux
- Extreme stress

Clinical Notes

There are two primary types of chronic gastritis:

1. **Type A gastritis** is the less common form and is found in the fundus and body of the stomach. It is usually associated with pernicious anemia and the presence of circulating antibodies against parietal cells.

2. **Type B gastritis** is the more common form and results from an *H. pylori* infection. The incidence of this form of gastritis increases with age, being present in up to 100% of the people over 70.

DIAGNOSIS

The patient presentation, history, physical examination, laboratory tests, and procedures outcomes support a diagnosis of **perforated gastric ulcer**.

Gastric Ulcer

Peptic ulcers are breaks in the mucosal lining of the stomach or duodenum. Gastric and duodenal ulcers share many pathophysiology features.

- Endoscopy revealed an ulcer in the posterior wall of the body of the stomach in this patient. The ulcer exposes the wall of the stomach to gastric secretions. This stimulates pain afferents and results in the chronic, epigastric pain described by this patient.

- In this patient, the perforated ulcer resulted in hemorrhage and spilling of gastric contents into the lesser peritoneal sac. This irritated the parietal peritoneum lining this portion of the peritoneal cavity and accounts for the left upper quadrant pain and tenderness in the epigastric and left hypochondriac regions on deep palpation, as well as guarding of the anterior wall.

- This patient also described referred pain to the left shoulder region. Irritation of parietal peritoneum on the abdominal surface of the diaphragm (in the superior recess of the lesser sac) stimulated afferent fibers in the phrenic nerve, which accounts for the referred pain.

- The patient described "coffee-grounds" vomitus, a classic sign for upper gastro-intestinal tract bleeding. Blood contains iron and, when the iron is oxidized by gastric acid, it has the appearance of coffee grounds.

Clinical Note

Hematemesis suggests that the upper gastrointestinal bleeding is more acute or more severe, or originates from the proximal stomach.

Duodenal Ulcer

Epigastric pain is the most common symptom of both gastric and duodenal ulcers.

- Subtle differences in the timing of some of the signs and symptoms could be used to suggest the location of the ulcer.

- The location of the ulcer can only be confirmed by endoscopy.

Endoscopy revealed an ulcer in the posterior wall of the body of the stomach on this patient.

Gastritis

Gastritis covers a broad spectrum of inflammatory changes in the gastric mucosa that may result in moderate-to-severe epigastric pain.

- Patients may experience epigastric pain, loss or appetite, nausea, and/or vomiting. Patients suffering from gastritis frequently have asymptomatic periods.

- This diagnosis was not confirmed by endoscopy in this patient.

Anemia Reduced erythrocytes, hemoglobin, or blood volume

Hematemesis Vomiting blood

Biliary Obstruction

Patient Presentation

A 48-year-old female is admitted to the emergency department complaining of moderate abdominal pain that has been increasing in intensity over the last several days.

Relevant Clinical Findings

History

The patient reports that her pain began just below her ribs in the midline about 5 days ago. It has been getting worse each day and is now located across the upper part of her abdomen and extends to her back between her shoulders. She reports having painful "attacks" an hour or so after meals. On the previous day, she was feverish, had chills, nausea, and vomited twice. She describes her urine as the color of "Coca Cola." She has had no prior significant illness, is not taking any medications, and may drink a beer with her husband on the weekends.

Physical Examination

Noteworthy vital signs:

- Height: 5′ 7″
- Weight: 189 lb
- BMI: 30 (normal: 18.5–24.9; obese: >30)
- Temperature: 39°C (102.8°F)
  Normal: 36.0–37.5°C (96.5–99.5°F)
- Pulse: 117 bpm
  Adult resting rate: 60–100 bpm
- Respiratory rate: 21 cycles/min
  Normal Adult: 14–18 cycles/min; women slightly higher

The following findings were noted on physical examination:

- Jaundice

- Localized tenderness in right upper quadrant (RUQ) with light and deep palpation
- Firm, regular, smooth but tender liver edge, palpable below right costal margin
- Positive Murphy sign

Clinical Note

Murphy sign. This test is used during an abdominal examination. It is performed by asking the patient to breathe out, and then gently placing the fingers below the right costal margin at the midclavicular line. The patient is then instructed to inspire. During inspiration, abdominal contents are normally pushed downward with the movement of the diaphragm. If the patient expresses pain with inspiration, the test is considered positive. The pain is caused when an inflamed gallbladder that moved downward on inspiration is compressed by the position of the examiner's fingers. A positive test requires a similar maneuver on the patient's left hand side with no pain.

Laboratory Tests

Test	Value	Reference value
Serum bilirubin (direct)	3.4	0.3–1.3 mg/dL
Alkaline phosphatase	65	30–95 U/L
Serum albumin	4.5	4.0–5.0 mg/dL
Serum amylase	45	20–96 U/L
Serum lipase	35	3–43 U/L
Lactate dehydrogenase (LDH)	105	115–221 U/L
Aspartate aminotransferase (AST)	35	12–48 U/L

Jaundice Yellowish color of the skin, mucous membranes, and/or conjunctiva. The term jaundice is synonymous with icterus

Imaging Studies

- The presence of gallstones was not confirmed by abdominal ultrasound, although there was an indication of dilated bile ducts.
- CT cholangiography (helical CT) revealed biliary pathology with a radiolucent stone in the bile duct.

Clinical Problems to Consider

- Biliary obstruction
- Chronic pancreatitis
- Liver cirrhosis

LEARNING OBJECTIVES

1. Describe the anatomy of the gallbladder and biliary ducts.
2. Describe the anatomy of the pancreas.
3. Describe the anatomy of the liver.
4. Explain the anatomical basis for the signs and symptoms associated with this case.

RELEVANT ANATOMY

Gallbladder and Biliary Ducts

The gallbladder is a hollow organ that lies in a fossa between the right and quadrate lobes on the visceral surface of the liver. It receives bile from the liver, then stores and concentrates it. The gallbladder has three parts.

1. The **fundus** is its rounded end. It usually projects from the inferior margin of the liver and is located along the costal margin on the midclavicular line.

2. The **body** forms the majority of the gallbladder. It is in contact with the liver, transverse colon, and superior part of the duodenum.

3. The **neck** is the narrow part that leads to the cystic duct. The mucosa in this part has a distinct **spiral fold** (spiral valve) that aids in directing the flow of bile.

Bile is a dark green to yellow-brown fluid produced by the liver. It is composed of water, bile salts, cholesterol, and other molecules. The functions of bile include:

- Emulsification of fats in food
- Excretion of bilirubin, a by-product of normal heme catabolism during red blood cell degradation
- Neutralizing excess gastric acid
- Bactericide

Hepatocytes secrete bile into **bile canaliculi**, microscopic collecting ducts within the liver parenchyma. These canaliculi drain into successively larger ducts (**Fig. 3.3.1**):

- The **right and left hepatic ducts** arise from the respective lobes of the liver.
- The **common hepatic duct** is formed by the union of the right and left hepatic ducts.
- The **cystic duct** connects the neck of the gallbladder to the common hepatic duct.
- The **bile duct** (common bile duct) begins at the junction of the cystic and common hepatic ducts. It is located in the margin of the **hepatoduodenal ligament** and descends posterior to the superior part of the duodenum. A thickened circular muscle at the distal end of the duct forms the **sphincter of the bile duct**. This sphincter regulates the flow of bile along the biliary ducts. The bile duct usually joins the **main pancreatic duct** on the posterior surface of the head of the pancreas.
- The **hepatopancreatic ampulla** (Vater) is formed when the main pancreatic and bile ducts unite. The distal end of this ampulla opens on the **major duodenal papilla** in the descending portion of the duodenum.

Blood Supply

The **cystic artery** is the primary blood supply to the gallbladder, cystic duct, and hepatic ducts

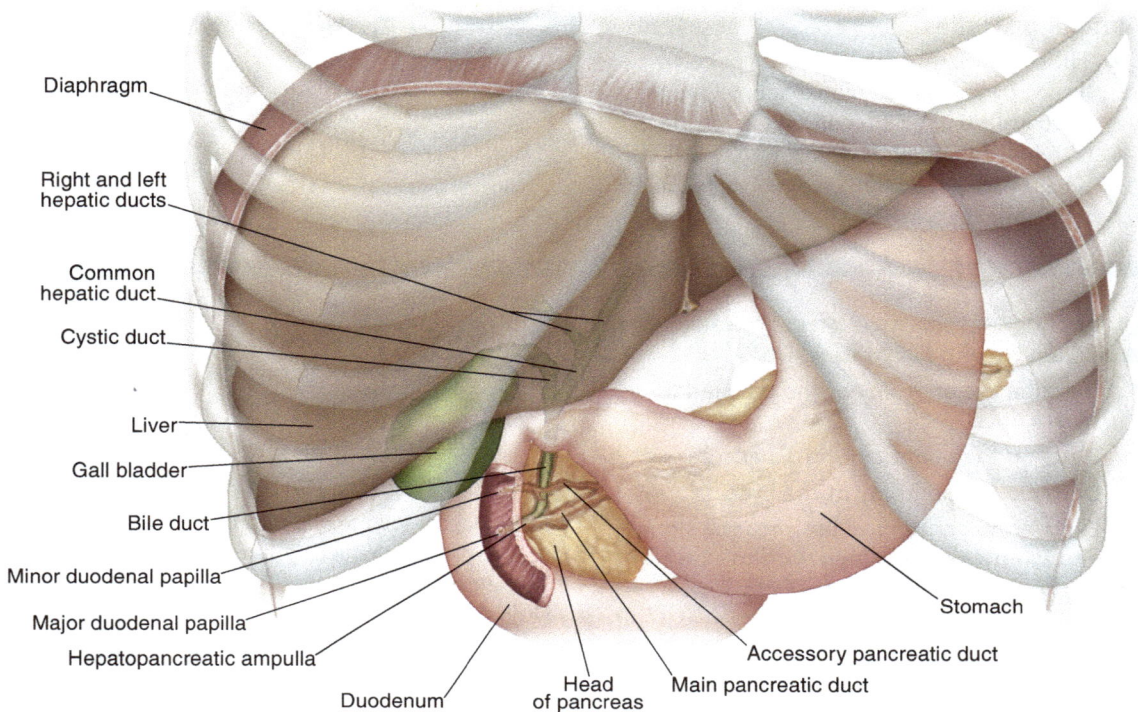

FIGURE 3.3.1 Anterior view of the gallbladder and biliary ducts.

(**Fig. 3.3.2A**). This artery usually arises from the **right hepatic artery**, although variations in its origin and course are common. **Direct branches** from the right hepatic artery supply the intermediate part of the biliary duct system. The **posterior superior pancreaticoduodenal artery** supplies the distal biliary tree including the hepatopancreatic ampulla. The **cystic vein** drains the gallbladder, cystic duct, and hepatic ducts, as well as the proximal part of the bile duct (**Fig. 3.3.2B**). This vein may enter the liver or drain into the hepatic portal vein. Small veins that join the portal vein drain the distal part of the biliary duct system.

Clinical Note

Ligation of the **posterior superior pancreaticoduodenal artery** can result in necrosis of the terminal portion of the biliary duct system.

Lymphatics

Lymph vessels from the gallbladder, cystic duct, and hepatic ducts drain primarily into **cystic nodes** located at the junction of the cystic and common hepatic ducts. Lymph from the remainder of the biliary system drains into **hepatic lymph nodes** located along hepatic arteries in the hepatoduodenal ligament. Lymph in these nodes passes to **celiac nodes** and then to **abdominal lymph trunks**.

Nerve Supply

The gallbladder and biliary system are innervated primarily by the **celiac plexus**. The nerve fibers from this plexus pass along branches of the celiac trunk to portions of the biliary apparatus (gallbladder and ducts). Neuronal fibers in this plexus include:

- **Sympathetic (visceral efferent).** Preganglionic axons originate from cell bodies in the T5-T9 lateral horns of the spinal cord. These

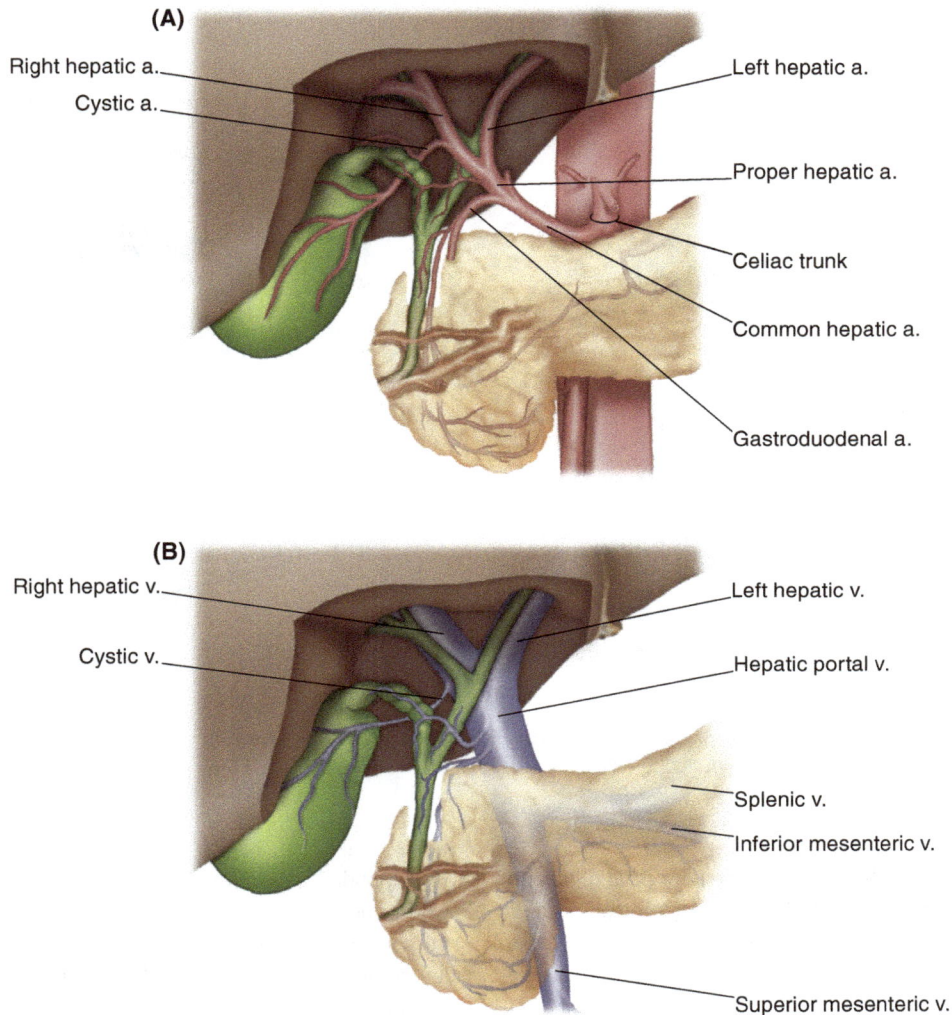

(A)

Right hepatic a.
Cystic a.

Left hepatic a.

Proper hepatic a.

Celiac trunk

Common hepatic a.

Gastroduodenal a.

(B)

Right hepatic v.
Cystic v.

Left hepatic v.

Hepatic portal v.

Splenic v.
Inferior mesenteric v.

Superior mesenteric v.

FIGURE 3.3.2 Anterior view showing the blood supply of the biliary tree.
(A) Arteries and **(B)** veins.

axons are carried by the **greater splanchnic nerves** to the **celiac ganglia** (prevertebral ganglia). Postganglionic axons are distributed along branches of the celiac trunk.

- **Parasympathetic (visceral efferent).** Preganglionic parasympathetic axons originate from cell bodies in the brainstem and travel in the **anterior and posterior vagal trunks.** These trunks pass into the abdominal cavity through the esophageal hiatus of the diaphragm. Parasympathetic preganglionic fibers are distributed

with sympathetic fibers along the branches of the celiac trunk to the biliary system.

- **Visceral afferents.** Afferents associated with **pain** receptors in the wall of the gallbladder and biliary ducts travel with the **greater splanchnic (sympathetic) nerves.** The afferent cell bodies associated with these receptors are located in the T5-T9 dorsal root ganglia. Visceral pain sensations from the biliary system will be felt primarily in the RUQ and epigastic region and referred along the T5-T9 dermatomes (**Fig. 3.3.3**).

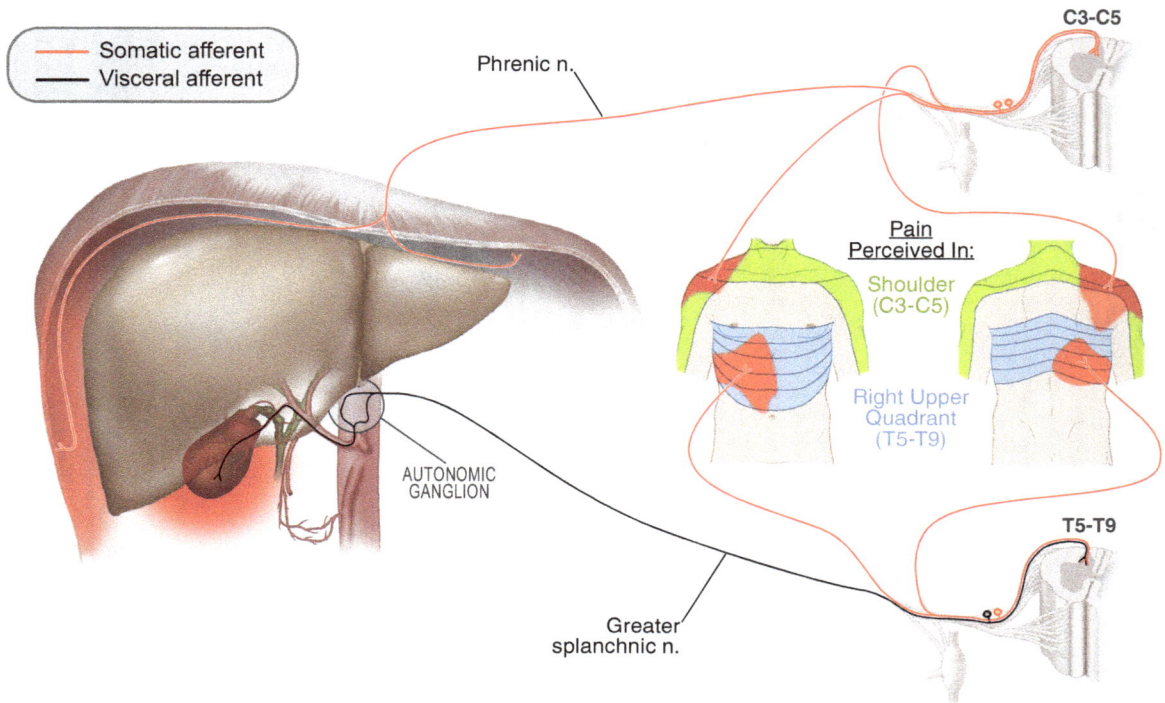

FIGURE 3.3.3 Diagrammatic representation of visceral and somatic referred pain from the gallbladder and biliary ducts.

An inflamed gallbladder may irritate parietal perito-neum on the adjacent abdominal (inferior) surface of the diaphragm. Pain impulses are transmitted in **somatic afferent** fibers of the phrenic nerve (C3-C5). Sensation from this irritated peritoneum pres-ents as aching or sharp pain referred to the shoulder and/or base of the neck, along the C3-C5 derma-tomes (**Fig. 3.3.3**).

For additional information about the autonomic ner-vous system and referred pain, see Chapter 1: Auto-nomic Nervous System.

Pancreas

The pancreas is an accessory gland of the digestive system. It is retroperitoneal, located in the posterior wall of the lesser sac. The pancreas extends from the medial margin of the duodenum to the hilum of the spleen. It is crossed anteriorly by the transverse mesocolon.

Its expanded, medial part is the **head**, which is fixed to the medial aspect of the descending and horizon-tal parts of the duodenum. The **uncinate process** is a medial projection from the head that extends poste-rior to the superior mesenteric vessels. The pancreatic **neck** is the short segment between the head and body; it is anterior to the superior mesenteric vessels. The **body** of the pancreas is the largest part of the gland and leads to the **tail**. The tail lies anterior to the left kidney and its tip is adjacent to the hilum of the spleen. There are two ducts in the substance of the pancreas (**Fig. 3.3.1**) that collect pancreatic enzymes and deliver them to the descending (second) part of the duodenum.

1. The **main pancreatic duct** extends from the tail of the gland to the head, where it meets the bile duct. The main pancreatic duct and bile duct unite to form the hepatopancreatic ampulla, which opens onto the **major duodenal papilla**.

2. The **accessory pancreatic duct** collects pancreatic enzymes from the uncinate process and the head of the gland. This duct opens at the **minor duodenal papilla**.

These two pancreatic ducts are usually connected.

Blood Supply

The **splenic artery** is the primary blood supply to the pancreas. This artery courses along the superior border of the pancreas, its serpentine course causing it to randomly "dip" into pancreatic tissue. It gives rises to as many as 10 small branches to the body and tail. The head and neck of the pancreas receive blood from the **superior and inferior pancreaticoduodenal arteries**, branches of the **gastroduodenal**, and **superior mesenteric** arteries, respectively. Pancreatic veins have a similar pattern as the arteries. They drain into the **splenic and superior mesenteric veins**, which join to form the **hepatic portal vein**.

Nerve Supply

The innervation of the pancreas is from the celiac and superior mesenteric plexuses. Nerve fibers from these plexuses pass along branches of the celiac trunk and superior mesenteric arteries. These plexuses include:

- **Sympathetic (visceral efferent).** Preganglionic sympathetic axons originate from cell bodies in the T5-T9 lateral horns of the spinal cord. These axons are carried by the **greater splanchnic nerves** to **celiac ganglia** (prevertebral ganglia). Postganglionic axons are distributed along branches of the celiac trunk (i.e., splenic and gastroduodenal). In addition, preganglionic axons originate from cells in the T10-T11 lateral horns travel in the **lesser splanchnic nerves** to **superior mesenteric ganglia** (prevertebral ganglia). These postganglionic axons are distributed along branches of the superior mesenteric artery (i.e., inferior pancreaticoduodenal).

- **Parasympathetic (visceral efferent).** Preganglionic parasympathetic axons originate from cell bodies in the brainstem. They are carried in the **CN X** and their respective **anterior and posterior vagal trunks**. These trunks pass through the esophageal hiatus of the diaphragm into the abdominal cavity. Preganglionic parasympathetic fibers are distributed with sympathetic fibers along the branches of the celiac trunk and superior mesenteric artery. They synapse in terminal ganglia and postganglionic fibers supply the gland.

Liver

The liver is the largest gland in the body. It is located just inferior to the diaphragm, primarily in the RUQ of the abdomen. Its **diaphragmatic surface** (superior) is in contact with the thoracic diaphragm, and its **visceral surface** (inferior) is related to structures of the abdominal cavity.

The liver has four **anatomical lobes**:

1. The **right lobe** is separated from the left lobe on the diaphragmatic surface by the falciform ligament. On the visceral surface, it lies to the right of the inferior vena cava.

2. The **left lobe** is located to the left of the falciform ligament. On the visceral surface, it lies to the left of the ligamentum venosum and ligamentum teres.

3. The **quadrate lobe** is only visible on the visceral surface. It is bounded by the fossa for the gallbladder and the ligamentum teres.

4. The **caudate lobe** is also only visible on the visceral surface. It is limited by the groove for the inferior vena cava and ligamentum venosum.

Functionally, the liver is divided into right and left lobes by its blood supply and biliary drainage. The left lobe includes the caudate, a portion of the quadrate, and left anatomical lobes. The right functional lobe is formed by the right anatomical lobe, with a portion of the quadrate.

The liver is intraperitoneal and has several associated peritoneal ligaments (**Table 3.3.1**).

An important function of the liver is the production of **bile**. This fluid aids in emulsifying fats in the small intestine. Bile leaves the liver in the **right and left hepatic ducts** (the biliary duct system is described above). These ducts join to form the

TABLE 3.3.1	Peritoneal ligaments associated with the liver.
Peritoneal ligament	**Description**
Falciform	▪ Connects liver to anterior abdominal wall ▪ Divides right and left anatomical lobes on diaphragmatic surface
Coronary	▪ Connects superior aspect of liver to diaphragm ▪ Has anterior and posterior layers
Left triangular	▪ Connects superior aspect of left lobe to diaphragm ▪ Formed by merged anterior and posterior layers of coronary ligament
Right triangular	▪ Connects posterolateral aspect of right lobe to diaphragm ▪ Formed by merged anterior and posterior layers of coronary ligament
Hepatoduodenal	▪ Connects liver to superior margin of proximal duodenum ▪ Part of lesser omentum
Hepatogastric	▪ Connects liver to lesser curvature of stomach ▪ Part of lesser omentum

common hepatic duct. The **common hepatic duct**, hepatic portal vein, and proper hepatic artery are considered the **portal triad**. The portal triad extends between the quadrate and caudate lobes.

Blood Supply

The **proper hepatic artery** is the primary blood supply to the liver. This artery arises from the common hepatic artery, a branch of the celiac trunk, and travels in the hepatoduodenal ligament. It divides into right and left hepatic branches that supply respective functional lobes.

Venous blood that drains from the liver parenchyma collects in central veins that converge to form hepatic veins. These veins open directly into the **inferior vena cava**.

Nerve Supply

The innervation to the liver is from the **celiac plexus**. Nerve fibers from this plexus pass along branches of the celiac trunk to reach the liver. Fibers in this plexus include:

▪ **Sympathetic (visceral efferent).** Preganglionic sympathetic axons originate from cell bodies in the T5-T9 lateral horns of the spinal cord. These axons are carried in the **greater splanchnic nerves** to the **celiac ganglia** (prevertebral ganglia). Postganglionic sympathetic axons are distributed along branches of the celiac trunk (common, proper, and right and left hepatic).

▪ **Parasympathetic (visceral efferent).** Preganglionic parasympathetic axons originate from cell bodies in the brainstem. They are carried in the **CN X** and their respective **anterior and posterior vagal trunks**. These trunks pass through the esophageal hiatus of the diaphragm into the abdominal cavity. Preganglionic parasympathetic fibers are distributed with sympathetic fibers along the branches of the celiac trunk. They synapse in terminal ganglia and postganglionic fibers supply the liver.

CLINICAL REASONING

This patient presents with signs and symptoms indicating a clinical condition involving the **liver, biliary system**, and/or **pancreas**.

Biliary Obstruction

Cholestasis refers to the blockage of any duct that carries bile from the liver to the gallbladder, or from the gallbladder to the duodenum. The major signs and symptoms result directly from the failure of bile to reach its proper destination.

Clinical Notes

- Cholelithiasis are the most common cause of biliary obstruction. Gallstones are formed by accretion of crystalized components of bile. In the United States, 80% of gallstones are composed primarily of cholesterol. Cholesterol gallstones result when bile in the gallbladder becomes supersaturated with cholesterol. Cholesterol then precipitates from solution as microscopic crystals. Over time, the crystals grow, collect, and may fuse to form macroscopic stones. Most gallstones remain in the gallbladder; however, some may enter the biliary duct system.

- Gallstones usually become impacted at the distal end of the hepatopancreatic duct, in the neck of the gallbladder, or in the cystic duct. If a gallstone blocks the cystic duct, cholecystitis may occur due to the stasis of bile within the gallbladder.

- The biliary ducts may also be occluded secondary to inflammation and/or malignancy (e.g., hepatobilliary or pancreatic). Physical obstruction of the biliary duct system results in hyperbilirubinemia.

Signs and Symptoms

- RUQ and epigastric pain
- Jaundice
- Dark urine
- Pale stools
- Nausea or vomiting
- Loss of appetite
- Itching

Predisposing Factors

Since the majority of gallstones are composed of cholesterol, most of the predisposing factors result in increased cholesterol secretion from the liver, which increases cholesterol in bile.

- Sex: women are twice as likely as men to develop gallstones.

- Age: individuals >60 years of age have higher risk for gallstones. With aging, the body tends to secrete more cholesterol into bile.

- Weight: being even moderately overweight increases the risk for developing gallstones. Obesity in women is a major risk factor for gallstones. With increased weight, the amount of bile salts in bile is reduced and, therefore, the relative levels of cholesterol are increased.

- Rapid weight loss: as the body metabolizes fat during prolonged fasting and rapid weight loss—such as "crash diets"—the liver secretes extra cholesterol into bile.

- Diet: diets high in fat and cholesterol and low in fiber increase the risk.

- Cholesterol-lowering drugs: drugs that lower cholesterol levels in blood actually increase the amount of cholesterol secreted into bile.

- Diabetes

- Ethnicity: Native Americans have a genetic predisposition to secrete high levels of cholesterol in bile and have the highest rate of gallstones in the United States.

- Family history of gallstones

Chronic Pancreatitis

Pancreatitis is an inflammation of the pancreas. Acute pancreatitis appears suddenly and lasts for days. Chronic pancreatitis may persist for years. Mild cases of pancreatitis may resolve without treatment, but severe cases can cause life-threatening complications.

Signs and Symptoms

Acute pancreatitis

- RUQ and epigastric pain
- Pain and tenderness with deep palpation of RUQ

Cholelithiasis Gallstones
Cholecystitis Gallbladder inflammation

Hyperbilirubinemia Elevated serum bilirubin

- Abdominal pain that radiates to the back and is worse after eating
- Nausea and vomiting

Chronic pancreatitis
- RUQ and epigastric pain
- Indigestion
- Unexplained weight loss
- Steatorrhea

Predisposing Factors
- Alcohol abuse
- Cholelithiasis
- Response to specific medication
- Chemical exposure
- Abdominal trauma
- Surgery and certain medical procedures
- Infection
- Abnormalities of the pancreas or intestine

Clinical Notes
- Most cases of chronic pancreatitis are due to alcohol abuse and are found in patients with at least a 5- to 7-year history of abuse.
- Gallstones are the second most likely cause of pancreatitis, most often in women >50 years of age. Pancreatitis results when the stone blocks the pancreatic duct, which traps digestive enzymes in the pancreas. In chronic cases, these enzymes may destroy the local pancreatic tissue.

Liver Cirrhosis

Cirrhosis is a complication of liver disease that involves loss of function liver cells and irreversible scarring of the liver. It is characterized by replacement of normal liver tissue with fibrotic scars and regenerative nodules.

Steatorrhea Presence of excess fat in feces

Signs and Symptoms
- Jaundice
- Fatigue and weakness
- Loss of appetite
- Itching
- Easy bruising

Predisposing Factors
- Alcohol abuse
- Viral hepatitis B and C
- Genetic disorders (resulting in accumulation of toxic substances in the liver)
- Nonalcoholic fatty liver disease
- Primary biliary cirrhosis
- Cryptogenic cirrhosis
- Autoimmune liver disease
- Extended exposure to medications or toxins

Clinical Note

Complications of cirrhosis include edema and ascites, spontaneous bacterial peritonitis, bleeding from varices, hepatic encephalopathy, hepatorenal syndrome, hepatopulmonary syndrome, hypersplenism, and liver cancer.

DIAGNOSIS

The patient presentation, history, physical examination, lab results, and imaging studies support a diagnosis of **biliary obstruction due to a gallstone**.

Biliary Obstruction

Blockage in the biliary tree can occur anywhere along the course of the hepatic, cystic, bile, or hepatopancreatic ducts.

- The presence of a gallstone large enough to block bile flow was confirmed by the radiologic studies completed as part of the workup in this patient.
- If the bile ducts remain blocked for a significant time, inflammation/infection can occur in

the gallbladder (cholecystitis) and duct system. This may lead to chills, fever, and epigastric pain similar to that reported by this patient.

- The inflamed gallbladder can irritate adjacent parietal peritoneum on the inferior diaphragmatic surface. This results in referred somatic pain to the region of the right shoulder, as reported in the patient.

- The obstructed bile duct also causes accumulation of bilirubin in the blood, as evidenced in the lab tests. Deposition of this compound in skin, conjunctiva, and oral mucous membranes causes jaundice.

- The dark urine is an indication of high levels of bilirubin in the urine, while the lack of bilirubin in the intestine results in pale-colored stool.

The imaging studies give conclusive evidence of biliary obstruction in this patient.

Pancreatitis

The pancreas has a dual function as an endocrine (insulin production) and exocrine gland (digestive enzymes). In pancreatitis, cellular injury impairs exocrine secretion.

- Serum amylase and lipase are the most widely used tests for pancreatitis. Approximately 85% of patients with acute pancreatitis have threefold greater than normal values for these enzymes.

- When the lab data is inconsistent with the diagnosis of pancreatitis, it may be confirmed by (sequentially by cost and invasiveness) abdominal ultrasound, CT, or endoscopic ultrasound (EUS).

- The normal amylase and lipase values in this patient, coupled with the results of imaging studies, rule out the diagnosis of pancreatitis.

Liver Cirrhosis

The liver has numerous functions, including bile production. Cirrhosis of the liver results in fibrosis that significantly reduces hepatic function and altered hepatic blood flow.

- Patients with liver cirrhosis have reduced white blood cell and platelet counts and prolonged prothrombin time.

- Cirrhosis also leads to reduced serum albumin, while serum bilirubin, transaminases, and alkaline phosphatase will be elevated.

- These patients frequently present with symptoms of portal hypertension, including ascites.

- Patient history, laboratory values, and radiographic findings are usually sufficient for diagnosis, although liver biopsy may be used for confirmation.

- The normal liver function values, together with the results of imaging studies, rule out a diagnosis of liver cirrhosis in this patient.

Patient Presentation

A 28-year-old female is admitted to the emergency department with severe right lower quadrant abdominal pain.

Relevant Clinical Findings

History

The patient reports that she awoke in the middle of the night with what she thought was indigestion. The pain was vague and accompanied at times by nausea. She took antacids and returned to sleep. Several hours later she awoke with excruciating pain in the right lower quadrant. She was nauseated and could find relief from the pain only if she laid quietly in the fetal position.

Physical Examination

Noteworthy vital signs:

- Temperature: 37.9°C (100.3°F)

  Normal: 36.0–37.5°C (96.5–99.5°F)

The following findings were noted on physical examination:

- Hypoactive bowel sounds
- Guarding of inferior, anterolateral abdominal wall
- Right lower quadrant extremely tender to palpation, with rebound tenderness

Laboratory Tests

Test	Value	Reference value
Erythrocytes (count)	4.3	$4.3–5.6 \times 10^6/mm^3$
Leukocytes (count)	13.5	$3.54–9.06 \times 10^3/mm^3$
Human chorionic gonadotrophin (hCG)	1	<5 mIU/mL

Urinalysis was normal.

Clinical Note

Human chorionic gonadotrophin (hCG) is a hormone produced during pregnancy. Levels of hCG may be measured in the blood or urine as a pregnancy test. High levels of the hormone indicate the presence of an implanted embryo. Levels rise from <5 mIU/mL in a nonpregnant female to over 30,000 mIU/mL 4–6 weeks into a pregnancy.

Imaging Studies

- Abdominal and pelvic ultrasounds were inconclusive.
- Contrast-enhanced abdominal CT revealed a markedly thickened appendicular wall and distend lumen.

Clinical Problems to Consider

- Appendicitis
- Pelvic inflammatory disease (PID)
- Tubo-ovarian abscess (TOA)
- Viral gastroenteritis

Guarding Muscle spasm (especially of anterior abdominal wall) to minimize movement at site, at or near injury or disease (e.g., inflammation associated with appendicitis or diverticulitis). May be detected with palpation during physical examination.

Palpitation Forcible or irregular pulsation of the heart that is perceptible to the patient

RELEVANT ANATOMY

Cecum and Vermiform Appendix

The **large intestine** is divided into several anatomical regions. It begins at the **cecum**, a dilated, blind pouch that lies in the right iliac fossa inferior to the ileocecal valve. The "saclike" nature of the cecum allows it to store large volumes of semiliquid chyme that enters from the ileum through the ileocecal valve. This valve lies at the junction of the cecum with the **ascending colon**.

The **vermiform appendix** is a narrow, fingerlike blind-ended tube that arises from the posteromedial wall of the cecum. The tenia coli (longitudinal, muscular bands characteristic of large intestine) on the cecum merge at the base of the appendix and form its

outer longitudinal muscle. The muscular wall of the appendix is similar to the small intestine in that it has complete outer longitudinal and inner circular layers. Unlike the small intestine, it lacks plicae circulares and villi. The appendix usually contains numerous lymphoid nodules, suggesting an immunologic function.

The appendix varies in length (2–20 cm) and position. It is usually longer in children and may shorten and atrophy with age. The appendix is most frequently retrocecal, that is, posterior to the cecum (**Fig. 3.4.1**). It may also be in a pelvic (descending) position, in which it extends over the pelvic brim. Less commonly, it may be directed superomedially toward the root of the mesentery of the small intestine, directly medial toward the rectum, or postileal (posterior to the terminal ileum).

Blood Supply

The blood supply to the cecum and appendix (**Fig. 3.4.2**) arises from branches of the **ileocolic artery**, which supplies the cecum, appendix, and terminal ileum. The cecum and appendix are supported by mesenteries, the mesocecum, and mesoappendix, respectively. The **appendicular branch of the ileocolic artery** is the primary arterial supply to the appendix. It courses posterior to the

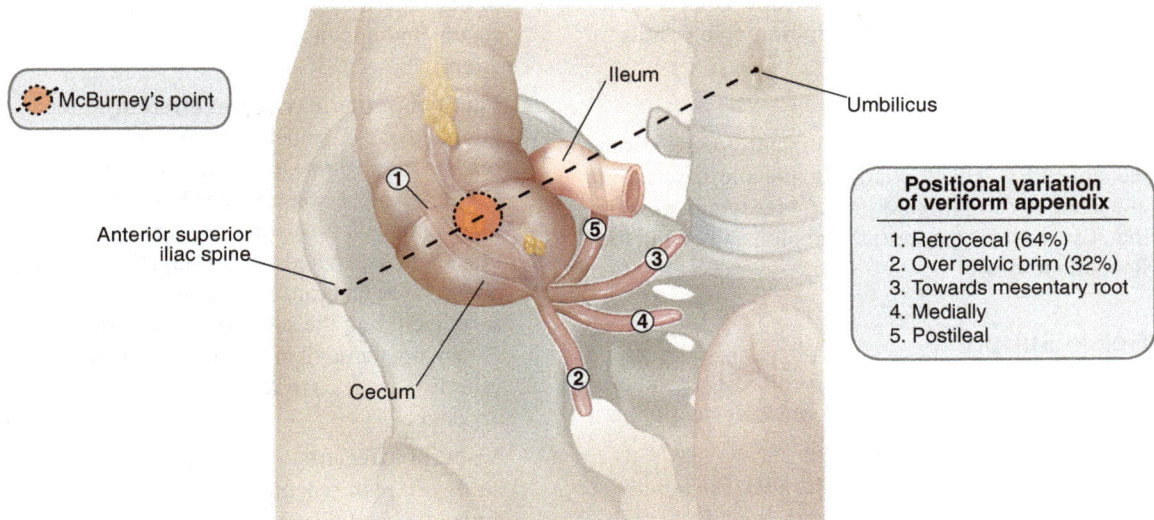

FIGURE 3.4.1 Anterior view of right lower quadrant showing positional variation in the appendix.

(A)

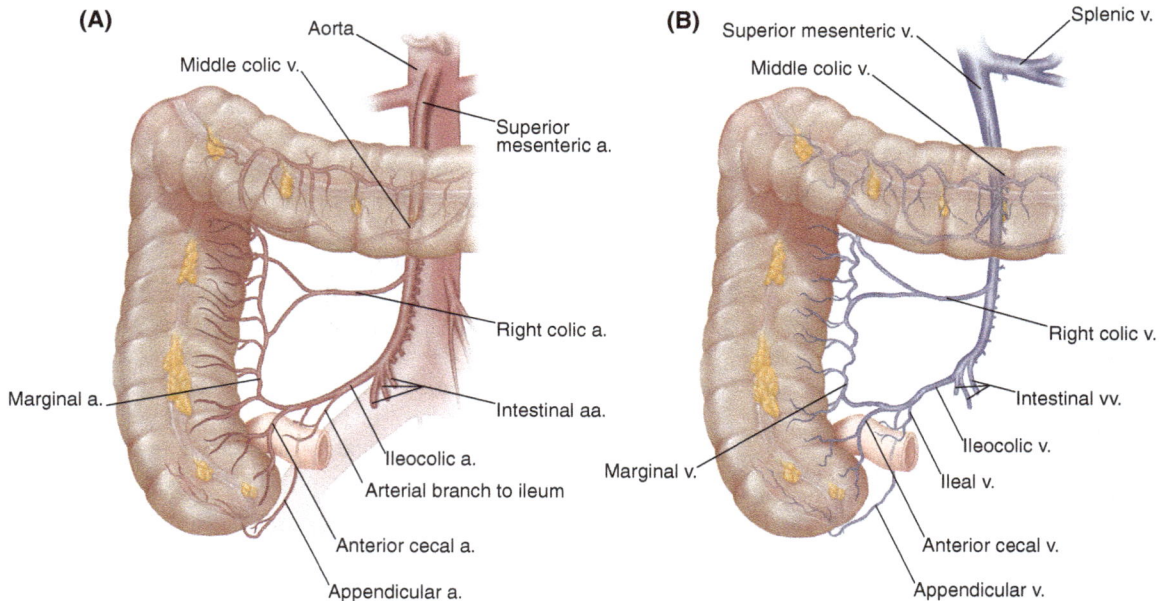

(B)

FIGURE 3.4.2 Anterior view of the blood supply of the terminal ileum and proximal colon. (**A**) Arteries and (**B**) veins.

terminal ileum and enters the meso-appendix. Venous blood from the cecum, appendix, and terminal ileum drains through **tributaries of the ileocolic vein**. These tributaries follow courses similar to the accompanying arteries. The **ileocolic vessels** arise from or terminate in **superior mesenteric** vessels.

Lymphatics

There may be as many as 15 lymphatic vessels from the appendix. These course in the mesoappendix and unite to form three or four larger lymph vessels that join those of the ascending colon and end in the **ileocolic nodes**.

Nerve Supply

The cecum and appendix received their autonomic innervation via the **superior mesenteric plexus**. Neuronal fibers in this plexus include:

- **Sympathetic (visceral efferent).** Preganglionic sympathetic neurons originate from cell bodies in T10-T11 lateral horns of the spinal cord. Preganglionic axons are carried by the

lesser splanchnic nerves to the **superior mesenteric ganglion** (prevertebral ganglia). Postganglionic axons are distributed in the **superior mesenteric plexus** along branches of the superior mesenteric artery and to the cecum and appendix along the iliocecal artery.

- **Parasympathetic (visceral efferent).** Preganglionic parasympathetic neurons located in the brain stem give rise to axons that travel in the **CN X**. The CN X enters the abdomen through the esophageal hiatus in the diaphragm. Vagal fibers are distributed (with sympathetic fibers) along the main branches of the abdominal aorta including the superior mesenteric. Parasympathetic fibers follow along specific vessels (iliocecal) to reach the cecum and appendix.

- **Visceral afferents.** Afferents fibers associated with pain receptors in the wall of the appendix would be located in the **lesser splanchnic nerves**. The afferent neuronal cell bodies are positioned in the T10-T11

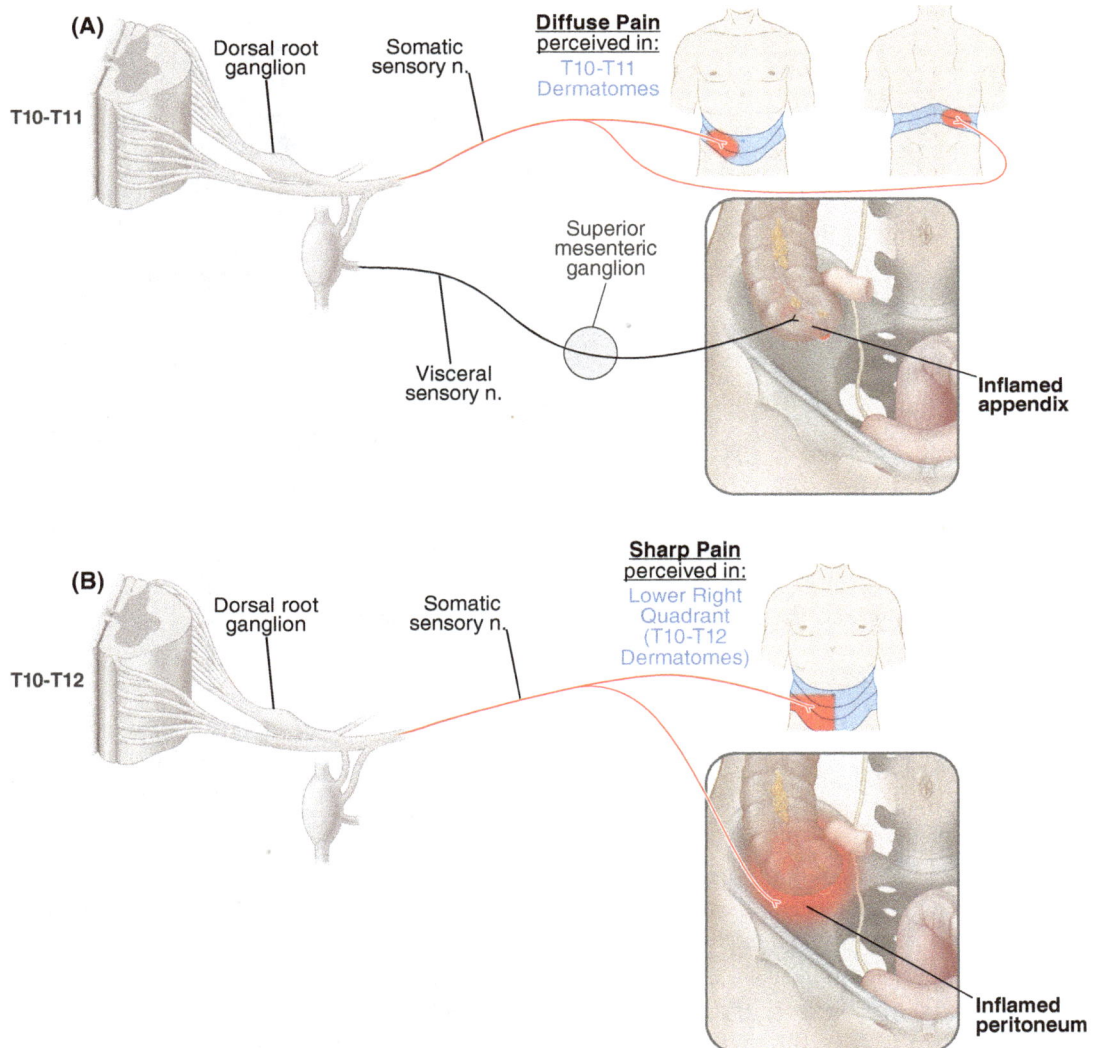

FIGURE 3.4.3 Diagrammatic representation of (**A**) visceral and (**B**) somatic referred pain from the appendix.

dorsal root ganglia. Visceral pain sensations from the appendix will be perceived in the paraumbilical region and right upper quadrant (RUQ) and referred along the T10-T11 dermatomes.

Over time the inflamed appendix may also irritate parietal peritoneum on the adjacent anterior abdominal wall. Parietal peritoneum lining the anterior abdominal wall receives sensory innervation from **somatic afferent** fibers in the **thoraco-abdominal (T7-T11)** and **subcostal (T12) nerves**. Sensation from this irritated peritoneum presents sharp pain in the RLQ (**Fig. 3.4.3**).

For additional information about the autonomic nervous system and referred pain, see Chapter 1: Autonomic Nervous System.

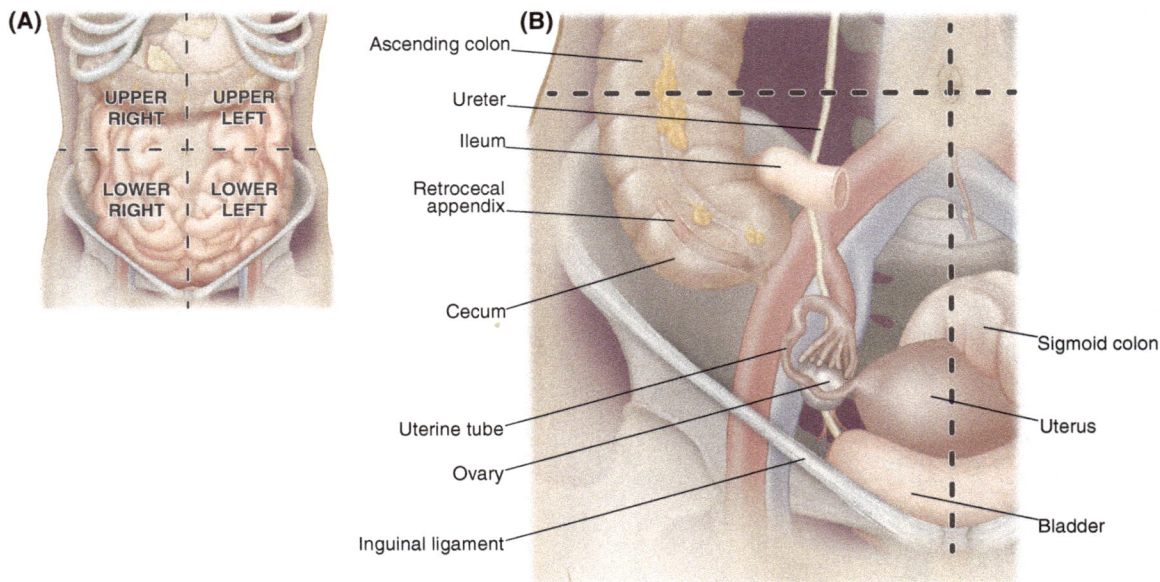

FIGURE 3.4.4 Anterior view of (**A**) abdominal quadrants and (**B**) right lower quadrant viscera.

Viscera of the Right Lower Quadrant (RLQ)

The abdominal cavity is divided into four quadrants, defined by two planes (**Fig. 3.4.4A**):

1. The **median (median sagittal) plane** passes vertically through the center of the body. This plane divides the body into right and left halves.

2. The **transumbilical plane** passes horizontally through the body at the level of the umbilicus.

In order to interpret abdominal signs and symptoms, it is important to know which organs are located in each quadrant (**Table 3.4.1**).

Abdominal Pain

The organs associated with the RLQ (as indicated **Table 3.4.1**) are the most likely source of this patient's pain (**Fig. 3.4.4B**). However, there are three distinct forms of abdominal pain:

1. Somatic
2. Visceral
3. Referred

Somatic abdominal pain can arise from the skin, fascia, skeletal muscles, and parietal peritoneum. This type of pain is localized. Parietal peritoneum is innervated by the same nerves that supply overlying muscles. The specific nerves that innervate parietal peritoneum are given in **Table 3.4.2**.

Visceral abdominal pain may occur with ischemia, chemical or mechanical insult, or over-distension of abdominal organs or mesenteries. Pain receptors are associated with visceral afferent neurons that connect with the spinal cord. For additional information about the autonomic nervous system and referred pain, see Chapter 1: Visceral Afferent Pathways.

Pain from abdominal viscera may be perceived in somatic structures, such as skin or underlying muscle. This is considered **referred pain**. Visceral pain from the appendix may be produced by over-distension, spasm of its smooth muscle wall, or irritation of its mucosa. Visceral afferent fibers accompany sympathetic nerves that supply the appendix. These fibers, whose cell bodies are located in the T10-T11 dorsal root ganglia, travel through the superior mesenteric plexus and lesser splanchnic

TABLE 3.4.1	Organs associated with abdominal quadrants.

Right abdominal quadrants	Left abdominal quadrants
Right upper quadrant (RUQ) *RUQ extends from the median plane, superior to the transumbilical plane* ▪ Liver ▪ Gallbladder ▪ Pancreas (head) ▪ Right suprarenal gland ▪ Right kidney (upper lobe) ▪ Ascending colon (distal) ▪ Hepatic flexure ▪ Transverse colon (proximal)	**Left upper quadrant (LUQ)** *LUQ extends from the median plane, superior to the transumbilical plane* ▪ Liver (left lobe) ▪ Pancreas (body and tail) ▪ Left suprarenal gland ▪ Left kidney (upper lobe) ▪ Descending colon (proximal) ▪ Splenic flexure ▪ Transverse colon (distal)
Right lower quadrant (RLQ) *RLQ extends from the median plane, inferior to the transumbilical plane* ▪ Right kidney (lower lobe) ▪ Cecum ▪ Appendix ▪ Most of ileum ▪ Ascending colon (proximal) ▪ Right uterine tube ▪ Right ovary ▪ Right ureter ▪ Uterus (if enlarged) ▪ Urinary bladder (if full)	**Left lower quadrant (LLQ)** *LLQ extends from the median plane, inferior to the transumbilical plane* ▪ Left kidney (lower lobe) ▪ Most of jejunum ▪ Descending colon ▪ Sigmoid colon ▪ Left uterine tube ▪ Left ovary ▪ Left ureter ▪ Uterus (if enlarged) ▪ Urinary bladder (if full)

nerves. Therefore, referred pain from the appendix is likely to be a vague, and localized around the umbilicus (T10 dermatome).

Later in the inflammatory process, the appendix may irritate adjacent parietal peritoneum and result in somatic pain in the RLQ (**Fig. 3.4.4**).

TABLE 3.4.2	Innervation of parietal peritoneum.

Location of parietal peritoneum	Nerve supply
Central part of diaphragm	Phrenic (C3-C5)
Peripheral part of diaphragm	Intercostal (T7-T11)
Anterior abdominal wall	Thoracoabdominal (T7-T11) Subcostal (T12) Iliohypogastric (L1)
Pelvic wall	Obturator (L2-L4)

Clinical Notes

▪ **Guarding** is a protective phenomenon in which the muscles of the abdominal wall increase in tone in response to irritation of the underlying parietal peritoneum.

▪ **Rebound tenderness** is elicited during palpation of the abdomen. If the parietal peritoneum is inflamed, when the examiner removes the hand following palpation, movement ("rebound") of the peritoneum will elicit pain.

▪ **Colic** (sometimes referred to as "abdominal cramps") is a form of visceral pain that results from vigorous, spasmodic contraction of the smooth muscle of the gastrointestinal tract.

▪ **McBurney's point** refers to the point on the skin of the RLQ that is one-third of the distance from the anterior superior iliac spine to the umbilicus (**Fig. 3.4.1**). This point represents the surface projection of the most common location of the base of the appendix.

- Deep tenderness at McBurney's point is known as **McBurney sign**. This sign is clinical indication of acute appendicitis with inflammation that is no longer limited to the appendix and is irritating parietal peritoneum.

CLINICAL REASONING

This patient presents with signs and symptoms of **RLQ pain**.

Appendicitis

Appendicitis most commonly results following obstruction of the lumen of the appendix. The blockage is caused by a fecalith. Enlarged lymphoid nodules associated with viral infections, worms (e.g., pinworms), and tumors (e.g., carcinoid or carcinoma) may also obstruct the lumen. Once the lumen is obstructed, local bacteria multiply and invade the appendiceal wall. Venous engorgement and subsequent arterial compromise result from the high intraluminal pressures. Finally, gangrene and perforation may occur. If this process evolves slowly, adjacent organs, such as the terminal ileum, cecum, and greater omentum may wall off the area and a localized abscess will develop. Alternatively, rapid progression of vascular impairment may cause perforation of the appendiceal wall, providing free access of luminal contents to the peritoneal cavity.

Signs and Symptoms

- Initial vague paraumbilical pain or discomfort
- Later pain shifts to RLQ over McBurney's point
- Rebound tenderness with guarding in RLQ
- Pain worsened by walking or coughing
- Nausea and vomiting
- Constipation or diarrhea
- Low-grade fever

Fecalith Fecal stone
Gangrene Necrosis due to loss of blood supply
Rebound tenderness Pain elicited during abdominal examination when the examiner removes pressure suddenly during palpation. This clinical sign is associated with peritoneal inflammation (e.g., peritonitis, appendicitis)

- Psoas sign may be positive
- Obturator sign may be positive

Predisposing Factors

- Age: more frequent in adolescents and young adults
- Sex: males predominate 3:2

Clinical Notes

- **Psoas sign** refers to pain that is elicited with passive extension of the hip.
- **Obturator sign** refers to pain elicited with passive flexion and internal rotation of the hip.

Pelvic Inflammatory Disease

Pelvic inflammatory disease (PID) refers to an infection that begins in the vagina or cervix and ascends to involve the remainder of the female reproductive tract. This type of infection can extend beyond the uterine tubes to cause pelvic peritonitis, generalized peritonitis, or a pelvic abscess. PID is most commonly due to sexually transmitted pathogens (e.g., *Chlamydia trachomatis*, *Neisseria gonorrhoeae*, and *Gardnerella vaginalis*).

Clinical Note

The primary signs and symptoms of peritonitis are acute abdominal pain, abdominal tenderness, and abdominal guarding, which are made worse by movement of the peritoneum.

Signs and Symptoms

- Abdominal and/or low back pain
- Dyspareunia
- Fever and chills
- Fatigue
- Diarrhea or vomiting
- Dysuria
- Purulent vaginal discharge

Peritonitis Inflammation of peritoneum
Dyspareunia Pain during sexual intercourse
Dysuria Pain during urination
Purulent Containing, discharging, or causing the production of pus

Predisposing Factors

- Sexually active and <25 years of age
- Unprotected sexual activity with multiple partners
- Previous history of sexually transmitted diseases (STDs)
- Previous history of PID
- Recent gynecologic procedures (e.g., IUD)

Tubo-Ovarian Abscess

A **tubo-ovarian abscess (TOA)** is an inflammatory mass involving the uterine tube, ovary, and occasionally, adjacent pelvic viscera. These abscesses typically result from recurrent infections of the upper reproductive tract. A single infectious episode (e.g., acute salpingitis), however, may also precipitate TOA. Sometimes, an ovulation site is the point of entry for the infectious agent and location of the abscess. The exudate produced by the abscess may lead to rupture, which results in peritonitis.

Signs and Symptoms

- Lower abdominal and pelvic pain
- Rebound tenderness in both lower quadrants
- Nausea and vomiting
- Fever
- Tachycardia
- Feeling of fullness

Predisposing Factors

- Age: reproductive age women
- Low parity
- History of PID

Clinical Note

The duration of symptoms is approximately 1 week, with onset usually 2 or more weeks after a menstrual period.

Tachycardia Increased heart rate: >100 bpm (normal adult heart rate: 55–100 bpm)
Parity Having given birth

Viral Gastroenteritis

Viral gastroenteritis is an inflammation of the epithelium of the colon such that there is a loss of absorptive function. The stools consist of water and fecal remnants, but do not contain blood, pus, or mucus. Common causes are noroviruses, rotavirus, adenovirus, caliciviruses, enterovirus, and coronavirus.

Signs and Symptoms

- Lower abdominal pain
- Nausea and vomiting
- Abdominal cramping
- Explosive diarrhea
- General malaise without fever

Predisposing Factors

- Age: children or elderly
- Group settings, such as schools, cruise ships, and social gatherings
- Conditions that result in weakened immune systems (e.g., HIV/AIDS and patients undergoing chemotherapy)

DIAGNOSIS

The patient presentation, history, physical examination, lab tests, and procedures outcomes support a diagnosis of **acute appendicitis**.

Acute Appendicitis

Appendicitis is an inflammation of the appendicular mucosa, usually following obstruction of the lumen of the appendix. The appendix is blind-ended, so blockage of its lumen will increase the intraluminal pressure, which may compromise its blood supply.

- The patient's initial, diffuse paraumbilical discomfort is consistent with the distribution of referred pain from the appendix. The pain progressed, becoming more severe and precisely located in the RLQ, indicating that the inflamed appendix has irritated the adjacent parietal peritoneum. This accounts for the RLQ somatic pain, guarding, and rebound tenderness.

Malaise Feeling of general body weakness or discomfort, often marking the onset of an illness

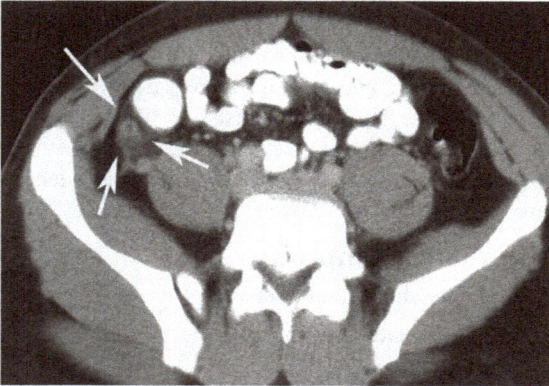

FIGURE 3.4.5 Contrast-enhanced abdominal CT image shows hyperenhanced wall of the appendix. *Source:* Fig. 9-16 in *CURRENT Diagnosis & Treatment: Gastroenterology, Hepatology, & Endoscopy.*

- The positive McBurney, psoas, and obturator signs are a further indication of the involvement of the peritoneum.

- Abdominal imagining studies confirm a thickened appendicular wall with an enlarged lumen, which is consistent with this condition (**Fig. 3.4.5**).

Pelvic Inflammatory Disease

PID is initiated by infection, usually sexually transmitted, that moves superior from the vagina and cervix into the uterus and uterine tubes.

- Women with PID present with a wide variety of symptoms, but the most common presenting complaint is lower abdominal pain. Most women (75%) also exhibit abnormal vaginal discharge.

- The diagnosis of acute PID is primarily based on patient history of previous PID and clinical findings and is not consistent with this patient's information.

Tubo-Ovarian Abscess

TOAs are very frequently the result of untreated PID. An enlarging pelvic abscess may bleed due to adjacent vessel erosion or rupture of the abscess.

- The typical patient with TOA is a young female of low parity, with a history of PID.

- Symptoms include pelvic and abdominal pain, fever, nausea and vomiting, and four-quadrant abdominal tenderness and guarding.

- The absence of an adnexal mass (i.e., abscess) on pelvic ultrasound eliminates this as a possible diagnosis.

Viral Gastroenteritis

Gastroenteritis is an inflammation of the large intestine caused most commonly by a norovirus infection. The virus is typically spread in contaminated food and water, but may also be airborne.

- Viral gastroenteritis has a sudden onset and generally lasts for 2–4 days.

- The illness is usually characterized by diarrhea and abdominal cramps/pain, nausea, and vomiting. The stools are typically loose, watery, and without blood or mucus.

- This patient did not report diarrhea as a symptom, so viral gastroenteritis is not likely to be the cause of her abdominal pain.

Adnexa Structures accessory to an organ; used commonly with reference to the uterus to refer to uterine tube, ovaries, and uterine ligaments

Ureterolithiasis

Patient Presentation

A 43-year-old male is brought to the emergency department complaining of intermittent, extreme pain in his right side. The pain radiates to his groin, inner thigh, and testicle, and is accompanied by sweating, fever, and nausea.

Relevant Clinical Findings

History

The patient reported that he had been experiencing minor "back" pain for a few days, but recently the pain increased in intensity. Over the last few hours, the pain extended to his groin. Over-the-counter medications did not relieve the pain. He recently had urinary urgency with dysuria and hematuria.

Physical Examination

During physical examination, the patient was restless and switched positions frequently.
   Noteworthy vital signs:

- Temperature: 37.6°C (99.8°F)

   Normal: 36.0–37.5°C (96.5–99.5°F)

The following findings were noted on physical examination:

- Tenderness at the right costovertebral angle (a region of the back between rib 12 and the lumbar vertebrae)
- Guarding in the right lower quadrant (RLQ)
- No rebound tenderness

Laboratory Tests

Test	Value	Reference value
Erythrocytes (count)	4.6	$4.3–5.6 \times 10^6/mm^3$
Leukocytes (count)	10.2	$3.54–9.06 \times 10^3/mm^3$
Serum calcium	11.4	8.7–10.2 mg/dL
Urinalysis Erythrocytes	8	<5 cells/high power field
Leukocytes	12	<10 cells/high power field
Crystals	Present	None

Imaging Studies

- Helical CT scan showed a radiopaque mass in the upper urinary tract.

Clinical Problems to Consider

- Pyelonephritis
- Nephrolithiasis
- Ureterolithiasis

LEARNING OBJECTIVES

1. Describe the anatomy of the kidneys.

2. Describe the anatomy of the ureters.
3. Explain the anatomical basis for the signs and symptoms associated with this case.

Dysuria Pain during urination
Hematuria Blood in urine
Rebound tenderness Pain elicited during abdominal examination when the examiner removes pressure suddenly during palpation. This clinical sign is associated with peritoneal inflammation (e.g., peritonitis, appendicitis)

Guarding Muscle spasm (especially of anterior abdominal wall) to minimize movement at site, at or near injury or disease (e.g., inflammation associated with appendicitis or diverticulitis). May be detected with palpation during physical examination.

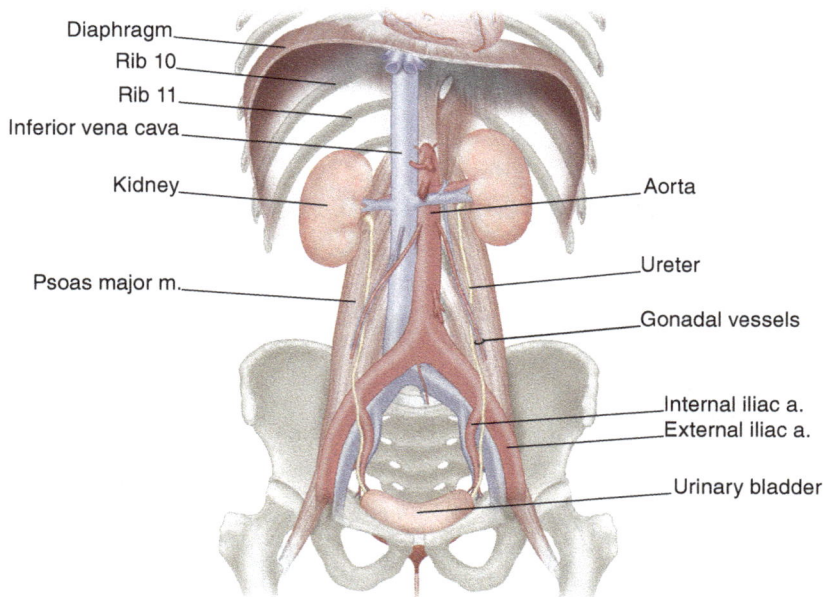

FIGURE 3.5.1 Anterior view of urinary system.

RELEVANT ANATOMY

The urinary system consists of two **kidneys**, two **ureters**, **urinary bladder**, and **urethra** (**Fig. 3.5.1**). The urinary system can be divided into upper and lower parts:

- The **upper urinary system** includes the kidneys and abdominal (superior) part of the ureter.
- The **lower urinary system** includes the pelvic (inferior) part of the ureter, urinary bladder, and urethra.

Kidney

The kidneys are retroperitoneal and lie on the posterior abdominal wall, adjacent to the T12-L3 vertebrae. The **right kidney** lies slightly inferior to the left because of the space occupied by the right lobe of the liver.

The kidneys are covered by a fibrous **renal capsule** (**Fig 3.5.2**). The concave, medial margin of each kidney has a vertical slit known as the **renal hilum**, which is the passageway for renal vessels, lymphatics, nerves, and the renal pelvis. Within the kidney, the hilum opens into the **renal sinus**. This space is occupied by the renal pelvis, calices, vessels, nerves, and perirenal fat.

- The **hilum of the right kidney** is located at the L1-2 intervertebral disc.

- The **hilum of the left kidney** is slightly higher at the L1 vertebral body.

The parenchyma of the kidney is divided into two zones (**Fig 3.5.2**):

1. Outer **renal cortex**
2. Inner **renal medulla**

Extensions of the cortex, the **renal columns**, divide the medulla into **renal pyramids**. The blunt tip of each pyramid, the renal papilla, projects into the renal sinus. Each papilla is surrounded by a **minor calyx**, and two or three minor calices converge to form a **major calyx**. The funnel-shaped **renal pelvis** is formed by the union of the two or three major calyces. The renal pelvis exits the renal hilum and is continuous with the abdominal portion of the **ureter**. The kidneys are protected by layers of connective tissue and fat (**Fig. 3.5.3**).

- **Pararenal fat** is a layer of adipose tissue located just deep to parietal peritoneum. It is most obvious posterior to the kidneys.

- **Renal (Gerota) fascia** is a thin, connective tissue envelop. Its posterior layer fuses with investing fascia of psoas major muscle, and its anterior layer blends with the adventitia of the renal vessels.

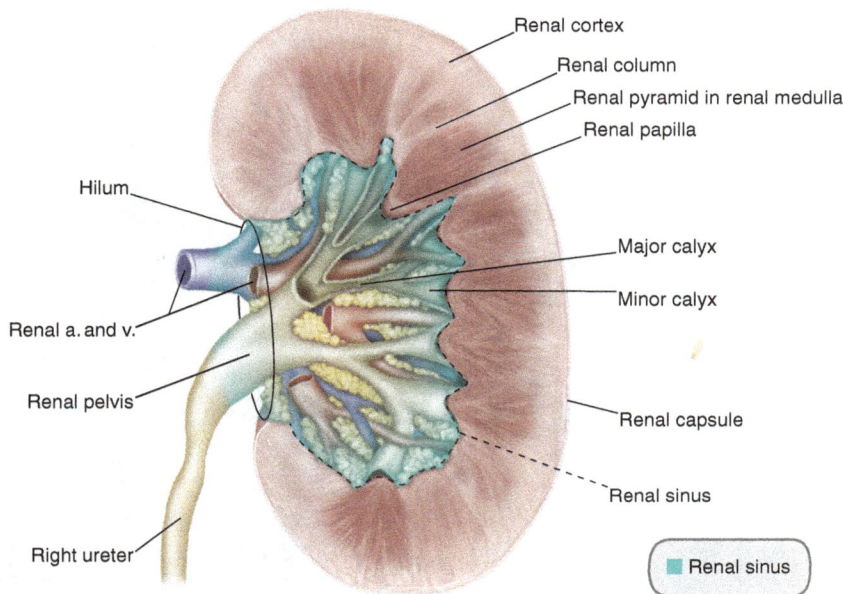

FIGURE 3.5.2 Coronal section through the kidney.

- **Perirenal fat** is a layer of adipose tissue deep to renal fascia that is continuous with the fat within the renal sinus.

Blood Supply

Each kidney is supplied by a **renal artery** that arises from the abdominal aorta at the level of the L2 vertebral body. Near the renal hilum, each renal artery typically divides into several segmental arteries that are distributed to the kidney. The **right renal artery is longer than the left** and passes posterior to the inferior vena cava (IVC). Both renal arteries lie posterior to the renal veins in the renal hilum.

The venous drainage of the kidney follows a similar pattern to the arterial supply. Both renal veins drain into the IVC. The **left renal vein is longer than the right** and passes anterior to the abdominal aorta. Inferior phrenic, suprarenal, and gonadal veins drain into the left renal vein.

Nerve Supply

The innervation to each kidney arises from a renal plexus located along renal vessels. Neuronal fibers in this plexus include:

- **Sympathetic (visceral efferent).** Preganglionic axons originate from cell bodies in the T-12 lateral horn of the spinal cord. These axons are carried by the **least splanchnic nerves** to the **aorticorenal ganglia** (prevertebral ganglia). Postganglionic axons are distributed along branches of the renal arteries.

- **Parasympathetic (visceral efferent).** Preganglionic parasympathetic axons originate from cell bodies in the brain stem and travel in the CN X. The CN X pass into the abdominal cavity as the anterior and posterior vagal trunks through the esophageal haitus of the diaphragm. Parasympathetic preganglionic fibers are distributed with sympathetic fibers along the renal vessels to synapse in terminal ganglia near the kidneys.

Anatomical Relationships

The kidneys are related to the posterior abdominal wall, as well as to organs of the abdominal cavity. **Table 3.5.1** summarizes the anatomical relationships of the kidneys.

Ureter

The **ureters** are paired, retroperitoneal muscular ducts that carry urine from the kidneys to the urinary bladder (**Fig. 3.5.1**). The ureter is continuous with the apex of the **renal pelvis**. The junction between the

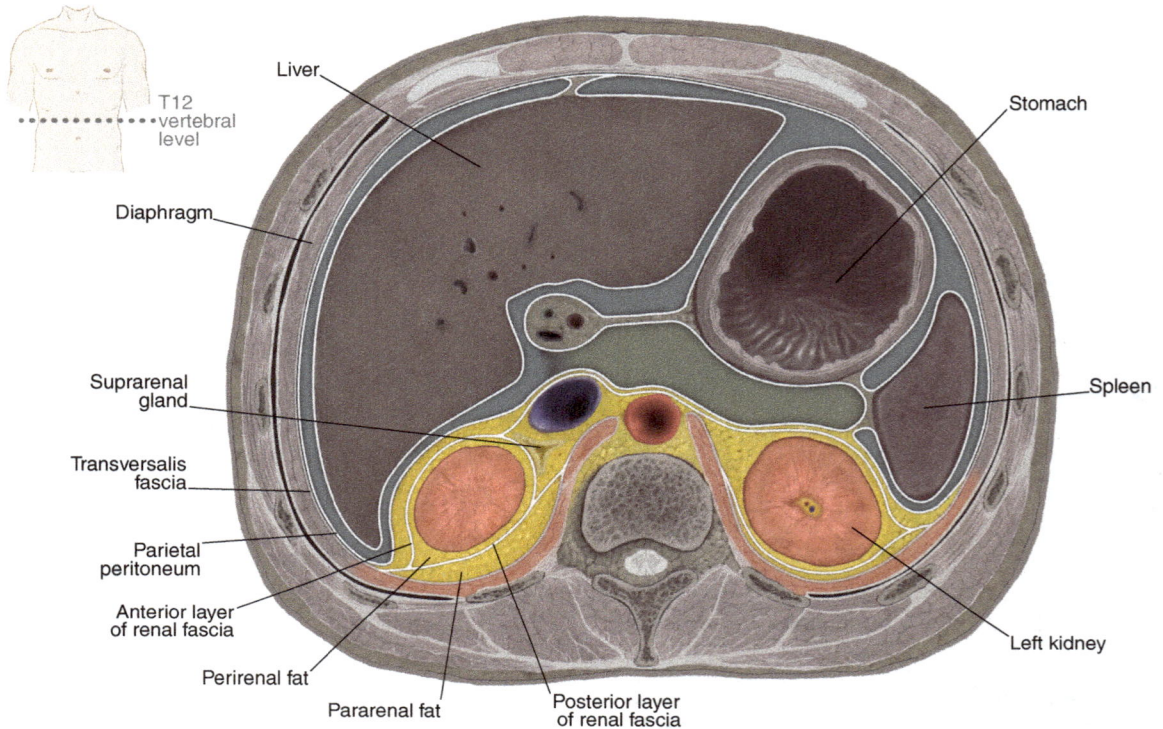

FIGURE 3.5.3 Horizontal section through the abdominal cavity at the level of the kidneys showing layers of renal fat and fascia.

TABLE 3.5.1	Anatomical relationships of right and left kidneys.	
Direction	**Structures**	
Superiomedial	▪ Suprarenal gland	
Anteriorly		
Right kidney	▪ Liver	
	▪ Duodenum	
	▪ Ascending colon	
Left kidney	▪ Stomach	
	▪ Spleen	
	▪ Pancreas	
	▪ Jejunum	
	▪ Descending colon	
Posteriorly	▪ Diaphragm	
	▪ Transversus abdominis muscle	
	▪ Quadratus lumborum muscle	
	▪ Psoas muscles	

renal pelvis and ureter is marked by a slight constriction. The ureter has two parts, **abdominal and pelvic**, of approximately equal length. The **abdominal part** descends along the medial aspect of the psoas major muscle and is crossed by the gonadal vessels. The **pelvic part** of the ureters begins where they cross the pelvic inlet. Here, the ureters pass anterior to the common iliac vessels, near their bifurcation. Each ureter descends in extraperitoneal areolar tissue along the lateral pelvic wall. Deep in the pelvis, the ureters move anterior and medial to reach the posteroinferior aspect of the urinary bladder. Each ureter has an oblique course through the bladder wall. Therefore, as the bladder fills and becomes distended, the ureter is compressed, limiting the reflux of urine into the ureter. The ureter is normally constricted at three places along its course:

1. At the junction of the ureter and renal pelvis

2. Where the ureter crosses the pelvic brim

3. As it passes through the wall of the urinary bladder

Blood Supply

Arteries to the **abdominal part** of the ureter arise from the **renal and gonadal arteries**, as well as direct branches from the **abdominal aorta**. The **pelvic part** may be supplied by branches from the **common and internal iliac arteries**, and by the **inferior vescial (male)** and **uterine (female) arteries**. There are arterial anastomoses along the entire extent of the ureter.

Veins of the ureter have a similar pattern to the arterial supply. They drain primarily into the **renal, gonadal, and internal iliac veins**.

Nerve Supply

The innervation to the ureters is derived from **renal, aortic, superior hypogastric, and inferior hypogastric plexuses**. Sympathetic and parasympathetic fibers in these plexuses follow blood vessels to the ureters. The density of these fibers increases in the pelvic ureter. Peristaltic contractions of the ureter originate in the proximal part and propagate toward the bladder.

Visceral afferent fibers conveying pain sensations from the ureter primarily follow sympathetic fibers to T11-L2 dorsal root ganglia. Ureteric pain is usually referred to the ipsilateral lower abdominal quadrant and groin region.

CLINICAL REASONING

This patient presents with signs and symptoms of **acute, severe right flank, and groin pain.**

Pyelonephritis

Pyelonephritis is a localized inflammation of the renal pelvis and kidney. This usually results from bacterial infection that spreads from the lower urinary tract. *Escherichia coli* accounts for 70–90% of urinary tract infections (UTIs), including pyelone-

phritis. **Acute pyelonephritis** is characterized by the sudden onset of inflammation. **Chronic pyelonephritis** results from a long-standing infection, or a series of frequent, repeated UTIs. Uncontrolled cystitis or prostatitis can progress to overt pyelonephritis, sepsis, and renal failure.

Signs and Symptoms

- Fever
- Flank and/or abdominal pain
- Nausea and vomiting
- Hematuria
- Urinary frequency and urgency
- Dysuria
- Pyuria

Predisposing Factors

- Sex: female
- Chronic or recurrent UTIs
- Mechanical or structural abnormality of urinary tract
- Kidney stones
- Urinary tract catheterization
- Ureteral stents
- Neurogenic bladder
- Prostate disease
- Diabetes mellitus
- Family history for UTIs

Nephrolithiasis

Nephrolithiasis is also known as renal calculi or kidney stones. A kidney stone is a solid crystalline mass that forms in the renal calyces or pelvis. Kidney stones result when urine becomes supersaturated and substances in the urine crystallize. One or more "stones" may be present at any given time.

Symptoms usually arise when the stone(s) begin to move through the urinary tract.

Cystitis Inflammation of the urinary bladder
Prostatitis Inflammation of the prostate
Sepsis Presence of pathogenic organisms, or their toxins, in blood or tissues

Pyuria Pus in urine
Calculus An abnormal concretion, usually of mineral salts

Signs and Symptoms

- Initial flank and/or abdominal pain
- Pain moves to groin region and testicle (male)
- Fever
- Chills
- Nausea and vomiting
- Hematuria
- Dysuria
- Urinary frequency and urgency

Predisposing Factors

- Age: increasing to the sixth decade
- Sex: male
- Polycystic kidney disease
- Hyperparathyroidism
- Cystinuria
- Dietary factors
- Hypercalciuria
- Low urine volume
- Certain medications (e.g., some diuretics and laxative overuse)

Clinical Notes

There are four types of kidney stones.

1. **Calcium stones** are most common, particularly in men 20–30 years of age. Calcium can combine with other substances, such as oxalate, phosphate, or carbonate, to form a stone. Oxalate is present in certain foods, such as spinach and vitamin C supplements. Diseases of the small intestine also increase risk of this type of stone.

2. **Cystine stones** form in people who have cystinuria. This is an autosomal-recessive disease characterized by defective transepithelial transport of cystine that results in the buildup of cystine in the urine.

3. **Struvite stones** form in patients infected with ammonia-producing organisms. They are potentiated by alkaline urine and high magnesium/plant-based diets. These stones are most frequently found in women with a history of UTIs. They can become very large.

4. **Uric acid stones** form when there is an excess amount of uric acid in urine. Uric acid is the product of purine metabolism, and therefore may be related to a high-protein diet. They most frequently occur in patients with gout or those undergoing chemotherapy.

Ureterolithiasis

Ureterolithiasis is the formation or presence of a calculus in one or both ureters. This is usually a stone that has moved from the kidney. It can cause extensive distension of the ureter and may block the flow of urine. As a ureteric stone is forced along the ureter by peristaltic contractions, it usually causes severe, rhythmic, sharp pain known as **ureteric colic**.

Clinical Note

Ureteric colic starts suddenly in the flank. The pain radiates to the anterior abdomen and moves inferiorly to the groin. It can also radiate into the testicle (male) or to the labia (female). The pain is severe and the patient may be very restless and uncomfortable. Episodes typically last 30–60 minutes and may reoccur every hour or two.

Signs and Symptoms

- Initial flank and/or abdominal pain
- Pain moves to groin region and testicle or labia
- Fever
- Chills
- Nausea and vomiting
- Hematuria
- Dysuria
- Urinary frequency and urgency

Predisposing Factors

- Age: increasing to the sixth decade
- Sex: male
- Polycystic kidney disease
- Hyperparathyroidism

Cystinuria Elevated urinary secretion of cysteine, lysine, arginine, and ornithine

Hypercalciuria Elevated urinary secretion of calcium

- Cystinuria
- Dietary factors
- Hypercalciuria
- Low urine volume
- Certain medications (e.g., some diuretics and laxative overuse)

DIAGNOSIS

The patient presentation, history, physical examination, and lab results support a diagnosis of **ureterolithiasis (ureteric stone)**.

Ureterolithiasis

A ureteric stone is most likely to obstruct the ureter at one of three narrow areas along its course. Pain afferents that supply the ureter have cell bodies located in T11-L2 dorsal root (sensory) ganglia. Therefore, ureteric pain is referred to the T11-L2 dermatomes. This includes abdominal skin overlying the course of the ureter (including the flank), the skin of the inguinal region, anterior and medial thigh, and the scrotum/labia.

- Severe, rhythmic pain beginning in the flank and radiating to the abdomen, groin, and scrotum in this patient is consistent with ureteric colic (i.e., pain from with a ureteric stone).

- Laboratory test results provide further support for this diagnosis, including elevated white blood cells and high levels of serum calcium. Urinalysis showed both red and white blood cells and crystals. The patient is likely to have a UTI, which could be confirmed with urine culture.

- Helical CT confirms the presence of a stone in the abdominal part of the ureter, just distal to the renal pelvis (**Fig. 3.5.4**).

Nephrolithiasis

Kidney stones form either on the surfaces of the renal papillae or within the renal collecting system.

- A stone located in the renal pelvis can produce interstitial renal edema, stretch the renal capsule, and enlarge the kidney (i.e., nephromegaly).

- Kidney stones are a common cause of hematuria

- The colicky-type pain known as renal colic usually begins in the superolateral back over the costovertebral angle, and is occasionally

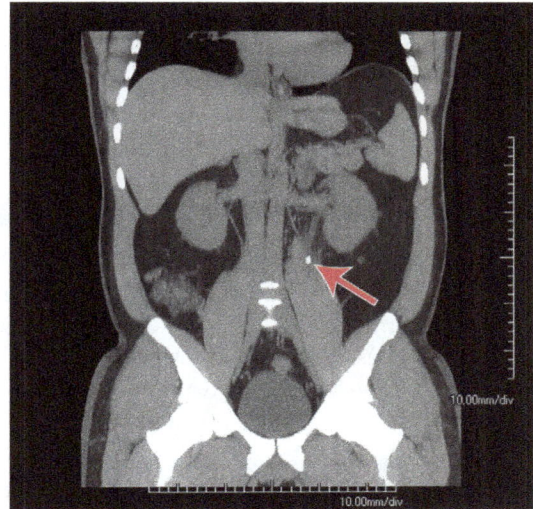

FIGURE 3.5.4 Noncontrast CT of coronal section through abdominal cavity. Red arrow shows 6-mm stone within the proximal third of the left ureter. *Source:* Fig. 97-1A in *Tintinalli's Emergency Medicine*, 7e.

subcostal. This pain tends to remain constant, and the pattern depends on the individual's pain threshold. This is not consistent with the pain described by this patient.

- Renal imaging studies (noncontrast abdominal CT scan) confirmed a ureteric stone, rather than one in the kidney.

Pyelonephritis

Pyelonephritis is a symptomatic infection of the kidney that is considered an upper UTI.

- Pyelonephritis is usually suggested by the history and physical examination, and supported by urinalysis.

- Results from laboratory studies may show pyuria, leucocyte esterase [dipstick leukocyte esterase test (LET)], and hematuria. The nitrite production test (NPT) would provide evidence for bacteriuria and proteinuria.

- While this patient has a UTI, it is not the source of the pain as described.

Edema Swelling of skin due to abnormal accumulation of fluid in subcutaneous tissue

Bacteriuria Bacteria in urine

Proteinuria Elevated protein in urine

3.6 | Ruptured Spleen

Patient Presentation

A 32-year-old male is admitted to the emergency department with shortness of breath. He also complains of pain when sitting or lying down. When standing, the breathing difficulty and pain are less.

Relevant Clinical Findings

History

The patient reports that earlier in the day he fell from his all-terrain vehicle. He reports falling "hard" on his left side, but did not have any pain following the crash so he completed the ride and did not seek medical help. On returning home, he showered, ate a meal, and fell asleep. Hours later, he awoke with shortness of breath and severe abdominal pain that improved when he stood up. Sitting or lying down exacerbated both symptoms.

Physical Examination

Noteworthy vital signs:

- BP: 105/70 mm Hg
  Systolic: 100–140 mm Hg
  Diastolic: 60–90 mm Hg

- Pulse: 112 bpm
  Adult resting rate: 60–100 bpm

The following findings were noted on chest and abdominal physical examination:

- Bilateral wheezing
- Diffuse abdominal discomfort
- Left upper quadrant (LUQ) pain with rebound tenderness
- Positive Kehr sign

Clinical Note

The **Kehr sign** results when blood from an injured spleen irritates the diaphragm and creates referred pain in the region of the left shoulder.

Imaging Studies

- Chest radiograph was unremarkable.
- Abdominal CT scan showed hemoperitoneum.

Clinical Problems to Consider

- Rib fracture
- Retroperitoneal hemorrhage
- Ruptured spleen

LEARNING OBJECTIVES

1. Describe the anatomy of the spleen.
2. Describe the retroperitoneal space and its contents.
3. Describe the anatomy of the ribs.
4. Explain the anatomical basis for the signs and symptoms associated with this case.

Rebound tenderness Pain elicited during abdominal examination when the examiner removes pressure suddenly during palpation. This clinical sign is associated with peritoneal inflammation (e.g., peritonitis, appendicitis)

Hemoperitoneum Blood in peritoneal fluid within the peritoneal cavity

Hemorrhage Escape of blood from vessels:
- Petechia: <2 mm diameter
- Ecchymosis (bruise): >2 mm
- Purpura: a group of petechiae or ecchymoses
- Hematoma: hemorrhage resulting in elevation of skin or mucosa

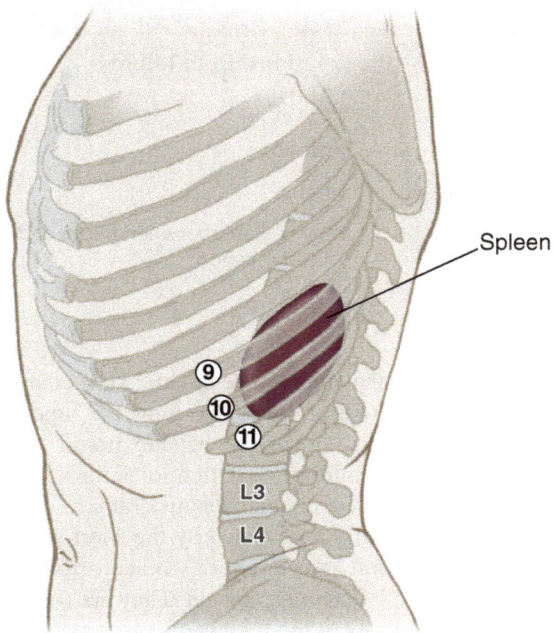

FIGURE 3.6.1 Left lateral view showing the position of spleen relative to ribs 9–11.

RELEVANT ANATOMY

Spleen

The **spleen** lies in the LUQ of the abdomen, just inferior to the diaphragm and lateral to the stomach. It is associated with ribs 9–11 and is normally not palpable on physical examination (**Fig. 3.6.1**).

The spleen is the largest lymphoid organ in the body. Its primary functions are to filter blood, phagocytose aging red blood cells and circulating microorganisms, supply immunocompetent T and B lymphocytes, and manufacture antibodies. The spleen has a thick connective tissue capsule. Septae extend from the capsule into the substance of the spleen. These septae convey blood vessels toward the center of the organ. The spleen is highly vascular and it is estimated that it filters 10–15% of the total blood volume every minute. It may also hold in reserve 40–50 mL of red blood cells and 25% of circulating platelets at any time.

The spleen is invested with peritoneum except at the hilum, where the splenic vessels enter and exit. It is connected to the stomach and the left kidney by peritoneal folds, the gastrosplenic and splenorenal ligaments, respectively (**Fig. 3.6.2**). These

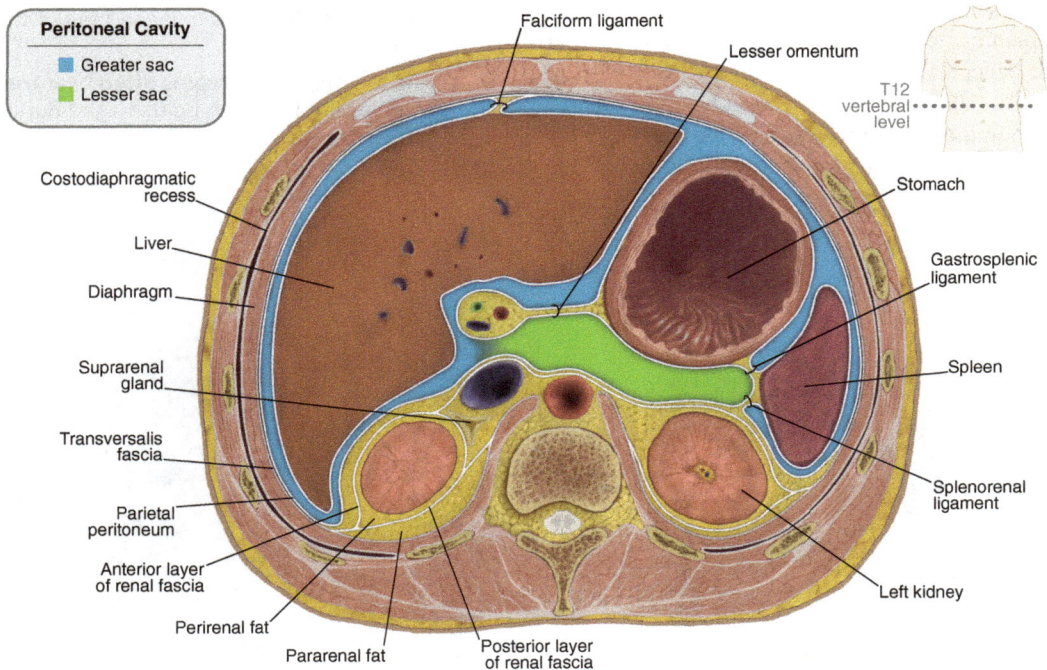

FIGURE 3.6.2 Horizontal section of the abdominal cavity at the level of the spleen.

TABLE 3.6.1	Impressions on the visceral surface of the spleen.	
Impression	**Viscera**	**Relationship to hilum**
Gastric (largest impression)	Stomach	Anterior
Renal	Left kidney	Posterior
Colic	Splenic flexure of large intestine	Inferior
Pancreatic	Tail of pancreas	Intermediate, between hilum and colic impression

ligaments contribute to the mobility of the spleen. Their length and thickness vary greatly: short and thick ligaments decrease spleen mobility.

The spleen has two surfaces:

1. The **diaphragmatic surface** is smooth, convex, and occupies the left hemidiaphragm.
2. The **visceral surface** is irregular. It has multiple impressions formed by adjacent viscera. **Table 3.6.1** summarizes the visceral impressions on the spleen.

Blood Supply

The spleen is supplied by the **splenic artery (Fig. 3.6.3A)**. It is the largest branch of the celiac trunk. The splenic artery courses along the superior border of the pancreas, its serpentine course causing it to randomly "dip" into pancreatic tissue. The splenic artery commonly bifurcates near the hilum of the spleen, forming branches that supply the upper and lower poles of the spleen. The **splenic vein** courses with the artery (**Fig. 3.6.3B**) and joins the superior mesenteric vein to form the hepatic portal vein. The inferior mesenteric vein often joins the splenic vein. The arterial supply and venous drainage of the spleen is frequently augmented by the short gastric vessels. The short gastric arteries branch from the splenic before it enters the spleen. They are the primary blood supply to the fundus of the stomach.

Clinical Note

- Splenic vessels have multiple small branches that distribute to the body and tail of the pancreas.
- The tail of the pancreas is often near the splenic hilum and can be damaged during splenectomy.

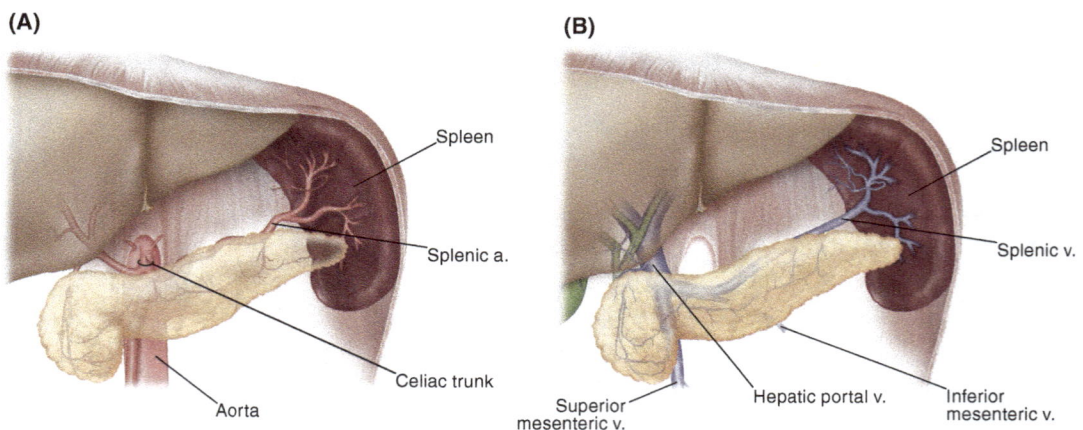

FIGURE 3.6.3 Anterior view of the spleen and its blood supply. (**A**) Arteries and (**B**) veins.

TABLE 3.6.2 | Structures located in subdivisions of the retroperitoneal space.

Anterior pararenal space	Perirenal space	Posterior pararenal space
Abdominal aorta	Suprarenal glands	Fat
Inferior vena cava		
Pancreas	Kidneys	
Ascending colon	Renal vessels	
Descending colon		

Duodenum
  Descending (2nd) part
  Horizontal (3rd) part
  Ascending (4th) part		

Retroperitoneal Space

The **retroperitoneal space** (retroperitoneum) is an anatomical area in the abdominal cavity located posterior (retro) to the parietal peritoneum on the posterior abdominal wall (**Fig. 3.6.2**). The retroperitoneal space is bounded by the diaphragm superiorly and posteriorly by the vertebral column, psoas, iliacus, iliopsoas and quadratus lumborum muscles. The retroperitoneal space can be subdivided into three parts (from anterior to posterior):

1. The **anterior pararenal space** lies between parietal peritoneum and the anterior layer of the renal fascia.

2. The **perirenal space** is limited by the anterior and posterior layers of renal fascia.

3. The **posterior pararenal space** lies between the posterior lamina of renal fascia and the muscles of the posterior abdominal wall.

The structures located within subdivisions of the retroperitoneal space are summarized in **Table 3.6.2**.

Retroperitoneal Structures

Retroperitoneal structures lie on the posterior abdominal wall and have a single layer of parietal peritoneum on their anterior surface (**Table 3.6.3**). Organs are considered secondarily retroperitoneal if they had a mesentery early in development but subsequently lost that mesentery as they assumed their permanent position on the posterior abdominal wall.

Ribs

There are 12 pairs of ribs; narrow curved, flat bones, that artculate with thoracic vertebrae posteriorly. Anteriorly, most ribs connect with the sternum via costal cartilages. There are three classifications of ribs:

1. **True ribs** (1–7) attach to the sternum directly by costal cartilage.

2. **False ribs** (8–10) attach to the sternum indirectly via shared costal cartilages.

3. **Floating ribs** (11–12) end in posterior abdominal musculature and are not attached to the sternum. Some sources include floating ribs as a subcategory of false ribs.

TABLE 3.6.3 | Retroperitoneal organs.

Primarily retroperitoneal	Secondarily retroperitoneal

Suprarenal glands
Kidneys
Ureters
Urinary bladder
Abdominal aorta
Inferior vena cava
Esophagus
  (abdominal part)
Rectum (superior
  part)
	Pancreas (head, neck, body)
	  Tail is intraperitoneal,
	  within splenorenal
	  ligament
	Duodenum (descending,
	  horizontal, ascending)
	  Superior part is
	  intraperitoneal
	Ascending colon
	Descending colon

The ribs, and other parts of the thoracic skeleton and wall, provide projection for organs of the thorax and abdomen.

Clinical Notes

- Ribs 4–9 are the most commonly fractured.
- A rib is most likely to fracture either at the point of impact or at the angle of the rib.

CLINICAL REASONING

This patient presents with signs and symptoms related to **blunt abdominal trauma**.

Rib Fracture

Simple rib fractures (i.e., not comminuted) are the most common form of chest injury. A direct trauma to the ribs may result in fracture of one or more ribs and/or costal cartilages. The ribs are particularly vulnerable because of their subcutaneous location.

Clinical Notes

- Costochondral separation(s) result when the costal cartilage of a rib separates from the sternum and can cause pain similar to rib fractures.
- A severe blow to the chest may fracture the sternum.

Signs and Symptoms
- Pain on inspiration
- Parasternal pain
- Dyspnea
- Local pain at site of injury

Predisposing Factors
- Age: more common with increasing age
- Trauma or history of trauma
- Contact sports

Comminuted fracture A fracture with three or more pieces
Dyspnea Difficulty breathing and shortness of breath

- Osteoporosis
- Cancerous lesion

Clinical Notes

- Approximately 25% of rib fractures are not visible with radiographic imaging and must be diagnosed by physical exam. Patients may experience bruising, muscle spasms over the ribs, and stabbing pain if the rib is touched at the point of injury.
- Sharp bone associated with a rib fracture may injure membranes, blood vessels, and/ or organs of the thoracic or abdominal cavities.

Retroperitoneal Hemorrhage and Hematoma

A hemorrhage in the retroperitoneal space results in a hematoma that may be localized to the inferior aspect of this space. This hematoma may not be observed during physical examination. A retroperitoneal hematoma is most commonly due to traumatic impact or from a penetrating wound to the abdomen. The injury results in retroperitoneal bleeding following damage to the inferior vena cava and/or abdominal aorta or to vessels that supply retroperitoneal organs.

Signs and Symptoms
- Abdominal pain
- Back pain
- Positive psoas sign
- Positive Cullen sign
- Positive Grey Turner sign

Predisposing Factors
- Major trauma
- Minor trauma in patients with defective clotting factors
- History of invasive femoral vascular procedures

Hematoma Localized extravasation of blood, usually clotted

Clinical Notes

- **Psoas sign** refers to pain that is elicited with passive extension of the hip.
- **Cullen sign** is superficial edema and purplish discoloration of the subcutaneous fatty tissue around the umbilicus. This sign is frequently seen in conjunction with Grey Turner sign.
- **Grey Turner sign** is a purplish discoloration of the skin and subcutaneous tissue of the flank(s).

The primary causes for Cullen and Grey Turner signs include:

- Bleeding from blunt trauma to abdomen, from a ruptured abdominal aortic aneurysm, and from a ruptured ectopic pregnancy.
- Acute pancreatitis with damage to local blood vessels

Ruptured Spleen

A ruptured spleen is any disruption of the capsule of the spleen. Splenic rupture potentially allows a large volume of blood to leak into the peritoneal cavity. Patients with this type of injury may require emergency surgery since the blood escaping the spleen is at arterial pressure and a large volume may be lost in a short period. With a small rupture, the bleeding may stop without surgical intervention. Rupture of a normal spleen is most likely the result of trauma. Spontaneous ruptures may occur if the spleen is enlarged (e.g., when the immune system is compromised, as in mononucleosis).

Clinical Notes

- Although protected by the ribcage, the spleen remains the most commonly injured organ in all age groups with blunt trauma to the abdomen.
- The situations in which blunt abdominal trauma occurs include motor vehicle accidents, domestic violence, sporting events, and accidents involving bicycle handlebars.

Signs and Symptoms

- Tachycardia
- Hypotension
- Falling hematocrit
- Shock
- Tenderness of LUQ
- Splenomegaly

Predisposing Factors

- Trauma or history of trauma
- Pathologic enlargement (e.g., infectious mononucleosis, AIDS, and leukemia)

DIAGNOSIS

The patient presentation, medical history, physical examination, and imaging studies support a diagnosis of a **ruptured spleen** due to abdominal trauma.

Ruptured Spleen

The spleen is the most commonly injured organ in blunt abdominal trauma. The spleen lies lateral to the stomach in the left upper abdominal quadrant, protected by the thoracic cage. Blood from a ruptured spleen most frequently enters the peritoneal cavity (**Fig. 3.6.4**).

- Hemoperitoneum is a cardinal sign, and was confirmed by abdominal CT. Hemoperitoneum can cause diffuse abdominal pain, but is more likely to irritate parietal peritoneum adjacent

---

Edema Swelling of skin due to abnormal accumulation of fluid in subcutaneous tissue

Aneurysm Circumscribed dilation of an artery, in direct communication lumen Tachycardia Increased heart rate: >100 bpm (normal adult heart rate: 55–100 bpm)

Tachycardia Increased heart rate: >100 bpm (normal adult heart rate: 55–100 bpm)

Hypotension Abnormal decrease in arterial pressure

Hematocrit A blood test that measures the percentage of erythrocytes in whole blood.

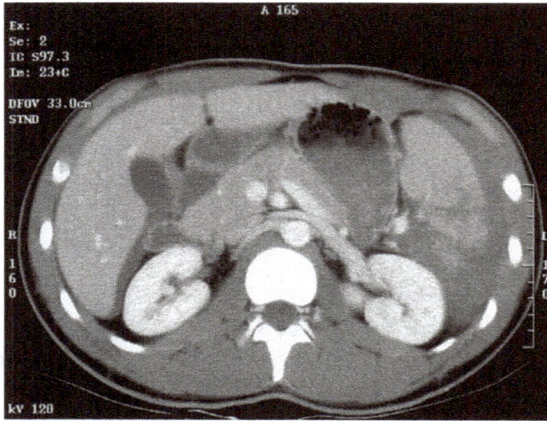

FIGURE 3.6.4 CT scan showing hemoperitoneum following injury to the spleen. *Source:* Fig. 299.4.1A in *Tintinalli's Emergency Medicine*, 7e.

to the spleen. This results in significant **LUQ pain and rebound tenderness**.

- **Left shoulder tenderness** (**Kehr sign**) may also be present due to irritation of diaphragmatic parietal peritoneum and referred pain to C3-5 dermatomes.
- Blood filling the subphrenic recess of the peritoneal cavity can also put pressure on the diaphragm causing dyspnea.
- If the intra-abdominal bleeding exceeds 5–10% of circulating blood volume, clinical signs of early shock may manifest. Signs include tachycardia, tachypnea, restlessness, and anxiety.
- Although not observed in this patient, increased hemoperitoneum will lead to abdominal distension with guarding of the abdominal wall and overt shock.

Rib Fracture

The ribs form the majority of the thoracic skeleton. Fragments of fractured ribs can penetrate the abdominal wall and injure the peritoneum and underlying viscera (especially the liver and spleen). This can lead to hemothorax, pneumothorax, or hemoperitoneum. Ribs 4–9 are the most commonly injured.

- In patients with multiple fractured ribs (especially ribs 9–11), unexplained hypotension may be the result of intra-abdominal bleeding from the liver or spleen.
- Patients with multiple fractured ribs will often have difficulty taking a deep breath or coughing.
- Radiographic imaging did not indicate rib fracture(s) in this patient.

Retroperitoneal Hemorrhage

The retroperitoneal space is located between the parietal peritoneum and the posterior abdominal wall. The anterior boundary of this space is distensible and, therefore, any pathology of retroperitoneal viscera or accumulating blood in this space will expand anteriorly toward the peritoneal cavity.

- Retroperitoneal hemorrhage may result from major trauma. It may also occur secondary to minor trauma in individuals with clotting disorders.
- This type of hemorrhage may occur as a result of invasive procedures involving the femoral vessels, such as coronary artery catheterization.
- Back pain and abdominal pains are typical symptoms and patients frequently exhibit a positive psoas sign.
- Abdominal CT scan did not indicate retroperitoneal hematoma in this case.

Tachypnea Increased respiration rate (normal adult respiration rate: 14–18 cycles/min)

Hemothorax Blood in pleural cavity
Pneumothorax Air or gas in pleural cavity

REVIEW QUESTIONS

1. A patient presents with a gastric ulcer located in the posterosuperior part of the pyloric canal. The ulcer has perforated the stomach wall and a vessel supplying the region. What artery is most at risk from this ulcer?
   A. Gastroduodenal
   B. Left gastric
   C. Right gastric
   D. Right gastro-omental
   E. Short gastric

2. A 58-year-old male is diagnosed with a tumor on the upper pole of his right kidney. During a posterior approach to the tumor, the surgeon would first encounter:
   A. Pararenal fat
   B. Parietal peritoneum
   C. Perirenal fat
   D. Renal capsule
   E. Renal fascia

3. A 63-year-old male presents to the clinic with chronic abdominal pain that has increased in intensity over the last month. He has a long-standing history of alcohol abuse. Imaging studies confirm a diagnosis of chronic pancreatitis with erosion of the pancreas and damage to adjacent structures. Which artery is most at risk in this patient?
   A. Common hepatic
   B. Gastroduodenal
   C. Left gastro-omental
   D. Short gastric
   E. Splenic

4. A 29-year-old male presents to the clinic with diffuse abdominal pain that originated in the groin. During the physical examination, a small reducible mass is observed in the inguinal region. The physician explained it was likely he had an inguinal hernia. During surgical repair, a diagnosis of indirect inguinal hernia is confirmed. How is the surgeon able to make this diagnosis?
   A. The hernial sac extended beyond the superficial inguinal ring.
   B. The hernial sac originated lateral to the inferior epigastric artery.
   C. The hernial sac ruptured the conjoint tendon.
   D. The patient was younger than 45 years of age.
   E. The hernia was reducible.

5. A perforated peptic ulcer is diagnosed in a 41-year-old male patient. During surgical repair, a small amount of blood is noted posterior to the stomach. This part of the peritoneal cavity is the:
   A. Supracolic space
   B. Infracolic space
   C. Greater sac
   D. Lesser sac (omental bursa)
   E. Retroperitoneal space

6. During a bowel resection, the surgeon mobilizes the descending colon from the posterior abdominal wall. What is the blood supply to this part of the colon?
   A. Left colic
   B. Left gastro-omental
   C. Left gonadal
   D. Left renal
   E. Middle colic

Questions 7–8 refer to the following clinical case.

7. A female patient is admitted to the hospital with acute abdominal pain. She states that she had experienced nausea and vomiting the previous day. Physical examination reveals marked tenderness in the right upper quadrant and abdominal ultrasound confirms the presence of a large gallstone in the cystic duct. Pain afferents from the cystic duct are carried in the:
   A. Greater splanchnic nerves
   B. Intercostal nerves
   C. Lesser splanchnic nerves
   D. Phrenic nerves
   E. Vagus nerves (CN X)

8. During laparoscopic surgery performed to remove the stone, the surgeon notes a dilated structure in the free margin of the lesser omentum. This structure is most likely the:
   A. Bile duct
   B. Common hepatic artery
   C. Common hepatic duct
   D. Main pancreatic duct
   E. Proper hepatic artery

9. A female patient is admitted to the emergency department following an accident during which she experienced blunt force trauma to her abdomen and damage to her spleen. During emergency

splenectomy, the splenic artery is ligated proximal to its branching near the hilum of the spleen. Which structure, also supplied by the splenic artery, would be at risk if the artery is ligated improperly?

A. Body of pancreas
B. Fundus of stomach
C. Left kidney
D. Left lobe of liver
E. Lesser curvature of stomach

10. A patient is diagnosed with a malignant tumor in the head of the pancreas. During surgery to remove the tumor, the surgeon identifies and protects structures associated with this part of the pancreas. Which structure would **not** be at risk during this surgery?

A. Accessory pancreatic duct
B. Bile duct
C. Common hepatic duct
D. Main pancreatic duct
E. Superior pancreaticoduodenal artery

11. A patient is diagnosed with an abscess on the superior aspect of the left kidney. He complains not only of left flank pain but also of pain in his left shoulder. Physical examination does not reveal musculoskeletal problems in his shoulder. The physician explains that his shoulder pain is referred pain, caused by the abscess irritating an adjacent

structure. Which structure is most likely the source of the pain?

A. Body of pancreas
B. Left lobe of liver
C. Parietal peritoneum on the thoracic diaphragm
D. Renal fascia on left suprarenal gland
E. Visceral peritoneum on the large intestine

12. A 69-year-old male with a history of hypercholesterolemia and advanced atherosclerosis is admitted to the emergency department with severe abdominal pain, nausea, and vomiting. Imaging studies reveal stenosis of the celiac trunk and its main branches, causing ischemia in organs they supply. Which organ is **least** likely to be affected?

A. Body of pancreas
B. Gallbladder
C. Head of pancreas
D. Quadrate lobe of liver
E. Spleen

13. During resection of the liver, the surgeon ligates the right hepatic artery. Which lobe(s) of the liver is/are supplied by this artery?

A. Left and right lobes
B. Right and caudate lobes
C. Right and quadrate lobes
D. Right lobe only
E. Right, caudate, and quadrate lobes

CHAPTER 4

Pelvis

4.1 Benign Prostatic Hyperplasia

Patient Presentation

A 65-year-old white male visits the family medicine clinic complaining of progressive difficulty urinating over the past 6 months.

Relevant Clinical Findings

History

The patient reports the frequent need to urinate. His urine stream is weak, it starts and stops, and it "dribbles" at the end. At times, he also has a burning sensation when he urinates.

Physical Examination

- **Digital rectal examination** (DRE) (**Fig. 4.1.1**) revealed an enlarged prostate, with a symmetrical, smooth, firm posterior surface that is without tenderness.

Laboratory Tests

Test	Value	Reference value
Erythrocytes (count)	4.5	$4.3–5.6 \times 10^6/mm^3$
Hematocrit	40	38.8–46.4%
Leukocytes (count)	7.9	$3.54–9.06 \times 10^3/mm^3$
Prostate-specific antigen (PSA)	7	0.0–4.0 ng/mL

Clinical Note

PSA is a protein produced by the cells of the prostate gland. PSA levels can be elevated with prostatitis, benign prostate hyperplasia, or prostate cancer. However, PSA levels are not diagnostic between hyperplasia and cancer.

Clinical Problems to Consider

- Benign prostatic hyperplasia (BPH)
- Prostate carcinoma

LEARNING OBJECTIVES

1. Describe the anatomy of the prostate.
2. Explain the anatomical basis for the signs and symptoms associated with this case.

RELEVANT ANATOMY

Prostate

The prostate is the largest accessory gland of the male reproductive system. In the adult, the prostate measures 4 cm (transverse), 3 cm (superoinferior), and 2 cm (anteroposterior). The healthy adult prostate weighs about 20 g and is symmetrical and lacks palpable nodules. A median sulcus lies between the two lateral lobes. With aging, the prostate may enlarge: by age 40, it may reach the size of an apricot; by age 60, it may be the size of a lemon.

The **prostatic part of the urethra** passes through the prostate; it receives not only the openings of the prostate glands but also those of the ejaculatory ducts.

Hyperplasia Increase in size of a tissue or organ due to increased cell numbers (antonym: hypertrophy)

Palpitation Forcible or irregular pulsation of the heart that is perceptible to the patient

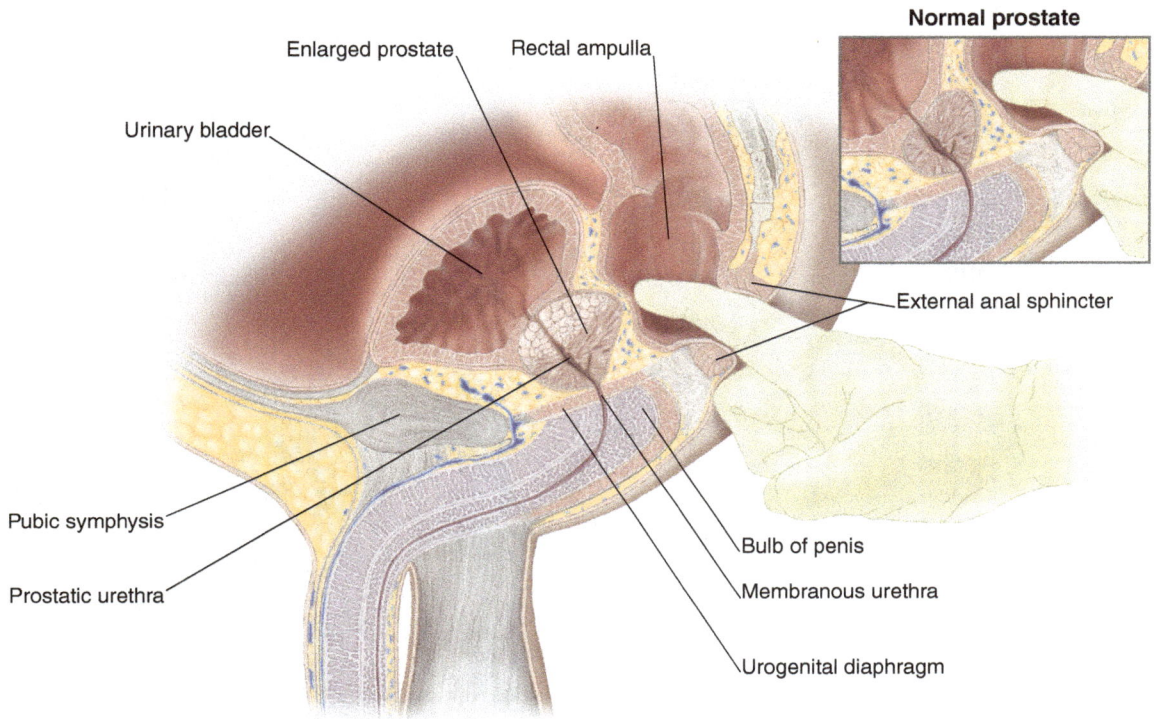

FIGURE 4.1.1 Median sagittal view of the male pelvis showing a digital rectal examination. The posterior surface of the prostate, as well as the seminal glands, may be palpated through the anterior rectal wall.

The prostate is contained within a fibrous capsule and is composed of numerous compound tubulo-alveolar glands arranged concentrically around the prostatic urethra:

- **Mucosal glands** are closest to the prostatic urethra
- **Submucosal glands** are peripheral to the mucosal glands
- **Main prostatic glands** are located toward the periphery of the prostate

Prostatic excretory ducts (usually 20–30) open into the prostatic urethra.

Anatomically, four prostatic *lobes* are defined (**Fig. 4.1.2**):

1. **Lateral** (right and left)
2. **Middle** (median)
3. **Anterior** (isthmus)
4. **Posterior**

Clinically, four prostatic *zones* are defined:

1. **Peripheral**
2. **Central**
3. **Transitional** (periurethral)
4. **Fibromuscular**

The fibromuscular zone contains smooth muscle that helps to expel prostatic fluid.

Prostatic secretion is a thin milky fluid that is slightly acidic (pH 6.1–6.5) that contributes approximately 30% (by volume) of the ejaculate. Among the secreted prostatic proteins, the proteolytic enzyme **PSA** helps liquefy the ejaculate to aid in sperm motility.

Anatomical Relationships

The anatomical relationships of the prostate are shown in **Figure 4.1.3** and outlined in **Table 4.1.1**.

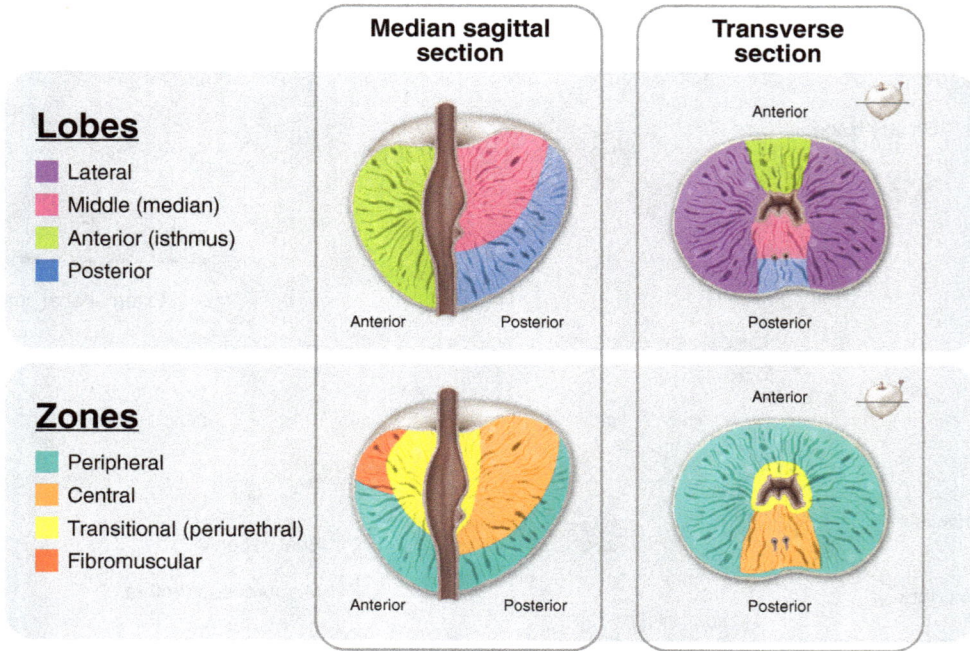

FIGURE 4.1.2 Lobes and zones of the prostate. Hyperplasia of mucosal glands in the transitional zone is responsible for BPH. In contrast, most prostate cancers originate in the peripheral and central zones.

FIGURE 4.1.3 Posterior view of prostate showing its anatomical relationships. A posterior median furrow indicates the separation of the right and left lateral lobes.

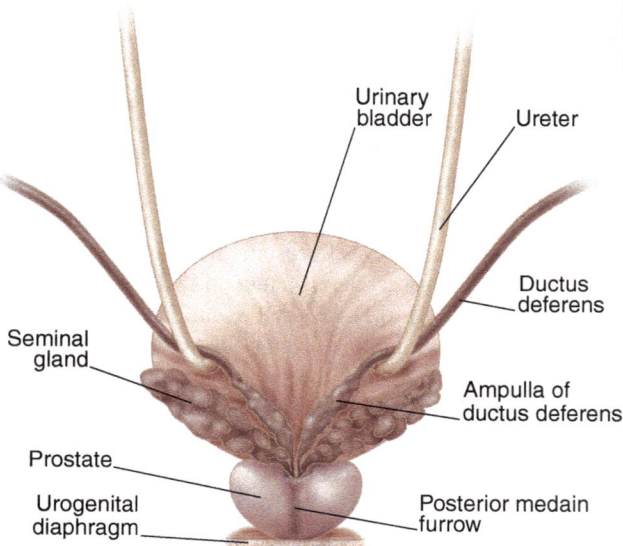

TABLE 4.1.1	Anatomical relationships of the prostate.	
Direction	**Related structure(s)**	**Note**
Anterior	▪ Pubic symphysis	Inferior part
Posterior	▪ Ampulla of rectum	
Superior	▪ Urinary bladder ▪ Seminal glands ▪ Ampulla of ductus deferens	Palpable if distended Palpable if enlarged
Inferior	▪ Urethral sphincter and deep perineal muscles	
Lateral	▪ Levator ani	

Because of its relationship to the rectum, the prostate can be palpated during a DRE.

Blood Supply and Lymphatics

The vasculature of the prostate is shown in **Figure 4.1.4**.

- **Arterial supply** is from branches of the **internal iliac artery**, including the **inferior vesical**, **internal pudendal,** and **middle rectal arteries**.

- **Venous drainage** of the prostate is through the **prostatic venous plexus**, which drains into the **internal iliac vein**. Blood from this plexus also anastomoses with the **vesical venous plexus**, as well as the **internal vertebral venous plexus** (not shown).

- **Lymph** from the prostate drains not only to **internal iliac nodes** but also to **sacral nodes** (not shown).

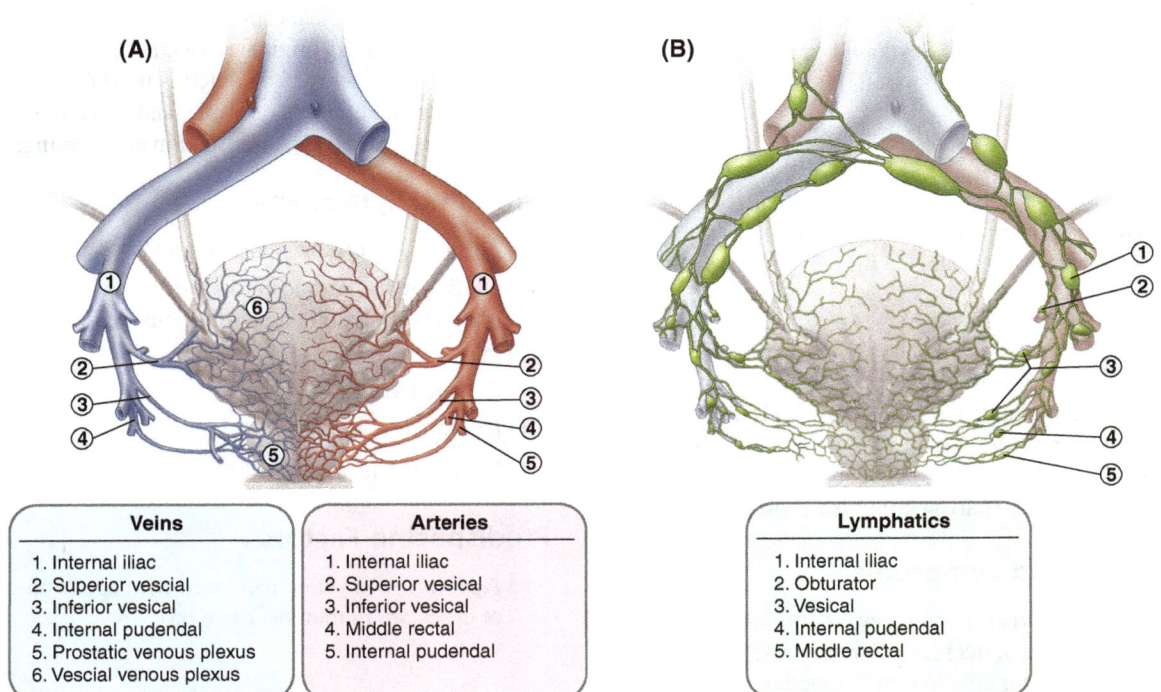

(A)

Veins
1. Internal iliac
2. Superior vescial
3. Inferior vesical
4. Internal pudendal
5. Prostatic venous plexus
6. Vescial venous plexus

Arteries
1. Internal iliac
2. Superior vesical
3. Inferior vesical
4. Middle rectal
5. Internal pudendal

(B)

Lymphatics
1. Internal iliac
2. Obturator
3. Vesical
4. Internal pudendal
5. Middle rectal

FIGURE 4.1.4 Vasculature of the prostate (posterior view). **(A)** Arterial supply (red) and venous drainage (blue) and **(B)** lymphatic drainage.

Nerve Supply

The prostate receives autonomic innervation from the **prostatic (nerve) plexus**.

- **Sympathetic (visceral efferent)**. Preganglionic sympathetic axons originate from cells in the T12-L1 lateral horn of the spinal cord. These fibers leave the spinal cord via **lumbar splanchnic nerves** and then travel through **aortic and superior hypogastric plexuses**. After synapsing in **ganglia** in the superior hypogastric plexus, postganglionic fibers enter a **hypogastric nerve** and then to the **inferior hypogastric plexus**. Fibers leave this plexus along periarterial plexuses to reach the prostatic plexus. Sympathetic fibers are vasomotor. They also control musculature of the preprostatic (urethral) sphincter, which is important to prevent ejaculate from entering the bladder.

- **Preganglionic parasympathetic** axons arise from S2 to S4 lateral horn cells. They emerge as pelvic splanchnic nerves and contribute to the inferior hypogastric plexus. These fibers follow periarterial plexuses to terminal ganglia near the gland. Postganglionic parasympathetic fibers innervate smooth muscle of the prostatic capsule.

CLINICAL REASONING

This patient presents with signs and symptoms of **prostatism**, consistent with narrowing of the prostatic urethra.

Benign Prostatic Hyperplasia

Benign prostatic hyperplasia (BPH) is the most common benign tumor in men. With age, mucosal glands in the transitional zone undergo cell proliferation, which leads to narrowing of the prostatic urethra.

Signs and Symptoms

- **Prostatism** (frequency, nocturia, urgency, hesitancy, decreased stream, and sensation of not being able to empty bladder)

- Enlarged prostate, with symmetrical, smooth, and firm posterior surface (median prostatic sulcus may be obliterated)
- Absence of prostatic tenderness or palpable nodules
- Elevated serum PSA

Predisposing Factors

- Age: 50% of men in their 50s; >90% in men after 80 years

Clinical Notes

- Negative DRE findings do not rule out BPH since obstructed urinary flow can occur without palpable prostate enlargement.
- Urinalysis is indicated to rule out hematuria or other evidence of a urinary tract infection.

Prostate Carcinoma

Prostate cancer is an adenocarcinoma (i.e., malignant epithelial neoplasm in glandular structures). Most prostate cancers originate in the peripheral zone (70%) or central zone (25%). DRE is used for initial staging when the tumor is palpable, and transurethral needle biopsy is used for confirmation and grading.

Signs and Symptoms

- Early stages often asymptomatic
- DRE reveals hard posterior surface, with nodule(s) palpable; median sulcus may not be evident
- Elevated serum PSA
- Prostatism
- Metastases in advanced stages

Predisposing Factors

- Age: rare <40 years; increases with age (>65% of cases are in men over age 65)

Nocturia Excessive urination at night

Hematuria Blood in urine

- Race: more common in African Americans; least common in Asian Americans and Native Americans
- Diet: high fat
- Family history

Clinical Notes

- BPH is not linked to prostate cancer and does not increase the chances of getting prostate cancer.
- Prostate cancer typically metastasizes initially to the seminal glands and urinary bladder. It may also metastasize to surrounding bone. Hematogenous metastasis to the spine via prostatic and vertebral venous plexuses is facilitated by the lack of valves in these venous channels.
- Metastasis to the rectum is less common because retroprostatic (Denonvillier) fascia separates the rectum from genitourinary structures.

DIAGNOSIS

The patient presentation, medical history, and physical examination support a diagnosis of **BPH**.

Hematogenous Spread via vasculature

Benign Prostatic Hyperplasia

BPH involves cellular proliferation in the **transitional (periurethral) zone** of the gland and results in prostatic enlargement. If the prostate is enlarged sufficiently, urine escape from the bladder may be restricted. BPH is a normal part of aging: 50% of men have BPH by age 60 and this increases to 90% by age 85.

- The patient is at risk for BPH due to his age.
- The lack of hardness and absence of nodules on the posterior surface of the prostate are consistent with BPH.

Prostate Carcinoma

In contrast to BPH, most (70%) prostate cancer develops in the **peripheral zone** of the prostate. Many patients with prostate carcinoma are asymptomatic and are diagnosed following routine screening (either for PSA levels or upon DRE) and biopsy.

- With prostate cancer, the posterior prostatic surface is hard, nodules are palpable, and the median sulcus may not be evident.

Both BPH and prostatic carcinoma may share signs and symptoms, including prostatism and elevated PSA levels.

Tubal Pregnancy

Patient Presentation

A 24-year-old white female is admitted to the emergency department complaining of severe abdominal pain.

Relevant Clinical Findings

History

The patient has been experiencing sharp pain in the right lower abdominal quadrant that she describes as 10 on a 1–10 scale. She has experienced nausea and vomiting over the last 4 hours. The patient's medical history includes the previous delivery of a healthy child. She suffered from pelvic inflammatory disease (PID) approximately 5 years ago.

Clinical Note

A notable symptom is localized, sharp abdominal pain. The pain described by this patient is due to stimulation of somatic afferent neurons with receptors located in the parietal peritoneum. These respond to stretching or irritation of parietal peritoneum adjacent to the affected organ. Stimulation of visceral afferent fibers with receptors in the wall of an organ would produce a vague, dull, poorly localized pain.

Physical Examination

Results of abdominal and pelvic examinations:

- Nondistended, tender abdomen with guarding
- Normal bowel sounds
- External genitalia are normal
- Uterus palpable but tender

- No vaginal bleeding or discharge
- No evidence of trauma

Laboratory Tests

Test	Value	Reference value
Erythrocyte (count)	4.5	$4.3–5.6 \times 10^6/mm^3$
Hematocrit	36	38.8–46.4%
Leukocyte (count)	6.7	$3.54–9.06 \times 10^3/mm^3$
Human chorionic hormone (hCG)	3,000	<5 mIU/mL

Imaging Studies

- Transvaginal ultrasound revealed an empty uterine cavity.

Clinical Notes

- **hCG** is secreted by the blastocyst and placenta. It can be detected in urine as early as 8 days after conception. hCG peaks between 10 and 12 weeks of pregnancy and falls to relatively low levels for the remainder of the pregnancy.

- During a **transvaginal ultrasound**, the transducer is placed in the vagina, which allows for clearer visualization of the uterus than would be seen with abdominopelvic ultrasound. This procedure can detect intrauterine pregnancies as early as 5 weeks of gestation.

Clinical Problems to Consider

- Appendicitis
- Ectopic pregnancy
- Ovarian cyst
- PID

Guarding Muscle spasm (especially of anterior abdominal wall) to minimize movement at site, at or near injury or disease (e.g., inflammation associated with

Palpitation Forcible or irregular pulsation of the heart that is perceptible to the patient

1. Describe the anatomy and vasculature of the uterine tubes.
2. Describe the anatomical relationships between the uterine tubes and other female pelvic organs.
3. Explain the anatomical basis for the signs and symptoms associated with this case.

RELEVANT ANATOMY

Uterine Tubes

Each uterine (Fallopian) tube is approximately 10 cm in length and extends posterolaterally from the body of the uterus. They facilitate, via peristaltic and ciliary actions, the transport of an ovum to the uterus. They also serve as the site for fertilization and transport of the zygote to the uterus. The uterine tube consists of four regions, from medial to lateral (**Fig. 4.2.1**):

1. **Uterine (intramural)**—lies within the uterine wall.

2. **Isthmus**—narrow, thick-walled portion between the uterine and ampullary parts.

3. **Ampulla**—dilated, thin-walled part between the isthmus and infundibulum. It is the longest segment of the tube.

4. **Infundibulum**—"funnel-shaped", distal end. It terminates with fimbriae (fingerlike extensions) and opens into the peritoneal cavity.

The isthmus, ampulla, and infundibulum are referred to as the **extramural parts** of the uterine tube.

Anatomical Relationships

An ovary lies adjacent to the infundibulum of each uterine tube. During ovulation, the fimbriae sweep over the ovarian surface to direct the ovum toward the distal (peritoneal) opening of the uterine tube.

The **broad ligament** (**Fig. 4.2.1**) is the double layer of peritoneum that reflects from the lateral pelvic wall and suspends and supports the uterus, uterine tubes, and ovaries. The uterine tubes are located in the superior margin of the broad ligament. The broad ligament is divided into parts that are named according to the reproductive organs they support (**Table 4.2.1**).

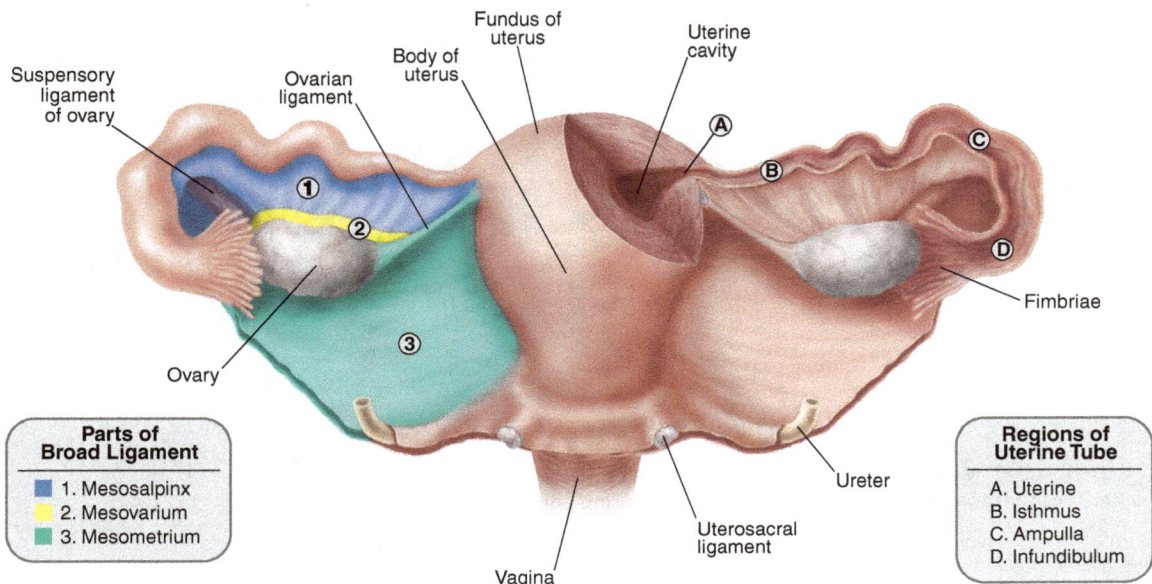

Parts of Broad Ligament
- ■ 1. Mesosalpinx
- ■ 2. Mesovarium
- ■ 3. Mesometrium

Regions of Uterine Tube
- A. Uterine
- B. Isthmus
- C. Ampulla
- D. Infundibulum

FIGURE 4.2.1 Posterior view of the female reproductive organs and broad ligament.

TABLE 4.2.1	Parts of the broad ligament.	
Part	**Organ supported**	**Description**
Mesosalpinx	Uterine tube	Suspends uterine tube in superior limit of broad ligament
Mesovarium	Ovary	Suspends ovary from posterior layer of broad ligament
Mesometrium	Uterus	Connects lateral margin of uterus with lateral pelvic wall

Peritoneum covers the anterior and superior aspect of the uterine body and is known as the epimetrium (serosa) of the uterus (**Fig. 4.2.2**). From the anterior surface of the uterus, the peritoneum reflects onto the superior surface of the urinary bladder. At the point of reflection, this creates a recess in the peritoneal cavity called the **vesico-uterine pouch**. Posteriorly, the peritoneum is reflected from the uterus onto the anterior surface of the rectum, forming the **recto-uterine pouch (Douglas)**, also known as the cul-de-sac.

The female reproductive organs are closely related to portions of the intestinal tract, including loops of small intestine and portions of the colon, including the vermiform appendix.

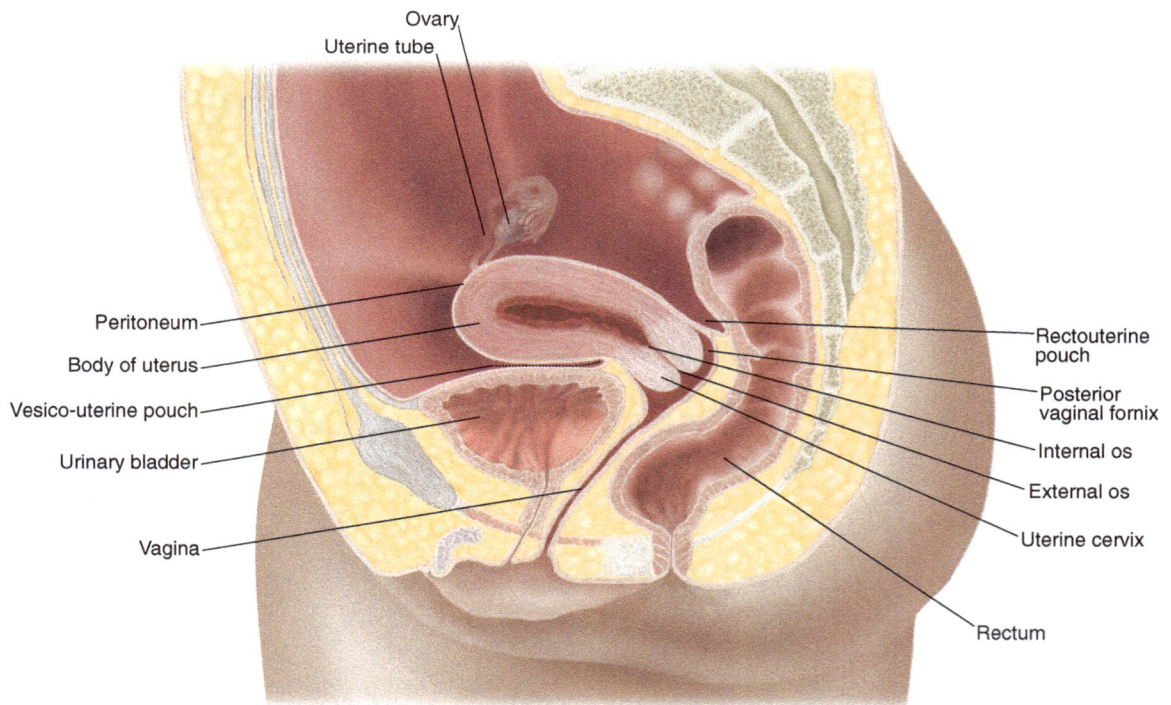

FIGURE 4.2.2 Median sagittal view of the female pelvis. Note the peritoneal reflections that form the vesico–uterine and rectouterine pouches.

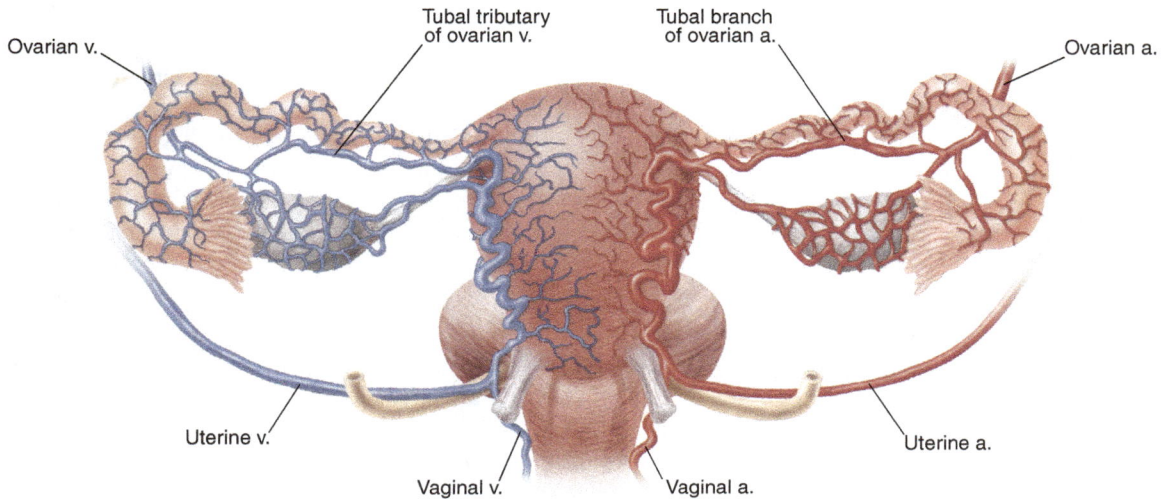

FIGURE 4.2.3 Posterior view of the vasculature of the uterine tube (arterial supply shown in red, venous drainage shown in blue).

Blood Supply

- **Arterial supply** is provided from anastomoses between tubal branches of **uterine artery** (branch of **internal iliac artery**) and **ovarian artery** (branch of abdominal aorta) (**Fig. 4.2.3**).

- **Venous drainage** is through tributaries of **internal iliac vein** (via uterine vein) and **renal or inferior vena cava** (via ovarian veins) (**Fig. 4.2.3**).

- **Lymph** from the uterine tubes drains principally to lateral **aortic and para-aortic nodes**, but also along the round ligament of the uterus to **superficial inguinal nodes**.

CLINICAL REASONING

This patient has severe pain in the right lower abdominal quadrant. This is consistent with pathology of viscera located in this quadrant of the abdominopelvic cavity including the cecum and appendix, right ureter, uterine tube, or ovary.

Appendicitis

Inflammation of the appendix is known as appendicitis. The cause of appendicitis is not always clear. It may be due to an obstruction (50–80% of cases) of the lumen by a fecal stone (i.e., a hard piece of fecal material). It may also occur following a gastrointestinal virus or other local inflammation.

Clinical Note

Deep tenderness on palpation at **McBurney's point** is known as **McBurney sign**. McBurney's point is a surface landmark located in the lower right quadrant about one-third of the distance from the anterior superior iliac spine to the umbilicus. Tenderness at this site indicates direct contact of an inflamed abdominal organ with parietal peritoneum. This is most commonly involves the appendix and, therefore, McBurney sign most frequently suggests acute appendicitis.

Signs and Symptoms

- Initial pain is vague and in umbilical region
- Later pain is more intense and localized to lower abdominal right quadrant
- Nausea and vomiting
- Loss of appetite
- Low-grade fever
- Constipation or diarrhea
- Abdominal bloating or swelling
- Psoas sign may be positive
- Obturator sign may be positive

Predisposing Factors

- Age: more frequent in adolescents and young adults
- Sex: males predominate 3:2

Clinical Notes

- Complications related to appendicitis include peritonitis following ruptured or perforated appendix or abdominal abscess.
- **Psoas sign** refers to pain that is elicited with passive extension of the hip.
- **Obturator sign** refers to pain elicited with passive flexion and internal rotation of the hip.

Ectopic Pregnancy

The term ectopic pregnancy is used to describe zygote implantation at any site other than the uterine lining.

The uterine tube is the most common site (90%) for an ectopic pregnancy; this is referred to as a **tubal pregnancy**. Within the uterine tube, the most fre-quent site for implantation is the ampulla (75–90%) followed by the isthmus (5–10%). Other sites for ectopic pregnancies include the ovary, peritoneum, and cervical canal.

In all of these locations, the fertilized ovum undergoes the initial stages of development, including the formation of placental tissue and fetal membranes. In a tubal pregnancy, the placenta may be partially separated from the wall and result in intra-tubal bleeding. More frequently, the placental tissue invades the wall of the uterine tube and causes a rupture that results in hemoperitoneum. This can be detected by culdocentesis, a procedure used to sample fluid that collects in the rectouterine pouch.

Signs and Symptoms

Early symptoms mimic a normal pregnancy

- Missed menstrual period
- Positive pregnancy test
- Breast tenderness
- Nausea
- Fatigue

  Additional symptoms

- Light vaginal bleeding
- Lower abdominal pain
- Unilateral "cramping" or sharp pelvic pain

  Symptoms related to uterine tube rupture

- Sharp stabbing pelvic pain
- Dizziness
- Low blood pressure

Predisposing Factors

- Previous ectopic pregnancy
- Previous or current PID
- Salpingitis
- Endometriosis

Hemoperitoneum Blood in peritoneal fluid within the peritoneal cavity

Culdocentesis Transvaginal aspiration of fluid from rectouterine pouch (of Douglas); also known as the cul-de-sac. The procedure involves inserting a needle through the posterior vaginal fornix to access the recto-uterine pouch

Salpingitis Inflammation of the uterine tube

- Drug treatment for infertility
- Stenosis of uterine tube
- Previous pelvic surgery

Clinical Notes

- About 5-6 weeks following a normal menstrual period, the symptoms of a tubal pregnancy may become acute. Pelvic examination, serum levels of hCG, and pelvic ultrasound studies should be conducted to confirm the diagnosis. Rupture of a tubal pregnancy constitutes a medical emergency.
- Infertility may be a consequence of tubal pregnancy.

Ovarian Cyst

An ovarian cyst is a fluid-filled sac (follicle) on the surface of the ovary. When one or more of the normal monthly follicles continues to grow beyond the typical size, it is considered a **functional cyst**. There are two types of functional cysts:

Follicular cyst

- Normally, luteinizing hormone (LH) serves as a signal for the follicle to rupture and release the ovum. A **follicular cyst** develops when the normal monthly surge of LH fails to occur and the ovum is not released. In this case, the follicle continues to expand into a cyst. These cysts are rarely painful and typically resolve within a few weeks.

Corpus luteum cyst

- The corpus luteum normally forms following the release of the ovum from a follicle and in response to increased estrogen and progesterone levels. Occasionally, the rupture site of the follicle seals itself and fluid is trapped in the follicle. In response to elevated hormone levels, the follicle continues to accumulate fluid

and forms a **corpus luteum cyst**. These cysts may increase in size to several centimeters in diameter, bleed, and/or rupture causing acute abdominal pain.

Signs and Symptoms

- Pelvic pain that is constant, dull, and aching
- Dyspareunia
- Abnormal uterine bleeding
- Dysmenorrhea
- Abdominal bloating or swelling

Predisposing Factors

- No predisposing factors have been established

Clinical Notes

Complications related to ovarian cysts include:

- Abdominal discomfort
- Pressure on urinary bladder, which may result in urinary frequency
- Ovarian torsion—turning of the ovary

Pelvic Inflammatory Disease

PID is a bacterial infection of the female reproductive tract. Causative bacteria may be introduced by sexual activity, during placement of contraceptive devices, or during gynecologic or obstetric procedures. PIDs most frequently result from sexually transmitted diseases (STD) such as *Neisseria gonorrhoeae* and *Chlamydia trachomatis*.

Signs and Symptoms

- Abdominal and/or low back pain
- Dyspareunia

Endometriosis is a disorder in which the endometrial cells grow outside the uterus. It is commonly painful and may involve the ovaries, bowel, or bladder; it rarely extends beyond the pelvis.

Stenosis Narrowing a canal (e.g., blood vessel and vertebral canal)
Dyspareunia Pain during sexual intercourse
Dysmenorrhea Difficult or painful menstruation

- Fever and chills
- Fatigue
- Diarrhea or vomiting
- Dysuria
- Purulent vaginal discharge

Predisposing Factors

- Sexually active and <25 years of age
- Unprotected sexual activity with multiple partners
- Previous history of STDs
- Previous history of PID
- Recent gynecologic procedures (e.g., IUD)

Clinical Notes

- Patients infected with *chlamydia* may be asymptomatic.
- Genital sores, vaginal discharge with odor, and bleeding between menstrual periods can also be indications of PID.
- PID diagnosis is based on presenting symptoms, pelvic examination, and vaginal and cervical cultures. Evidence from laparoscopy and endometrial biopsy may be used to confirm the diagnosis.
- Ectopic pregnancy, infertility, and/or chronic pelvic pain may result following one or more episodes of PID.

DIAGNOSIS

The patient presentation, medical history, physical examination, laboratory tests, and imaging studies support a diagnosis of **tubal (ectopic) pregnancy**.

Ectopic Pregnancy

Ectopic pregnancy refers to the implantation of a fertilized egg in a location other than the uterine cavity, usually in the uterine tube. The uterine tube is not designed to expand to accommodate fetal development. Ectopic pregnancy creates the potential for rupture of the tube that can lead to massive hemorrhage and death.

- The patient presents with acute, sharp pain in the right lower abdominal quadrant. For women of reproductive age, the two most likely diagnoses are acute appendicitis and tubal pregnancy.
- Prior PID puts the patient at higher risk for tubal pregnancy.
- A diagnosis of tubal pregnancy in this patient is consistent with the high levels of serum hCG.
- Ultrasound imaging confirmed that the implantation was not in the uterine cavity.

Clinical Note

Hemoperitoneum would confirm a rupture of the tubal pregnancy and establish the need for immediate surgical intervention.

Appendicitis

Appendicitis is an inflammation of the appendicular mucosa, usually following obstruction of its lumen. This results in increased intraluminal pressure, which can lead to vascular impairment, perforation of the appendiceal wall, and peritonitis.

- Patients with acute appendicitis complain of intense abdominal pain in the right lower quadrant similar to that describe by this patient.
- However, these patients would most likely have a fever, elevated white cell counts, and tenderness at McBurney's point, not seen in this patient, making this an unlikely diagnosis.

Dysuria Pain during urination

Purulent Containing, discharging, or causing the production of pus

Ovarian Cyst

Ovarian cysts form on the surface of the ovary as part of the normal menstrual cycle. However, one of these cysts may persist, enlarge, and eventually rupture causing acute abdominal pain.

- Sudden onset of unilateral pelvic pain suggests rupture of an ovarian cyst in this patient, but increased serum hCG is not consistent with diagnosis.

- Pelvic ultrasound could be used to confirm the presence of an ovarian cyst.

Pelvic Inflammatory Disease

PID is initiated by infection, usually sexually transmitted, that spreads from the vagina to other reproductive organs.

- The most common presenting complaint with PID is lower abdominal pain.

- Most patients also exhibit abnormal vaginal discharge.

- The diagnosis of acute PID is primarily based on patient history of previous PID and clinical findings and is not consistent with this patient's information.

Patient Presentation

A 56-year-old postmenopausal female presents at the OB/GYN clinic with a complaint of episodes of vaginal bleeding.

Relevant Clinical Findings

History

The patient reports mild pain during sexual intercourse, and that she limits the frequency of intercourse because bleeding often follows. A Pap smear (Papanicolaou test) 10 years ago revealed human papillomavirus (HPV) and dysplastic cells. She has had several male sex partners over the past 20 years. She has smoked cigarettes for most of her adult life.

Physical Examination

Results of pelvic examination and colposcopy:

- Uterus anteverted and anteflexed
- Roughened and mottled area on the vaginal cervix at the external os and on the posterior part of the vaginal fornix
- No abnormal masses detected by digital vaginal and rectal evaluation

Procedures

Cells from the area of the external os were harvested by curettage and an endocervical punch biopsy was taken.

The pathology report indicates:

- Dysplastic and cancerous cells in the curettage sample
- Squamous cancer cells approximately 2 mm into the biopsy tissue

Clinical Problems to Consider

- Cervical carcinoma in situ
- Cervical carcinoma with metastases
- Endometrial carcinoma

LEARNING OBJECTIVES

1. Describe the anatomy of the uterus.
2. Describe the anatomy of the vagina.
3. Describe the anatomy related to a pelvic examination.
4. Explain the anatomical basis for the signs and symptoms associated with this case.

RELEVANT ANATOMY

Uterus

The **uterus** is composed of two parts: the **body** and **cervix** (**Fig. 4.3.1**). A **fundus** and **isthmus** are subdivisions of the body of the uterus. The cervix is subdivided into **supravaginal and vaginal parts**. The slit-like **uterine cavity** lies in the body. This cavity is continuous with the **cervical canal**, which connects the vaginal canal with the uterine cavity. The **internal os** (*Latin*: mouth) is at the junction of the uterine cavity and the cervical canal (**Fig. 4.3.2**). On the external surface of the uterus, the position of the internal os is indicated by the isthmus, a shallow notch where the body joins the supravaginal cervix. The cervical canal opens onto the blunt vaginal cervix as the **external os** (**Fig. 4.3.2**). This area has a **transitional**

Curettage Removal of material (e.g., growths) from the wall of a body/organ cavity or surface with a curet (spoon-shaped instrument)

Dysplasia Abnormal tissue development

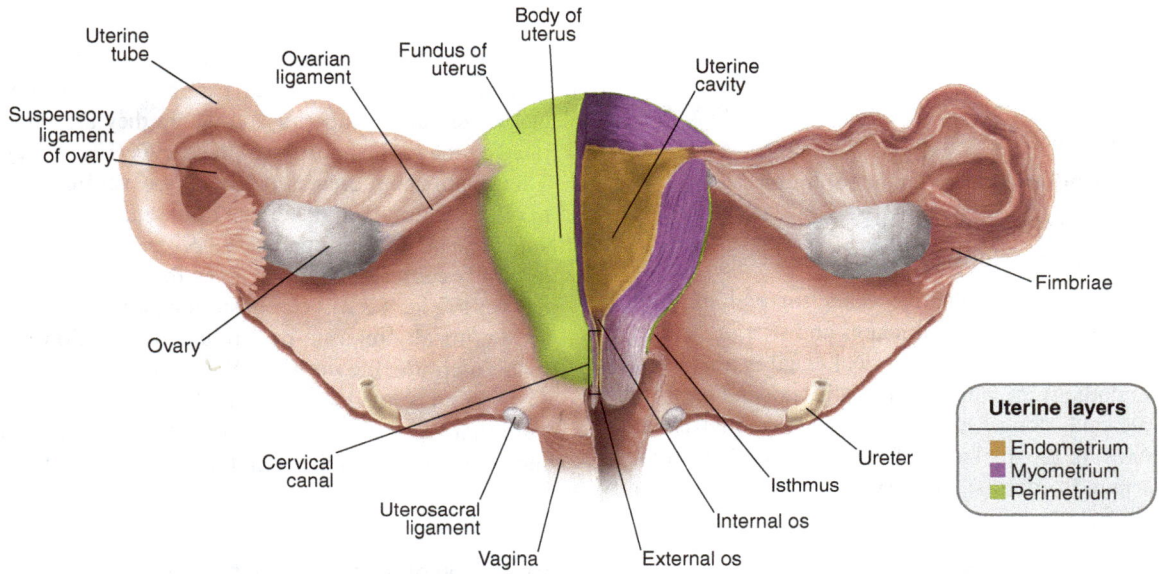

FIGURE 4.3.1 Posterior view of female reproductive organs and broad ligament. The right uterine tube and the right side of the uterus and superior vagina are shown in coronal section.

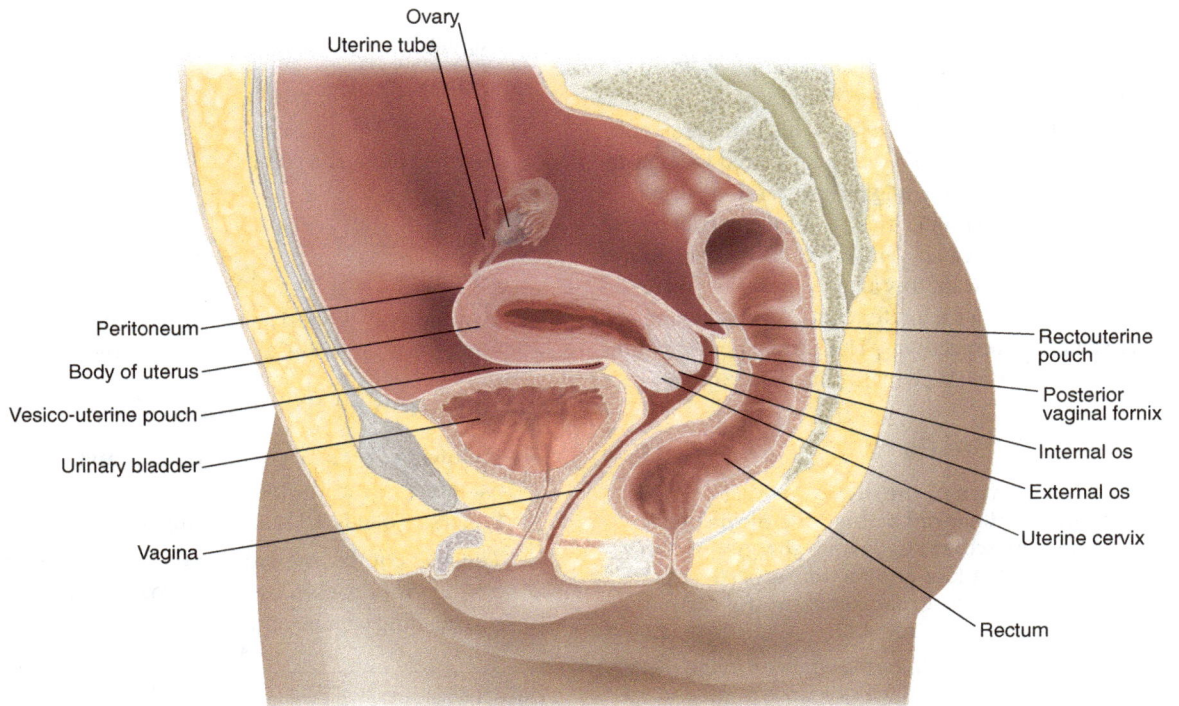

FIGURE 4.3.2 Median sagittal view of female pelvis.

zone of epithelium between the columnar cells of the cervical canal and the squamous cells of the vaginal surface of the cervix and the vaginal lining.

The wall of the nonpregnant uterine body is thick and divided into:

- **Endometrium:** epithelial lining
- **Myometrium:** middle smooth muscle layer
- **Perimetrium:** external serosal coat

During the reproductive years, the endometrium is in constant flux as it thickens and sloughs with each menstrual cycle. The cervical canal is also lined by endometrium, but this epithelium does not proliferate and slough in response to hormonal changes. Most of the thickness of the uterine body is myometrium. In contrast, the cervical wall is predominantly fibroelastic tissue.

Anatomical Position

In its anatomical position, the anterior (vesical) surface of the body of the uterus rests on the superior surface of the urinary bladder (**Fig. 4.3.2**). This is most conspicuous when the urinary bladder and rectum are empty. At the level of the internal os, where the uterine cavity and cervical canal are continuous, there is a slight angulation between the isthmus of the body and supravaginal part of the cervix. This anterior angulation is approximately 170° and is referred to an **anteflexion**. A more acute angle (**anteversion**) occurs at the external os with the junction of the cervical canal and the vaginal canal. Since the cervix pierces the anterior vaginal wall, this angle is near 90°. Increases in these two angles are referred to as **retroflexion** and **retroversion**, respectively. The normal position of the uterus is important in reducing the chance of prolapse.

Clinical Notes

- The position of the uterine body is dynamic and is influenced by the tone of the pelvic floor musculature, volume changes of the urinary bladder and rectum, and to a lesser degree by the broad ligament (for further discussion, see Case 4.4).
- Implantation of a fertilized ovum in the endometrium and maintenance of pregnancy are influenced by the position of the uterus.

Vagina

The vagina is a musculofascial sheath, 7–9 cm long. Its anterior and posterior walls are normally in contact except at the superior end where they are separated by the vaginal portion of the **cervix** (**Fig. 4.3.2**). This relationship creates a recess, the **vaginal fornix**. The fornix is divided into right and left lateral, anterior, and posterior parts. The posterior part is the largest because the cervix pierces the anterior vaginal wall (**Fig. 4.3.2**). The posterior part is also the most clinically relevant due to its close relationship to the **rectouterine pouch** of the peritoneal cavity. The vagina is lined by a stratified squamous epithelium that lacks glands. The vaginal wall releases a serous exudate to moisten and lubricate the surface.

Anatomy of the Pelvic Examination

Evaluation of the pelvic organs is an important part of the female physical examination. It is used to assess female reproductive and other organs in the perineum and pelvic cavity:

- Vulva
- Vagina
- Uterus
- Uterine cervix
- Uterine (Fallopian) tubes
- Ovaries
- Urinary bladder
- Rectum

The gowned and properly draped patient is in the lithotomy position on the examination table.

Examination of the Vaginal Canal

- Following visual evaluation of the **vulva**, the examiner places the gloved index and middle fingers of one hand into the vagina.
- The walls of the **vaginal canal** are assessed for irregularities caused by the encroachment of surrounding organs or pathological masses.

Vulva Female external genitalia
Lithotomy position Supine with buttocks at edge of examination table, hips and knees flexed, feet in stirrups

Bimanual Examination

With the fingers still in the vagina, the fingers of the other hand press deeply against the suprapubic portion of the anterior abdominal wall. This "bimanual palpation" compresses midline structures between the fingers of the two hands.

- This allows evaluation of the size and position of the **uterine cervix**, **uterine body**, and **urinary bladder**.
- This may also allow for the detection of pelvic masses and/or cysts.

At any time during the examination, the examiner may use a speculum to separate the vaginal walls for visual inspection.

The digital vaginal examination can also be used to determine pelvic dimensions since several bony landmarks of the pelvis are palpable transvaginally.

- These include the **ischial spines** and, in most women, the **sacral promontory**.

Rectovaginal Examination

A second type of evaluation involves the examiner placing one finger in the vagina and another finger of the same hand in the anal canal and rectum. This procedure may allow palpation of the **ovaries** and **uterine ligaments**, as well as a **retroverted and/or retroflexed uterus**.

CLINICAL REASONING

This patient presents with vaginal bleeding and a history of HPV. This is consistent with **pathology of the uterus and/or vagina**.

Cervical Carcinoma In Situ

Most cervical cancer is malignancy of the squamous epithelial cells of the vaginal part of the cervix (squamous cell carcinoma). Cervical carcinoma in situ, the earliest stage of cervical cancer, is characterized by malignant cells that are restricted to the cervix and extend less than 3 mm into the cervical wall. This stage also is referred to as stage I cervical carcinoma.

Palpitation Forcible or irregular pulsation of the heart that is perceptible to the patient

TABLE 4.3.1	Staging of cervical carcinoma.
Stage	**Metastasis**
Stage I	Cells restricted to cervix, <3 mm into cervical wall
Stage II	Cells in other parts of the uterus and in upper vagina
Stage III	Cells in the lower vagina and possibly the pelvic walls
Stage IV	Cells in the bladder or rectum, or more distant organs (especially lung)

Signs and Symptoms

- Often asymptomatic in early stages
- Vaginal bleeding unrelated to menstrual cycle or postmenopausal vaginal bleeding
- Dyspareunia

Predisposing Factors

- History of HPV infection
- Multiple sex partners
- Smoking

Clinical Notes

- Detection of cervical carcinoma at stage I results in a high survival rate.
- Cervical carcinoma is no longer the most common cancer of the female reproductive tract due largely to early detection with the Pap smear.

Cervical Carcinoma with Metastases

Metastatic carcinoma of the cervix implies that the cancer cells have invaded more than 3 mm into the cervical tissue and that the cells have invaded other tissues and/or organs. Stages II–IV are used to grade the degree of spread of cancer cells (**Table 4.3.1**).

Signs and Symptoms

- Vaginal bleeding unrelated to menstrual cycle or postmenopausal vaginal bleeding

---

Dyspareunia Pain during sexual intercourse

TABLE 4.3.2	Staging of endometrial carcinoma.
Stage	**Metastasis**
Stage I	Cells confined to the body of the uterus
Stage II	Cells extend to the uterine cervix
Stage III	Cells invade other pelvic organs
Stage IV	Cells in distant organs

- Pelvic pain
- Dyspareunia
- Hematochezia
- Oliguria due to compressed ureters

Predisposing Factors

- History of HPV infection
- Multiple sex partners
- Smoking
- Failure to have regular Pap smear

Endometrial Carcinoma

Most endometrial cancers develop from the glands of the endometrium (adenocarcinoma). Stages I–IV are used to grade the degree of spread of cancer cells (**Table 4.3.2**).

Signs and Symptoms

- Often asymptomatic in early stages
- Postmenopausal vaginal bleeding (vaginal bleeding unrelated to menstrual cycle in premenopausal women)

Predisposing Factors

- Age: 50–69 years
- Postmenopausal
- Obesity
- Estrogen therapy (without progesterone) for symptoms of menopause
- Other medical conditions, including breast cancer and hereditary nonpolyposis colon cancer

---

Hematochezia Bloody stool
Oliguria Decreased urine output

Clinical Notes

- The majority (75%) of women with endometrial cancer are postmenopausal.
- This is the most common cancer of the female reproductive tract in developed countries.

DIAGNOSIS

The patient presentation, medical history, physical examination, and laboratory tests support a diagnosis of **cervical carcinoma in situ**.

Cervical Carcinoma In Situ

The external os of the vaginal part of the cervix is a zone of epithelial transition. At this point, the simple columnar/cuboidal cells lining the uterine tubes, uterine cavity, and cervical canal change to the stratified squamous epithelium of the vaginal canal. Cyclic hormonal influence (normal or therapeutic), infections, and modifiable behaviors make the epithelium at the external os susceptible to dysplasia and malignancy.

- This patient is at risk for cervical cancer because of her history of HPV infection, multiple sex partners, and smoking.
- While not diagnostic for cervical carcinoma, a common complaint in postmenopausal women is vaginal bleeding ("spotting").
- The finding of squamous cells invading deeper tissues of the cervix strongly supports the diagnosis of cervical carcinoma.
- This is most likely stage I (in situ) of the disease as revealed by the 2-mm invasion of the cervical tissue by the malignant cells. It might also be considered an early stage II carcinoma of the cervix because of the involvement of the vagina.

Cervical Carcinoma with Metastases

This condition implies that malignant cells have invaded deep into the tissues of the cervix, and may well have spread to other adjacent or distant organs.

- Cervical carcinoma with metastases is ruled out by the fact that the punch biopsy shows the malignant cells are less than 3 mm into the cervical tissue.

Clinical Note

Nearly all women who develop cervical carcinoma have prior infections of HPV. The most common method of infection with HPV is sexual intercourse. Intercourse with multiple partners without the use of latex condoms increases the risk for infection. Women with HPV who smoke have the highest probability of developing cervical carcinoma. The causal factor(s) that increases the incidence of cervical carcinoma in women with HPV who smoke is unknown.

Endometrial Carcinoma

The endometrium lines the uterine cavity and its superficial layers are sloughed as part of menstrual fluid. The endometrium contains a rich network of glands.

- **Endometrial carcinoma** is unlikely because the cancer cells are squamous. Nearly all endometrial carcinomas are adenocarcinomas involving glandular cells of the endometrium (simple cuboid/columnar epithelium).

CASE 4.4 | Urethrocele with Stress Incontinence

Patient Presentation

A 59-year-old multiparous patient complains of difficulty controlling urine flow.

Relevant Clinical Findings

History

The patient describes "dribbling" of urine when she sneezes, coughs, laughs, or does bench press exercise at the gym. She indicates that over the past year the volume of urine lost during an "episode" has progressively increased to the point of causing embarrassment. She also describes a mild discomfort during sexual intercourse.

Physical Examination

Results of pelvic examination:

- Soft bulge in lower anterior wall of vagina
- Q-tip test indicates urethra has a change in angulation of 40° (**Fig. 4.4.1**)

Clinical Note

The **Q-tip test** is used to assess adequate support for the urethra. A sterile Q-tip is inserted into the urethra and the deviation of shaft of the Q-tip as it protrudes from the external urethral orifice is compared in a relaxed and straining situation. A deviation of >30° is considered a positive Q-tip test. This indicates poor pelvic floor support and a change in the orientation of the urethra.

Laboratory Tests

- Urine culture is negative for infectious agents.

Clinical Problems to Consider

- Bacterial cystitis
- Cystocele
- Urethral syndrome (urethritis without evidence of bacterial/viral infection)
- Urethrocele

LEARNING OBJECTIVES

LEARNING OBJECTIVES

1. Describe the anatomy of the female pelvic organs and pelvic floor.
2. List factors that may contribute to pelvic floor weakness.
3. Explain exercises that help tone and strengthen pelvic floor musculature.
4. Explain the anatomical basis for the signs and symptoms associated with this case.

RELEVANT ANATOMY

Female Urethra

The female urethra (**Fig. 4.3.2**) is about 4 cm long and extends from the internal urethral orifice of the urinary bladder to the external urethral orifice in the vestibule of the perineum. It may be embedded in the anterior wall of the vagina along all, or part, of its course. The ducts of many small, mucous-producing, urethral glands open into the lumen of the urethra. The female urethra,

---

Cystitis Inflammation of the urinary bladder
Multiparous Having borne multiple children

Urethritis Inflammation of the urethra

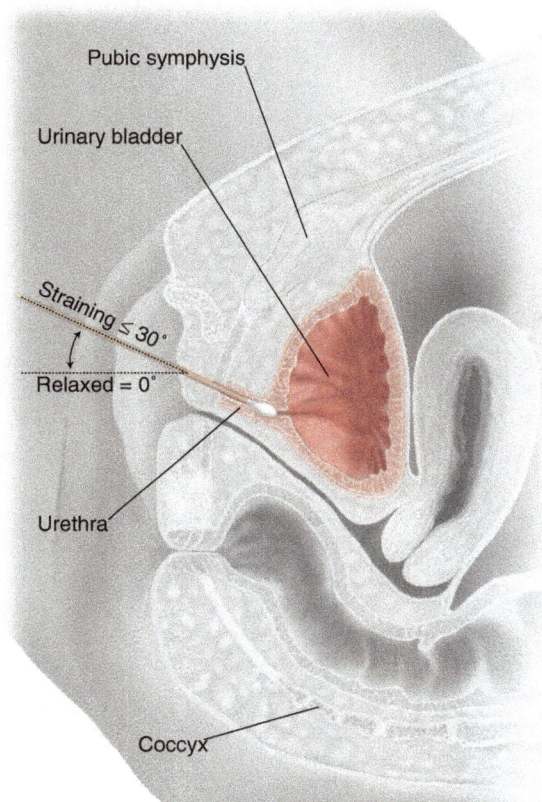

FIGURE 4.4.1 Median sagittal view of the female pelvis. The angulation of a Q-tip as it protrudes from the external urethral orifice is compared in relaxed and straining situations to assess support for the urethra.

contrasted to the male urethra, is not divided into anatomical regions.

Urogenital and Pelvic Diaphragms

The pelvic outlet is closed by two diaphragms (urogenital and pelvic) that are composed of skeletal muscles and their investing fascia (**Fig. 4.4.2**). These form the pelvic floor.

1. The **urogenital (UG) diaphragm** is the smaller and is positioned anterior and superior to the pelvic diaphragm. It is so named because the female urethra and the vagina pierce it to enter the vestibule of the perineum. It lies on the horizontal plane

connecting the **ischiopubic rami**. The primary muscle of the UG diaphragm is the **urethral sphincter**, which compresses the urethra to permit voluntarily control of urine flow.

2. The larger **pelvic diaphragm** forms the remainder of the pelvic floor. It is comprised of four paired muscles that attach to pelvic walls. They are from anterior to posterior: **puborectalis**, **pubococcygeus**, **iliococcygeus**, and **ischiococcygeus (coccygeus)**. The puborectalis, pubococcygeus, and iliococcygeus are collectively termed the **levator ani** muscle. Fibers of these muscles decussate with the same muscle of the opposite side, insert on a midline fibrous raphe, or attach to the coccyx. The anal canal passes through the pelvic diaphragm.

The pelvic diaphragm is concave on its pelvic perspective, in contrast to the flat UG diaphragm. There is a gap in the pelvic diaphragm between the two puborectalis muscles. This gap is the **genital hiatus**. The urethra and vagina exit the pelvis by passing through the genital hiatus to reach and pierce the UG diaphragm. *Thus, neither diaphragm completely closes the pelvic outlet.* Despite their different orientations, collectively they form a "floor" that provides the primary support for pelvic viscera.

Pelvic Fascia

Endopelvic fascia fills the interstices between organs and other structures of the pelvis. It is thickened at specific points to form **pelvic ligaments** (e.g., transverse cervical or cardinal ligament, uterosacral ligament), musculofibrous bands that contain fascicles of smooth muscle. These ligaments extend between the pelvic walls and the surfaces of organs, especially the urinary bladder and the supravaginal part of the uterine cervix. These ligaments offer some support to pelvic organs.

In summary, the muscles of the **pelvic floor** are primary in maintaining the normal positions and relationships of female pelvic organs (uterus, vagina, urinary bladder, urethra, and rectum). A weakened pelvic floor may result in altered positions of pelvic viscera that commonly herniate the walls of the vagina.

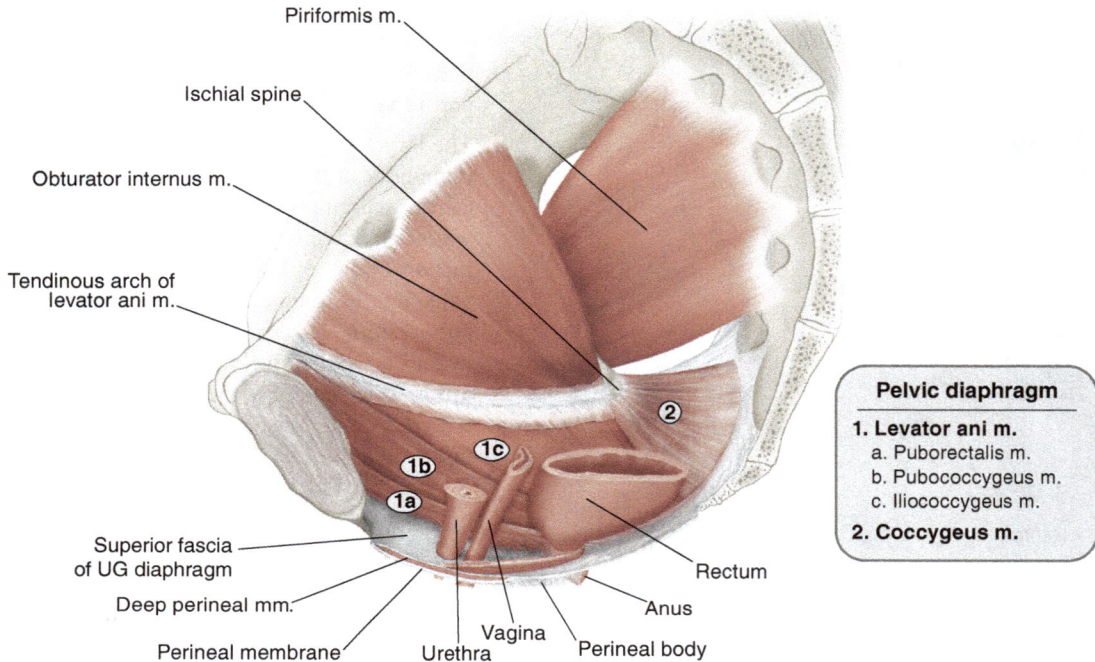

Pelvic diaphragm
1. **Levator ani m.**
   a. Puborectalis m.
   b. Pubococcygeus m.
   c. Iliococcygeus m.
2. **Coccygeus m.**

FIGURE 4.4.2 Parasagittal section of the female pelvis showing the urogenital and pelvic diaphragms.

CLINICAL REASONING

This patient presents with involuntary loss of urine, a bulge in her anterior vaginal wall, and discomfort during intercourse. These signs and symptoms are consistent with a genitourinary condition.

Bacterial Cystitis

Bacterial cystitis is a common infection of the urinary bladder that may involve the urethra. Urine will be positive for gram-negative bacteria, usually *Escherichia coli*. Bacterial cystitis is part of a spectrum of infections known as urinary tract infections (UTIs). These infections usually resolve within 2–3 days with over-the-counter therapeutics that contain citrate.

Signs and Symptoms

- Urgency and increased frequency of urination
- Dysuria
- Pyuria
- Hematuria
- Foul-smelling urine
- Positive urine culture
- Low-grade systemic fever
- Suprapubic tenderness
- Dyspareunia

Predisposing Factors

- Length of female urethra
- Close relationship of urethral orifice to anus
- Pregnancy and related hormonal changes
- Menopause and related hormonal changes
- Diabetes
- Diet high in sugar
- Use of scented soaps and lubricants

Dyspareunia Pain during sexual intercourse
Dysuria Pain during urination

Hematuria Blood in urine
Pyuria Pus in urine

Clinical Notes

- Sexually transmitted bacterial infections such as *Neisseria gonorrhoeae* and *Chlamydia trachomatis* may present with symptoms similar to UTIs and it may be necessary to rule these out prior to developing a treatment plan.
- Often referred to as "Honeymoon cystitis" since the infection frequently occurs following frequent or vigorous sexual intercourse.

Cystocele

This condition exists when the urinary bladder herniates into the anterior vaginal wall (**Fig. 4.4.3**). While it may be the result of trauma or surgery, the most common cause is a stretching and subsequent weakening of the pelvic floor musculature and pubovesical ligaments secondary to vaginal delivery. As a fetus enters the lower birth canal, the muscles of the pelvic and UG diaphragms relax to allow maximum stretching as the baby passes through the pelvic floor. Overstretching, tearing, or other trauma to these structures during childbirth will have long-term effects on the position and function of the female pelvic organs. Prenatal and postnatal exercises (e.g., Kegel) that tone the pelvic floor muscles significantly reduce the likelihood of trauma to these muscles during childbirth.

Signs and Symptoms

- Stress incontinence—involuntary leakage of urine during increased intra-abdominal pressure (sneezing, coughing, laughing, straining, lifting)
- Dyspareunia

Predisposing Factors

- Childbirth
- Coughing (chronic bronchitis or asthma)
- Increasing age
- Sedentary lifestyle

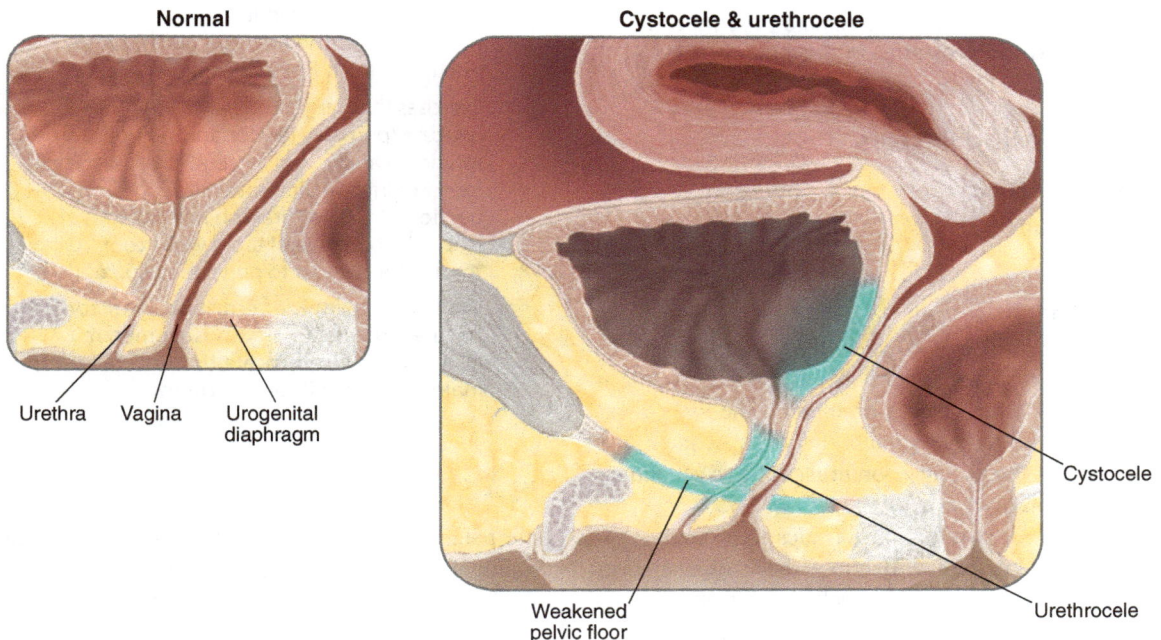

Normal

Urethra   Vagina   Urogenital diaphragm

Cystocele & urethrocele

Cystocele

Urethrocele

Weakened pelvic floor

FIGURE 4.4.3 Median sagittal view of the female pelvis showing normal anatomy and prolapse of the urinary bladder (cystocele) and urethra (urethrocele) into the vagina. A weakened pelvic floor musculature is the major factor in these conditions.

Clinical Note

Cystocele is frequently associated with urethrocele and imaging may be required to distinguish them, or to confirm that both exist.

Urethral Syndrome

Urethral syndrome is urethritis without evidence of bacterial or viral infection (by urinalysis).

Signs and Symptoms

- Dysuria
- Increased frequency of urination
- Constant urgency
- No pyuria or urethral discharge

Clinical Note

Current data suggests that urethral syndrome involves a low-grade infection of the paraurethral (Skene's) glands. These glands are homologous to the prostate gland and, thus, urethral syndrome in the female thought to be equivalent to prostatitis in the male.

Predisposing Factors

- Age: 30–50 years
- Sexually active

Urethrocele

This condition exists when the urethra herniates into the anterior vaginal wall (**Fig. 4.4.3**). While it may result from trauma or surgery, the most common cause is an overstretching and subsequent **loss of tone in the pelvic floor musculature** secondary to vaginal delivery.

Failure of the pelvic floor muscles to properly support pelvic viscera will allow these organs, over time, to "sag" toward the pelvic floor, most often encroaching on the vagina. In the case of the urethra, decreased physical support and its close relationship to the anterior vaginal wall may cause it to herniate into the vagina. Herniation alters the alignment of the urethra with the bladder floor and its passage through the UG diaphragm. This compromises the effectiveness of muscles and elastic tissues that contribute to urinary continence.

Signs and Symptoms and Predisposing Factors

The signs and symptoms and predisposing factors for urethrocele are the same as for cystocele (see above). Indeed, these two conditions are often present together and it may require imaging to determine which one or if both are present.

Clinical Notes

- "Stress incontinence" is used to describe the cardinal symptom of urethrocele. Increased intra-abdominal pressure ("stress") will compress the walls of the urinary bladder and cause unwanted leakage ("incontinence") of urine into the urethra and discharge from the external urethral orifice.
- Exercises that target the pelvic floor musculature (primarily levator ani and urethral sphincter) are known as Kegel exercises. These exercises involve alternate relaxation and contraction of the urethral sphincter to start and stop urination several times while contracting the levator ani as if to prevent escape of flatus.
- While childbirth is a contributing factor for urethrocele, nulliparous women may also develop urethrocele through failure to maintain pelvic floor muscle tone.

DIAGNOSIS

The patient presentation, medical history, physical examination, and laboratory tests support a diagnosis of **urethrocele with possible cystocele**.

Flatus Gas or air in gastrointestinal tract that is expelled through the anus

Nulliparous Having never borne children

Urethrocele with Possible Cystocele

Loss of tone for muscles of the pelvic floor is a leading cause of prolapse of genitourinary organs in the female. Vaginal deliveries stretch the muscles of the pelvic outlet, and without proper postnatal exercises, they may lose the ability to maintain pelvic organs in their proper position and relationships.

- Stress incontinence is a cardinal sign of a weakened pelvic floor.

- The woman's age and the fact that she is multiparous predispose her to this condition.

- A soft bulge in the lower anterior vaginal wall supports the diagnosis of a herniated urethra and potentially the urinary bladder as well.

- The positive Q-tip test indicates poor pelvic floor support and a change in the orientation of the urethra.

- Differentiation between urethrocele and cystocele is difficult by physical examination and these two conditions frequently co-exist.

- Imaging studies are required to make an accurate diagnosis.

Urethral Syndrome or Cystitis

Both of these conditions are considered to be caused by infectious agents. Urethral syndrome patients present with symptoms of urethral infection (urethritis), although this would not be detected by urinalysis.

- The lack of dysuria and urethral discharge, and failure to detect a bacterium or virus in urinalysis, helps rule out urethral syndrome and cystitis.

Patient Presentation

A 54-year-old African American male visits the family medicine clinic complaining of constipation and bright red blood in his stool. He is referred to a gastroenterology clinic for colonoscopy.

Relevant Clinical Findings

History

The patient reports a recent change in bowel habits together with constipation and cramping. He has smoked one and half packs of cigarettes a day for the past 35 years. He has no family history of cancer.

Physical Examination

Noteworthy vital signs:

- Height: 5′ 8″
- Weight: 220 lb
- BMI: 31.9 (normal: 18.5–24.9; obese: >30)

Results of abdominal examination:

- No masses were noted on deep palpation
- No guarding or tenderness
- Hypoactive bowel sounds

Results of digital rectal examination (DRE):

- Bright red blood

Laboratory Tests

Test	Value	Reference value
Hemoglobin	10.6	14–17 gm/dL
Carcinoembryonic antigen (CEA)	560	0–3.0 ng/mL

Clinical Note

CEA is a protein expressed in the fetus. It may show increased levels in blood as a tumor marker, particularly in association with colorectal, pancreatic, stomach, breast, and lung cancer. CEA may also be elevated with "benign" conditions including smoking, infection, inflammatory bowel disease (IBD), pancreatitis, and cirrhosis of the liver (in these conditions, serum CEA is typically <10 ng/mL).

Imaging Studies

- Contrast imaging revealed a distended sigmoid colon, with luminal narrowing and marked wall thickening.

Colonoscopy and Biopsy

- Colonoscopy with biopsy indicated an adenocarcinoma.

Clinical Problems to Consider

- Colorectal cancer
- Diverticulosis
- Sigmoid volvulus
- Ulcerative colitis

1. Describe the anatomy of the large intestine.

2. Explain the anatomical basis for the signs and symptoms associated with this case.

RELEVANT ANATOMY

Large Intestine

The large intestine is approximately 1.5 m in length. It begins on the right side of the abdominopelvic cavity, in the right lower quadrant, at the ileocecal junction and ends at the anus (**Fig. 4.5.1**). The large intestine has three divisions (proximal to distal):

1. **Cecum** and **vermiform appendix**
2. **Colon** (ascending, transverse, descending, sigmoid)
3. **Rectum** and **anal canal**

Cecum and Appendix

The cecum, the most proximal part of the large intestine, is a blind pouch. It extends inferiorly from the ileocecal junction and is continuous superiorly with the ascending colon. The appendix, a diverticulum of the cecum, is slender and variable in length (2–20 cm). It is most frequently retrocecal. The cecum and appendix each have a mesentery.

Colon

The colon has four divisions:

1. The retroperitoneal **ascending colon** begins at its junction with the cecum and extends superiorly to the right colic (hepatic) flexure. This flexure lies just inferior to the right lobe of the liver, deep to ribs 9 and 10. At this flexure, the ascending colon turns sharply to the left and continues as the transverse colon.

2. The **transverse colon** is the longest portion of the large intestine. It is suspended by a mesentery and may loop inferiorly into the pelvic cavity. It extends from the right colic flexure to the left colic (splenic) flexure. The left colic flexure lies more superiorly (deep to ribs 8 and 9) and is more acute. At the left colic flexure, the transverse colon turns inferiorly to continue as the descending colon.

3. The **descending colon** is retroperitoneal. It extends inferiorly to the left iliac fossa, where it is continuous with the sigmoid colon.

4. The **sigmoid colon** is S-shaped, ends at the rectum (anterior to the S3 vertebra), and has a mesentery.

Characteristic features of the large intestine include:

- **Teniae coli** are longitudinal bands of smooth muscle. The three teniae expand on the surfaces of the appendix and rectum to form a complete muscular layer.

- **Haustra** are recurring sacculations (pouches) along colon.

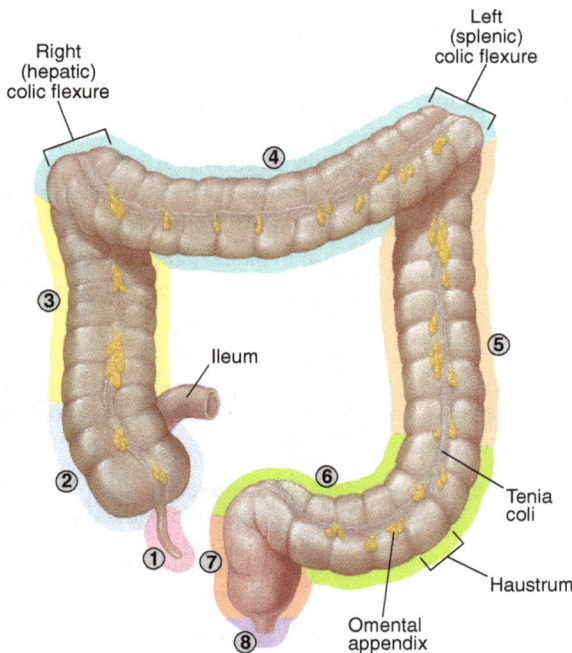

Right (hepatic) colic flexure

Left (splenic) colic flexure

Ileum

Tenia coli

Haustrum

Omental appendix

Regions of the large intestine

1. Appendix
2. Cecum
3. Ascending colon
4. Transverse colon
5. Descending colon
6. Sigmoid colon
7. Rectum
8. Anal canal

FIGURE 4.5.1 Anatomy of the large intestine.

- **Omental appendices** are projections of fat, attached to teniae coli.

Rectum and Anal Canal

The **rectum** is approximately 12 cm long. It begins in the pelvic cavity opposite the S3 vertebral body and is continuous with the anal canal at the pelvic floor, anterior to the coccyx. The rectum lacks haustra, teniae coli, and omental appendages.

The **anal canal** is the distal 4 cm of the large intestine. It begins superiorly where the rectum passes through the pelvic floor and opens inferiorly as the **anus**.

The large intestine develops from three regions of the embryonic alimentary canal:

1. **Right colon** (cecum to mid-transverse colon) is derived from embryonic midgut.

2. **Left colon** (mid-transverse colon through upper anal canal) is derived from embryonic hindgut.

3. **Lower anal canal** is derived from proctodeum.

Blood and Nerve Supply of the Large Intestine

The vasculature and innervation of the large intestine (**Table 4.5.1**) are determined by the embryonic origin of each segment.

Blood vessels of the large intestine are shown in **Figure 4.5.2**. Blood is supplied to most of the large intestine by branches and tributaries of the superior and inferior mesenteric arteries and veins respectively. Right, middle, and left colic arteries supply the large intestine through anastomotic arcades that derive from the marginal (juxtacolic) artery (Drummond). This artery is located along the mesenteric aspect of the large intestine. Superior and inferior mesenteric veins join the splenic vein to form the hepatic portal vein. Branches and tributaries of the internal pudendal vessels (i.e., inferior rectal) supply the lower anal canal.

CLINICAL REASONING

This patient presents with obstructive symptoms and clinical signs of blood in the stool. This is consistent with pathology that narrows the lumen of the distal large intestine and bleeding from the bowel wall (**Fig. 4.5.3**).

Colorectal Carcinoma

The vast majority (98%) of cancer of the large intestine develops from adenomatous polyps (also known as adenomas). An adenoma is a small, benign tumor of the intestinal glands (crypts) that may protrude into the intestinal lumen. They are more common in the distal colon and rectum than in proximal large bowel. These colonic polyps have malignant potential; particularly those that are flat and have invaded the submucosa of the intestinal wall.

TABLE 4.5.1 | **Vasculature and nerve supply of the large intestine.**

Large intestine	Blood supply	Lymph nodes	Innervation
Embryonic midgut Cecum Appendix Ascending colon Transverse colon (proximal)	Superior mesenteric vessels	▪ Superior mesenteric	**Sympathetic** ▪ Superior mesenteric plexus **Parasympathetic** ▪ Vagus nerve (CN X)
Embryonic hindgut Transverse colon (distal) Descending colon Sigmoid colon Rectum Anal canal (upper)	Inferior mesenteric vessels	▪ Superior mesenteric ▪ Inferior mesenteric ▪ Internal iliac	**Sympathetic** ▪ Inferior mesenteric plexus **Parasympathetic** ▪ Pelvic splanchnic nerves
Proctodeum Anal canal (lower)	Inferior rectal vessels	▪ Superficial inguinal	**Somatic**[a] ▪ Inferior rectal nerve

[a]Somatic nerves to the lower anal canal also contain sympathetic fibers.

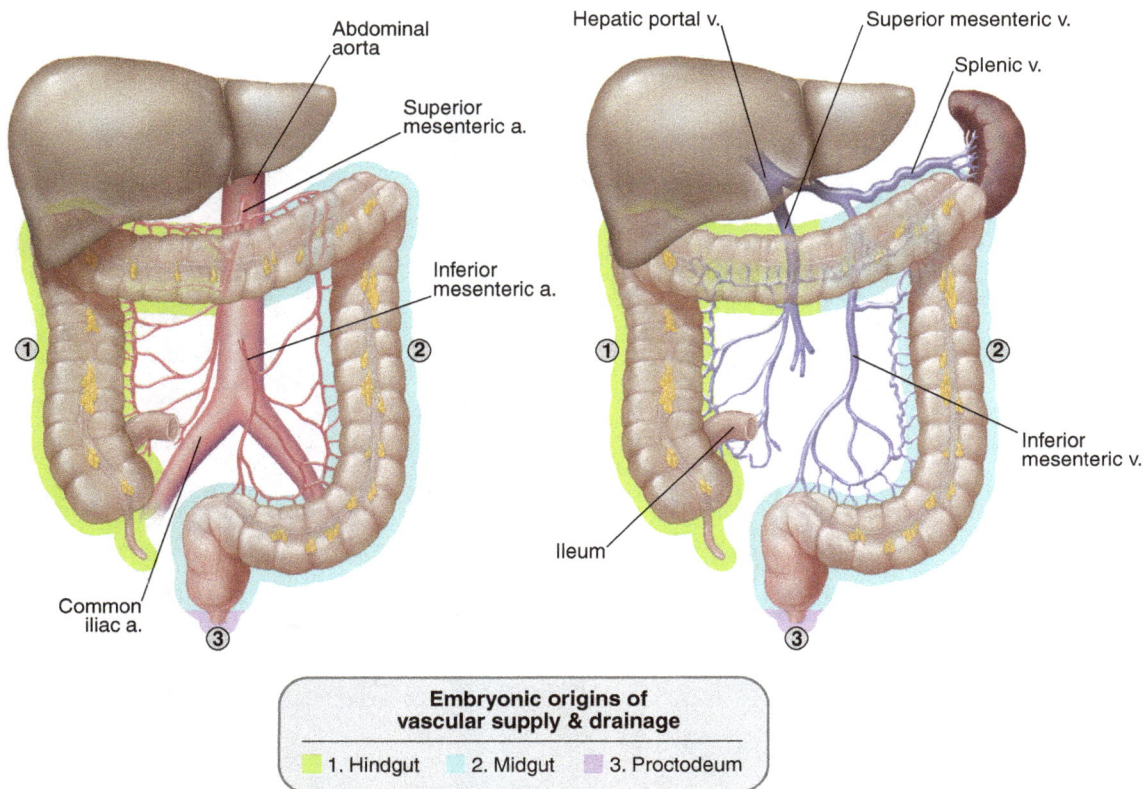

FIGURE 4.5.2 Blood supply of the large intestine.

Clinical Note

Colon cancer is second leading cause of cancer death—after lung cancer—in the United States. The majority (>90%) of colon cancers occur after age 50.

Signs and Symptoms

Clinical signs and symptoms vary depending on tumor location.

Right colon (cecum through proximal transverse colon)

- Unexpected or unexplained weight loss
- Persistent right abdominal discomfort

- Dyspepsia
- Occult fecal blood
- Fatigue and/or anemia
- Palpable abdominal mass (if tumor large enough)

Stool in the proximal colon is liquid and, therefore, a lesion does not impede its passage. As a result, these tumors frequently enlarge without symptoms of bowel obstruction or changes in bowel habits. Lesions of the cecum and ascending colon may ulcerate and cause chronic, intermittent bleeding. While this bleeding may not be detected by fecal occult blood tests, patients may present with symptoms of fatigue and anemia.

Dyspepsia Indigestion
Occult Hidden

Anemia Reduced erythrocytes, hemoglobin, or blood volume

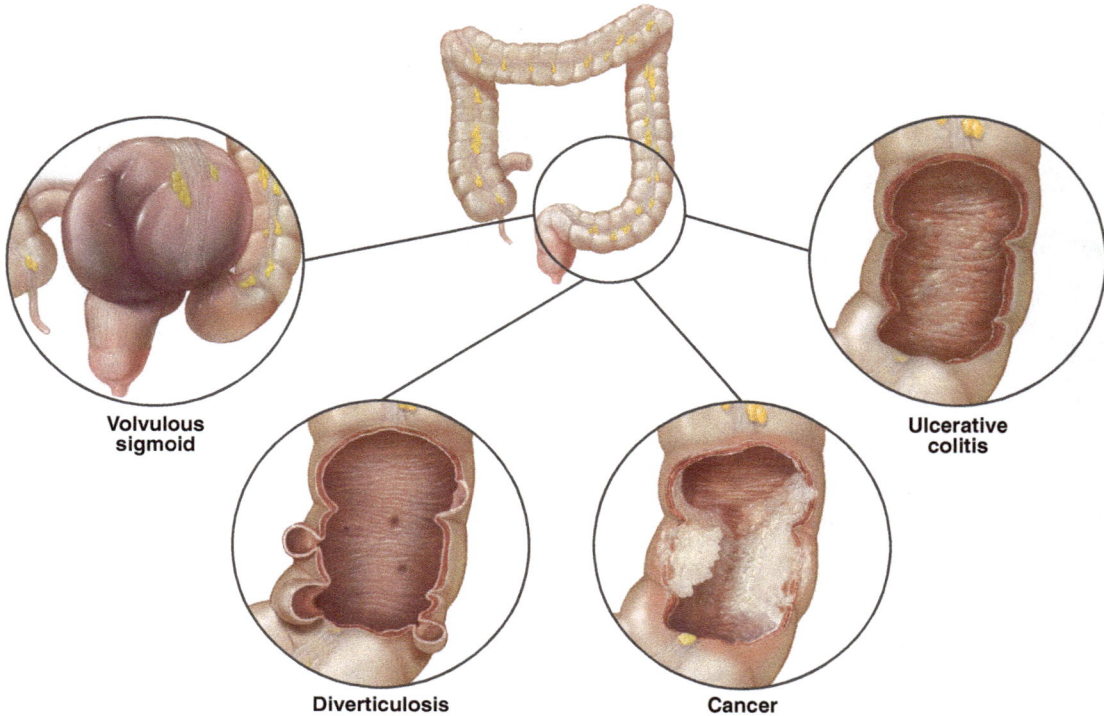

FIGURE 4.5.3 Pathologies of the large intestine.

Left colon (distal transverse through sigmoid colon)

- Change in bowel habits
- Obstructive symptoms
- Bleeding common but rarely massive
- Stool may have gross blood and mucus

Stool becomes more solid in this part of the colon and tumors may impede its passage. This results in **obstructive symptoms** (e.g., abdominal cramping, straining, and constipation). Radiographic imaging, colonoscopy, or sigmoidoscopy is used to detect and diagnose annular constricting ("apple-core" or "napkin-ring") lesions (**Fig. 4.5.4**).

Rectum

- Change in bowel habits
- Hematochezia
- Rectal tenesmus
- Tumor palpable on DRE

Anemia is not common and these symptoms are also consistent with hemorrhoids.

Predisposing Factors

- Age: >90% occur in individuals >50 years of age
- Sex: higher incidence in men
- Race: African Americans have the highest incidence, morbidity, and mortality
- Diet: high fat, low fiber, red meat
- Smoking

Tenesmus Straining and painful sphincter spasm related to the sensation of incomplete evacuation of either bowel or bladder

Palpitation Forcible or irregular pulsation of the heart that is perceptible to the patient

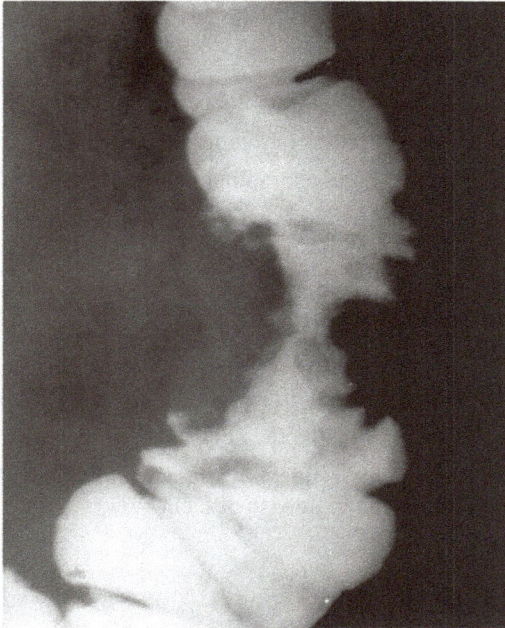

FIGURE 4.5.4 Annular, constricting adenocarcinoma of the descending colon. This radiographic appearance is referred to as an "apple-core" lesion and is always highly suggestive of malignancy. *Source*: Fig. 91-2 in Longo et al. *Harrison's Principles of Internal Medicine*, 18 e. www.accessmedicine.com.

FIT and **FOBT** detect blood in stool that is not visible with the naked eye. FIT is the more sensitive test because it uses antigen–antibody reactions. Fecal blood may be the only symptom of colorectal cancer, although it is also associated with other gastrointestinal conditions, for example, hemorrhoids, anal fissures, ulcers, or Crohn disease.

With early detection, and removal of adenomatous polyps that have not spread through the mucosa, the current 5-year survival rate is >90%. If the tumor has invaded the muscularis, the 5-year survival is 85%, and if regional lymph nodes are involved, it is 35–65%. Metastasis through nodes (lymphatogenous) is the most common pathway; however, tumor cells may also spread though blood vasculature (hematogenous), or directly to, and along, contiguous structures.

- The American Cancer Society (ACS) recommends that at-risk individuals begin screening at 50 years of age; for African Americans at age 45.

- Colorectal cancer has higher incidence in economically prosperous populations, perhaps related to environmental or dietary factors.

- Excessive alcohol use (according to the Center for Disease Control and Prevention, consuming an average of more than two drinks per day)

- Familial history, especially familial adenomatous polyposis

Clinical Notes

Colon cancer commonly does not produce clinical signs or symptoms until the disease is advanced. It is often detected through routine screening, including:

- Fecal immunochemical test (FIT) or fecal occult blood test (FOBT)

- DRE

- Sigmoidoscopy or colonoscopy (the latter has higher efficacy because the scope is longer and, therefore, examines more of the colon)

Diverticulosis

Diverticulosis is an abnormal evagination (outpouching) of the colon, most commonly (95%) in the sigmoid colon. These diverticula usually consist of mucosa and submucosa that herniate through the muscularis layers.

Signs and Symptoms

Diverticulosis is usually asymptomatic (80%). Complications of diverticulosis are diverticulitis (infection, inflammation, rupture) and bleeding. Symptoms of diverticulitis include:

- Acute, episodic abdominal pain

- Mild left lower quadrant tenderness

Hematogenous Spread via vasculature
Lymphatogenous Spread via lymphatic vasculature

- Changes in bowel habits (constipation, diarrhea)
- Leukocytosis

Predisposing Factors

- Age: present in 10% of Americans >40 years; incidence increases with age
- Low-fiber diet
- Constipation
- Obesity

Sigmoid Volvulus

A volvulus is a twisting of part of the intestine (most frequently in the sigmoid colon) either around itself or of its mesentery.

Signs and Symptoms

- Colicky abdominal pain
- Abdominal distention
- Obstipation
- Nausea and vomiting (late symptoms)

Predisposing Factors

- Age: 60% are neonates <2 months of age; 40% are elderly >70 years old
- Elongated sigmoid colon and rectum

Clinical Note

If blood flow to the volvulus is obstructed (strangulation), the tissue becomes gangrenous and necrotic. Because gangrene can develop rapidly and lead to bowel perforation, a volvulus can become a surgical emergency.

Ulcerative Colitis

Ulcerative colitis is a severe inflammatory disease that most commonly affects the sigmoid colon and rectum.

Gangrene Necrosis due to loss of blood supply
Obstipation Intestinal obstruction; severe
  constipation

Signs and Symptoms

- Rectal bleeding
- Diarrhea containing blood, pus, and mucus
- Rectal tenesmus (straining) and urgency
- Cramping and abdominal pain
- Systemic symptoms (fever, vomiting, weight loss, dehydration)

Predisposing Factors

- Age: onset between 15 and 30 years, and then between 50 and 70
- Sex: women affected slightly more than men
- Race: more frequent in whites and individuals of Ashkenazic Jewish descent

DIAGNOSIS

The patient presentation, medical history, physical examination, laboratory tests, and procedures support a diagnosis of **adenocarcinoma of the sigmoid colon**.

Adenocarcinoma of the Sigmoid Colon

Most colorectal carcinomas begin in glandular tissue. They usually begin as small, benign adenomatous polyps that may develop into cancer.

- This patient is at risk for colon cancer because of his race, age, sex, obesity, and tobacco use.
- Right colon tumors are generally not associated with changes in bowel habits because they do not cause obstruction, and blood in stool is typically occult.
- Changes in bowel habits, obstructive symptoms, and gross blood in the stool are consistent with a tumor in the left colon.
- Tumors in the rectum that are associated with changes in bowel habits are typically palpable on DRE.
- Diagnosis is confirmed by the finding of gross tumor obstruction of the sigmoid colon and histopathology after colonoscopy.

Diverticulosis

Diverticulosis, characterized by small mucosal herniations of the gastrointestinal tract, most commonly affects the sigmoid colon (*inflammation of these diverticula is known as **diverticulitis***).

- As with some cases of colon cancer, Diverticulosis also results in changes in bowel habits.

Sigmoid Volvulus

An intestinal volvulus is formed by twisting of a portion of the gastrointestinal tract, including its mesentery. Consequently, this may lead to obstruction and vascular necrosis. Volvulus most commonly occurs in the sigmoid colon and cecum.

- Volvulus is accompanied by colicky abdominal pain, distension, and obstipation. In addition, volvulus is found most frequently in the neonates and the elderly.

Ulcerative Colitis

Ulcerative colitis is an IBD that commonly affects the colon, particularly the rectum.

- Ulcerative colitis causes diarrhea and systemic symptoms.

Obstipation Intestinal obstruction; severe constipation

REVIEW QUESTIONS

1. During a pelvic examination, an Ob/Gyn resident distinguishes a pulse adjacent to the lateral vaginal fornix. This is the pulse in the:
   A. Internal iliac artery
   B. Internal pudendal artery
   C. Ovarian artery
   D. Uterine artery
   E. Vaginal artery

2. During digital rectal examination of a male, an urologist evaluates structures on the postero-inferior surface of the urinary bladder. Which paired structures are closest to the midline?
   A. Ampullae of the ducta deferens (vasa deferentia)
   B. Superior vesicle arteries
   C. Seminal glands
   D. Superior rectal (anal) arteries
   E. Ureters

3. Imaging reveals a collection of fluid in the pelvic portion of the peritoneal cavity of a 39-year-old female. To sample the fluid, a sterile needle is introduced into the peritoneal cavity by piercing the wall of the posterior vaginal fornix. Properly placed, the tip of the needle should be in which part of the peritoneal cavity?
   A. Left paracolic gutter
   B. Lesser sac (omental bursa)
   C. Rectouterine pouch (cul-de-sac)
   D. Rectovesical pouch
   E. Vesico-uterine pouch

4. During a follow-up appointment 30 days after transurethral resection of his prostate, a 63-year-old male relates that he is now impotent. The perivascular nerve plexus that lies just outside the capsule of the prostate gland contains fibers responsible for penile erection. These nerve fibers would have their preganglionic parasympathetic cell bodies in the:
   A. Penis
   B. Sacral dorsal root ganglia
   C. Sacral plexus
   D. Sacral spinal cord
   E. Sympathetic trunk

5. A 44-year-old female undergoes a total hysterectomy in which the uterus, including the cervix, is removed. During the procedure, the surgeon identifies and transects all structures associated with the uterus. Which structure is not transected?
   A. Mesometrium
   B. Mesovarium
   C. Round ligament of the uterus
   D. Transverse cervical (cardinal) ligament
   E. Uterine artery

6. During an annual physical examination of 45-year-old female, the fundus of her uterus can only be

palpated by digital rectal examination. This situation is *most likely* to occur when the:

A. Uterine cervix is anteverted and the body is anteflexed
B. Uterine cervix is anteverted and the body is retroflexed
C. Uterine cervix is retroverted and the body is anteflexed
D. Uterine cervix is retroverted and the body is retroflexed
E. Uterus is displaced by a partially full urinary bladder

7. In 1948, gynecologist Arnold Kegel proposed a series of exercises to strengthen the pelvic floor muscles of women with urinary incontinence after childbirth. Which muscle would *not* show an increase in tone when these Kegel exercises are properly done?

A. External urethral sphincter
B. Iliococcygeus
C. Obturator internus
D. Pubococcygeus
E. Puborectalis

8. Two weeks ago, a 43-year-old woman had an elective hysterectomy. She is in the clinic today complaining of increased vaginal discharge. Physical examination establishes uncontrolled leakage of urine into the vagina, and radiologic imaging with contrast confirms a vesicovaginal fistula. During the surgery, the superior vesical artery was inadvertently ligated, which led to necrosis of the bladder wall and formation of the fistula. What artery usually gives rise to the superior vesical artery?

A. Internal pudendal
B. Obturator
C. Superior rectal
D. Umbilical
E. Uterine

9. A 24-year-old female patient indicates that she and her partner are planning a pregnancy. Pelvic evaluation indicates her diagonal conjugate dimension is 12 cm. This represents the distance between the:

A. Inferior aspect of the pubic symphysis and the ischial spine
B. Inferior aspect of the pubic symphysis and the sacral promontory
C. Ischial spines
D. Ischial tuberosities
E. Superior aspect of the pubic symphysis and the sacral promontory

10. During his first colonoscopy, a 58-year-old male has three polyps removed from the sigmoid colon. Pathologic evaluation reveals that two of the polyps contain malignant cells. Which group of lymph nodes should be evaluated in determining treatment?

A. External iliac
B. Femoral
C. Left deep inguinal
D. Left superficial inguinal
E. Lumbar

CHAPTER 5

Perineum

Ruptured Urethra

Patient Presentation

A 12-year-old boy is admitted to the emergency department with complaints of a painful, enlarged scrotum and mild distention of the lower abdomen.

Relevant Clinical Findings

History

The patient relates that 30 hours earlier, while racing with friends, his foot slipped from the pedal of his bicycle and he came down hard, straddling the bicycle frame. He has noticed the following, subsequent to the accident:

- Weak urine stream
- Mild penile pain during urination
- Hematuria

Physical Examination

Results of physical examination of the external genitalia, anterior abdominal wall, and anal region:

- Contusions to the scrotum
- Subcutaneous fluid infiltration of the scrotum and shaft of the penis
- Subcutaneous fluid infiltration of the inferior anterior abdominal wall
- Normal anal canal and peri-anal region

Laboratory Tests

- Urinalysis reveals erythrocytes and leucocytes.

Clinical Problems to Consider

- Inguinal hernia
- Rupture of spongy urethra with extravasation of urine
- Sexual abuse

LEARNING OBJECTIVES

1. Describe the fascial layers of the anterior abdominal wall and male urogenital (UG) region.
2. Describe the subdivisions of the male urethra.
3. Explain the contents of the superficial pouch of the male UG region.
4. Explain the anatomical basis for the signs and symptoms associated with this case.

RELEVANT ANATOMY

Male Perineum

The **male perineum** is divided into two regions (**Fig. 5.1.1**):

1. **Anal triangle** (posterior)
2. **Urogenital** (UG) **triangle** (anterior)

Hematuria Blood in urine
Contusion Mechanical injury beneath skin that results in subcutaneous hemorrhage (i.e., a bruise)
Extravasation Escape of body fluid into surrounding tissues

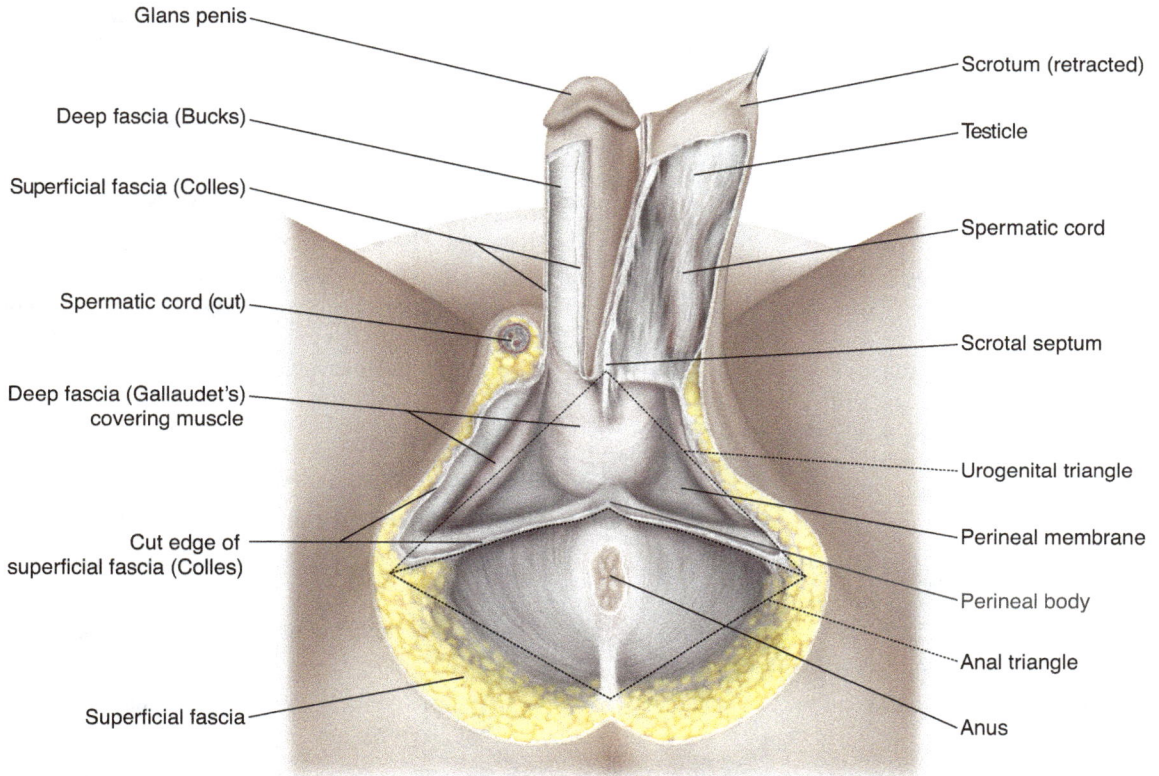

FIGURE 5.1.1 Inferior view of the male perineum showing the urogenital and anal triangles. The superficial pouch is opened to reveal its contents.

The plane separating these regions passes through the perineal body at the midline and the anterior aspects of the ischial tuberosities laterally. The anus is the prominent feature for the anal triangle, while the external genitalia dominate the UG triangle. The body (shaft) of the penis and the scrotum are considered part of the UG triangle.

Fascia of the Male Urogenital Triangle

Superficial Fascia

Two distinct layers of the **superficial (subcutaneous) fascia** can be distinguished on the anterior abdominal wall, and are continuous into the UG triangle (**Table 5.1.1**). The terminology for these layers is different in each region.

There is a noteworthy difference between superficial layers on the anterior abdominal wall and the UG triangle:

- On the **anterior abdominal wall**, the superficial layer (Camper) is composed primarily of fat.
- On the scrotum, the superficial layer (dartos) contains numerous smooth muscle fascicles that insert on the dermis.

TABLE 5.1.1	**Layers of superficial fascia.**	
Layer	**Anterior abdominal wall**	**Male UG triangle**
Superficial	Fatty layer (Camper)	Muscular layer (dartos)
Deep	Membranous layer (Scarpa)	Membranous layer (Colles)

- On the shaft of the penis, the superficial layer is mostly loose connective with little or no smooth muscle. This layer is also frequently called the dartos layer.

Contraction of the dartos muscle, which inserts into the dermis, changes the scrotal dimensions and, thereby, draws the testis toward the body wall. This affects the temperature of the testis and contributes to the regulation of spermatogenesis.

Deep Fascia

The deep fascia (investing or muscular) forms a continuous, membranous layer of variable thickness on the anterior abdominal wall and in the UG triangle. Different terms are applied to deep fascia in the UG triangle (**Table 5.1.2**).

Perineal Pouches in the Male

The male UG triangle is divided into two potential pouches or spaces (**Fig. 5.1.1**):

1. The **superficial perineal pouch** is located between Colles fascia and the perineal membrane.
2. The **deep perineal space** is superior to the perineal membrane.

Thus, the perineal membrane separates the contents of the superficial and deep perineal spaces.
At the posterior edge of the UG diaphragm, Colles fascia fuses with the perineal membrane. This creates a barrier that separates the contents of the superficial perineal pouch from those of the anal triangle. Important structures within this pouch include:

- Erectile bodies and associated muscles
- Penis and spongy (penile) urethra
- Scrotal contents
- Perineal nerves and blood vessels

The erectile bodies include the midline, **bulb of the penis** (and its extension into the shaft of the penis as the **corpus spongiosum**), and right and left **crura of the penis** (which extend into the shaft of the penis as paired **corpora cavernosa**).

Male Urethra

The **male urethra** is divided into four parts:

1. **Preprostatic:** within neck of urinary bladder
2. **Prostatic:** passes through the core of prostate gland
3. **Membranous** (intermediate): passes through pelvic floor
4. **Spongy** (penile): within the bulb and corpus spongiosum

The spongy urethra is the longest part and lies entirely within the superficial perineal space. It ends at the tip of the **glans penis** as the **external urethral orifice**.

CLINICAL REASONING

This patient presents with subcutaneous fluid accumulation in the external genitalia (scrotum and penis) and inferior anterior abdominal wall. This is consistent with compromise of the urinary tract.

Indirect Inguinal Hernia

An **indirect inguinal hernia** usually involves a portion of small bowel and associated peritoneum. This hernia extends through the **deep and/or superficial inguinal rings**, areas of structural weakness of the anterior abdominal wall.
The underlying cause for this type of hernia is a failure of the **processus vaginalis** to close at the deep inguinal ring during the fetal period. Therefore, an indirect inguinal hernia is considered **congenital**, even though

TABLE 5.1.2	Deep fascia of the urogenital triangle.
Deep fascia of UG triangle	**Anatomical relationships**
Perineal membrane	Forms inferior fascia of UG diaphragm
Investing fascia (Gallaudet)	Envelops bulbospongiosus, ischiocavernosus, superficial transverse perineal muscles
Deep fascia of penis (Buck)	Envelops shaft of penis

it may not develop until the second or third decade. Intra-abdominal pressure will eventually cause a loop of small bowel to enter the patent processus vaginalis, forcing the hernial sac into the inguinal canal.

An indirect inguinal hernia enters the deep inguinal ring, lateral to the inferior epigastric vessels, and passes along the inguinal canal (see Fig. 3.1.4B). It may protrude through the superficial inguinal ring and extend into the scrotum.

Signs and Symptoms

- Subcutaneous bulge at superficial ring of inguinal canal
- Asymmetric enlargement of scrotum
- Bulge may be more conspicuous with straining
- A range of pain and tenderness: none; dull, aching, regional pain; acute localized pain

Predisposing Factors

- Sex: male (9:1)
- Age: <25 years
- Family history
- Past history of inguinal hernia
- Chronic cough
- Smoking
- Excess weight

Clinical Notes

- An inguinal hernia in a male under age 25 is nearly always indirect.
- Bowel incarcerated in the inguinal canal may undergo strangulation and necrosis, which is a medical emergency.
- Differentiation between indirect and direct inguinal hernias by physical examination is highly inaccurate.

Rupture of Spongy Urethra

The spongy (penile) urethra enters the superior (deep) surface of the bulb of the penis. The urethra then curves to follow the long axis of the bulb of the penis and corpus spongiosum. The midline bulb of the penis and the urethra within are vulnerable to blunt trauma to the UG triangle. Most commonly, the trauma results from falls that end with the victim straddling an object such as a tree limb, fence rail, or bicycle frame ("straddle" injury). The trauma may lacerate one or more of the following:

- Bulbospongiosus muscle and its investing fascia (Gallaudet)
- Bulb of the penis
- Spongy urethra in the bulb of penis

If trauma lacerates all of the above, urine will extravasate from the spongy urethra during urination and occupy the potential space between the deep fascia (Gallaudet, perineal membrane, Buck) and superficial (Colles) fascia (**Fig. 5.1.2**). These fascial layers are continuous into the scrotal wall and onto the anterior abdominal wall. Thus, extravasated urine, limited by the fascial planes, can track into the scrotum and onto the anterior abdominal wall. On the anterior abdominal wall, the urine will occupy the plane between the deep (investing) fascia of external abdominal oblique muscle and the membranous layer of superficial fascia (Scarpa).

Signs and Symptoms

- Progressive expansion of subcutaneous "puffiness" in the UG triangle and anterior abdominal wall
- Weakened urine stream
- Contusions to scrotum
- Hematuria
- Dysuria

Incarcerated Trapped
Strangulated Constricted or twisted to prevent air or blood flow

Necrosis Pathologic death of cells, tissues, or organs
Lacerated Torn
Dysuria Pain during urination

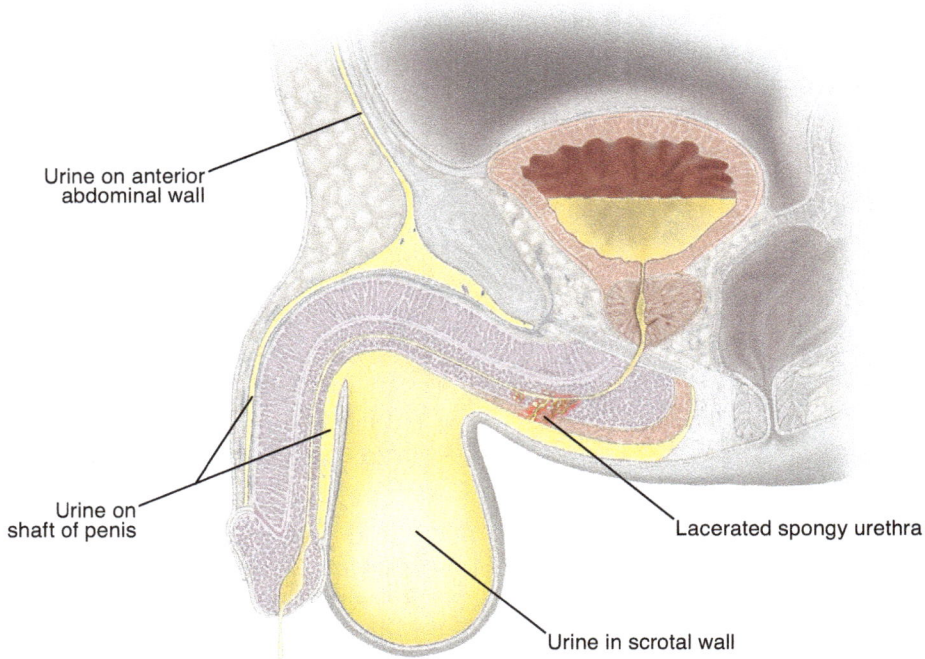

Urine on anterior abdominal wall

Urine on shaft of penis

Lacerated spongy urethra

Urine in scrotal wall

FIGURE 5.1.2 Lacerated urethra with urine leakage into superficial pouch.

Predisposing Factor

- Sex: male

Clinical Notes

- Risk of infection from extravasated urine is low since urine is sterile

- A straddling, traumatic injury in a female will not result in extravasation of urine between the fascial planes because the female urethra does not course with an erectile body, but empties directly into the vestibule of the UG triangle.

- The membranous layer of superficial fascia of the anterior abdominal wall fuses with the deep fascia of the thigh (fascia lata) approximately 2 cm inferior to the inguinal ligament. This obliterates the potential space between the two fascial layers and prevents extravasated urine from entering the thigh.

Sexual Abuse

In general, sexual abuse should be considered in any child that presents with bruising or bleeding in the anorectal area, proximal thigh, or external genitalia. These patients should be evaluated by a credentialed health care provider.

Signs and Symptoms

- Bruising/bleeding to perineum
- Lubricant residue
- Socially withdrawn child

Predisposing Factors

- Social and/or geographic isolation of family
- Substance abuse or addiction in household

Clinical Note

The list of signs and symptoms for sexual abuse is long and discovery may require long-term intervention by professionals.

DIAGNOSIS

The patient presentation, medical history, physical examination, and laboratory tests support a diagnosis of a **traumatic rupture of spongy urethra**.

Traumatic Rupture of Spongy Urethra

The urethra in the male enters the bulb of the penis upon passing through the perineal membrane of the UG diaphragm. This spongy portion of the urethra, with the other components of the external genitalia, lies in the superficial perineal pouch of the UG triangle of the perineum. Blunt trauma to the bulb of the penis may rupture the spongy urethra, allowing urine to escape into this pouch. Urine extravasation will be limited by the superficial and deep fasciae that define the superficial pouch.

- The patient is at risk because he is male and because of the blunt force trauma during a straddle accident on a bicycle.

- The trauma has produced a tear in spongy urethra that extends through the erectile tissue of the bulb of the penis, the bulbospongiosus muscle, and the deep fascia (Gallaudet) covering bulbospongiosus muscle.

- The subcutaneous distension is distributed to the scrotum, shaft of the penis, and anterior abdominal wall. This is consistent with extravasation of urine into the interfascial spaces due to traumatic rupture of spongy urethra.

Indirect Inguinal Hernia

Indirect inguinal hernias result from failure of the processus vaginalis to close during the fetal period. This peritoneal diverticulum extends from the deep inguinal ring, along the inguinal canal, to the superficial inguinal ring. From the superficial ring, the processus extends into the developing scrotum/labium majus. A patent processus vaginalis is nine times more likely in males, although the actual hernia may not develop until adolescence or young adulthood. Typically, a loop of small intestine enters the patency at the deep inguinal ring and extends for a variable distance along the inguinal canal. Potentially, the hernia can extend into the scrotum.

- The scrotal distension associated with **indirect inguinal hernia** would be asymmetric.

- This hernia can often be reduced (at least temporarily) by digital manipulation at the superficial inguinal ring.

- Contusions to the scrotum would not be expected with an indirect inguinal hernia.

Sexual Abuse

Health care providers must always be alert to the possibility of abuse whenever a child presents with perineal trauma. In this patient, the examiner did not note any physical or behavioral signs suggesting abuse.

Pudendal Neuralgia

Patient Presentation

A 27-year-old white male presents in clinic complaining of constant genital pain over the past 5 months. He also indicates urinary and sexual dysfunction.

Relevant Clinical Findings

History

The patient reports he rides a stationary bike and lifts weights 5 days a week. This is a continuation of the training program he followed in high school and college, where he was an all-American wrestler and football player.

He describes progressive symptoms that include burning, numbness, and stabbing pains in the penis, scrotum, and peri-anal region. The pain is aggravated by sitting, relieved by standing, and absent when lying down.

Physical Examination

- Results of digital rectal examination (DRE) were normal.
- External genitalia were normal.

Laboratory Tests

Test	Value	Reference value
Urinalysis (WBC)	0	0
Prostate-specific antigen (PSA)	2.1	0.0–4.0 ng/mL
Prostate secretion analysis (WBC)	0	0–1/high power field

Clinical Notes

- **PSA** is a protein produced by cells of the prostate gland. PSA levels can be elevated with prostatitis, benign prostate hyperplasia, or prostate cancer. Serum PSA levels, however, cannot distinguish diagnostically between hyperplasia and cancer.
- Prostate secretion analysis microscopically examines prostatic fluid for signs of infection or inflammation. Samples are cultured to analyze for bacteria.

Clinical Problems to Consider

- Chronic prostatitis
- Interstitial cystitis (IC)
- Pudendal neuralgia (PN)

LEARNING OBJECTIVES

1. Describe the anatomy of the pudendal nerve.
2. Describe the anatomy of the urinary bladder.
3. Describe the anatomy of the prostate gland.
4. Explain the anatomical basis for the signs and symptoms associated with this case.

RELEVANT ANATOMY

Pudendal Nerve

The **pudendal nerve** is a branch of the sacral plexus. It arises within the pelvis from the ventral rami of the second, third, and fourth sacral spinal nerves (**Fig. 5.2.1**).

Hyperplasia Increase in size of a tissue or organ due to increased cell numbers (antonym: hypertrophy)

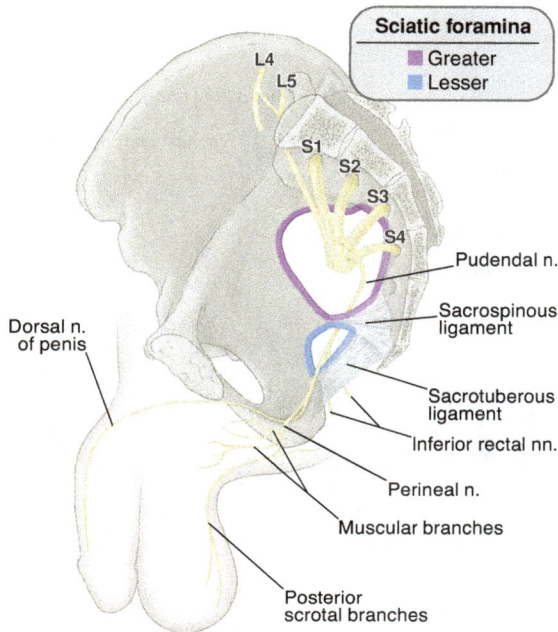

FIGURE 5.2.1 Medial view of the pudendal nerve, its course, and its relationship to the sacrotuberous and sacrospinous ligaments.

Course of the Pudendal Nerve

The pudendal nerve leaves the pelvis by passing between the piriformis and coccygeus muscles to enter the gluteal region at the inferior part of the **greater sciatic foramen** (**Figs. 5.2.1 and 5.2.2**). It crosses the posterior surface of the **ischial spine** and enters the **lesser sciatic foramen**.

Sciatic Foramina

The boundaries of the **greater sciatic foramen** are:

- **Greater sciatic notch** of the ilium
- **Sacrotuberous** ligament
- **Sacrospinous** ligament

The boundaries of the **lesser sciatic foramen** are (**Fig. 5.2.2**):

- **Lesser sciatic notch** of the ischium
- **Sacrotuberous** ligament
- **Sacrospinous** ligament

The lesser sciatic foramen directs the nerve into the perineum, along the lateral wall of the ischioanal fossa. Here, the nerve lies within the

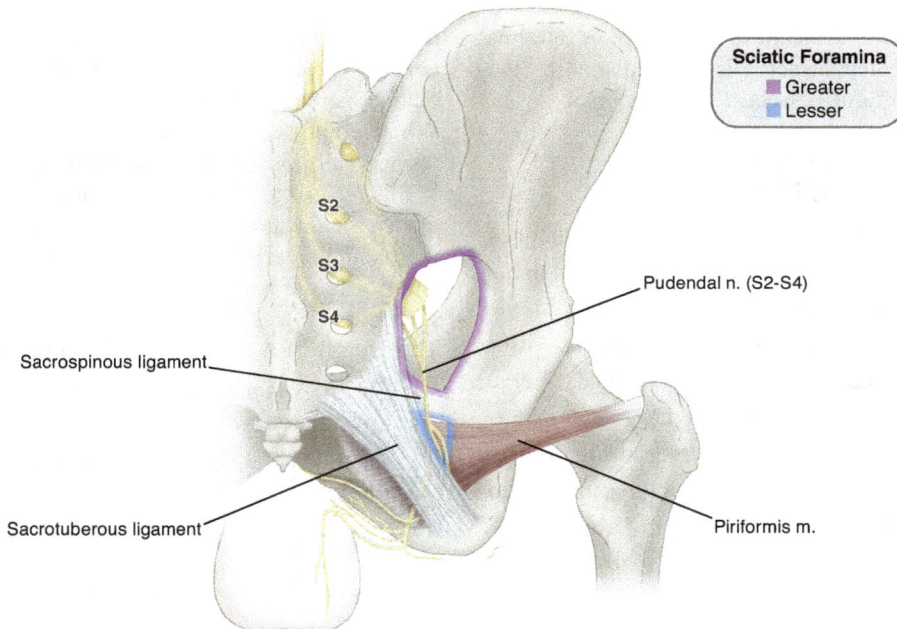

FIGURE 5.2.2 Posterior view of the pudendal nerve and its relationship to the sciatic foramina.

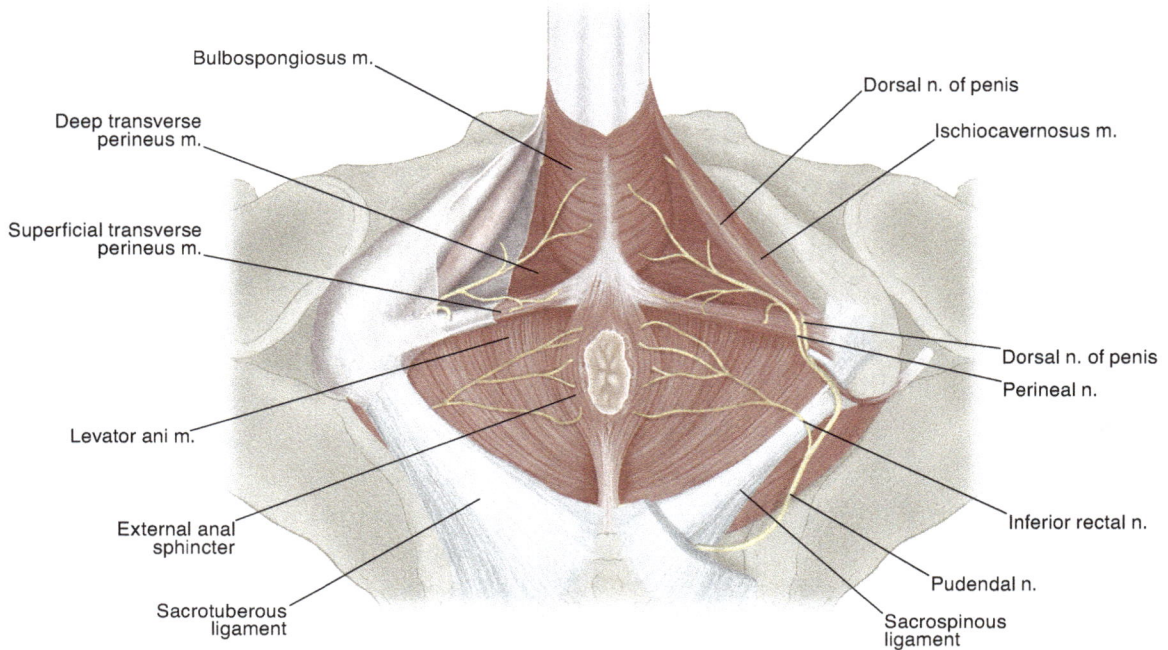

FIGURE 5.2.3 Inferior view of pudendal nerve course and relationship to sacrotuberous and sacrospinous ligaments.

pudendal canal, a sheath of the obturator internus fascia. The pudendal nerve is accompanied by the **internal pudendal vessels** along its course.

Branches and Distribution of the Pudendal Nerve

The pudendal nerve supplies most of the structures of the perineum (**Figs. 5.2.1** and **5.2.3**). A notable exception is the testis. The first branches of the pudendal nerve are the inferior rectal nerves—usually two or three in number. The pudendal nerve ends in the pudendal canal by giving rise to its terminal branches: the perineal nerve and the dorsal nerve of the penis/clitoris. The structures innervated by the pudendal nerve are outlined in **Table 5.2.1**.

Urinary Bladder

The urinary bladder is a distensible muscular reservoir. The smooth muscle of the bladder wall is the **detrusor muscle**. The superior surface of the bladder is covered by peritoneum. Therefore, the urinary bladder is retroperitoneal.

When empty, it is located in the pelvic cavity, immediately posterior to the pubic symphysis. As it fills, it expands superiorly into the abdominal cavity. In the female, it lies anteroinferior to the uterus and anterior to the vagina. In the male, it is anterior to the rectum and superior to the prostate gland.

The urinary bladder consists of four parts:

1. The **apex** is the pointed anterior end that is directed toward the pubic symphysis.

2. The **body** lies between the apex and the fundus.

3. The **fundus** (base) forms the posterior wall of the bladder. The ureters enter the bladder at the superior margin of the fundus.

4. The **neck** is the most inferior part and is continuous with the urethra. The neck is supported

TABLE 5.2.1	Sensory and motor distribution of the pudendal nerve.	
Branch	**Sensory distribution**	**Motor distribution**
Nerve to levator ani		▪ Levator ani
Inferior rectal	▪ Peri-anal skin	▪ External anal sphincter
Perineal Posterior scrotal/labial	▪ Skin of posterior scrotum/labium majus ▪ Skin of labium minus	
Muscular branches		▪ External urethral sphincter ▪ Superficial and deep transverse perineus muscles ▪ Bulbospongiosus muscle ▪ Ischiocavernosus muscle
Dorsal nerve of penis/clitoris	▪ Skin of penis/clitoris	

by the pubovesical ligament that attaches the neck to the pubic bones.

Blood Supply

The blood supply to the urinary bladder is derived from branches of the **internal iliac artery**. One or more **superior vesical arteries** branch from the umbilical artery and supply the anterior and superior parts of the bladder. In the male, the **inferior vesical artery** provides blood to the bladder fundus and neck. In the female, this area is supplied by branches of the **vaginal artery**. The obturator and inferior gluteal arteries may also supply blood to the bladder.

The veins that drain the bladder have a pattern similar to the arteries. In the male, these veins drain into the **prostatic venous plexus** and then to the **internal iliac vein**. Vesical veins in the female drain into the **vaginal** or **uterovaginal venous plexus** before entering the **internal iliac vein**.

Nerve Supply

Autonomic innervation to the urinary bladder includes sympathetic and parasympathetic fibers.

- **Sympathetic (visceral efferent).** Preganglionic sympathetic axons originate from cells in the T12-L2 lateral horns of the spinal cord. These fibers leave the spinal cord via **lumbar splanchnic nerves** and then travel through **aortic and superior hypogastric plexuses**. After synapsing in **ganglia** in the **superior hypogastric plexus**, postganglionic fibers enter the **hypogastric nerve** and then pass into the **inferior hypogastric plexus**. Fibers leave this plexus along peri-arterial plexuses to innervate the bladder.

- **Parasympathetic (visceral efferent).** Preganglionic parasympathetic axons arise from cells in the S2 to S4 lateral horns. They emerge as pelvic splanchnic nerves and contribute to the inferior hypogastric plexus. These fibers follow periarterial plexuses to terminal ganglia near the bladder. Postganglionic fibers enter the urinary and innervate the detrusor muscle.

- Visceral afferent impulses are carried in the **pelvic splanchnic nerves**. Receptors in the bladder that detect **stretch and distension or pain** have their **afferent fibers in pelvic splanchnic nerves**. Their cell bodies are located in S2-S4 dorsal root ganglia. Pain from the urinary bladder is referred to the S2-S4 dermatomes.

Prostate

The prostate is the largest accessory gland of the male reproductive system. In the adult, the prostate measures 4 cm (transverse), 3 cm (supero-inferior),

TABLE 5.2.2	Anatomical relationships of the prostate.	
Direction	**Related structure(s)**	**Note**
Anterior	• Pubic symphysis	Inferior part
Posterior	• Ampulla of rectum • Sacrum and coccyx	
Superior	• Urinary bladder	Palpable if distended
	• Seminal glands • Ampulla of ductus deferens	Palpable if enlarged
Inferior	• Deep perineal muscles	
Lateral	• Levator ani	

and 2 cm (anteroposterior). The healthy adult prostate weighs about 20 g, is symmetrical, and lacks palpable nodules. A median sulcus lies between the two lateral lobes. With aging, the prostate may enlarge: by age 40, it may reach the size of an apricot; by age 60, it may be the size of a lemon. The **prostatic part of the urethra** passes through the prostate; it receives the ducts of the prostate glands and ejaculatory ducts.

Anatomical Relationships

The anatomical relationships of the prostate are outlined in **Table 5.2.2** (see also Fig. 4.1.3). Because of its relationship to the rectum, the prostate can be palpated during a DRE.

Blood Supply

Arterial supply to the prostate is from branches of the **internal iliac artery**, including the **inferior vesical**, **internal pudendal**, and **middle rectal arteries**. **Venous drainage** is through the **prostatic venous plexus**, which drains into the **internal iliac vein**. Blood from this plexus also anastomoses with the **vesical venous plexus**, as well as the **internal vertebral venous plexus**.

Nerve Supply

The prostate receives autonomic innervation from the **prostatic (nerve) plexus**.

- **Sympathetic (visceral efferent).** Preganglionic sympathetic axons originate from

cells in the T12-L1 lateral horns of the spinal cord. These fibers leave the spinal cord via **lumbar splanchnic nerves** and then travel through **aortic and superior hypogastric** plexuses where they synapse. Postganglionic fibers join a hypogastric nerve and pass to the **inferior hypogastric plexus**. Fibers leave this plexus along periarterial plexuses to reach the prostate. Sympathetic fibers are vasomotor. They are also motor to the preprostatic (urethral) sphincter, which is important to prevent ejaculate from entering the bladder.

- **Preganglionic parasympathetic** axons arise from cells in the S2 to S4 lateral horns. They emerge as pelvic splanchnic nerves and contribute to the inferior hypogastric plexus. These fibers follow periarterial plexuses to terminal ganglia near the gland. Postganglionic parasympathetic fibers innervate smooth muscle of the prostate and its capsule.

CLINICAL REASONING

This patient presents with chronic perineal pain that is consistent with pathology of the structures of the inferior pelvis or perineum.

Chronic Prostatitis

Chronic prostatitis is an inflammation of the prostate that can affect men of any age. It is considered chronic if it lasts more than 3 months. It exists in bacterial and nonbacterial forms:

- **Chronic nonbacterial prostatitis** is the most common form, accounting for 90% of cases.

- **Chronic bacterial prostatitis** results from recurrent urinary tract infections (UTI) that have entered the prostate gland.

Clinical Note

According to the National Institutes of Health, chronic prostatitis is the number one reason men under the age of 50 visit a urologist.

Signs and Symptoms

- Severe pelvic pain
- Perineal pain
- Pain on ejaculation
- Hematuria
- Blood in semen
- Prostatism (frequency, nocturia, urgency, hesitancy, decreased stream, sensation of not being able to empty bladder)

Predisposing Factors

- History of recurrent UTI
- History of acute prostatitis
- Nerve disorder involving lower urinary tract
- Exposure to pathogens
- Sexual abuse

Interstitial Cystitis

Interstitial cystitis (IC) is a persistent, painful form of chronic cystitis related to inflammation and fibrosis of the bladder wall. IC has no clear etiology or pathophysiology. It is characterized most commonly by fissures in the bladder mucosa (detected by cystoscopy during bladder distension). Symptoms of IC resemble those of bacterial infection, but urinalysis is negative for organisms.

Signs and Symptoms

- Mild-to-severe pain in the bladder and pelvic area
- Daytime and nighttime urgency and increased frequency of urination
- Hematuria
- Dysuria without evidence of bacterial infection
- Symptoms worse during menstruation

Predisposing Factors

- Sex: female more common than male
- History of irritable bowel syndrome
- History of fibromyalgia

- History of lupus erythematosus or other autoimmune disorders

Pudendal Neuralgia

Pudendal neuralgia (PN), is a painful neuropathic syndrome caused by inflammation of the pudendal nerve. Irritation and trauma to the nerve are the two main causes for an inflammatory response, which usually results in scarring or thickening of the nerve. Once damaged, the pudendal nerve may "misreport" pain from the perineal region, which can lead to urination, defecation, and/or sexual problems similar to those seen in chronic prostatic syndromes.

Clinical Notes

- One cause of PN is pudendal nerve entrapment (PNE), which occurs when the nerve becomes compressed and/or stretched along its course. It is hypothesized that hypertrophy of the muscles of the pelvic floor causes remodeling of the ischial spine and significant narrowing of the lesser sciatic foramen. As a result, the pudendal nerve is compressed as it traverses between the sacrotuberous and sacrospinous ligaments.
- Remodeling of the ischial spine may also change the course of the nerve such that it is stretched over the sacrospinous ligament or ischial spine during repetitive hip flexion (e.g., during cycling) or hip flexion over extended periods of time (e.g., long-distance truck driving).

Signs and Symptoms

- Severe pelvic and/or perineal pain
- Constipation
- Dysuria
- Erectile dysfunction

Predisposing Factors

- Trauma secondary to childbirth
- Sacroiliac joint and/or pelvic floor dysfunction
- Competitive cycling or rowing

Nocturia Excessive urination at night
Cystitis Inflammation of the urinary bladder

Hematuria Blood in urine
Dysuria Pain during urination

- Weight training
- Prolonged sitting

DIAGNOSIS

The patient presentation, medical history, physical examination, and laboratory tests support a diagnosis of **PN**.

Pudendal Neuralgia

The pudendal nerve is at risk for irritation or entrapment as it passes through the lesser sciatic foramen. Here the ligaments are tightly attached to the adjacent bones and the nerve may become stretched or compressed between the ligaments resulting in inflammation. The pain related to this neuralgia is positional and is worsened by sitting.

- Distribution of perineal pain in this patient is consistent with pudendal nerve involvement.

- The patient is at risk for nerve "entrapment" because of his long-term athletic training and activities and resulting physical changes to his pelvic floor.

Cystitis or Prostatitis

Cystitis and prostatitis are the inflammation of the urinary bladder and prostate; respectively. Inflammation due to infection in either of these organs could result in pelvic and perineal pain similar to that described by this patient.

- There was no laboratory or physical evidence of infection or inflammation to support diagnosis of **cystitis** or **prostatitis**.

Prostatitis Inflammation of the prostate

Patient Presentation

A 54-year-old Latina complains of bouts of pain, itching, and burning sensations in the anal region.

Relevant Clinical Findings

History

The patient had four pregnancies during her third and fourth decades. Periodically over the past 5 years, she has observed bright red blood streaks on stools. She experiences frequent episodes of constipation and the hard stools are often blood-streaked and have areas of thick mucus. Her job requires long periods of sitting at a desk and it is often then that the pain, itching, and burning develop.

Physical Examination

Results of pelvic examination:

The physician assistant examines the anus and surrounding tissues and makes the following observations:

- Peri-anal skin has a series of subcutaneous, dark-colored, hard lumps consistent with venous thromboses.
- Anal aperture is occupied by a series of moist, globular, dark-colored swellings.

Clinical Problems to Consider

Based on the patient's description of symptoms and the physical examination observations, the most likely problems to consider are:

- Anal fissure
- Hemorrhoids
- Peri-anal abscess

LEARNING OBJECTIVES

1. Describe the anatomy of the anorectum.
2. Explain the anatomical basis for the signs and symptoms associated with this case.

RELEVANT ANATOMY

Anal Canal and Rectum

The **anal canal** is the distal 4 cm of the large intestine. It begins superiorly where the rectum passes through the pelvic diaphragm and opens inferiorly as the **anus**. The superior and inferior portions of the anal canal differ in terms of embryologic origin, internal features, vasculature, and innervation (**Fig. 5.3.1**).

Superior Portion of the Anal Canal

This portion is **hindgut-derived** (endoderm) and lined by a **simple columnar epithelium**, similar to that of more proximal regions of the gastrointestinal tract. It is characterized by **anal columns** (Morgagni), 6–10 vertical ridges formed by underlying blood vessels. Each column contains a terminal branch of the superior rectal artery and vein. The superior ends of the anal columns are located at the **anorectal junction**. The inferior ends of the columns are joined by thin, crescentic folds of mucosa to form the **anal valves**. The recess superior to each

---

Thrombus A fixed mass of platelets and/or fibrin (clot) that partially or totally occludes a blood vessel or heart chamber. An embolism is a mobile clot in the cardiovascular system.

(A)

Inferior transverse rectal fold

Anal column

Anal valve

Anal sinus

Internal anal sphincter

External anal sphincter

(B)

① 1. Rectal ampulla
② 2. Anorectal junction
③ 3. Columnar zone
④ 4. Pectinate line
⑤ 5. Anal pecten
⑥ 6. Anocutaneous line
⑦ 7. Cutaneous zone
⑧ 8. Anal verge

Anal regions

1. Rectal ampulla
2. Anorectal junction
3. Columnar zone
4. Pectinate line
5. Anal pecten
6. Anocutaneous line
7. Cutaneous zone
8. Anal verge

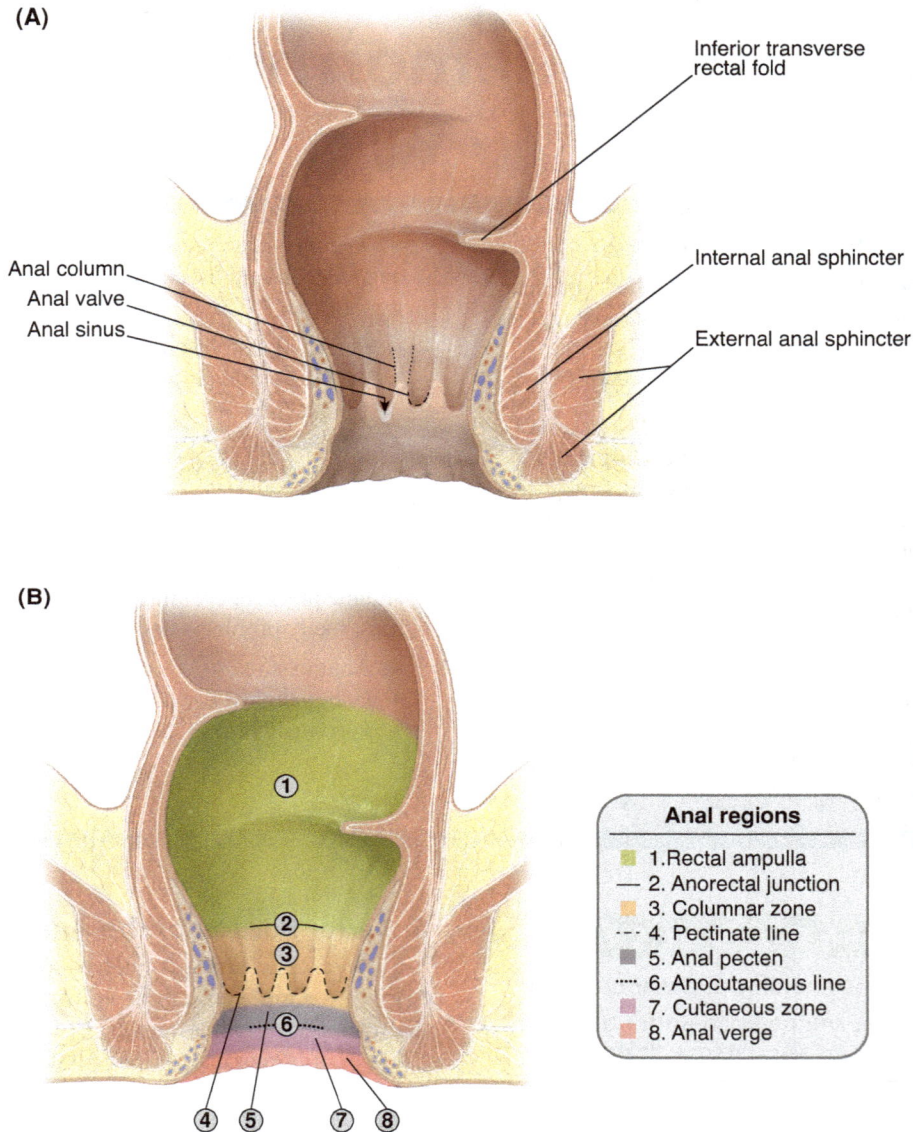

FIGURE 5.3.1 Coronal sections through the anal canal and surrounding tissue. **(A)** Anatomy of the anal canal and **(B)** regions of the anal canal.

valve is known as an **anal sinus**. The sinuses contain the openings of anal crypts (Morgagni), which are ducts of anal glands that produce mucus to facilitate passage of stool. The anal valves and sinuses form a circumferential, scalloped **pectinate** (or **dentate**) **line**; this delineates the junction of the superior and inferior portions of the anal canal.

The blood supply for the superior anal canal is provided by the **superior rectal artery** and **vein** (**Fig. 5.3.2**). Lymphatic drainage is into **internal iliac nodes**. The nerve supply is provided through the inferior hypogastric plexus (see Chapter 1). Thus, the superior anal canal is similar to other regions of the gastrointestinal tract in that overdistension,

SUPERIOR to Pectinate Line

ARTERIAL	VENOUS	LYMPHATIC
Aorta	Internal rectal venous plexus	Internal iliac nodes
↳Inferior mesenteric a.	↳Superior rectal v.	
↳Superior rectal a.	↳Inferior mesenteric v.	
	↳Splenic v.	
	↳Hepatic portal v.	

AUTONOMIC INNERVATION: Rectum and proximal anal canal

SYMPATHETIC
Preganglionic:
T12-L2 Lateral horns → Splanchnic nn. → Intermesenteric (aortic) plexus → Prevertebral ganglia
(least thoracic/lumbar)

Postganglionic:
Proximal rectum: *Superior rectal artery (vasomotor)*
Distal rectum and proximal anal canal: Hypogastric plexuses → *Middle rectal artery (vasomotor)*

PARASYMPATHETIC
Preganglionic:
S2-S4 Lateral horns → Pelvic splanchnic nn. → Inferior hypogastric plexus → Rectal plexus → Terminal ganglia
(nervi eregentes)

Postganglionic:
Enteric plexus → *Smooth muscle of rectal wall and internal anal sphincter*

VISCERAL SENSORY
Stretch receptors → Axons travel with parasympathetics

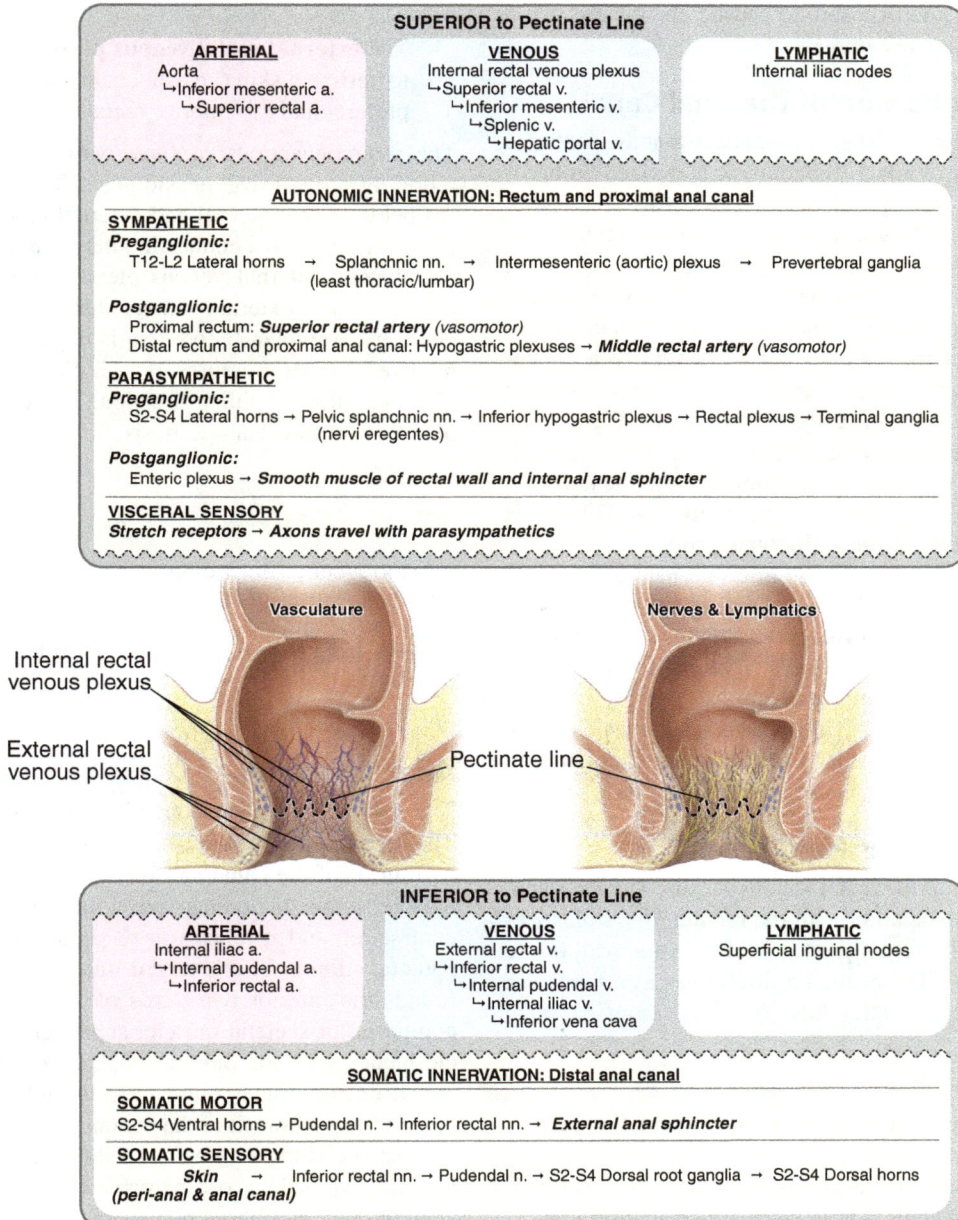

Vasculature

Nerves & Lymphatics

Internal rectal venous plexus

External rectal venous plexus

Pectinate line

INFERIOR to Pectinate Line

ARTERIAL	VENOUS	LYMPHATIC
Internal iliac a.	External rectal v.	Superficial inguinal nodes
↳Internal pudendal a.	↳Inferior rectal v.	
↳Inferior rectal a.	↳Internal pudendal v.	
	↳Internal iliac v.	
	↳Inferior vena cava	

SOMATIC INNERVATION: Distal anal canal

SOMATIC MOTOR
S2-S4 Ventral horns → Pudendal n. → Inferior rectal nn. → *External anal sphincter*

SOMATIC SENSORY
Skin → Inferior rectal nn. → Pudendal n. → S2-S4 Dorsal root ganglia → S2-S4 Dorsal horns
(peri-anal & anal canal)

FIGURE 5.3.2 Coronal sections through the rectum, anal canal, and surrounding tissues showing vasculature and innervation with relation to the pectinate line. Arterial, venous, lymphatic, and innervation pathways are outlined.

ischemia, and exposure to caustic chemicals elicit a visceral pain response.

Inferior Portion of the Anal Canal

This portion of the anal canal begins at the pectinate line and extends to the anus. It is derived embryologically from proctodeum (ectoderm) and, therefore, is lined by a stratified squamous epithelium. Its lining is smooth and lacks anal columns. Just inferior to the pectinate line is a narrow transition zone—the **pecten**—where the simple columnar epithelium of the superior canal changes to a nonkeratinized, stratified squamous epithelium. A second transition zone, the **anocutaneous line** (**Fig. 5.3.1**), marks the junction of nonkeratinized to keratinized, stratified squamous epithelium of skin; the anocutaneous line, also known as the **white line** (**Hilton**), is located at the **intersphincteric groove**.

The blood supply for the inferior anal canal is provided by the **inferior rectal artery** and **vein**; if present, **middle rectal vessels** anastomose with superior and inferior rectal vessels. Lymphatic drainage is into **superficial inguinal nodes**. Sensory innervation is provided by inferior rectal nerves (S2-S4). Thus, the lower anal canal has receptors for all modes of general sensation, similar to skin.

Rectal Venous Plexuses

While the term "rectal venous plexus" suggests anatomical association with the rectum, these plexuses are actually found predominantly in the wall of the anal canal. The pectinate line serves as a landmark for the preferential flow of blood from the anal canal:

Superior to the Pectinate Line

- The **internal rectal venous plexus** lies deep to the epithelium (i.e., submucosal) of the anal columns and drains preferentially to superior rectal veins. These veins are normally somewhat varicotic and lack valves. This distension allows the anal columns to act as appositional cushions, assisting the anal sphincters to keep the anal canal closed.

Inferior to the Pectinate Line

- The **external rectal venous plexus** lies deep to peri-anal skin (i.e., subcutaneous) and drains preferentially to inferior rectal veins.

The interconnections of these valveless plexuses represent a portacaval anastomosis. The anastomoses between the internal and external rectal venous plexuses establish a connection between the hepatic portal vein (internal venous plexus) and the inferior vena cava (external venous plexus). Since these veins are avalvular, blood flow is dependent on the relative vascular pressures in each plexus and, thus, blood can flow in either direction. This is an example of a portacaval anastomosis.

Clinical Note

Arteriovenous anastomoses also exist in the rectal plexuses of the anal canal. This accounts for the bright red blood when there is hemorrhage from over-distended veins of the anal columns.

Anal Sphincters

The anal canal is surrounded by two sphincters. The **internal anal sphincter** surrounds the superior two-thirds of the canal and is composed of smooth muscle; sympathetic innervation maintains its tonic contraction and it relaxes in response to parasympathetic stimuli. The **external anal sphincter** surrounds the inferior two-thirds of the canal and is composed of skeletal muscle; superiorly, this muscle blends with the puborectalis portion of levator ani. It receives innervation from the inferior rectal nerve (S4)—a branch of the pudendal nerve. A shallow groove in the anal mucosa, the intersphincteric groove (**Fig. 5.3.1**), is a palpable landmark for the zone of overlap of these two sphincters.

Varico- Relating to dilated or distended vein

Hemorrhage Escape of blood from vessels:
- Petechia: <2 mm diameter
- Echymosis (bruise): >2 mm
- Purpura: a group of petechiae or ecchymoses
- Hematoma: hemorrhage resulting in elevation of skin or mucosa

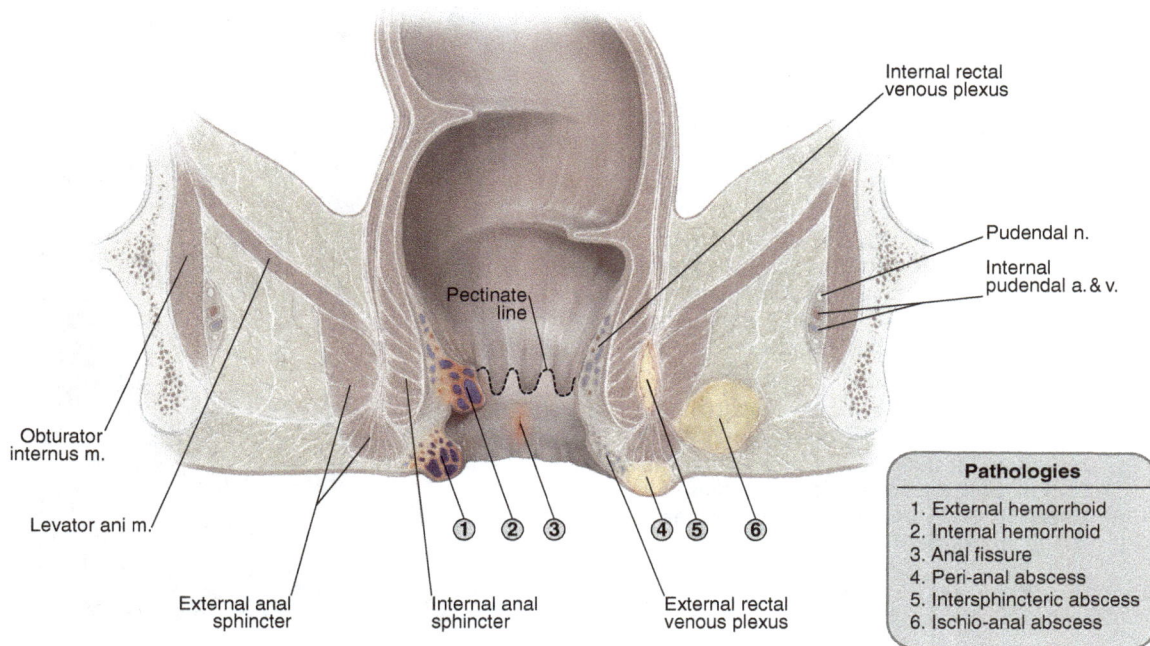

FIGURE 5.3.3 Pathologies of the anal canal.

CLINICAL REASONING

This patient presents with signs and symptoms indicating a clinical condition of the terminal part of the large intestine.

Anal Fissure

An **anal fissure** is a short longitudinal tear in the anal mucosa that may be restricted to the epithelium but can involve the entire mucosal thickness. Most fissures occur in the posterior midline between the pectinate (dentate) line and the anal verge (**Fig. 5.3.3**).

Signs and Symptoms

- Severe pain (burning sensation) during defecation
- Bright red blood on stool or toilet tissue
- Pruritis

Pruritis Itching

Predisposing Factors

- Age: most common between ages 30 and 50 years
- Constipation
- Low intake of dietary fiber

Anal fissures are commonly caused by hard stools as they pass through the lower anal canal. This region is sensitive to general sensation (carried by the inferior rectal nerve) and accounts for the pain and pruritus. The risk of developing an anal fissure is increased by constipation and straining during bowel movements as the mucosa is stretched and scraped by compacted fecal matter. Less frequently, fissures result from foreign objects being inserted into the anal canal and rectum.

The recurring pain of the unhealed fissure is exacerbated with each bowel movement and leads to a cycle of worsening symptoms. To compensate, the patient may attempt to reduce the frequency of bowel movements through decreased dietary intake, leading to weight loss.

Clinical Notes

- Anal fissures have equal incidence in both sexes.

- In patients with an anal fissure, there is a good correlation with physiological changes in the anal canal (e.g., increased hypertonicity and hypertrophy of the internal anal sphincter; increased anal canal and sphincter resting pressures). With constipation, the hypertrophied internal sphincters and increased pressures further compress the stool against the epithelium, compounding the pain with each bowel movement.

- More than 90% of anal tears heal within a few days to weeks, typically aided by topical creams. Patients are usually counseled regarding lifestyle modifications to reduce the risk of recurrence (increase in dietary fiber and water intake, exercise, and the use of stool softeners).

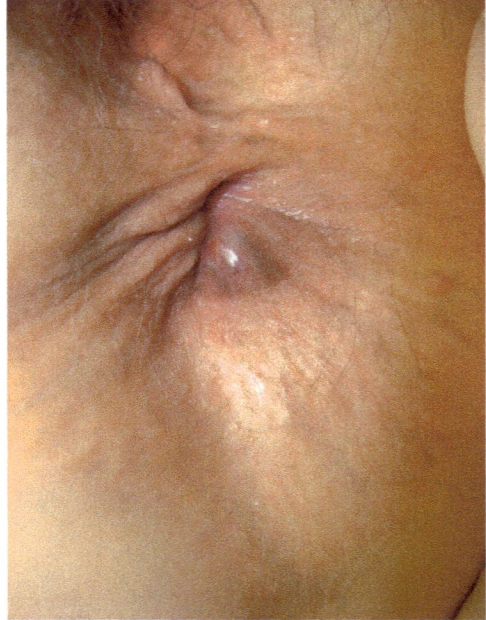

FIGURE 5.3.4 External hemorrhoid, with a small skin break that resulted in bleeding. *Source:* Plate 28 in *DeGowin's Diagnostic Examination,* 9e.

Hemorrhoids

Hemorrhoids are classified as internal and external (**Fig. 5.3.2**). **Internal hemorrhoids** involve the internal rectal venous plexus. These veins become overdistended and their mucosal coating is vulnerable to abrasion and ulceration during defecation. They may enlarge to the point of prolapse from the anus and may resemble a cluster of small grapes. Patients often describe only vague anal discomfort consistent with their visceral innervation.

External hemorrhoids are associated with the external venous plexus of the peri-anal skin. They are characterized by deep purple, hard, nodular masses that are highly susceptible to thrombosis (**Fig. 5.3.4**). Patients frequently discover these hemorrhoids while cleaning following a bowel movement. They rarely become ulcerated from abrasions by feces. Peri-anal

skin is richly supplied with somatic pain receptors, and sharp hemorrhoidal pain is most often related to external hemorrhoids.

Internal and external hemorrhoids may exist simultaneously since these venous plexuses have extensive anastomoses and are valveless.

Signs and Symptoms

Internal Hemorrhoids

- Blood on stool or toilet paper
- Pruritus
- Painful (visceral) when prolapsed

External Hemorrhoids

- Peri-anal pain (somatic)
- Peri-anal inflammation and swelling
- Bleeding
- Hard peri-anal mass (thrombosed external venous plexus)

Hypertonicity of muscle Abnormal increase in muscle tone
Hypertrophy Increase in size of a tissue or organ due to increased cell size, that is, without increased cell number (antonym: hyperplasia)
Ulcer A lesion on skin or mucous membrane

Predisposing Factors

- Chronic constipation or diarrhea
- Low-fiber diet
- Pregnancy
- Liver disease (portal hypertension)

Clinical Notes

- Hemorrhoids have equal incidence in both sexes.
- Most cases of hemorrhoids resolve themselves once the factor causing increased pelvic venous pressure is resolved (i.e., change in diet; end of pregnancy). The goal of any treatment is to reduce the venous distention in the plexuses so that the symptoms are alleviated.

Peri-anal Abscess

A **peri-anal** abscess is an infection in soft tissues that surround the anal canal, usually in a discrete cavity (**Fig. 5.3.2**). Peri-anal abscesses typically result from **cryptitis**, blockage and infection of ducts that drain the anal mucous glands. These ducts (crypts of Morgagni) open into the anal sinuses, just superior to the pectinate (dentate) line. Because these glands extend into the plane between internal and external sphincters, an infection that becomes suppurative and breaks through the internal anal sphincter may spread to adjacent potential spaces, such as the ischio-anal fossa, to create an ischio-anal abscess.

Signs and Symptoms

- Dull, peri-anal discomfort and, pruritus
- Peri-anal pain often exacerbated by long periods of sitting or during defecation
- Small, subcutaneous, fluctuant, erythematous mass near anus

---

Abscess Collection of purulent exudate (pus)
Suppurate To form pus
Fluctuance Indication of pus in a bacterial infection in which redness and induration (hardening) develop in infected skin
Erythematous Reddened skin

Predisposing Factors

- Age: peak incidence in adults between 20 and 40 years; infants also frequently affected
- Sex: male (2:1 to 3:1)
- Chronic constipation
- Previous history of anorectal abscess

Clinical Notes

- Most anorectal abscesses are peri-anal (60%) or ischiorectal (20%). The remainder of abscesses are in intersphincteric, supralevator, or submucosal locations.
- Most abscesses resolve spontaneously, though some may require surgical intervention.

DIAGNOSIS

The patient presentation, medical history, and physical examination support a diagnosis of **external and internal hemorrhoids**.

The anal canal is divided into superior and inferior portions because of its developmental origins. Thus, the blood supply, lymphatic drainage, and nerve supply are different for the two parts. In the case of the venous system, there are extensive interconnections along the entire length of the anal canal, forming a portacaval anastomosis. Veins associated with the superior part of the anal canal are submucosal and drain preferentially into the hepatic portal vein, while veins of the inferior part are subcutaneous and preferentially drain into the inferior vena cava.

External Hemorrhoids

- The hard, subcutaneous, dark enlargements around the anus are diagnostic of thrombosis of blood in the external venous plexus, that is, external hemorrhoids. The pain described is consistent with this type of hemorrhoid.

Internal Hemorrhoids

- The moist, dark colored, globular enlargements that occupy the anus are prolapsed veins of the internal venous plexus, that is, internal hemorrhoids. The blood and mucus on the stool, and the anal discomfort is consistent with this type of hemorrhoid.

Peri-anal Abscess

Abscesses of the anal canal or rectum are infections in surrounding soft tissues that usually evolve from blockage and subsequent infection of the ducts that drain anal mucous glands. They are classified on the basis of location: peri-anal (60%), ischiorectal (20%), intersphincteric, supralevator, and submucosal. Clinical signs of a peri-anal abscess are absent with this patient. These would include:

- Small, subcutaneous, fluctuant, erythematous mass near the anus

Anal Fissure

Anal fissures are short tears in the anal mucosa. In contrast to peri-anal fistula, they do not extend to the peri-anal skin. Most fissures occur in the posterior midline of the anal canal. They may develop with the passage of hard stool, often associated with constipation.

- Severe pain with defecation.
- Bright red blood on stool or toilet tissue.
- An anal fissure is ruled out since no evidence of a tear was found during the physical examination.

Peri-anal Fistula

Patient Presentation

A 45-year-old white male visits the family medicine clinic with chief complaints of itching and irritation around the anus, and pain when defecating. He has had a fever of 37.8°C (100°F) for the past 2 days.

Relevant Clinical Findings

History

The patient indicates having been constipated for several months and that with every bowel movement he experiences severe pain. He indicates that this has made him afraid to have a bowel movement. He also reports that he sometimes finds bright red blood on the toilet paper and on the stool; he also noticed pus after wiping about 4 weeks ago. He reports being active and maintaining a relatively stable weight.

Physical Examination

Noteworthy vital signs:

- Temperature: 37.8°C (100°F) [normal: 36.0–37.5°C (96.5–99.5°F)]

Results of physical examination:

- Peri-anal swelling and irritation on the right, including a small, erythematous subcutaneous mass, excoriation, fluctuance, and inflammation
- A small opening on the right peri-anal region, from which purulent discharge can be expressed by digital rectal exam.

Laboratory Tests

Test	Value	Reference value
Leukocytes (count)	15	3.54–9.06 × 10³/mm³
C-reactive protein (CRP)	2.7	0–10 mg/dL

Clinical Note

C-reactive protein is a general marker for infection and inflammation.

Clinical Problems to Consider

- Anal fissure
- Peri-anal abscess
- Peri-anal fistula

LEARNING OBJECTIVES

1. Describe the anatomy of the anal canal and ischioanal fossa.
2. Explain the anatomical basis for the signs and symptoms associated with this case.

Erythematous Reddened skin
Excoriation Scratched or abraded area of skin

RELEVANT ANATOMY

Anal Canal

The **anal canal** is the distal 4 cm of the large intestine. It begins superiorly where the rectum

Fluctuance Indication of pus in a bacterial infection in which redness and induration (hardening) develop in infected skin
Purulent Containing, discharging, or causing the production of pus

(A)

Anal regions
■ 1. Rectal ampulla
— 2. Anorectal junction
■ 3. Columnar zone
- - - 4. Pectinate line
■ 5. Anal pecten
••••• 6. Anocutaneous line
■ 7. Cutaneous zone
■ 8. Anal verge

(B)

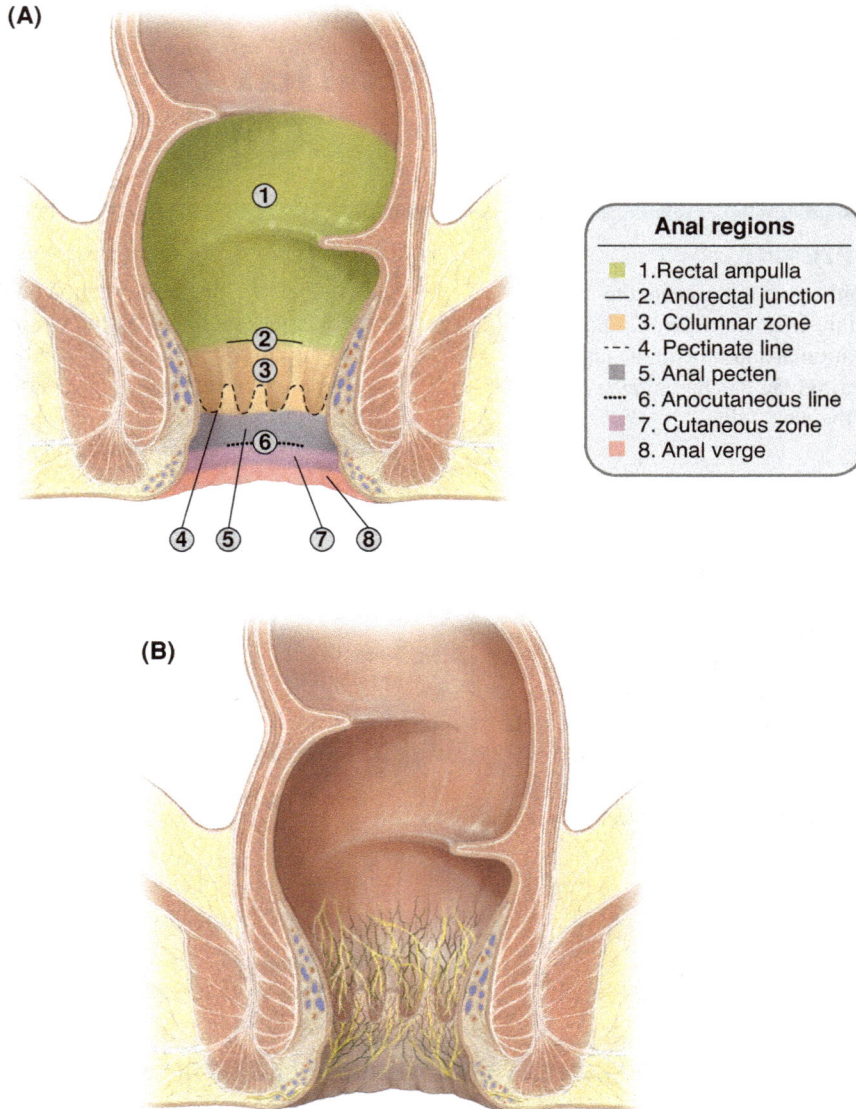

FIGURE 5.4.1 Coronal sections through the anal canal and surrounding tissue. **(A)** Regions of the anal canal and **(B)** innervation (yellow) and lymphatic vasculature (green) of the anal canal.

passes through the pelvic diaphragm and opens inferiorly as the **anus**. The superior and inferior portions of the anal canal differ in terms of embryologic origin, internal features, vasculature, and innervation.

Superior Portion of the Anal Canal

This portion is **hindgut-derived** (endoderm) and lined by a **simple columnar epithelium**, similar to that of more proximal regions of the gastrointestinal tract. It is characterized by **anal columns**

(Morgagni), 6–10 vertical ridges formed by under-lying blood vessels (**Fig. 5.4.1A**). Each column contains a terminal branch of the superior rectal artery and vein. The superior ends of the anal columns are located at the **anorectal junction**. The inferior ends of the columns are joined by thin, crescentic folds of mucosa to form the **anal valves**. The recess superior to each valve is known as an **anal sinus**. The sinuses contain the openings of anal crypts (Morgagni), which are ducts of anal glands that produce mucus to facilitate passage of stool. The anal valves and sinuses form a circumferential, scalloped **pectinate** (or **dentate**) **line**; this delineates the junction of the superior and inferior portions of the anal canal.

The blood supply for the superior anal canal is provided by the **superior rectal artery** and **vein**. Lymphatic drainage is into **internal iliac nodes** (**Fig. 5.4.1B**). The nerve supply is provided through the inferior hypogastric plexus (see Chapter 1). Thus, the superior anal canal is similar to other regions of the gastrointestinal tract in that overdistension, ischemia, and exposure to caustic chemicals elicit a visceral pain response.

Inferior Portion of the Anal Canal

This portion of the anal canal begins at the pectinate line and extends to the anus. It is derived embryologically from proctodeum (ectoderm) and, therefore, is lined by a stratified squamous epithelium. Its lining is smooth and lacks anal columns (**Fig. 5.4.1A**). Just inferior to the pectinate line is a narrow transition zone—the **pecten**—where the simple columnar epithelium of the superior canal changes to a nonkeratinized, stratified squamous epithelium. A second transition zone, the **anocutaneous line** (**Fig. 5.4.1A**), marks the junction of nonkeratinized to keratinized, stratified squamous epithelium of skin; the anocutaneous line, also known as the **white line** (**Hilton**), is located at the **intersphincteric groove**.

The blood supply for the inferior anal canal is provided by the **inferior rectal artery** and **vein**; if present, **middle rectal vessels** anastomose with superior and inferior rectal vessels. Lymphatic drainage is into **superficial inguinal nodes** (**Fig. 5.4.1B**). Sensory innervation is provided by inferior rectal nerves, which have contribution from S2 to S4 spinal nerves. Thus, the lower anal canal has receptors for all modes of general sensation, similar to skin.

For a discussion of the rectal venous plexuses, see Case 5.3.

Anal Sphincters

The anal canal is surrounded by two sphincters. The **internal anal sphincter** surrounds the superior two-thirds of the canal and is composed of smooth muscle; sympathetic innervation maintains its tonic contraction and it relaxes in response to parasympathetic stimuli. The **external anal sphincter** surrounds the inferior two-thirds of the canal and is composed of skeletal muscle; superiorly, this muscle blends with the puborectalis portion of levator ani. It receives innervation from the inferior rectal nerve (S4)—a branch of the pudendal nerve. A shallow groove in the anal mucosa—the intersphincteric groove—is a palpable landmark for the zone of overlap of these two sphincters.

Ischio-anal Fossae

The ischio-anal (ischiorectal) fossae are paired, fat-filled, pyramidal recesses that flank the anal canal (**Fig. 5.4.2**). The borders for each fossa are:

- **Inferior (base)**—skin over the anal triangle
- **Medial**—external anal sphincter and levator ani muscles
- **Lateral**—ischial tuberosities; and obturator internus muscle and its fascia

The apex of this pyramidal region is formed where fibers of levator ani take origin from the fascia of obturator internus muscle.

The fossae communicate with each other posterior to the anal canal (via the deep postanal space) and have anterior recesses that extend superior

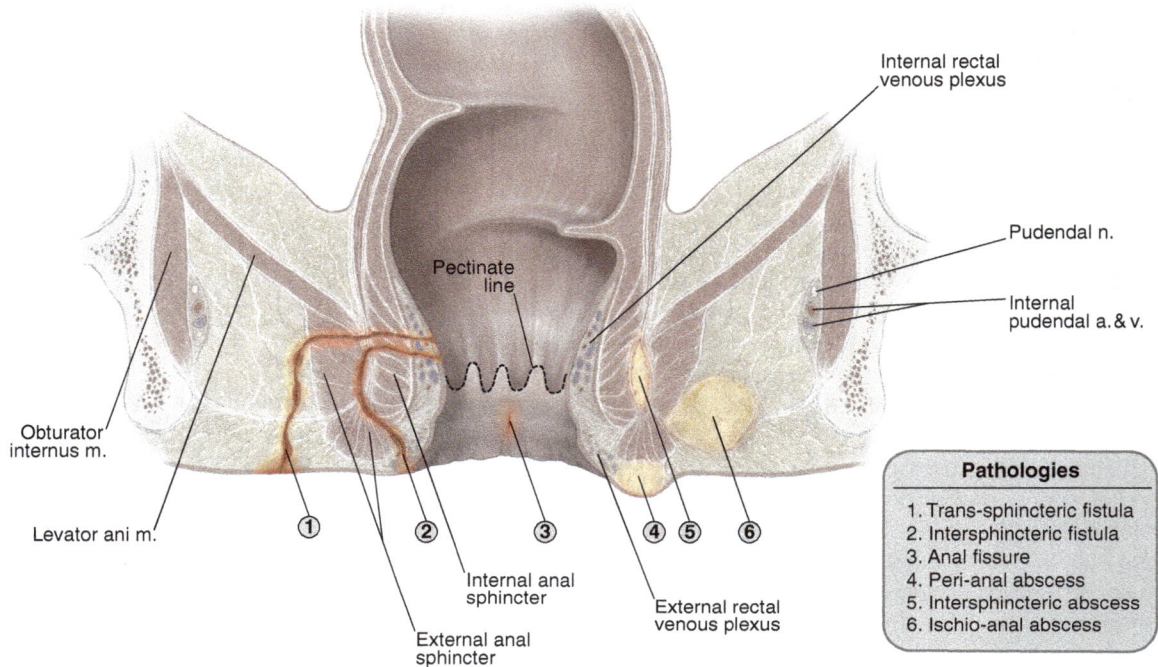

FIGURE 5.4.2 Pathologies of the anal canal.

to the urogenital diaphragm. The pudendal nerve and internal pudendal vessels are enclosed in the pudendal canal on the lateral wall of each fossa. Inferior rectal nerves (branches of the pudendal nerve) and vessels (branches of the internal pudendal artery and vein) traverse each fossa to reach the anal canal.

CLINICAL REASONING

This patient presents with signs and symptoms indicating a clinical condition of the terminal part of the large intestine.

Anal Fissure

An **anal fissure** is a short longitudinal tear in the anal mucosa that may be restricted to the epithelium but can involve the entire mucosal thickness. Most fissures occur in the posterior midline between the pectinate (dentate) line and the anal verge (**Fig. 5.4.2**).

Signs and Symptoms

- Severe pain (burning sensation) during defecation
- Bright red blood on stool or toilet tissue
- Pruritis

Predisposing Factors

- Age: most common 30–50 years
- Constipation
- Low intake of dietary fiber

Anal fissures are commonly caused by hard stools as they pass through the lower anal canal. This region is

---

Pruritis Itching

sensitive to general sensation (carried by the inferior rectal nerve) and accounts for the pain and pruritus. The risk of developing an anal fissure is increased by constipation and straining during bowel movements as the mucosa is stretched and scraped by compacted fecal matter. Less frequently, fissures result from foreign objects being inserted into the anal canal and rectum.

The recurring pain of the unhealed fissure is exacerbated with each bowel movement and leads to a cycle of worsening symptoms. To compensate, the patient will attempt to reduce the frequency of bowel movements through decreased dietary intake, leading to weight loss.

Clinical Notes

- Anal fissures have equal incidence in both sexes.

- In patients with an anal fissure, there is a good correlation with physiological changes in the anal canal (increased hypertonicity and hypertrophy of the internal anal sphincter; increased anal canal and sphincter resting pressures). With constipation, the hypertrophied internal sphincters and increased pressures further compress the stool against the epithelium, compounding the pain with each bowel movement.

- More than 90% of anal tears heal within a few days to weeks, typically aided by topical creams. Patients are usually counseled regarding lifestyle modifications to reduce the risk of recurrence (increase in dietary fiber and water intake, exercise, and the use of stool softeners).

Peri-anal Abscess

A **peri-anal** abscess is an infection in soft tissues that surround the anal canal, usually in a discrete cavity (**Fig. 5.4.2**).

Hypertonicity of muscle Abnormal increase in muscle tone

Hypertrophy Increase in size of a tissue or organ due to increased cell size, that is, without increased cell number (antonym: hyperplasia)

Signs and Symptoms

- Dull, peri-anal discomfort and pruritus
- Peri-anal pain often exacerbated by long periods of sitting or during defecation
- Small, subcutaneous, fluctuant, erythematous mass near anus

Predisposing Factors

- Age: peak incidence in adults between 20 and 40 years; infants also frequently affected
- Sex: men (2:1 to 3:1)
- Chronic constipation
- Previous history of anorectal abscess

Peri-anal abscesses typically result from **cryptitis**, blockage and infection of ducts that drain the anal mucous glands. These ducts (crypts of Morgagni) open into the anal sinuses, just superior to the pectinate (dentate) line. Because these glands extend into the plane between internal and external sphincters, an infection that becomes suppurative and breaks through the internal anal sphincter may spread to adjacent potential spaces, such as the ischio-anal fossa, to create an ischio-anal abscess.

Clinical Notes

- Most anorectal abscesses are peri-anal (60%) or ischiorectal (20%), with the remainder in intersphincteric, supralevator, and submucosal locations.

- Most abscesses resolve spontaneously, though some may require surgical intervention.

Peri-anal Fistula

A peri-anal **fistula** (*fistula in ano*) is a communication between the anal canal and the peri-anal skin (**Fig. 5.4.2**).

Abscess Collection of purulent exudate (pus)
Suppurate To form pus

Signs and Symptoms

- External peri-anal opening (a corresponding internal opening in the anal canal may be detected by DRE, anoscopy, or proctoscopy)
- Spontaneous or expressible purulent peri-anal discharge (may also include fecal material)
- Peri-anal itching, tenderness, pain and swelling

Predisposing Factors

- Sex: men (approximately 2:1)
- Previous peri-anal abscess or anorectal fistula
- Chronic constipation
- Crohn disease
- Less commonly, trauma, tuberculosis, carcinoma

Anal fistulae are named according to their relationship to the anal sphincters. The two most common types are:

1. **Intersphincteric** fistulae (70%) pass through the internal sphincter and then inferiorly between the sphincters to reach peri-anal skin.
2. **Transsphincteric** fistulae (25%) extend through both anal sphincters before entering the ischiorectal fossa and, only then, open onto the peri-anal skin.

Clinical Notes

- Most peri-anal fistulae develop secondary to a peri-anal abscess that develops a drainage "tunnel" to the peri-anal skin.
- Treatment of a simple (i.e., straight) peri-anal fistula typically involves a fistulotomy to open the fistulous tract, clean it out, and stitch it open so that it heals.

DIAGNOSIS

The patient presentation, medical history, physical examination, and laboratory tests support a diagnosis of **peri-anal fistula**.

Peri-anal Fistula

A peri-anal fistula is a communication between the anal canal and the peri-anal skin.

- The presence of a fistulous opening on the peri-anal skin from which purulent and fecal material can be expressed by DRE is diagnostic of this condition.

Anal Fissure

Anal fissures are short tears in the anal mucosa. In contrast to peri-anal fistula, they do not extend to the peri-anal skin. Most fissures occur in the posterior midline of the anal canal. They may develop with the passage of hard stool, often associated with constipation.

- Severe pain with defecation
- Bright red blood on stool or toilet tissue

Peri-anal Abscess

Abscesses of the anal canal or rectum are infections in surrounding soft tissues that usually evolve from blockage and subsequent infection of the ducts that drain anal mucous glands. They are classified on the basis of location: peri-anal (60%), ischiorectal (20%), intersphincteric, supralevator, and submucosal.

Clinical signs of a peri-anal abscess are absent with this patient. These would include small, subcutaneous, fluctuant, erythematous mass near the anus.

Fistula An abnormal passage between two epithelialized surfaces (e.g., anus and skin)

REVIEW QUESTIONS

1. As part of a preschool physical evaluation of a 5-year-old male, a normal cremasteric reflex is elicited. Which is *not* associated with the cremasteric reflex?

   A. Genitofemoral nerve
   B. Middle layer of spermatic fascia
   C. Obturator nerve
   D. Skin of the proximal, medial thigh
   E. Spinal cord segment L1

2. A 38-year-old male presents with complaints of peri-anal pain. Radiologic imaging reveals a large abscess in contact with the lateral wall of the left ischio-anal fossa. Which of the following contributes to the lateral wall of the ischio-anal fossa?

   A. Fascia of obturator internus muscle
   B. Levator ani muscle
   C. Perineal membrane
   D. Piriformis muscle
   E. Skin

3. A 13-year-old boy is brought to the pediatric clinic with concerns about the progressive enlargement of his scrotum. Transillumination of the scrotum reveals a testicular hydrocele. Which structure would *not* be penetrated by a needle used to drain the hydrocele?

   A. Cremasteric fascia
   B. Dartos fascia
   C. External spermatic fascia
   D. Parietal layer of tunica vaginalis
   E. Tunica albuginea

4. A female patient presents with complaints of severe anal pain during defecation. Physical examination reveals enlarged superficial inguinal lymph nodes, which drain the inferior part of the anal canal. Which structure would *not* have lymph drainage into these nodes?

   A. Camper fascia of right lower abdominal quadrant
   B. Clitoris
   C. Inferior one-third of rectum
   D. Labia minora
   E. Subcutaneous tissues over lateral thigh

5. During the annual gynecological examination of a 36-year-old patient, she indicates that the vaginal speculum "feels cold." Afferent impulses from temperature receptors in walls of the vaginal vestibule would travel in:

   A. Hypogastric nerves
   B. Pelvic splanchnic nerves
   C. Pudendal nerve
   D. Sacral splanchnic nerves
   E. Vaginal nerve plexus

6. A 28-year-old female in the third trimester of pregnancy complains of peri-anal pain. Physical examination reveals external hemorrhoids. Blood in these vessels will flow preferentially into the:

   A. Inferior rectal veins
   B. Middle rectal veins
   C. Superficial dorsal vein of the clitoris
   D. Superior rectal veins
   E. Uterovaginal venous plexus

7. A 3rd year medical student writes a note in the chart of an 81-year-old male that the scrotum is pendulous and its skin smooth. The internal medicine resident counsels the student that this is normal in elderly men because:

   A. Cremaster muscle loses it tone
   B. Dartos muscle loses it tone
   C. Impulse conduction in branches of the pudendal nerve is slower
   D. The perineal membrane atrophies
   E. The testes increase in size and weight with age

8. During labor, the obstetrical team decides that a mediolateral episiotomy should be performed. Which muscle is *least likely* to be damaged by this surgical incision?

   A. Bulbospongiosus
   B. Deep transverse perineal
   C. External anal sphincter
   D. Ischiocavernosus
   E. Superficial transverse perineal

9. A 37-year-old male complains of fecal incontinence. Digital rectal examination reveals that the right part of the external anal sphincter is flaccid. Which nerve supplies this muscle?

    A. Dorsal nerve of the penis
    B. Ilioinguinal
    C. Inferior rectal
    D. Perineal
    E. Pudendal

10. A 43-year-old female presents with pain in her external genitalia. Physical examination reveals an inflamed swelling on the right side of the vaginal vestibule, consistent with Bartholinitis. The greater vestibular (Bartholin) gland:

    A. Is modified erectile tissue
    B. Is a homologue of the prostate gland
    C. Is located deep to bulbospongiosus muscle
    D. Has a duct that opens into the deep perineal pouch
    E. Is ductless

CHAPTER 6

Neck

Internal Jugular Catheterization

Patient Presentation

A 40-year-old male is admitted to the emergency department complaining of severe diarrhea, vomiting, dehydration, weakness, and weight loss. He is malnourished and a central venous catheter is placed to provide parenteral nutrition.

Relevant Clinical Findings

History

The patient reported losing 25 lb during the past 9 months. The patient has a history of Crohn disease, with two previous resections of his small intestine. This has resulted in short-bowel syndrome, characterized by malabsorption, diarrhea, steatorrhea, fluid and electrolyte disturbances, and malnutrition.

Clinical Note

Crohn disease is a chronic, autoimmune, inflammatory disease that affects the gastrointestinal tract, usually the intestines. It results in transmural fibrosis and obstructive symptoms.

Physical Examination

Noteworthy vital signs:

- Height: 5′ 10″
- Weight: 120 lb

Results of physical examination:

- Diffuse abdominal tenderness
- Firmness in the right lower abdominal quadrant (consistent with appendicitis; or Crohn disease, which most commonly involves the terminal ileum)

Laboratory Tests

Test	Value	Reference value
Erythrocyte (count)	3.6	$4.3–5.6 \times 10^6/mm^3$
Hematocrit	32.2	38.8–46.4%
Hemoglobin	10	14–17 gm/dL
Serum albumin	3.2	4.0–5.0 mg/dL

Imaging Studies

- Placement of a central venous catheter, introduced via the right internal jugular vein, was confirmed radiographically.

Clinical Problems to Consider

- Internal jugular vein catheterization
- Subclavian vein catheterization

LEARNING OBJECTIVES

1. Describe the anatomy of the internal jugular vein.

2. Describe the anatomy of the subclavian vein.

3. Explain the anatomical basis for these procedures.

Parenteral A route other than the gastrointestinal tract (e.g., subcutaneous, intramuscular, and intravenous) to introduce nutrition, medication, or other substance into the body (Greek, para = around + enteron = bowel)

Steatorrhea Presence of excess fat in feces
Transmural Extending through, or affecting, the entire thickness of the wall of an organ or cavity

RELEVANT ANATOMY

Internal Jugular Vein

The **internal jugular vein** (IJV) is formed at the jugular foramen, where it is the continuation of the sigmoid venous dural sinus (**Fig. 6.1.1**). It ends in the root of the neck, where it joins the **subclavian vein** to form the **brachiocephalic vein**. The right brachiocephalic vein is shorter and more vertical. The right and left brachiocephalic veins unite posterior to the right sternoclavicular joint to form the superior vena cava (SVC).

In addition to the brain, the IJV drains blood from the face, neck viscera, and deep neck muscles. Its main tributaries include:

- Sigmoid dural venous sinus
- Inferior petrosal dural venous sinus
- Facial vein
- Lingual vein
- Pharyngeal vein
- Superior and middle thyroid veins

The IJV may also connect with the external jugular via the common facial and divisions of the retromandibular veins.

The **venous angle** is formed by the intersection of the IJV and subclavian. This usually lies posterior to the medial end of the clavicle. Lymph is returned to the venous circulation at this location.

- The **thoracic duct** joins the left venous angle.
- The **right lymphatic duct** joins the right venous angle.

The venous angle is associated with the apex of lung and cervical pleura, which extend into the root of the neck.

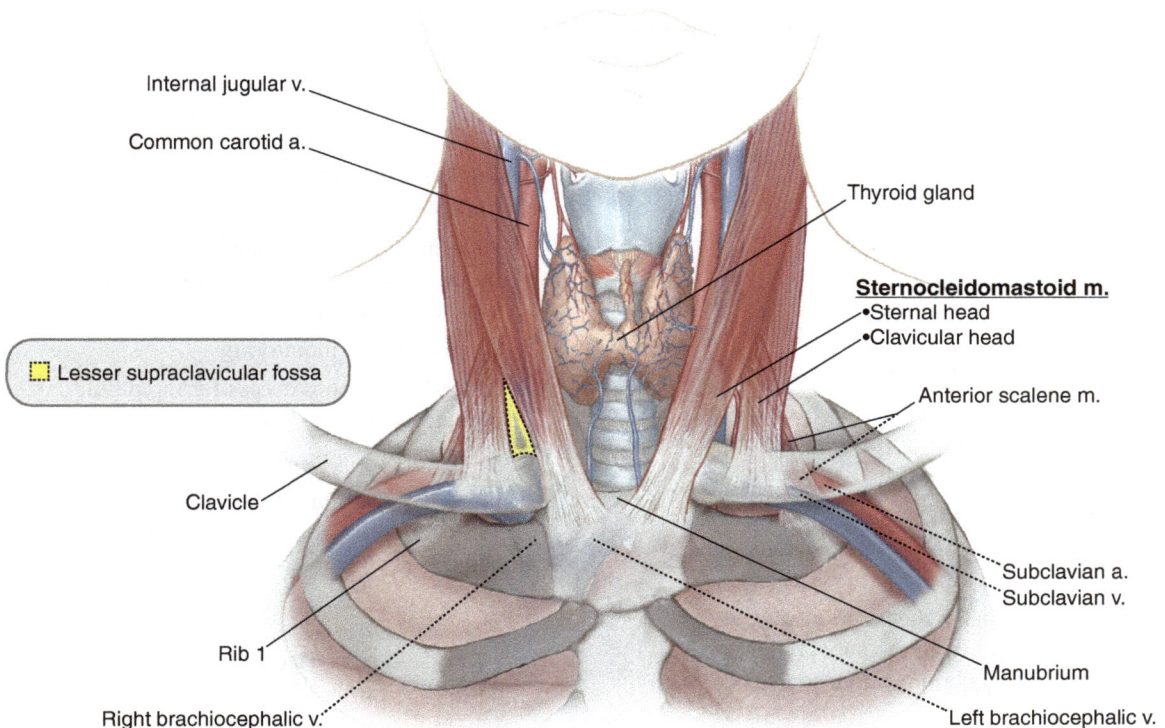

FIGURE 6.1.1 Anterior view of the anatomy of the internal jugular and subclavian veins.

Anatomical Relationships

The IJV lies in the carotid sheath, one of the cervical compartments defined by deep fascia. The sheath lies lateral to the visceral compartment. It also forms the lateral boundary of the retropharyngeal space. In the inferior part of the neck, the carotid sheath is deep to sternocleidomastoid muscle.

Within the carotid sheath, the IJV lies lateral and slightly anterior to the common/internal carotid artery (CCA/ICA). The vagus nerve (CN X) lies between and posterior to the vessels in the carotid sheath.

Surface Anatomy

The inferior part of the IJV lies in the **lesser supraclavicular fossa** (Sedillot's triangle). The clavicle and the sternal and clavicular heads of sternocleidomastoid border this narrow, shallow depression (**Fig. 6.1.1**).

Subclavian Vein

The **subclavian vein** is the direct continuation of the axillary vein, which drains the upper limb. The subclavian vein begins at the lateral border of rib 1 and ends where it joins the IJV to form the brachiocephalic vein (**Fig. 6.1.1**). Its tributaries include:

- Axillary vein
- External jugular vein
- Anterior jugular vein
- Dorsal scapular vein

Anatomical Relationships

As the subclavian vein arches over the rib 1, it lies posterior and inferior to the clavicle and subclavius muscle.

- The lateral portion of the subclavian vein is anterior and slightly inferior to the subclavian artery.

- More medially, the anterior scalene muscle separates it from the subclavian vein and artery. The phrenic nerve courses between the muscle and vein before entering the thorax on the anterior aspect of the muscle and between this muscle and the subclavian vein before it enters the thorax.

The patency of the subclavian vein is maintained by connections between its tunica adventitia and surrounding connective tissues:

- Investing fascia of subclavius and anterior scalene
- Costoclavicular ligament
- Periosteum on the posterior aspect of the clavicle

Clinical Note

The patency of the subclavian vein is maintained in hypovolemic and hypotensive conditions by its connective tissue supports with adjacent structures. This is advantageous for venous catheterization.

CLINICAL REASONING

Based on the physical examination and medical history, the patient required the administration of a nutritional formula and medications via a **central venous catheter**. Indications for central venous catheterization include:

- Administration of vasoactive drugs, chemotherapeutic and other caustic agents, and parental nutrition
- Hemodynamic monitoring
- Large-bore venous access for rapid fluid administration
- Long-term venous access

The internal jugular, subclavian, and femoral veins are common sites for the placement of these catheters.

Hypovolemic Low or inadequate blood volume
Hypotension Abnormal increase in arterial or venous
   pressure

Clinical Notes

Major types of central venous catheters include:

- A **peripherally inserted central catheter (PICC)** is a catheter that extends from a vein in the upper limb to the SVC. It provides central access for weeks to months.

- A **nontunneled catheter** is commonly placed into the internal jugular, subclavian, or femoral. The point of entry for this catheter through the skin is close to its entry into the vein.

- A **tunneled catheter** passes under the skin for some distance before it enters the vein. This helps secure the catheter, reduces the rate of infection, and permits free movement of the catheter port. It provides central access for weeks to months.

- A **subcutaneous port** is a permanent device composed of a catheter with a small reservoir. The patient's skin is punctured each time the catheter is used.

Internal Jugular Vein Catheterization

Advantages of using the **right IJV** include:

- It is usually larger than the left.
- Its course to the SVC is straight.

Disadvantages and risks of using the **left IJV** include:

- It turns to joins the subclavian to form brachiocephalic vein and again to enter the SVC.
- The **thoracic duct** joins the left venous angle. A misplaced catheter may damage this duct and result in chylothorax.
- **Cervical pleura (cupula)** extends further into the neck on the left. A misplaced catheter may puncture this pleura and result in pneumothorax.

Procedure for Right IJV Catheterization

The procedure for catheterization of the right internal jugular veins is outlined in Figure **6.1.2**.

- The body is placed in a Trendelenburg position (supine with head lower by 10–15°). This distends the IJV and decreases the risk of air embolism.

- The head is turned slightly to the left because:
  - It flattens the IJV slightly, which increases its cross-sectional area.
  - It stretches the sternocleidomastoid muscle and accentuates the lesser supraclavicular fossa.
  - The pulse of the CCA is more easily palpable in the apex of the fossa.

- The **key anatomical relationships are:**
  - The position of the carotid sheath in the **lesser supraclavicular fossa**.
  - The position within the carotid sheath of the IJV and CCA: **the IJV is lateral to the CCA**.

- After sterile preparation and draping of the patient, local anesthetic is infiltrated subcutaneously into the lesser supraclavicular fossa.

- An introducer needle, angled downward 30–45°, is inserted at the apex of the lesser supraclavicular fossa. The IJV and CCA are located by ultrasound or by palpation. The needle is advanced just lateral to the artery.

- Gentle aspiration during needle insertion is applied until dark, nonpulsatile venous blood is evident. Aspiration of bright-red, pulsatile blood indicates arterial puncture and an alternative site should be chosen.

- Once venous access is achieved, the following steps are required for catheter placement:
  1. A guidewire is introduced through the needle.
  2. The needle is removed.
  3. A catheter is inserted over the guidewire.
  4. The guidewire is removed.
  5. The catheter is connected to an intravenous line.

Clinical Note

Current evidence indicates the use of ultrasound to guide catheterization of the IJV may significantly reduce the number of attempts required as well as the risk of complications.

Chylothorax Accumulation of lymph in the pleural cavity
Pneumothorax Air or gas in pleural cavity
Embolus A mobile clot in the cardiovascular system, often derived from a thrombus, that is obstructive

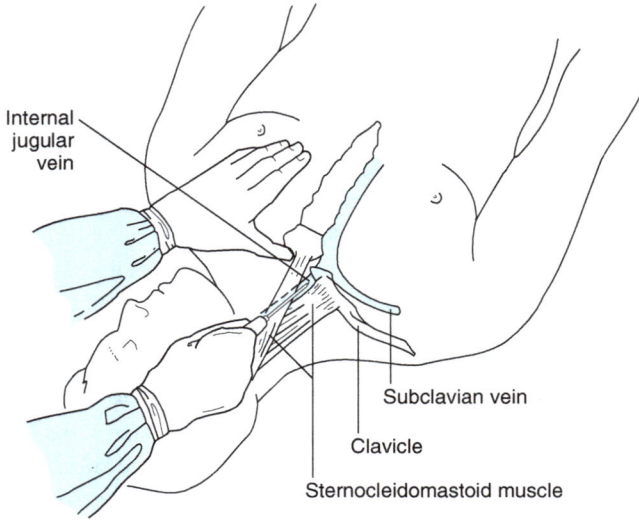

Internal jugular vein

Subclavian vein

Clavicle

Sternocleidomastoid muscle

FIGURE 6.1.2 Catheterization of the right internal jugular vein. *Source:* Fig. 13-7 in *Clinician's Pocket Reference.*

Subclavian Vein Catheterization

An infraclavicular approach is commonly used for insertion of a catheter into the subclavian vein (**Fig. 6.1.3**).

Outline of the Procedure

The procedure for catheterization of the subclavian vein is outlined in Figure 6.1.3.

- Patient positioning is similar to that used for IJV catheterization: Trendelenburg position, with the head turned to the opposite side.
- After sterile preparation and draping of the patient, local anesthetic is infiltrated subcutaneously 3 cm lateral to the midpoint of the clavicle.

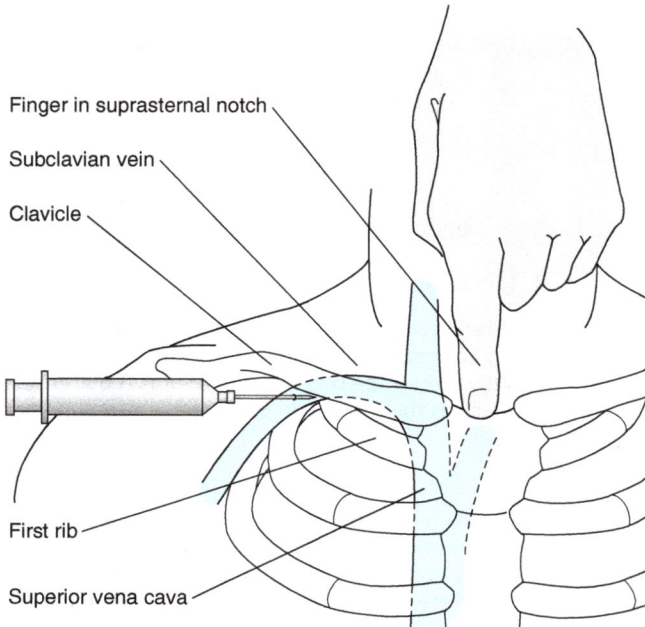

Finger in suprasternal notch

Subclavian vein

Clavicle

First rib

Superior vena cava

FIGURE 6.1.3 Catheterization of the right subclavian vein using an infraclavicular approach. *Source:* Fig. 13-6 in *Clinician's Pocket Reference.* www.accessmedicine.com.

TABLE 6.1.1	Risk of complications with central venous catheterization.[a]		
Complication	**Internal jugular**	**Subclavian**	**Femoral**
Pneumothorax (%)	<0.1–0.2	1.5–3.1	NA
Hemothorax (%)	NA	0.4–0.6	NA
Infection (per 1000 catheter-days)	8.6	4	15.3
Thrombosis (per 1000 catheter-days)	1.2–3.0	0–13	8–34
Arterial puncture (%)	3	0.5	6.25
Malposition (risk)	Low	High	Low

[a]Adapted from Graham et al (2007).

- **Key anatomical landmarks:**
  - **Midpoint of the clavicle**
  - **Suprasternal notch**
- The introducer needle is inserted along the inferior edge of the clavicle, 2–4 cm lateral to its midpoint. The needle is advanced toward the sternal notch.
- Gentle aspiration during needle insertion is applied until dark, nonpulsatile venous blood is evident.
- Placement of the catheter is similar to that described for IJV catheterization.

POTENTIAL COMPLICATIONS

Infectious, mechanical, and thrombotic risks are associated with central venous catheterization.

- **Infectious complications** may be reduced using an appropriate sterile technique, selection of an optimal catheter site, and maintenance of the catheter only as long as needed.
- **Mechanical complications** involve structures adjacent to or near the vein selected for catheterization. An inappropriately placed catheter may injure adjacent structures. Complications include:
- Arterial puncture
  - Hematoma
  - Pneumothorax
  - Hemothorax
- Chylothorax
- Arrhythmia

The internal jugular and subclavian veins are preferred because they are associated with lower risk of infection and fewer mechanical complications (**Table 6.1.1**).

Thrombotic complications. Central venous catheterization increases the risk of thrombosis and the resultant risk of venous thrombo-embolism. The subclavian vein has the lowest risk for thrombotic complications.

Clinical Note

If not contraindicated, the CDC recommends the subclavian vein be used for central venous catheterization in adults.

CONCLUSION

Understanding the anatomy of the internal jugular and subclavian veins, including their surface anatomy and anatomical relationships, is essential for their safe and successful catheterization. Careful sterile technique, as well as appropriate catheter management, may minimize catheter-related infections. Mechanical complications involve structures that are adjacent to these veins, either at the site of catheter insertion or along the course of the vessels.

Hematoma Localized extravasation of blood, usually clotted
Hemothorax Blood in pleural cavity
Arrhythmia Irregular heart beat

Thrombus A fixed mass of platelets and/or fibrin (clot) that partially or totally occludes a blood vessel or heart chamber. An embolism is a mobile clot in the cardiovascular system

6.2 | Hyperthyroidism

Patient Presentation

A 41-year-old woman visits the family medicine clinic complaining of fatigue, eyestrain, and weight loss over the last 16 months.

Relevant Clinical Findings

History

The patient describes that she no longer tolerates heat, as in the past. She also recalls having heart palpitations.

The patient reports having no previous medical problems but does note that her older brother had similar symptoms and requires medication.

Physical Examination

Noteworthy vital signs:

- Height: 5′ 9″
- Weight: 115 lb—this represents a 50 lb loss since her last medical evaluation
- Pulse: 110 bpm
  Adult resting rate: 60–100 bpm
- Blood pressure: 166/72 mm Hg
  Normal adult: 120/80

Results of physical examination:

- Diffusely enlarged, nontender goiter
- Diaphoresis

- Exophthalmos (**Fig. 6.2.1**)
- Hyper-reflexia
- Tremor with arms outstretched

Laboratory Tests

Test	Value	Reference value
Thyrotropin (thyroid-stimulating hormone, TSH)	0.1	0.34–4.25 ml U/L
Total triiodothyronine (T3)	277	77–135 ng/dL
Free thyroxine (T4)	42.4	9.0–16.0 pmol/L

Imaging Studies

- Radioactive iodine scanning and measurements of iodine showed increased uptake, distributed diffusely over the thyroid gland.

Biopsy

- Fine-needle aspiration biopsy confirmed pathology of the thyroid gland, including hyperplastic follicular epithelium, follicular hypertrophy, and little colloid.

Clinical Problems to Consider

- Graves disease
- Supraclavicular lymphadenopathy

LEARNING OBJECTIVES

1. Describe the anatomy and relationships of the thyroid and parathyroid glands.
2. Describe the anatomy of the recurrent laryngeal nerve.
3. Describe the anatomy of the lymphatics of the head and neck.
4. Explain the anatomical basis for the signs and symptoms associated with this case.

Palpitation Forcible or irregular pulsation of the heart that is perceptible to the patient
Diaphoresis Sweating
Exophthalmos Protrusion of the eye (synonym: proptosis)
Hyper-reflexia Exaggerated deep tendon reflex
Hyperplasia Increase in size of a tissue or organ due to increased cell numbers (antonym: hypertrophy)
Hypertrophy Increase in size of a tissue or organ due to increased cell size, that is, without increased cell number (antonym: hyperplasia)
Lymphadenopathy Disease of the lymph nodes; used synonymously to mean swollen or enlarged lymph nodes

(A)

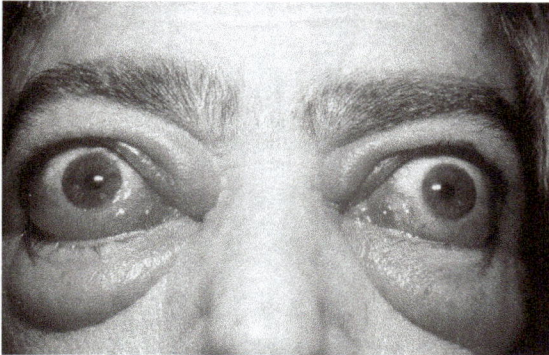

(B)

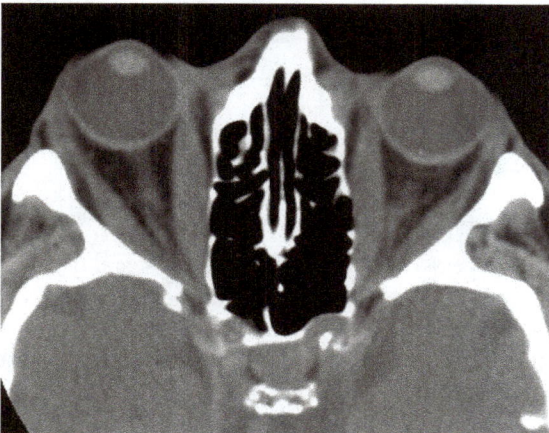

FIGURE 6.2.1 A patient with marked exophthalmos and periorbital edema. *Source:* Fig. 15-23 in *Vaughan & Asbury's General Ophthalmology,* 18e.

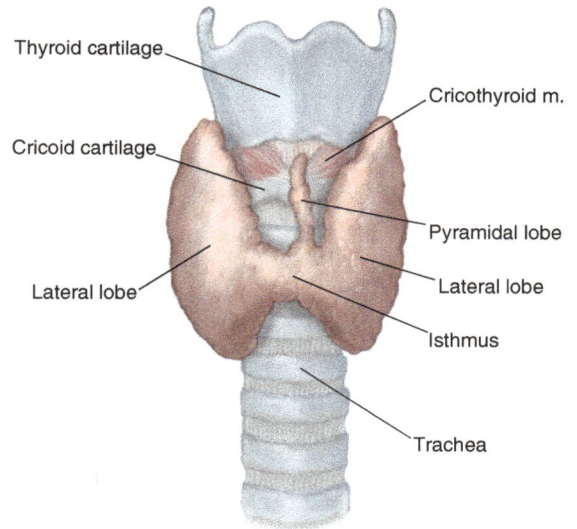

FIGURE 6.2.2 Anterior view of the thyroid gland.

approximately 50% of the population, projects superiorly from the isthmus just left of midline and represents a developmental remnant of the thyroglossal duct.

Thyroid Hormones

Under the stimulation of thyrotropin (thyroid stimulating hormone, TSH), the **thyroid gland** produces two **hormones**: **triiodothyronine (T3)** and **thyroxine (T4)**. T4 is the predominant form in terms of its relative level in the blood (T4:T3 ratio is 20:1). When released from the gland, however, T4 is converted to T3, which is 3–4 times more potent than T4. These hormones play an important role in body metabolism. Thyroid follicles are composed of a layer of epithelial cells that surround protein material (predominantly thyroglobulin) called **colloid.** The amount of follicular colloid reflects the overall activity of the gland:

- **When the gland is inactive**: follicles are large, colloid is abundant, and epithelial cells are flattened.
- **When the gland is active**: follicles are small, relatively little colloid is present, and epithelial cells are cuboid or columnar.

Anatomical Relationships

The thyroid gland is closely related to structures in the antero-inferior aspect of the neck (**Table 6.2.1**).

RELEVANT ANATOMY

Thyroid Gland

The **thyroid gland** is located in the antero-inferior neck. It extends from vertebral levels C5-T1. It is in the visceral compartment of the neck and, therefore, is enclosed in the visceral part of the prevertebral fascia. A fibrous capsule also envelops the gland.

The gland consists of paired lateral lobes that are connected across the midline by the thyroid isthmus (**Fig. 6.2.2**). The lateral lobes extend from the laminae of the thyroid cartilage to the level of the 4th or 5th tracheal cartilages. The isthmus crosses the midline just inferior to the cricoid cartilage. A pyramidal lobe, present in

TABLE 6.2.1	Anatomical relationships of the thyroid gland.
Direction	**Structure**
Anterior	Sternothyroid and sternohyoid muscles
Posterior	Parathyroid glands Tracheal rings 2–3 Recurrent laryngeal nerve
Posteromedial	Trachea and larynx
Posterolateral	Carotid sheath (associated closely with common carotid artery)

The thyroid gland is attached to the underlying laryngeal cartilages and trachea by loose connective tissue. The sternothyroid muscle, which overlies the thyroid gland, is attached to the oblique line of the thyroid cartilage. This muscle helps to hold the thyroid gland against the trachea and larynx, and limits the ability of the gland to move independent of the larynx. Consequently, the thyroid gland moves upward and downward with deglutition.

Blood Supply

The thyroid gland is supplied by two arteries and drained by three veins (**Fig. 6.2.3A**). A third artery, the thyroid ima, is present in approximately 10% of individuals. The blood supply to the thyroid gland is outlined in **Table 6.2.2**.

Lymphatics

Lymph from the thyroid gland drains primarily to **deep cervical nodes** and, thereby, into the jugular lymph trunks. Lymph from the thyroid can also drain directly into the thoracic duct. Prelaryngeal, pretracheal, paratracheal, and brachiocephalic nodes are also involved.

Nerve Supply

Endocrine secretion from the thyroid gland is regulated *hormonally* by **thyrotropin** (thyroid-stimulating hormone, TSH) released from the pituitary gland. **Sympathetic**, vasoconstrictor fibers from cell bodies in cervical paravertebral ganglia follow the superior and inferior thyroid arteries. The recurrent laryngeal nerves lie immediately posterior to the thyroid gland

---

Deglutition Swallowing

TABLE 6.2.2	Blood supply of the thyroid gland.	
Vessel	**Origin/termination**	**Notes**
Arteries		
Superior thyroid	External carotid	
Inferior thyroid	Subclavian → Thyrocervical trunk	
Thyroid ima	Brachiocephalic trunk or aortic arch	Present in 10% of individuals
Veins		
Superior thyroid	Internal jugular	
Middle thyroid	Internal jugular	
Inferior thyroid	Left brachiocephalic	Veins from both sides form plexus

(**Fig. 6.2.3B**) but do not innervate the parenchyma of the gland.

Parathyroid Glands

Four, pea-sized **parathyroid glands** are typically found on the posterior aspect of the lateral lobes of the thyroid gland. They usually lie within 1 cm of the intersection of the inferior thyroid artery and the recurrent laryngeal nerve.

Blood Supply

The parathyroid glands are supplied primarily by the **inferior thyroid arteries**. They may receive contributions from the superior thyroid, thyroid ima (if present), laryngeal, tracheal, or esophageal arteries. Blood from the parathyroids drains to the **thyroid venous plexus** and the left brachiocephalic vein.

Nerve Supply

Innervation of the parathyroid glands is similar to that of the thyroid gland, that is, sympathetic and vasoconstrictor fibers. Parathyroid hormone secretion is regulated primarily by negative feedback through receptors on parathyroid cells.

Recurrent Laryngeal Nerve

The recurrent laryngeal nerve is a branch of the vagus nerve (CN X). In the thorax, the course is different on each side (**Fig. 6.2.3B**):

(A)

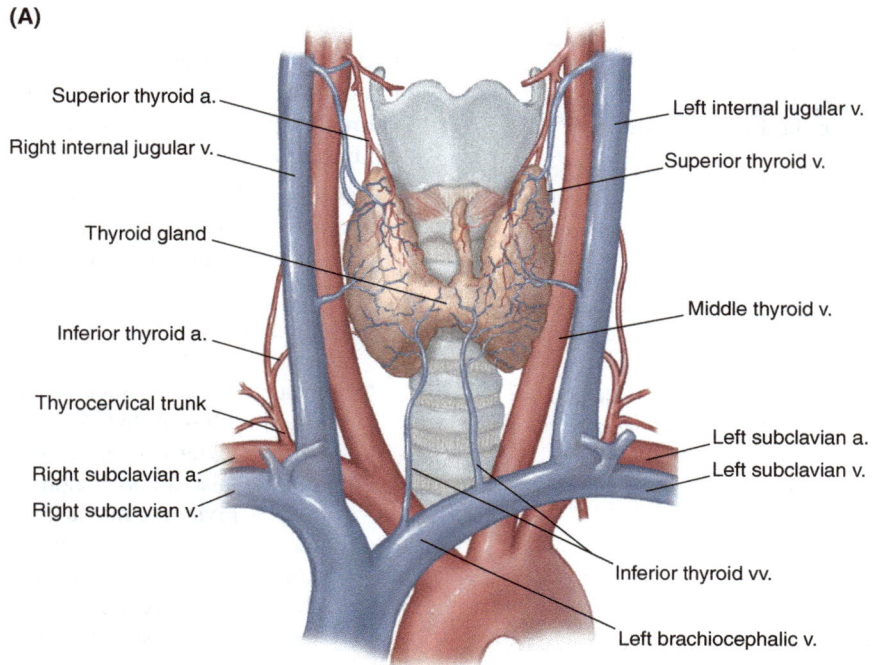

(B)

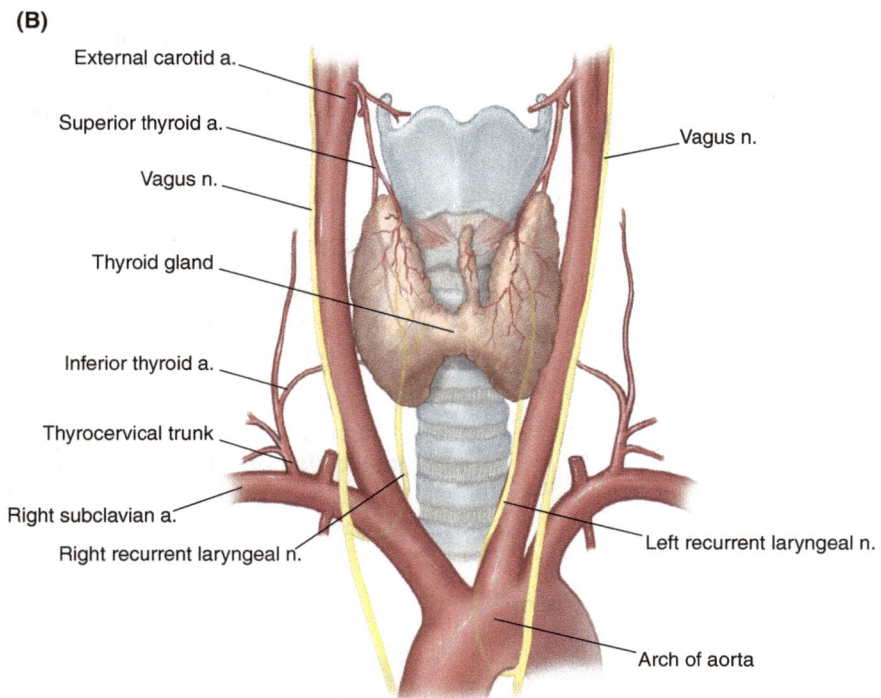

FIGURE 6.2.3 (**A**) Blood supply of the thyroid gland and (**B**) relationship of the thyroid gland to the left and right recurrent laryngeal nerves.

TABLE 6.2.3	Laryngeal distribution of the recurrent laryngeal nerve.	
Branch	**Structures supplied**	**Notes**
Anterior	All intrinsic laryngeal muscles (except cricothyroid)	Often called inferior laryngeal nerve Cricothyroid innervated by external branch of superior laryngeal nerve (branch of CN X)
Posterior	Sensory from infraglottic mucosa	Remaining laryngeal mucosa supplied by internal branch of superior laryngeal nerve

- The **left recurrent laryngeal nerve** branches as CN X crosses the arch of the aorta. It then passes posterior to ligamentum arteriosum, loops beneath the arch, and ascends into the neck in the tracheo-esophageal groove.

- The **right recurrent laryngeal nerve** leaves the CN X as it passes anterior to the first part of the right subclavian artery. The nerve loops the subclavian and, like the left, ascends in the tracheoesophageal groove.

The recurrent laryngeal nerves supply the superior part of the esophagus before terminating in the larynx; this nerve terminates as anterior and posterior branches that distribute differentially within the larynx (**Table 6.2.3**). Recurrent laryngeal and internal laryngeal nerves also supply parasympathetic innervation to laryngeal mucous glands.

Clinical Note

Proximity of the recurrent laryngeal nerve to the inferior thyroid artery places this nerve at risk for injury when this artery is dissected and branches are ligated during thyroid or parathyroid surgery.

Lymphatics of the Head and Neck

Lymph from the head and neck drains sequentially through **superficial cervical nodes** and **deep cervical nodes**. There are several groups of **deep cervical lymph nodes**:

- Nodes in the **superior group** are related to the superior part of the internal jugular vein. They include the **jugulodigastric nodes**, which drain part of the tongue.

- Nodes in the **inferior group** are related to the inferior part of the internal jugular vein, the brachial plexus, and subclavian vessels.
  - The **jugulo-omohyoid node**, associated with the intermediate tendon of omohyoid muscle, also drains lymph from the tongue.
  - **Supraclavicular nodes** are located posterior to the origin of sternocleidomastoid muscle. In addition to the head and neck, they also have lymphatic afferents from upper limb, chest wall, and breast.

- **Retropharyngeal nodes** receive lymph from the nasopharynx and auditory tube.

- **Paratracheal nodes** drain the trachea and esophagus.

- **Infrahyoid, prelaryngeal, pretracheal nodes** drain structures in the antero-inferior neck.

Clinical Note

Metastases from abdominal organs, lung, and breast frequently involve neck lymphatics, notably supraclavicular lymph nodes.

Lymph from deep cervical nodes is collected in **jugular lymph trunks** that terminate at the venous angle (the junction of the internal jugular and subclavian veins).

CLINICAL REASONING

Based on the patient history, physical examination, laboratory studies, and fine-needle aspiration biopsy, the patient has an **enlarged thyroid gland**.

Graves Disease

A goiter is an enlarged thyroid gland; the gland may be diffusely enlarged or may have isolated or multinodular goiter. Goiter may be associated with hypo- or hyperthyroidism. The presence of a goiter does not, however, reflect thyroid malignancy.

Hyperthyroidism results when the gland produces elevated levels of circulating thyroid hormones (T3 or T4). **Thyrotoxicosis** refers to the clinical manifestations associated with elevated thyroid hormone levels.

- **Primary hyperthyroidism** is caused by excess production of thyroid hormones from the thyroid gland.
  - **Graves disease**, a form of primary hyperthyroidism, is an autoimmune disorder in which the thyroid gland synthesizes and releases increased levels of thyroid hormones. It accounts for 60–80% of cases of thyrotoxicosis. Other forms of primary hyperthyroidism include toxic multinodular goiter is (second most common cause) and toxic nodular (adenoma) goiter.
- **Secondary hyperthyroidism** is caused by excess production of thyroid-releasing hormones from the hypothalamus or TSH from the pituitary.

Causes of secondary hyperthyroidism include:

- Thyrotropin-secreting pituitary adenoma—TSH stimulates thyroid gland to produce excess hormone.
- Hashimoto thyroiditis—the thyroid gland is overactive initially (hyperthyroidism) followed by a state of hypothyroidism.

Signs and Symptoms

Signs

- Tachycardia
- Tremor
- Goiter
- Warm, moist skin
- Hyper-reflexia
- Muscle weakness
- Eyelid retraction or lag (related to sympathetic overactivity)

Signs Specific for Graves Disease

- Ophthalmopathy (**Fig. 6.2.4A**): exophthalmos, upper eyelid retraction, peri-orbital edema, and conjunctival injection
- Thyroid acropachy (**Fig. 6.2.4B**)
- Dermopathy characterized by pretibial myxedema that resembles the color and texture of an orange peel (**Fig. 6.2.4C**)

Orbital inflammation and edema of the extra-ocular muscles are the primary cause of exophthalmos. Restriction of gaze, diplopia, and visual loss from optic nerve compression may occur if the muscles become sufficiently inflamed.

Symptoms

- Hyperactivity, irritability, dysphoria
- Heat intolerance and sweating
- Palpitations
- Fatigue and weakness
- Weight loss with increased appetite
- Frequent bowel movements or diarrhea

Tachycardia Increased heart rate: >100 bpm (normal adult heart rate: 55–100 bpm)
Ophthalmopathy A disease of the eye
Edema Swelling of skin due to abnormal accumulation of fluid in subcutaneous tissue
Acropachy Clubbing of fingers and/or toes caused by edema and periosteal changes

Dermopathy A disease of the skin
Myxedema A dermopathy associated with hypothyroidism caused by accumulation of subcutaneous mucoid material. It may be pronounced on the face and shin
Dysphoria Generalized mood depression
Diplopia Double vision

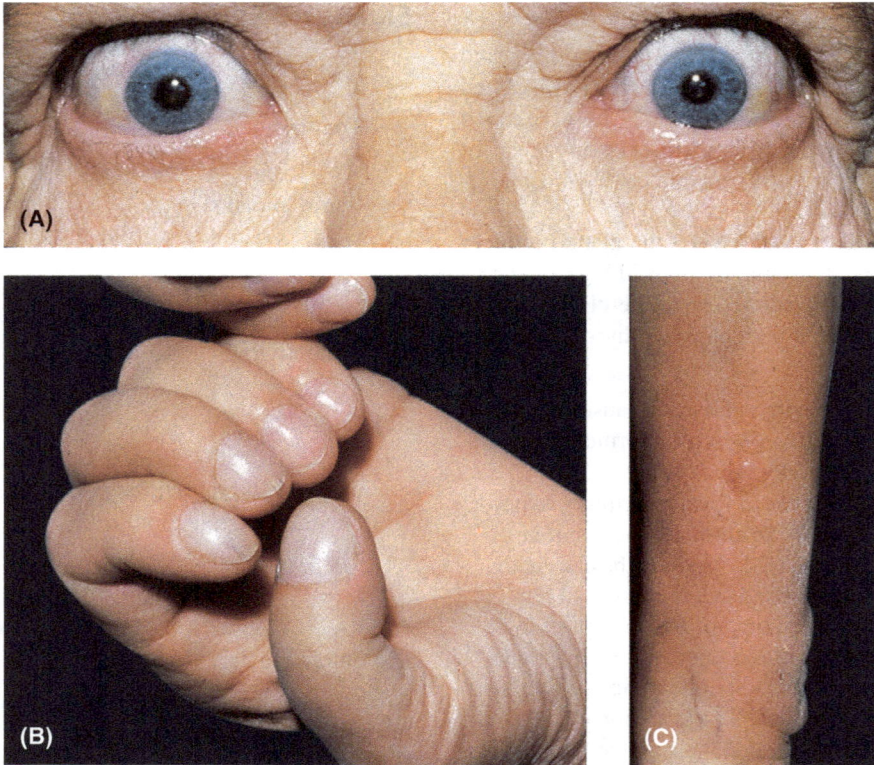

FIGURE 6.2.4 Graves disease. (**A**) Exophthalmos, (**B**) acropachy, and (**C**) dermopathy.
Source: Fig. 152-11 in *Fitzpatrick's Dermatology in General Medicine,* 7e. Accessmedicine.com.

- Polyuria
- Oligomenorrhea or amenorrhea (women); loss of libido in men

Predisposing Factors

- Sex: women (8:1)
- Age: 20–50 years
- Family history of hyperthyroidism, especially siblings
- High iodine intake
- Stress

Polyuria Excessive excretion of urine
Oligomenorrhea Reduced menses (periodic hemorrhage related to the menstrual cycle)
Amenorrhea Absence or abnormal cessation of menses (periodic hemorrhage related to the menstrual cycle)
Libido Sexual desire (female or male)

- Use of sex steroids
- Smoking

Supraclavicular Lymphadenopathy

The **supraclavicular lymph nodes,** located in the **greater supraclavicular fossa** (in the inferior part of the lateral cervical region), are not palpable under normal conditions. Supraclavicular lymph nodes are important clinically because they drain structures in the thorax and abdomen:

Clinical Note

- There are estimated 600 lymph nodes in the body, but only those in the submandibular, axillary, or inguinal groups are normally palpable.

Clinical Notes

- Enlarged **right supraclavicular lymph nodes** are suggestive of metastasis of lung and breast tumors.

- Enlarged **left supraclavicular lymph nodes** are most often related to the spread from malignancies from the abdomen (commonly the stomach) by way of the thoracic duct. One of these is particularly large and referred to as **Virchow node**. Detection of this node, referred to as a **Troisier sign**, is strongly suggestive of metastases in the cervical region. Consequently, this node is also called a **sentinel node** or **signal node**.

- Sites in the head and neck (notably, pharynx, tonsils, and tongue) may also metastasize to the inferior neck.

- Enlargement of supraclavicular lymph nodes is associated with metastasis in 54–85% of cases.

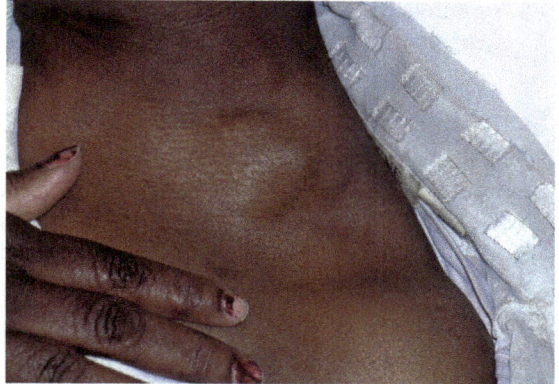

FIGURE 6.2.5 A patient with a lateral neck swelling revealed to be a Virchow node. *Source:* Fig. 7.34 in *The Atlas of Emergency Medicine.*

Signs and Symptoms

- Enlarged, palpable supraclavicular nodes (**Fig. 6.2.5**).

- Asking the patient to perform a Valsalva maneuver may enhance the ability to palpate these nodes.

Clinical Note

In a **Valsalva maneuver**, the patient exhales with moderate force with the airway closed (as if bearing down during a bowel movement). This increases intra-abdominal and intrathoracic pressures, thereby elevating the cervical pleural and structures—including supraclavicular nodes—further into the root of the neck. As a result, supraclavicular nodes may be palpable (**Fig 6.2.5**).

Predisposing Factors

- Risk factors are related to the primary tumor and its specific predilection to metastasize.

DIAGNOSIS

The patient presentation, medical history, physical examination, laboratory tests, and imaging studies support a diagnosis of **Graves disease**.

Graves Disease

The primary function of thyroid hormone is to regulate basal metabolic rate. Thyroid dysfunction that results in elevated or depressed thyroid hormone levels increases or decreases metabolism accordingly.

- Hyperthyroidism and increased basal metabolic rate account for most signs and symptoms (e.g., increased pulse and blood pressure, weight loss, diaphoresis, hyper-reflexia, and tremor).

- Goiter (an enlarged thyroid gland).

- Thyroid biopsy showed follicular hypertrophy and little colloid, which indicates increased glandular activity.

- A triad of signs and symptoms are pathognomonic for Graves disease: ophthalmopathy, dermopathy, and thyroid acropachy. Clinical

---

Pathognomonic Characteristic symptom that points unmistakably to a specific disease

signs related to the orbit are likely the result of lymphocytic infiltration of orbital tissues.

Supraclavicular Lymphadenopathy

Supraclavicular lymph nodes are important because they drain structures of the thorax and abdomen. When enlarged, they may be palpable and serve as an indicator of metastatic disease from the head and neck, as well as from distant primary tumors.

■ Supraclavicular lymphadenopathy, with palpable supraclavicular lymph node(s), presents as an anterolateral neck mass.

The presence of a palpable neck mass is a clinical sign shared by goiter and supraclavicular lymphadenopathy. The latter condition, however, is not associated with the wide constellation of additional signs and symptoms that characterize Graves disease.

6.3 Retropharyngeal Abscess

Patient Presentation

A 23-month-old boy was brought to the family medicine clinic with recurring fever, neck pain and swelling, and difficulty swallowing.

Relevant Clinical Findings

History

The parents report that the child has had repeated respiratory infections. They note that he seems to have difficulty eating and drinking, and that he pulls at his throat and ears.

Physical Examination

Noteworthy vital signs:

- Temperature (rectal): 40°C (104°F)
    Normal: 36.6–38°C (97.9–100.4°F)

Results of physical examination:

- Neck swelling
- Torticollis
- Dysphagia
- Tonsillar inflammation
- Cervical lymphadenopathy

Laboratory Tests

Test	Value	Reference value
Leukocytes (count)	17.4	$3.54–9.06 \times 10^3/mm^3$

Imaging Studies

- A lateral radiograph reveals anterior displacement of the trachea (**Fig. 6.3.1**).

Clinical Problems to Consider

- Acute bronchitis
- Epiglottitis
- Retropharyngeal abscess

LEARNING OBJECTIVES

1. Describe the cervical fasciae and compartments of the neck.
2. Describe the anatomy of the retropharyngeal space.
3. Describe the anatomy of the pharynx, esophagus, larynx, and tracheobronchial tree.
4. Explain the anatomical basis for the signs and symptoms associated with this case.

RELEVANT ANATOMY

Cervical Fasciae and Compartments of the Neck

Superficial Fascia

The subcutaneous tissue of the neck contains variable amounts of fat and structures outlined in **Table 6.3.1**.

Deep Fascia

Four subdivisions of cervical deep fascia are identified: **investing**, **pretracheal**, **prevertebral**, and

---

Torticollis Spasmodic contraction of neck muscles (often those innervated by CN XI)
Dysphagia Difficulty swallowing

Lymphadenopathy Disease of the lymph nodes; used synonymously to mean swollen or enlarged lymph nodes
Abscess Collection of purulent exudate (pus)

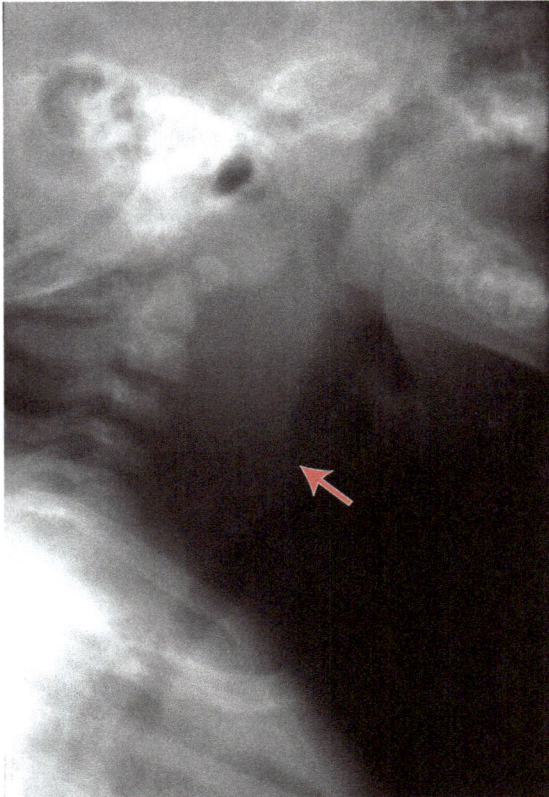

FIGURE 6.3.1 Lateral cervical radiograph showing the trachea (red arrow) that has been displaced anteriorly. *Source:* Fig. 119-9 in *Tintinalli's Emergency Medicine: A Comprehensive Study Guide, 7e.*

TABLE 6.3.1	Structures related to superficial cervical fascia.
Tissue	**Structure(s)**
Nerves	**Cutaneous branches of C2-C4 ventral rami (cervical plexus)** ■ Lesser occipital nerve (C2) ■ Great auricular nerve (C2, C3) ■ Transverse cervical nerve (C2, C3) ■ Supraclavicular nerves (C3, C4) **Cutaneous branches of C2-C8 dorsal rami** ■ Greater occipital nerve (C2) ■ Occipital cutaneous nerve (C3-C8) **Cervical branch of facial nerve (CN VII)**
Blood vessels	■ External jugular vein ■ Anterior jugular vein
Lymphatics	Superficial cervical nodes
Muscle	Platysma muscle

Nomenclature for this space is variable:

- In one view, the entire space, from cranium to mediastinum, is called *retropharyngeal*.
- Others subdivide this space at the pharyngo-esophageal junction:
  - **Posterior to pharynx:** *retropharyngeal space*
  - **Posterior to esophagus:** *retrovisceral space*

Clinical Note

The retropharyngeal space and the interfaces between all deep cervical fasciae are natural cleavage planes. These are useful for surgical dissection, but also serve as routes for the spread of infection.

Buccopharyngeal Fascia

The fascia that lines the posterior aspect of the pharynx and esophagus (i.e., the anterior wall of the retropharyngeal/retrovisceral space) is referred to commonly as **buccopharyngeal**. Some authors note that it extends anteriorly onto the buccinator muscles. The **alar fascia** is a thin layer that crosses the retropharyngeal space from the cranial base to the C7 vertebral level. It is attached laterally to the carotid

carotid sheath. Each fascia surrounds groups of structures and defines five cervical compartments (**Table 6.3.2**).

The interfaces created by cervical deep fasciae facilitate movement of adjacent structures in the neck. The cervical fasciae, the compartments they create, and structures associated with each compartment are shown in **Figure 6.3.2**.

Retropharyngeal Space

The **retropharyngeal space** is a potential space bounded by deep cervical fascia. It lies between the vertebral and visceral compartments and contains loose connective tissue (**Fig. 6.3.2**). The anatomical relationships of this space are outlined in **Table 6.3.3**.

TABLE 6.3.2	Cervical fasciae and compartments.	
Fascia	**Compartment**	**Location**
Investing	**Muscular compartment** ▪ Sternocleidomastoid muscle ▪ Trapezius muscle	Immediately deep to superficial fascia
Pretracheal *Visceral part* *Muscular part*	**Visceral compartment** ▪ Thyroid gland ▪ Parathyroid glands ▪ Trachea ▪ Esophagus ▪ Infrahyoid muscles	Surrounds structures anterior to cervical vertebrae and associated muscles
Prevertebral	**Vertebral compartment** ▪ Cervical vertebrae ▪ Longus colli and capitis muscles ▪ Scalene muscles ▪ Splenius capitis muscle ▪ Levator scapulae muscle ▪ Deep posterior neck muscles	Surrounds cervical vertebrae and associated muscles
Carotid sheaths (2)	**Vascular compartments (paired)** ▪ Common carotid artery ▪ Internal carotid artery ▪ Internal jugular vein ▪ Vagus nerve (CN X) ▪ Deep cervical lymph nodes	Lateral to visceral compartment Deep nodes lie along internal jugular vein

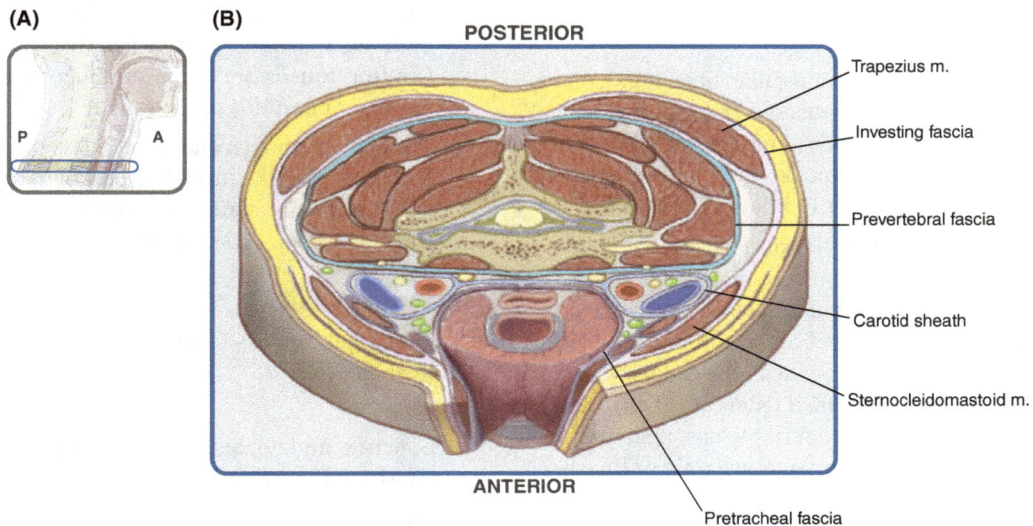

FIGURE 6.3.2 Cervical fasciae and neck compartments. (**A**) Midsagittal section and (**B**) transverse section.

TABLE 6.3.3	Anatomical relationships of the retropharyngeal space.		
Direction	**Fascia**	**Related structure(s)**	**Notes**
Anterior	Visceral part of pretracheal fascia (buccopharyngeal)	***Visceral compartment*** ▪ Pharynx ▪ Esophagus	
Posterior	Prevertebral	***Vertebral compartment*** ▪ Cervical vertebral bodies ▪ Longus capitis and colli muscles ▪ Sympathetic trunk	Sympathetic trunk embedded in prevertebral fascia, anterior to longus capitis and colli muscles
Lateral	Carotid sheaths	***Vascular compartments*** ▪ Common and internal carotid arteries (CCA/ICA) ▪ Internal jugular vein (IJV) ▪ Vagus nerve (CN X) ▪ Deep cervical lymph nodes	
Superior		Skull base	
Inferior		▪ Superior mediastinum ▪ Posterior mediastinum	▪ In superior mediastinum, pretracheal fascia extends along great vessels and blends with fibrous pericardium. ▪ In the thorax, prevertebral fascia blends with anterior longitudinal ligament.

sheaths and along the midline to the buccopharyngeal fascia.

Pharynx

The pharynx is a fibromuscular tube that extends from the cranial base (sphenoid and occipital bones) to the esophagus. It opens anteriorly into the nasal and oral cavities, and lies posterior to the larynx. It has three parts (superior to inferior) (**Fig. 6.3.3**):

1. The **nasopharynx** lies posterior to the nasal cavities, with which it communicates via the choanae (posterior nasal apertures). It extends inferiorly to the level of the soft palate, where it is continuous with the oropharynx. The **pharyngeal tonsil** (adenoids) lies in the submucosa of the roof of the nasopharynx. The **auditory tube** (pharyngotympanic or Eustachian) opens into its lateral wall.

2. The **oropharynx** lies posterior to the oral cavity and contains the posterior one-third of the tongue. It extends inferiorly to the level of the superior border of the epiglottis. It communicates with the oral cavity via the opening bordered laterally by the palatoglossal and palatopharyngeal arches. **Palatine tonsils** lie in the tonsillar bed between these arches. **Lingual tonsils** are in the submucosa of the posterior one-third of the tongue.

3. The **laryngopharynx** lies posterior to the larynx. It extends from the superior aspect of the epiglottis to the inferior border of cricoid cartilage, where it becomes continuous with the esophagus.

Muscular Wall

The pharyngeal wall is formed by circular and longitudinal muscles.

Circular
▪ Superior, middle, and inferior pharyngeal constrictor muscles

Longitudinal
▪ Palatopharyngeus
▪ Stylopharyngeus
▪ Salpingopharyngeus

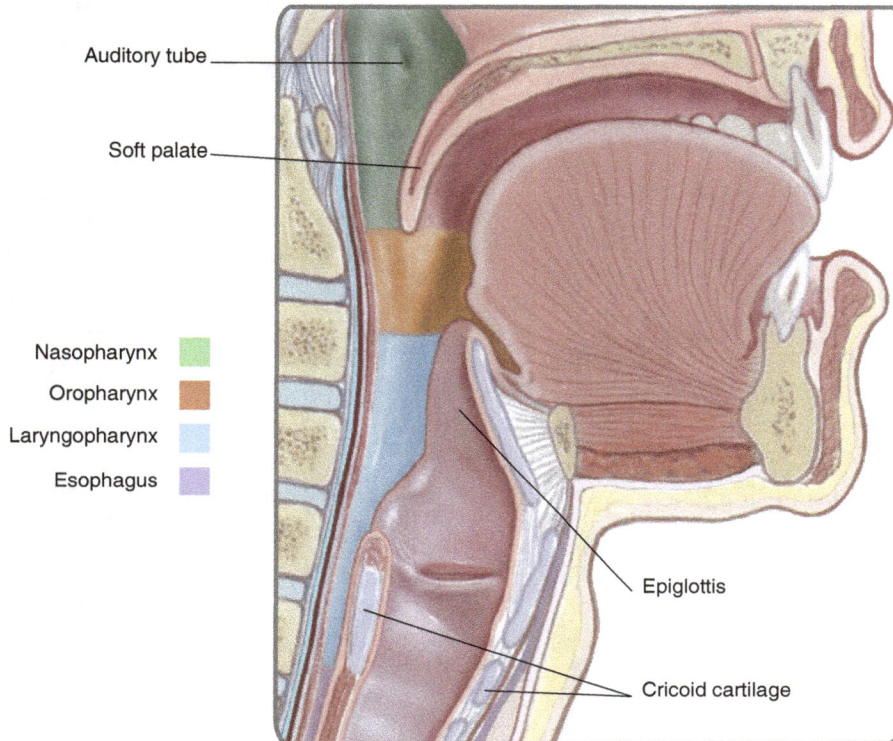

FIGURE 6.3.3 Midsagittal section of the pharynx.

Blood Supply and Lymphatics

The pharynx is supplied by branches of the **external carotid artery** and drains into the **internal jugular vein**. The blood and lymphatic supply of the pharynx are outlined in **Table 6.3.4**.

Nerve Supply

The pharynx receives somatic and visceral innervation, primarily via the **pharyngeal plexus**. This plexus is located on the external aspect of the posterior pharynx. It receives contributions from the glossopharyngeal nerve (CN IX), vagus nerve (CN X), and sympathetic nerves.

Sensory

- **Nasopharynx** is supplied by the **maxillary nerve (CN V2)**.
- **Oropharynx** is innervated by the **CN IX** via the pharyngeal plexus.

- **Laryngopharynx** is supplied by the **CN X** via the pharyngeal plexus.

Motor

- **Pharyngeal branches of the CN X** innervate most pharyngeal muscles.
- The **external branch of the superior laryngeal nerve** innervates the inferior pharyngeal constrictor.
- The **CN IX** innervates **stylopharyngeus**.

Autonomic

- **Sympathetic (visceral efferent).** Axons from postganglionic cell bodies in middle and inferior cervical ganglia follow vasculature to the pharynx.
- **Parasympathetic (visceral efferent).** Preganglionic parasympathetic fibers from the CN X synapse in ganglia in the pharyngeal plexus.

TABLE 6.3.4	Blood supply and lymphatics of the pharynx, esophagus, larynx, and tracheobronchial tree.		
	Arterial supply	**Venous drainage**	**Lymphatics**
Pharynx	**External carotid artery** ▪ Ascending pharyngeal ▪ Facial (ascending palatine, tonsillar) ▪ Maxillary (greater palatine, pharyngeal, artery of pterygoid canal) ▪ Lingual (dorsal lingual)	**Internal jugular vein** ▪ Pharyngeal venous plexus	**Nodes** ▪ Deep cervical ▪ Retropharyngeal ▪ Paratracheal **Pharyngeal tonsillar ring (Waldeyer)** ▪ Palatine, pharyngeal, and lingual tonsils form an incomplete ring of submucosal lymphoid tissue in walls of nasopharynx and oropharynx
Esophagus Cervical	**Subclavian artery** ▪ Inferior laryngeal (from inferior thyroid/thyrocervical trunk)	**Left brachiocephalic vein** ▪ Inferior thyroid plexus	**Nodes** ▪ Paratracheal ▪ Inferior deep cervical
Thoracic	**Thoracic aorta** ▪ Bronchial ▪ Esophageal **Abdominal aorta** ▪ Left gastric (from celiac trunk)	**Azygos** ▪ Hemiazygos ▪ Intercostal ▪ Bronchial	**Nodes** ▪ Posterior mediastinal
Abdominal	**Abdominal aorta** ▪ Esophageal (from left gastric/celiac trunk) ▪ Left inferior phrenic	**Hepatic portal vein** ▪ Left gastric ▪ Short gastric vv.	**Nodes** ▪ Left gastric (to celiac)
Larynx	**External carotid artery** ▪ Superior laryngeal (from superior thyroid) **Subclavian artery** ▪ Inferior laryngeal (from inferior thyroid/ thyrocervical trunk)	**Internal jugular vein** ▪ Superior laryngeal **Left brachiocephalic vein** ▪ Inferior laryngeal vv.	**Nodes** ▪ Deep cervical
Airway Trachea	**Subclavian artery** ▪ Inferior thyroid **Thoracic aorta** ▪ Bronchial (right or left)	**Left brachiocephalic vein** ▪ Inferior thyroid plexus	**Nodes** ▪ Pretracheal ▪ Paratracheal
Bronchi	Superior posterior intercostal or Left superior bronchial ▪ Right bronchial	**Pulmonary vv. or left atrium** ▪ Deep bronchial vv. **Right: Azygos** **Left: Hemiazygos and superior intercostal** ▪ Superficial bronchial vv.	**Nodes** ▪ Bronchopulmonary (hilar) ▪ Pulmonary ▪ Tracheobronchial (lymph drains to bronchomediastinal lymph trunks) ▪ Alveoli do not have lymphatic vessels

Postganglionic fibers innervate pharyngeal mucosal glands.

Esophagus

The **esophagus** is a muscular tube approximately 25 cm in length. It begins at the C6 vertebral level as the continuation of the laryngopharynx. At T10, it passes through the esophageal hiatus in the right crus of the diaphragm. It has **cervical, thoracic, and abdominal parts**. The abdominal part ends at the cardia of the stomach. The esophagus is divided into thirds based on muscle type:

- **Superior one-third: skeletal**
- **Middle one-third: skeletal and smooth**
- **Inferior one-third: smooth**

Blood Supply and Lymphatics

The esophagus is supplied by branches of the **subclavian artery** and **aorta**. Blood is drained primarily into the **left brachiocephalic vein** and **azygos venous system**. The abdominal esophageal veins are tributaries of the **hepatic portal vein**. The blood and lymphatic supply of the esophagus are outlined in **Table 6.3.4**.

Nerve Supply

The **superior one-third of the esophagus** is supplied by the **recurrent laryngeal nerves**. The **inferior *two-thirds* of the esophagus** is innervated by the **esophageal plexus**. This plexus receives contributions from both vagus nerves and both thoracic sympathetic trunks (T6-T10 spinal cord levels).

Sensory

- **Visceral afferent** fibers follow sympathetic nerves to T1-T4 spinal cord levels. Additional afferents are carried by CN X and have cell bodies in the inferior vagal ganglion.

Motor (somatic efferent)

- **Skeletal muscle** is supplied by the recurrent laryngeal nerve.

Parasympathetic (visceral efferent)

- **Smooth muscle** is supplied by CN X, via branches that synapse in, and pass through the esophageal plexus.

- Secretomotor fibers from CN X supply esophageal mucous glands.

Sympathetic (visceral efferent)

- **Cervical and thoracic parts of the esophagus**: Preganglionic sympathetic fibers arise from cell bodies in the lateral horn of the T1-T4/6 spinal cord. Postganglionic fibers are vasomotor.

- **Abdominal esophagus:** Preganglionic sympathetic fibers arise from the T5-T9 spinal cord lateral horns. These axons follow the **greater thoracic splanchnic nerve** to the **celiac ganglion**. Postganglionic fibers are vasomotor.

Clinical Note

It may be difficult to distinguish between esophageal and cardiac pain because sympathetic, visceral afferent fibers from both organs involve T1-T4 spinal cord.

Larynx

The **larynx** is part of the respiratory system. It is located between the oropharynx and the trachea, anterior to the laryngopharynx. The skeleton of the larynx is formed by nine cartilages (**Table 6.3.5**), and each of has a thick mucosa.

Laryngeal Cavity

The **laryngeal cavity** begins at the **laryngeal inlet** (**Table 6.3.6 and Fig. 6.3.4**). The cavity extends inferiorly to the inferior border of the cricoid cartilage, which marks the beginning of the trachea.

The laryngeal cavity is subdivided into three regions (**Table 6.3.6**).

Muscles

Movement of larynx is controlled by extrinsic and intrinsic muscles:

- **Extrinsic muscles** elevate (suprahyoid muscles) and depress (infrahyoid muscles) the larynx. Elevation of the larynx is essential for closure of the epiglottis during swallowing.

TABLE 6.3.5	Laryngeal cartilages.	
Cartilage	**Description**	**Notes**
Unpaired **Epiglottic**	Leaf-shaped; posterior to root of tongue, anterior to laryngeal inlet; attached to thyroid cartilage	Forms **epiglottis** with its mucous membrane; covers laryngeal inlet during swallowing
Thyroid	Largest cartilage; two laminae fused at anterior midline; has superior and inferior horns; has cricothyroid joints	Forms laryngeal prominence ("Adam's apple"); connected to hyoid bone and cricoid cartilage by thyrohyoid and cricothyroid membranes, respectively. Rotation (pivoting) of thyroid cartilage at cricothyroid joint regulates tension on vocal ligaments.
Cricoid	Signet ring shape; larger posteriorly; only circular laryngeal cartilage	Connected to trachea by cricotracheal ligament; articulates with arytenoid cartilages
Paired **Arytenoid**	Pyramidal, with two processes: 1. Vocal process—attachment of vocal ligament 2. Muscular process—attachment of cricoarytenoid muscles	Pivoting and gliding movements on cricoid cartilage adjust vocal ligaments
Corniculate	On apex of arytenoid cartilages	
Cuneiform	Nodules in aryepiglottic fold	

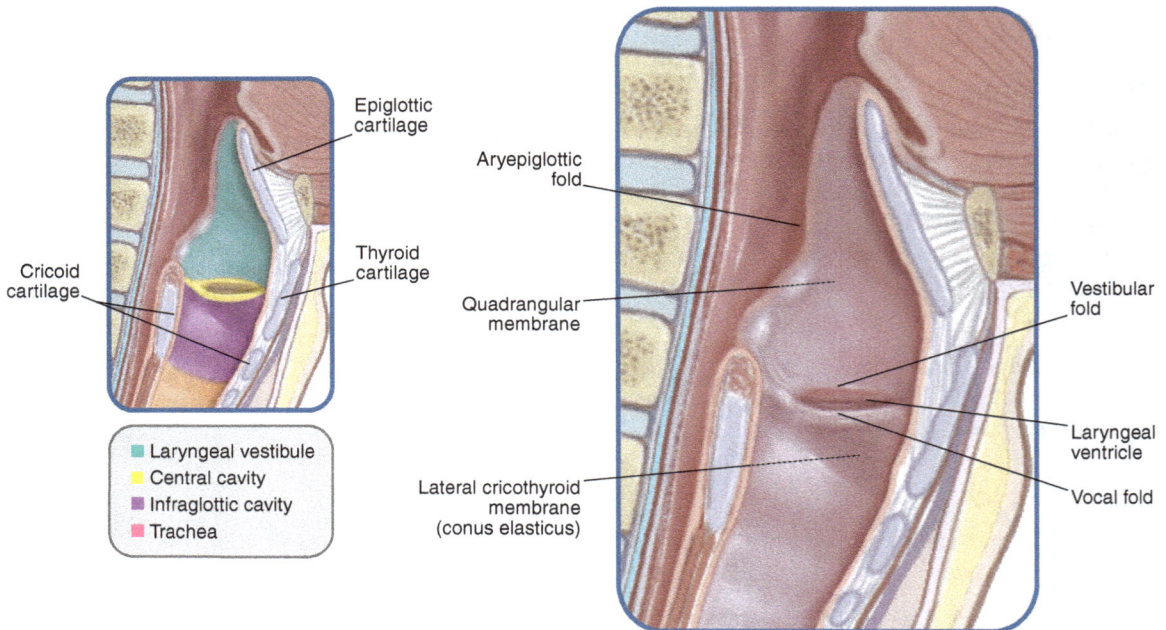

FIGURE 6.3.4 Midsagittal section of the larynx.

TABLE 6.3.6	**Regions of the larynx.**
Region	**Description**
Laryngeal vestibule **Laryngeal inlet** **Quadrangular membrane**	Space from laryngeal inlet (superiorly) to vestibular fold (inferiorly) Opening between laryngopharynx and larynx Fibroelastic membrane (covered by mucous membrane) ▪ **Anterior:** attached to lateral margin of epiglottic cartilage ▪ **Posterior:** attached to anterolateral margin of arytenoid cartilage ▪ **Superior (free) border** forms aryepiglottic fold ▪ **Inferior (free) border** forms vestibular fold
Central cavity **Vestibular fold** **Rima vestibuli** **Laryngeal ventricle** **Vocal fold** **Rima glottidis**	Space between vestibular and vocal folds Inferior (free) border of quadrangular membrane Space between *vestibular folds* Lateral diverticulum between vestibular and vocal folds Superior (free) border of conus elasticus; contains vocalis muscle Space between *vocal folds*
Infraglottic cavity **Lateral cricothyroid membrane (conus elasticus)**	Space from vocal folds to inferior border of cricoid cartilage Fibroelastic membrane (covered by mucous membrane); connects superior border of cricoid cartilage with thyroid cartilage ▪ **Superior (free) border** forms vocal ligament

▪ **Intrinsic muscles** include cricothyroid, thyroarytenoid, posterior and lateral crico-arytenoids, arytenoids, and vocalis. These muscles move laryngeal cartilages to adjust the length, tension, and position of the vocal ligaments.

Blood Supply and Lymphatics

Arterial supply to the larynx is derived from the **external carotid** and **subclavian arteries**. Venous drainage is into the **internal jugular** and **left brachiocephalic veins**. Lymphatic drainage of the larynx is through **deep cervical nodes**. The blood and lymphatic supply of the larynx are outlined in **Table 6.3.4**.

Nerve Supply

The larynx is innervated by the **superior and inferior laryngeal nerves**, which are branches of **CN X**.

Sensory

▪ The **internal branch of the superior laryngeal nerve** passes through the thyrohyoid membrane (with superior laryngeal blood vessels). It provides general sensation to the mucous membrane superior to the vocal folds. It also carries the afferent arm of the cough reflex.

▪ The **inferior laryngeal nerve**, a branch of **recurrent laryngeal**, provides general sensation inferior to the vocal folds.

▪ The distribution of the superior and inferior laryngeal nerves overlap in the region of the vocal folds.

Motor

▪ The **inferior laryngeal nerve** innervates all intrinsic laryngeal muscles except cricothyroid.

▪ The **external branch of the superior laryngeal nerve** supplies the cricothyroid muscle.

Autonomic

▪ **Sympathetic (visceral efferent).** Axons from postganglionic cell bodies in superior and middle cervical ganglia provide vasomotor innervation to vasculature of the larynx.

▪ **Parasympathetic (visceral efferent).** Preganglionic parasympathetic fibers in the superior and inferior laryngeal nerves synapse in ganglia in the pharyngeal plexus. Postganglionic parasympathetic fibers supply laryngeal mucosal glands.

Tracheobronchial Tree

The **trachea** begins at the inferior border of the cricoid cartilage. It ends in the mediastinum where it bifurcates into the **right and left main bronchi**. This hollow tube is approximately 2.5 cm in diameter in the adult. Its patency is maintained by a series of cartilaginous "rings" along its length. These rings are incomplete posteriorly and the gap in each cartilage is occupied by **trachealis** muscle.

Blood Supply and Lymphatics

Arterial supply to the trachea and bronchi is derived from the **subclavian arteries** and **thoracic aorta**. Many veins receive blood from the tracheobronchial tree (**Table 6.3.4**).
Pre- and paratracheal nodes drain lymph from the trachea. A deep pulmonary lymphatic plexus carries lymph from the bronchial tree to the bronchomediastinal lymph trunks (**Table 6.3.4**).

Nerve Supply of the Trachea

The **pulmonary plexus**, an extension of the cardiac plexus, supplies the trachea, bronchi, and lungs. The pulmonary plexus has sympathetic and parasympathetic components:

- Cardiac (cardiopulmonary splanchnic) nerves carry **postganglionic sympathetic fibers** from the cervical and upper four thoracic paravertebral ganglia to the cardiac plexus. Stimulation of these **visceral efferents** increases respiratory rate, and leads to bronchodilation and decreased glandular secretion. These fibers also conduct visceral afferent impulses (reflexive and nociceptive).
- **Preganglionic parasympathetic fibers** (cardiac branches) originate from the CN X. Parasympathetic stimulation decreases respiratory rate, and leads to bronchoconstriction and increased glandular secretion.

Nociception Nerve modality related to pain

CLINICAL REASONING

This pediatric patient presents with an **infection and inflammation of the respiratory tract**.

Acute Bronchitis

Bronchitis is an inflammation of the mucous membranes of the tracheobronchial tree that often follows an upper respiratory tract infection. The inflammatory response is accompanied by narrowed respiratory passages caused by increased mucus production and impaired clearance by respiratory cilia.

Signs and Symptoms

- Manifests initially as a common cold
- Cough may progress from nonproductive to productive
- Dyspnea
- Retrosternal pain in severe cases involving trachea
- Constitutional symptoms: malaise, chills, low-grade fever, sore throat, back and muscle pain

Predisposing Factors

- Age: most common <2 years and 9–15 years
- Associated most commonly with viral respiratory infections
- May occur in children with chronic sinusitis, allergies, and enlarged tonsils

Clinical Notes

- Young children have difficulty coughing up mucus and tend to swallow the mucus. This causes them to gag or vomit mucus.
- Acute pediatric bronchitis is generally self-limiting, 10–14 days after the onset of symptoms.

Epiglottitis

Epiglottitis (supraglottitis) is an inflammation of the epiglottis and supraglottic region. Epiglottitis may lead to respiratory arrest if the airway becomes obstructed.

Dyspnea Difficulty breathing and shortness of breath
Malaise Feeling of general body weakness or discomfort, often marking the onset of an illness

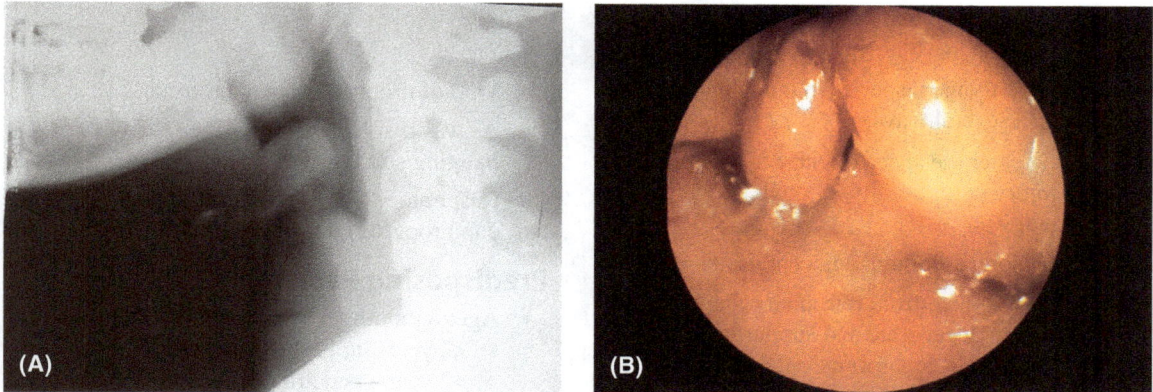

FIGURE 6.3.5 Imaging of the epiglottis in a pediatric patient with epiglottitis. **(A)** Lateral radiograph showing the "thumbprint" sign (arrow). *Source:* Fig. 50-4 in *Current Diagnosis and Treatment Emergency Medicine, 7e.* **(B)** Laryngoscopy of edematous, "cherry red" epiglottis. *Source:* Fig. 119-3 in *Tintinalli's Emergency Medicine.*

Signs and Symptoms

- Constitutional symptoms: high fever of sudden onset, severe sore throat, and systemic toxicity
- Edematous, "cherry red" epiglottis
- "Thumbprint" sign (refers to shape of epiglottis in a lateral neck x-ray).
- Tachycardia
- Drooling
- "Sniffing position"
- Signs and symptoms of respiratory obstruction (e.g., dyspnea and inspiratory stridor)
- Dysphagia
- Odynophagia
- Cyanosis

Tachycardia Increased heart rate: >100 bpm (normal adult heart rate: 55–100 bpm)

Stridor Noisy respiration, usually a sign of airway obstruction (especially involving the trachea or larynx)

Odynophagia Pain on swallowing

Cyanosis Bluish color of skin and mucous membranes from insufficient blood oxygen

Clinical Notes

- Diagnosis of epiglottitis may be confirmed by lateral radiographs (**"thumbprint sign"**) or direct fiberoptic laryngoscopy (edematous, "cherry-red" epiglottis) **(Fig. 6.3.5)**.

- The **"sniffing position"** in children with epiglottitis and a compromised airway is an assumed body posture to make breathing easier. In this position, the child is seated, with the body bent forward slightly and the neck hyperextended. The mouth may be held open and the tongue may protrude.

- Drooling is common due to difficultly and/or pain associated with swallowing.

Predisposing Factors

- Bacterial infection in children, most commonly *Streptococcus*

Clinical Note

Previously, *Haemophilus influenzae* type b (Hib) was the major infectious agent. Immunization with the Hib vaccine has significantly reduced the impact of this bacterium.

Retropharyngeal Abscess

A retropharyngeal abscess results from an infection of the retropharyngeal space. An inoculum can enter this space in several ways:

1. It may enter the space directly from a penetrating injury (e.g., usually, through the oropharyngeal wall).
2. It may spread through adjacent tissue walls.
3. It may originate from retropharyngeal lymph nodes that degenerate and suppurate, which can lead to abscess formation.

Fascia is normally tightly adherent to the tissue surfaces that limit the retropharyngeal space (e.g., posterior pharyngeal wall and anterior aspect of the cervical vertebral compartment), Infectious microorganisms, accompanied by inflammation and suppuration, can spread into this potential space. In addition, retropharyngeal lymph nodes can degenerate and suppurate, leading to abscess formation.

Clinical Note

Lateral retropharyngeal (Rouviere) nodes are the primary lymphatic drainage of the nasopharynx. As such, they may suppurate and lead to a retropharyngeal abscess.

Because of the potential for airway compromise, a retropharyngeal abscess can become life threatening.

Signs and Symptoms

- Constitutional symptoms: fever, chills, sore throat, malaise, decreased appetite, and irritability
- Cervical lymphadenopathy
- Palpable neck mass
- Dysphagia
- Odynophagia

- Trismus
- Torticollis
- Wall of oropharynx may protrude
- Airway displacement may be seen with imaging studies
- Respiratory distress (if the airway is compromised)

Predisposing Factors

- Age: 6 months-6 years (mean: 3–5 years)
  - <2 years: upper respiratory infection
  - >2 years: penetrating injury of oropharynx
- Predisposing infections: pharyngitis; adenitis, tonsillitis, and/or adenoiditis; otitis; sinusitis; nasal, salivary, and dental infections

Clinical Notes

- Pneumonia followed by asphyxiation may result with pharyngeal wall rupture and aspiration of pus
- Infection from the retropharyngeal space may extend into:
  - Mediastinum (inferiorly)
  - Carotid sheath (laterally)
  - Vertebral column (posteriorly)

DIAGNOSIS

The patient presentation, history, physical examination, laboratory tests, and imaging studies support a diagnosis of **retropharyngeal abscess**.

Retropharyngeal Abscess

Retropharyngeal abscesses are caused most commonly by a penetrating injury to the throat or secondary to infections that involve the neck and airways. Expansion of the retropharyngeal abscess, or its invasion of adjacent cervical compartments,

---

Suppurate To form pus
Trismus Jaw stiffnes
Tonsillitis Inflammation of the palatine tonsils

Adenitis Inflammation of a lymph node(s)
Adenoiditis Inflammation of the pharyngeal tonsils (adenoids)

may lead to signs and symptoms secondary to the infection:

- Dysphagia and odynophagia
- A palpable neck mass or cervical lymphadenopathy
- Trismus or torticollis
- Imaging may show an expanded retropharyngeal space

Epiglottitis

Signs and symptoms of epiglottitis are related to inflammation of the laryngeal inlet.

- Fever is usually the first symptom.
- Patients typically present a triad of signs and symptoms: drooling, dysphagia, and distress.
- Stridor, respiratory distress, and a "muffled" voice may develop. Most children appear toxic and anxious and may assume a sniffing position to maintain the airway.
- Diagnosis may be confirmed with lateral radiographs ("thumbprint sign") or direct fiberoptic laryngoscopy (edematous, "cherry-red" epiglottis). In a lateral neck x-ray, the "thumbprint

sign" refers to the shape of the epiglottis with epiglottitis.

Patients with epiglottitis or retropharyngeal abscess present similar constitutional symptoms and dysphagia related to infection and inflammation. In epiglottitis, the airway may be compromised due to inflammation and edema of the tissues at the laryngeal inlet. Patients may assume the sniffing position in order to compensate and maintain the airway. With retropharyngeal abscess, compromise of the airway may result when the abscess expands and compresses or displaces the pharynx, larynx, or trachea.

Acute Bronchitis

Acute bronchitis involves inflammation of the trachea and bronchial tree. The pediatric condition is usually associated with self-limiting viral infection of the respiratory tract (usually, the common cold).

- Bronchitis is usually diagnosed on the basis of the history and physical examination.

In bronchitis, the retropharyngeal space and epiglottis are not involved. Dysphagia and odynophagia would not develop.

6.4

Subclavian Steal Syndrome

Patient Presentation

A 45-year-old African American male visits his family physician complaining of pain in his left arm when working at his construction job.

Relevant Clinical Findings

History

The patient, a drywall installer, reports that his left upper extremity becomes painful when he works with his arms above his head. The patient reports frequent episodes of dizziness, presyncope, and blurred vision. These symptoms are not present at rest.

The patient has smoked since he was a teenager and currently smokes >25 cigarettes/day. There is no previous history of coronary or vascular disease.

Physical Examination

Noteworthy vital signs:

- Blood pressure (normal adult: 120/80)
  - Left arm: 50/70 mm Hg
  - Right arm: 115/80 mm Hg
- Pulses (adult resting rate: 60–100 bpm)
  - Weak left brachial and radial pulses
  - Carotid and femoral pulses normal

Results of physical examination:

- Left upper extremity was pale and cold.

Laboratory Tests

Test	Value	Reference value
Total cholesterol	220	<200 mg/dL
High-density lipoprotein (HDL)	30	Optimal: >60 mg/dL
Low-density lipoprotein (LDL)	140	Optimal: <100 mg/dL
Triglycerides	205	<150 mg/dL
Cholesterol ratio (total cholesterol:HDL)	7.3:1	≤4:1

Imaging Studies

- Contrast enhanced 3-D magnetic resonance angiography (MRA) showed an occlusion of the left subclavian artery.

Clinical Problems to Consider

- Spontaneous carotid artery dissection
- Subclavian steal syndrome

LEARNING OBJECTIVES

1. Describe the subclavian artery.
2. Describe the vertebral artery.
3. Describe the carotid artery.
4. Describe the cerebral arterial circle (Willis).
5. Explain the anatomical basis for the signs and symptoms associated with this case.

Presyncope Feeling faint or lightheaded (syncope is actually fainting)

Occlusion Blockage (e.g., of a blood vessel and canal)

RELEVANT ANATOMY

Subclavian Artery

The **left subclavian** arises as a direct branch of the **aortic arch**. The **right subclavian** is a branch of the **brachiocephalic trunk**. These arteries enter the root of the neck and arch over the superior surface of rib 1 (**Fig. 6.4.1**). At the lateral border of rib 1, the subclavian artery continues as the axillary artery.

For descriptive purposes, the subclavian artery is divided into three regions (**Table 6.4.1**).

Vertebral Artery

The vertebral artery is a branch of the first part of the subclavian, which arises opposite the internal thoracic artery (**Fig. 6.4.1**). Each vertebral artery helps supply the spinal cord, brainstem, cerebellum, and the posterior part of the cerebral hemisphere.

TABLE 6.4.1	The subclavian artery and its branches.	
Part	**Location**	**Branches**
Part 1	**Medial** to anterior scalene muscle	▪ Vertebral ▪ Internal thoracic (mammary) ▪ Thyrocervical trunk
Part 2	**Posterior** to anterior scalene muscle	▪ Costocervical trunk
Part 3	**Lateral** to anterior scalene muscle	▪ Dorsal scapular

For descriptive purposes, each vertebral artery is divided into four parts (**Table 6.4.2**). Near the medullary–pontine border, the vertebral arteries unite to form the **basilar artery**. The basilar artery terminates as paired **posterior cerebral arteries**.

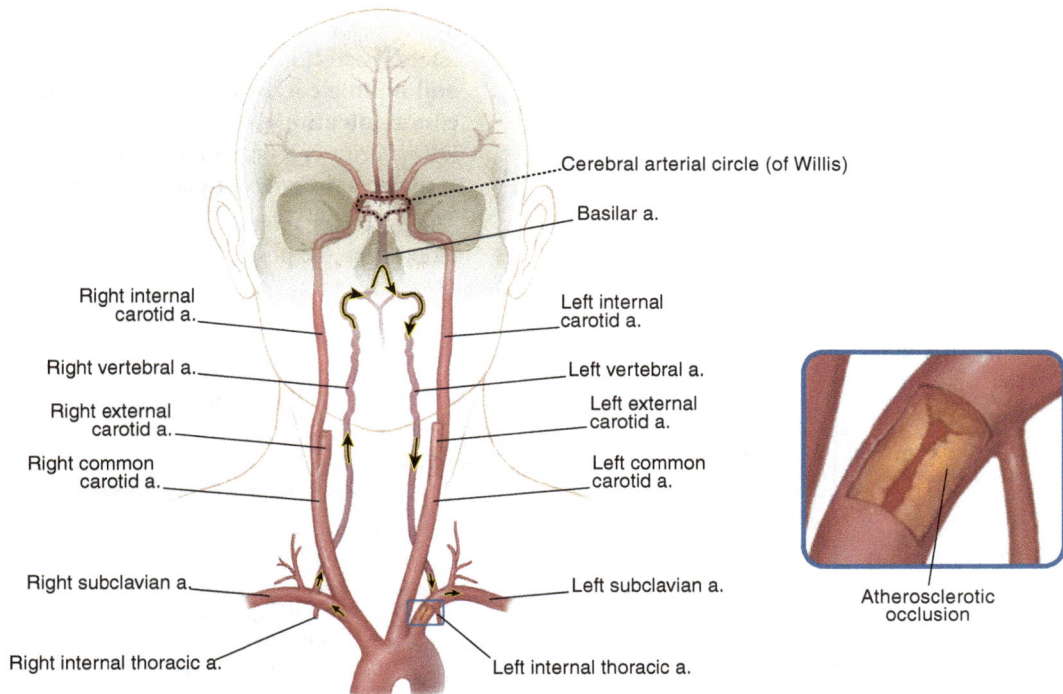

FIGURE 6.4.1 Anterior view of cervical arterial vasculature. An atherosclerotic occlusion of left subclavian artery (inset) is shown between the origins of the common carotid and vertebral arteries. The resulting directions of altered blood flow is indicated by the arrows.

TABLE 6.4.2	The vertebral artery.	
Part	**Alternate name**	**Description**
Prevertebral	Cervical	Proximal segment from origin to C6 transverse foramen (usually, the artery does not pass through C7 transverse foramen, although the vein and sympathetic nerves do)
Cervical	Vertebral	Segment that ascends in the neck through C6-C1 transverse foramina
Atlantic	Suboccipital	Segment between C1 transverse foramen and foramen magnum. It passes through posterior atlanto-occipital membrane to enter the posterior cranial fossa.
Intracranial		Distal segment that ascends along anterior aspect of medulla oblongata

Carotid Artery

The **common carotid artery** ascends in the carotid sheath and bifurcates at the C3 vertebral level to form **external and internal carotid arteries (Fig. 6.4.1)**.

- The **external carotid artery (ECA)** has eight branches that supply structures in the neck and head (**Table 6.4.3**).
- The **internal carotid artery (ICA)** has no branches in the neck. It passes through the **carotid canal** in the petrous part of the temporal bone to enter the cranial cavity. It terminates as the **anterior and middle cerebral arteries**. The ICA is divided into four parts (**Table 6.4.4**).

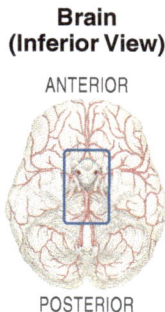

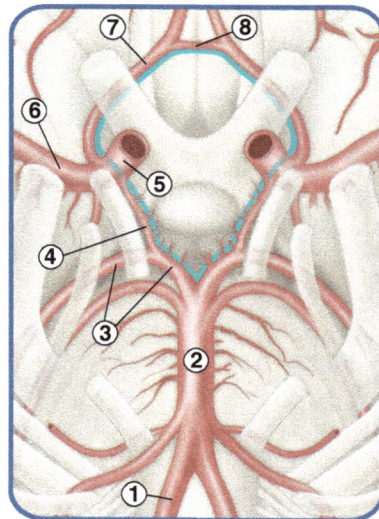

Cerebral Arterial Circle (Willis)

The **cerebral arterial circle (Willis)**, a vascular ring on the ventral aspect of the brain, represents anastomoses between the internal carotid and vertebrobasilar arterial systems (**Fig. 6.4.2**). The *circle of Willis* supplies most of the forebrain.

- The **anterior part of the circle** is formed by the cerebral parts of the right and left internal carotid arteries and their **anterior and middle cerebral** branches. The **anterior communicating artery** connects right and left anterior cerebral arteries and, thereby, completes the anterior part of the circle.
- The **posterior part of the circle** receives blood from the **posterior cerebral arteries**,

Brain (Inferior View)

ANTERIOR

POSTERIOR

■ Vascular ring

Arteries of the Circle of Willis

1. Vertebral a.
2. Basilar a.
3. Posterior cerebral a.
4. Posterior communicating a.
5. Internal carotid a.
6. Middle cerebral a.
7. Anterior cerebral a.
8. Anterior communicating a.

FIGURE 6.4.2 Cerebral arterial circle (of Willis).

TABLE 6.4.3	Branches of the external carotid artery.	
Branch	**Structures/area supplied**	**Comments**
Superior thyroid	▪ Thyroid gland ▪ Infrahyoid muscles ▪ Sternocleidomastoid muscle ▪ Larynx (via superior laryngeal artery)	Deep to infrahyoid muscles
Ascending pharyngeal	▪ Pharynx ▪ Prevertebral muscles ▪ Middle ear ▪ Cranial meninges	Only medial branch of ECA
Lingual	▪ Tongue ▪ Floor of oral cavity	Lies on middle pharyngeal constrictor Passes deep to posterior digastric, stylohyoid, and hyoglossus
Facial	▪ Face ▪ Soft palate ▪ Palatine tonsil	May have common origin with lingual Has similar course as lingual, but passes posterior to submandibular gland and then curves around inferior border of mandible (anterior to masseter muscle) to enter face Ends as angular artery
Occipital	▪ Posterior scalp ▪ Sternocleidomastoid muscle	
Posterior auricular	▪ Parotid gland ▪ Facial nerve (CN VII) ▪ Temporal bone ▪ Auricle ▪ Scalp	
Maxillary	▪ Tissues around the maxilla	Terminal branch of ECA; origin for branches to temporal, infratemporal, pterygopalatine fossae; meninges, teeth, nasal cavity, oral cavity, pharynx, face, scalp
Superficial temporal	▪ Parotid and temporal regions ▪ Scalp	Terminal branch of ECA

which are branches of the basilar. **Posterior communicating arteries** complete the posterior part of the circle by connecting these arteries with the internal carotid arteries.

CLINICAL REASONING

This patient presents with **activity-dependent upper limb pain and neurologic symptoms**.

Spontaneous Carotid Artery Dissection

Arterial dissection refers an oblique tear between laminae of the vessel wall. This allows blood to pass between layers and may result in an intramural hematoma or aneurysm. Dissection of the ICA occurs most frequently in its cervical part.

Signs and Symptoms

▪ Headache

▪ Neck and/or facial pain

▪ Amaurosis fugax: sudden, transient (seconds to minutes) unilateral vision loss; may be episodic

---

Intramural Within the wall of an organ or vessel
Aneurysm Circumscribed dilation of an artery, in direct communication lumen
Amaurosis fugax Transient blindness (Greek, amauros = dark; Latin, fugax = fleeting)

TABLE 6.4.4	The internal carotid artery.
Part	**Description**
Cervical	Proximal segment between carotid bifurcation and carotid canal
Petrous	Traverses carotid canal and superior portion of foramen lacerum (i.e., superior to its cartilage-filled, inferior opening)
Cavernous	Traverses cavernous dural venous sinus
Cerebral (intracranial)	Distal segment between cavernous sinus and cerebral arterial circle (Wills). Terminates as the **anterior and middle cerebral arteries**. ▪ The **ophthalmic artery,** one of its branches, joins the optic nerve (CN II) and passes through the optic canal.

- Other visual field disturbances (e.g., scintillating scotoma, a migraine-like visual aura with shimmering arcs of light).
- Partial Horner syndrome (ptosis with miosis); usually painful with ICA dissection
- Neck swelling
- Pulsatile tinnitus
- Hypo-ageusia

Predisposing Factors

- Age: most frequent in 40–50 years age group; approximately 25% suffer stroke

---

Scotoma Loss or absence of vision from an area of visual field
Aura Symptom that precedes partial epileptic seizure or migraine
Ptosis Drooping eyelid
Miosis Excessive pupillary constriction (opposite of mydriasis)
Tinnitus Sensation of ringing or other noises in the ears
Hypo-ageusia Diminished sensation of taste
Stenosis Narrowing a canal (e.g., blood vessel and vertebral canal)

Clinical Note

Carotid dissection may also result from trauma. A common scenario involves a motor vehicle accident in which cervical extension is combined with lateral flexion to the opposite side. It may also involve an incorrectly positioned shoulder belt that tightens across the neck. The forceful neck movements stretch the artery across cervical transverse processes, while seat belt may injure the artery directly.

Subclavian Steal Syndrome

Subclavian steal syndrome *may* occur with stenosis of the subclavian artery proximal to the origin of the vertebral artery. With occlusion, the syndrome *will* occur and blood flow to the upper limb (axillary) and brain (vertebral) on the ipsilateral side is compromised (see **Fig. 6.4.1**). This has two consequences:

1. Due to the pressure differences in the two vertebral arteries, blood reaching the basilar artery from the contralateral (unaffected) vertebral artery may enter the ipsilateral (affected) vertebral artery. Blood will flow, retrograde to reach the subclavian artery and perfuse the upper limb on the side of the occlusion (**Fig. 6.4.1**).

2. At the same time, flow from the contralateral vertebral artery retrogradely into the ipsilateral vertebral "steals" blood from the basilar artery and, thus, the vertebrobasilar vasculature. This can induce brain ischemia and lead to neurologic symptoms.

Signs and Symptoms

Signs and symptoms associated with this syndrome result from occlusion of the subclavian artery and may be exacerbated during activity involving the affected upper extremity.

Ipsilateral Upper Extremity due to reduced Blood Flow

- Diminished pulses and blood pressure
- Pallor and cold skin
- Pain and/or hemiparesis

Pallor Pale skin
Hemiparesis Unilateral weakness

Neurologic Symptoms due to Ischemia

- Vertigo (dizziness)
- Syncope or presyncope
- Dysarthria
- Visual dysfunction (e.g., diplopia) or loss

Predisposing Factors

- Atherosclerosis

Nonmodifiable Risk Factors for Atherosclerosis

- Age: >40 years
- Sex: more common in males (difference not apparent after menopause)
- Family history

Modifiable Risk Factors for Atherosclerosis

- Cigarette smoking
- Hypercholesterolemia
- Diabetes mellitus
- Hypertension
- Hyperhomocysteinemia

Clinical Note

Most patients with significant arterial occlusion of the proximal subclavian artery are asymptomatic. The term subclavian steal syndrome is applied to patients who have cerebral ischemia with neurologic symptoms that occur during, or immediately after, exercise or activity involving the ipsilateral upper extremity.

DIAGNOSIS

The patient history, physical examination, laboratory tests, and imaging studies support a diagnosis of **subclavian steal syndrome**.

Subclavian Steal Syndrome

The subclavian artery supplies the neck, head (including the brain), anterior thoracic and abdominal walls, and upper limb. Atherosclerotic lesions are a common cause of vascular stenosis or occlusion. Subclavian steal syndrome involves blockage of the subclavian artery proximal to the origin of the vertebral artery. As a result, blood is "stolen" from the contralateral vertebral artery and directed into the ipsilateral vertebral artery. This may result in neurologic symptoms.

- Physical examination may reveal a significantly lower blood pressure in the affected limb. Without occlusion of the subclavian, blood pressures in the upper extremities in a healthy individual should be similar.
- Stenosis or occlusion of the proximal vertebral artery (i.e., not involving the subclavian artery) may produce similar *neurologic symptoms*, but there would be no difference in upper limb blood pressures.
- Symptoms, when present, are commonly exacerbated upon activities or exercise that involve the affected upper extremity, particularly those in which the arm is raised over the head.

Spontaneous Carotid Artery Dissection

The ICA supplies the brain and structures in the orbit, including the eye.

- The most common presenting symptom of spontaneous internal carotid dissection is ipsilateral head pain.
- Amaurosis fugax suggests involvement of the ophthalmic artery.
- Partial Horner syndrome may be present, suggesting damage to the cervical sympathetic trunk or involvement of sympathetic, perivascular plexuses.
- Physical examination may reveal hemiparesis. This would, however, be activity independent and

---

Syncope Loss of consciousness (fainting)
Dysarthria Disturbance of speech
Diplopia Double vision

Hypercholesterolemia Elevated serum cholesterol (total cholesterol level >200 mg/dL)
Hypotension Abnormal decrease in arterial pressure
Hyperhomocysteinemia Elevated serum homocysteine

would not be restricted to the upper extremity (as would occur in subclavian steal syndrome).

■ Diagnosis is confirmed by computed tomographic or magnetic resonance imaging angiography, or Doppler ultrasound.

Clinical Note

Some physicians and hospitals may use injection angiography for diagnosis of carotid artery dissection.

REVIEW QUESTIONS

Questions 1–3 refer to the following clinical case.

A 44-year-old man was referred to the outpatient surgery clinic with a diagnosis of a nodular mass in his left thyroid lobe. After needle biopsy, thyroid cancer was diagnosed and the patient underwent total thyroidectomy.

1. Which nerve stimulates secretion of thyroid hormone?
   A. Motor fibers in ansa cervicalis
   B. Parasympathetic fibers traveling in the recurrent laryngeal nerve
   C. Parasympathetic fibers traveling in the superior laryngeal nerve
   D. Sympathetic fibers from the middle cervical ganglion
   E. None of the above

2. In addition to the internal jugular veins, which vein also commonly receives blood from the gland?
   A. Left brachiocephalic
   B. Left subclavian
   C. Right brachiocephalic
   D. Right subclavian
   E. None of the above

3. After surgery, the patient reported persistent hoarseness. Laryngoscopic examination reveals a paralyzed right vocal fold. Which nerve was most likely damaged during surgery?
   A. External laryngeal
   B. Internal laryngeal
   C. Recurrent laryngeal
   D. Superior laryngeal
   E. None of the above

4. A first-year emergency medicine resident inserts a subclavian venous catheter using an infraclavicular approach. Subsequently, the patient has difficulty breathing. What nervous structure may have been injured?
   A. External laryngeal nerve
   B. Phrenic nerve
   C. Recurrent laryngeal nerve

D. Sympathetic trunk
E. Vagus nerve (CN X)

5. A internal jugular venous catheter is usually introduced through the:
   A. Greater supraclavicular fossa
   B. Lesser supraclavicular fossa
   C. Occipital triangle
   D. Omoclavicular (subclavian) triangle
   E. None of the above

6. A 14-year-old boy is seen in the pediatric clinic complaining of a sore throat. Physical examination reveals enlargement and inflammation of the lymphoid tissue located between the palatoglossal and palatopharyngeal arches. Which tissue is inflamed?
   A. Deep cervical lymph nodes
   B. Lingual tonsil
   C. Palatine tonsil
   D. Pharyngeal tonsil

Questions 7–8 refer to the following clinical case.

A 49-year-old female undergoes fusion of the C4 and C5 vertebrae. The surgeon uses a lateral approach through the "safe" portion of the posterior cervical triangle (lateral cervical region).

7. Once through the skin and superficial fascia, which fascial layers did the surgeon dissect to reach the vertebral column?
   A. Investing only
   B. Investing and buccopharyngeal
   C. Investing and muscular part of pretracheal
   D. Investing and prevertebral
   E. Visceral part of pretracheal and prevertebral

8. Which structure would the surgeon use as a landmark to identify the "safe" portion of this cervical region?
   A. Accessory nerve (CN XI)
   B. External jugular vein
   C. Inferior belly of omohyoid
   D. Nerve point of the neck
   E. None of the above

CHAPTER 7

Head

CASE 7.1 | Skull Fracture

Patient Presentation

An 18-year-old male was brought to the emergency department after he was found unconscious on the ski slope. His friends told the ski patrol that while they were "boarding" on a difficult run he lost control and collided with a tree.

Relevant Clinical Findings

History

On the way to the hospital, the patient regained consciousness. When he arrived in the emergency department, he appeared dazed and complained of headache. He responded appropriately to questions (e.g., he knew time, date, and place) and confirmed that he lost control while skiing and "crashed" into a tree.

Physical Examination

Noteworthy vital signs include:

- Blood pressure: 120/80 mm Hg (normal adult: 120/80)
- Pulse: 75 bpm (adult resting rate: 60–100 bpm)
- Respiratory rate: 17 cycles/min (normal adult: 14–18 cycles/min; women slightly higher)

Noteworthy results of physical examination:

- Facial abrasions
- Swelling above the right ear

Noteworthy results of neurologic examination:

- Pupils were equal, round, and reactive to light (PERRL)
- Normal extraocular movements
- Numbness on right side of face

Imaging Studies

- Radiography upon admission to the emergency department confirmed a small, lateral skull fracture.
- Computed tomography (CT) revealed a hyperdense, biconvex (lens-shaped) mass between the brain and the skull.

The patient was kept for observation in the emergency department. Several hours later, the patient lost consciousness and his right pupil was dilated. He became bradycardic, hypertensive, with a decreased respiratory rate.

Clinical Problems to Consider

- Epidural hematoma
- Subdural hematoma

LEARNING OBJECTIVES

1. Describe the anatomy of the lateral skull.
2. Describe the anatomy of the cranial meninges and associated spaces.
3. Describe the anatomy of the dural folds and venous sinuses.
4. Explain the anatomical basis for the signs and symptoms associated with this case.

Bradycardia Decreased heart rate: <55 bpm (normal adult heart rate: 55–100 bpm)

Hypertension Abnormal increase in arterial pressure
Hematoma Localized extravasation of blood, usually clotted

RELEVANT ANATOMY

Lateral Skull

The skull, or cranium, forms the skeleton of the head (**Table 7.1.1**). It is divided into two parts:

1. The **neurocranium** is composed of eight "flat" bones that enclose and protect the brain.

2. The **viscerocranium**, or facial skeleton, is composed of 15 irregular bones that form the mouth, nose and nasal cavities, and most of the orbits.

The mandible is the only movable bone of the skull. All other bones articulate with each other at immovable joints known as **sutures**.

The **lateral aspect of the skull** is formed by parts of the neurocranium and viscerocranium (**Fig. 7.1.1**). The **parietal bone** and the **squamous part of the temporal bone** form the largest part of the

TABLE 7.1.1	Bones associated with neurocranium and viscerocranium.
Part of skull	**Contributing bones**
Neurocranium	<ul><li>Frontal (1)</li><li>Ethmoid (1)</li><li>Sphenoid (1)</li><li>Occipital (1)</li><li>Temporal (2)</li><li>Parietal (2)</li></ul>
Viscerocranium	<ul><li>Ethmoid (1)</li><li>Vomer (1)</li><li>Mandible (1)</li><li>Maxillae (2)</li><li>Zygomatic (2)</li><li>Inferior nasal conchae (2)</li><li>Nasal (2)</li><li>Lacrimal (2)</li><li>Palatine (2)</li></ul>

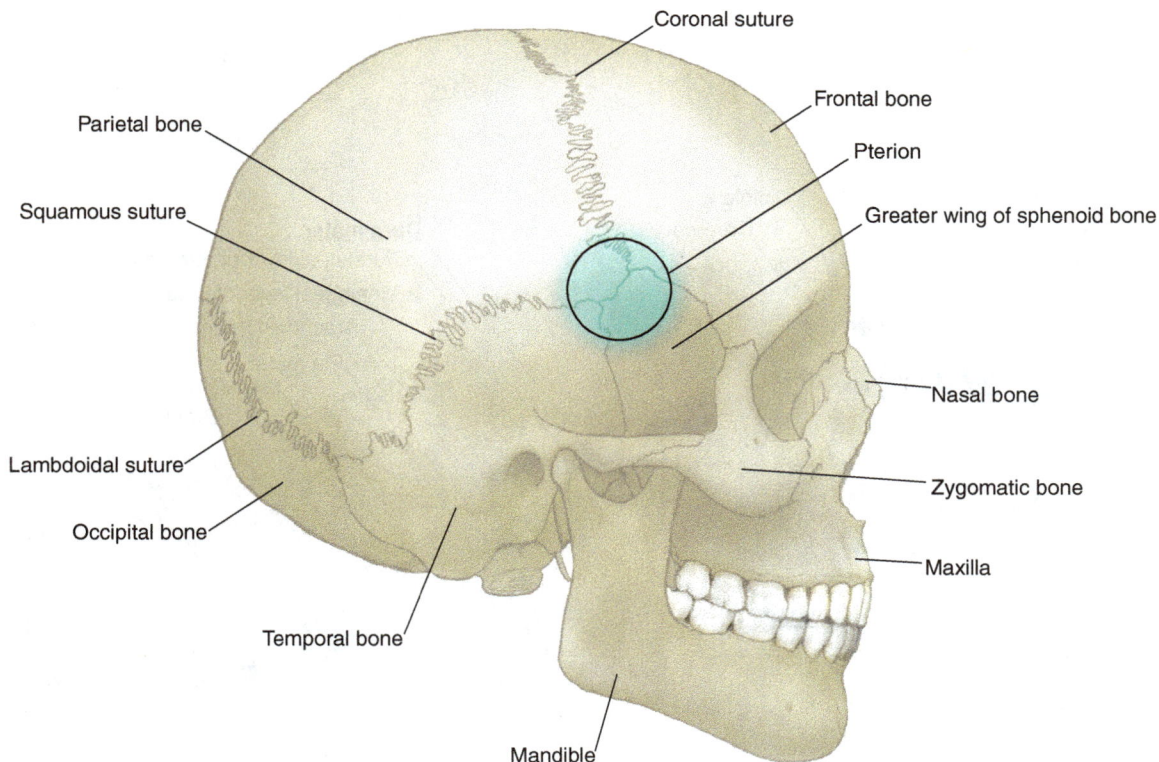

FIGURE 7.1.1 Lateral view of the skull. The pterion is indicated.

lateral skull and are joined at the **squamous suture**. The parietal bone also articulates with the frontal bone at the **coronal suture** and with the occipital bone at the **lambdoidal suture**.

The lateral surface of the **greater wing of the sphenoid bone** is located just anterior to the temporal bone. The **pterion**, a point of clinical significance, is a small region on the lateral skull where the frontal, parietal, temporal, and sphenoid (greater wing) bones articulate. The sutures involved in these articulations describe an "H" pattern.

Clinical Note

The **pterion** is one of the weakest parts of the skull as the bone is very thin. It is clinically important because the frontal (anterior) branch of the **middle meningeal artery** lies just deep to this point on the inner aspect of the skull. Trauma at this site may rupture this vessel and cause an epidural hematoma.

Cranial Meninges and Associated Spaces

The brain is protected by three layers of connective tissue, collectively known as **meninges**, located just inside the neurocranium (**Fig. 7.1.2**). From external-to-internal, they are:

1. The **dura mater** (Latin, *tough mother*) is the most external and is composed of a thick layer of dense connective tissue. The dura is composed of two laminae: an outer **periosteal layer** that is adherent to the inner surface of the skull and an inner **meningeal layer** that is contact with the arachnoid mater.

2. The **arachnoid mater** (Greek, arachne = *spider*) is a thin, transparent, avascular membrane that lines the internal surface of the dura mater. Numerous fine filaments, known as **arachnoid trabeculae** extend from this membrane to the pia mater. This filamentous network gives this layer the appearance of a spider's web.

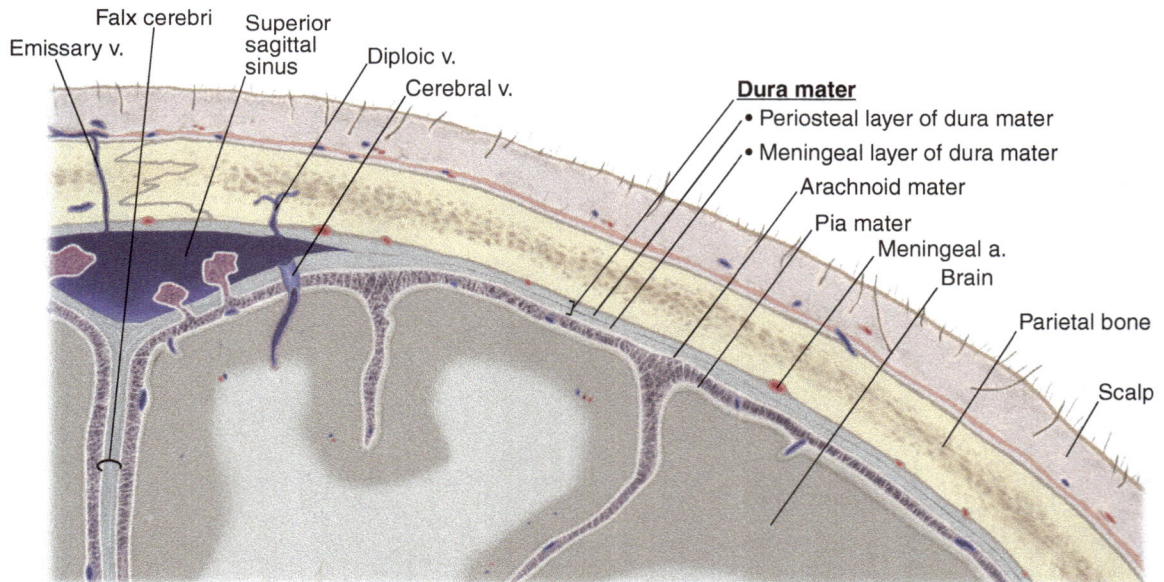

FIGURE 7.1.2 Coronal section through the skull showing the brain, cranial meninges, and associated spaces.

3. The **pia mater** (Latin, *delicate mother*) is composed of a thin membrane that is directly adherent to the surface of the brain.

Three spaces are formed between the meningeal layers (external-to-internal):

1. The **epidural space** is a potential space between bones of the neurocranium and the periosteal layer of dura mater.

2. The **subdural space** is a potential space between the meningeal layer of dura mater and the arachnoid mater.

3. The **subarachnoid space** is a located between the arachnoid mater and the pia mater. This space is traversed by the arachnoid trabeculae and is filled with cerebrospinal fluid (CSF). It also contains bloods vessels that supply the brain.

Blood Supply

The dura mater is the only layer of meninges that has a blood supply. In contrast, the arachnoid mater and pia mater are supported by the CSF.

The **middle meningeal artery** is the primary blood supply to the dura mater. It branches from the maxillary artery in the infratemporal fossa, passes between the roots of the auriculotemporal nerve, and enters the cranial cavity through **foramen spinosum** (**Fig. 7.1.3**). In the cranial cavity, it lies between the periosteal layer of dura mater and the internal surface of the neurocranium. The middle meningeal artery has two major branches, frontal (anterior) and parietal (posterior), that distribute across the lateral aspect of the dura. They anastomose with branches from the contralateral (opposite) side and other meningeal arteries.

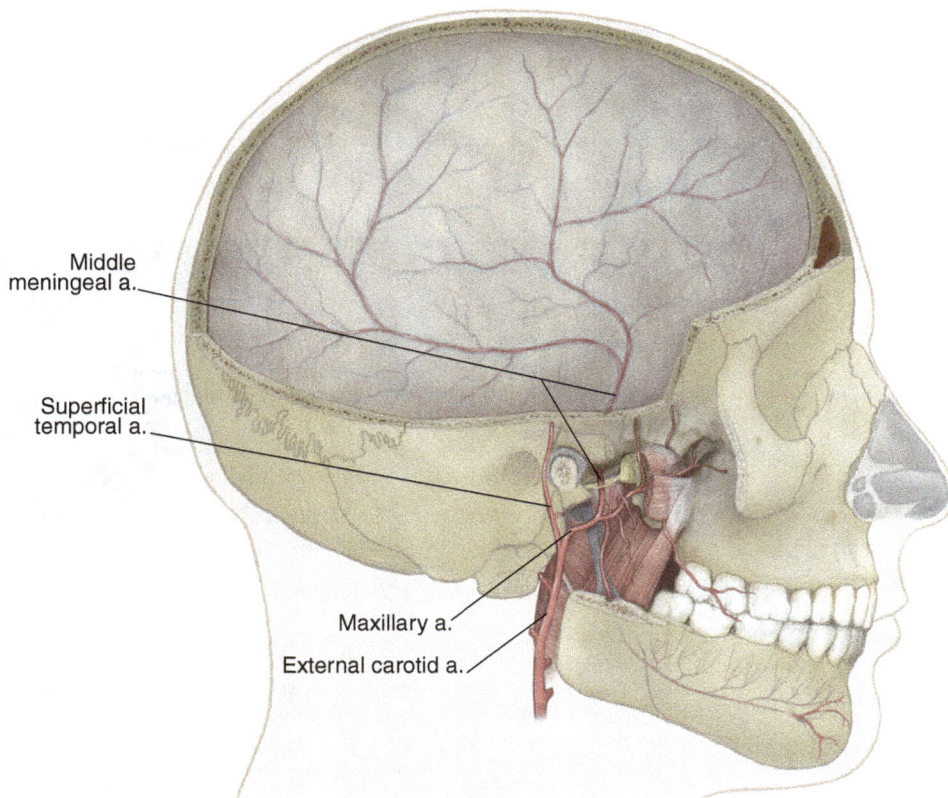

FIGURE 7.1.3 Lateral view of the middle meningeal artery.

The **accessory meningeal artery**, if present, is also a branch of the maxillary. It enters the cranial cavity through the **foramen ovale**. Other small, meningeal arteries arise from the anterior and posterior ethmoidal, internal carotid, ascending pharyngeal, and occipital arteries.

Meningeal veins that drain the dura mater collect in its periosteal layer. They open directly into the superior sagittal and other dural venous sinuses. A small component of the middle meningeal vein may traverse foramen spinosum and drain into the pterygoid venous plexus in the deep face.

Nerve Supply

The dura mater receives sensory innervation primarily from the three branches of the **trigeminal nerve (CN V)**. Other, small meningeal branches arise from the facial (CN VII), glossopharyngeal (CN IX), and vagus (CN X) nerves. The upper three spinal nerves (C1-C3) also contribute to the innervation of dura.

Dural Folds

The meningeal and periosteal layers of dura mater are fused except in four specific areas where the meningeal layer reflects into the cranial cavity to form partitions between brain regions. These folds support, stabilize, and separate parts of the brain (**Table 7.1.2**).

Dural Venous Sinuses

The **dural venous sinuses** (also called dural sinuses, cerebral sinuses, or cranial sinuses) are endothelium-lined spaces between the meningeal and periosteal layers of cranial dura mater (**Table 7.1.3**). They are created when the meningeal layer separates from the periosteal layer to form a dural fold. Dural sinuses differ from most veins (most veins superior to the heart lack valves) in that they do not have the characteristic layers of a vein.

Dural sinuses receive blood from cerebral veins (**Fig. 7.1.2**), cerebellar veins, and veins that drain the brainstem. They also receive blood from diploic veins (**Fig. 7.1.2**), which drain the spongy core (i.e., diploe) of the neurocranial bones and emissary veins that drain the scalp and other structures of the head. Dural sinuses also receive CSF from the subarachnoid space. Blood (and CSF) in the dural sinuses enters the **internal jugular vein**.

TABLE 7.1.2	Dural folds.
Dural fold	**Description**
Falx cerebri	Largest dural fold Sickle-shaped (crescentic) vertical fold in midsagittal plane Projects into longitudinal fissure (between right and left cerebral hemispheres) Anterior attachment: crista galli of ethmoid bone Posterior attachment: internal occipital protuberance
Falx cerebelli	Small, sickle-shaped (crescentic) vertical fold Divides right and left cerebellar hemispheres Inferior to tentorium cerebelli Posterior attachment: internal occipital crest
Tentorium cerebelli *"tent" over cerebellum*	Horizontal fold Separates cerebral hemispheres (occipital lobes) from cerebellum Anterior margin creates the tentorial incisure—an opening for brain stem Anterior attachment: petrous part of temporal bone Posterior attachment: occipital bone
Diaphragma sellae	Small, circular, horizontal fold Forms roof over sella turcica (pituitary fossa) and covers pituitary gland Has small opening for pituitary stalk

TABLE 7.1.3 | Dural venous sinuses.

Dural sinus	Location	Description
Unpaired sinuses		
Superior sagittal	Attached superior margin of falx cerebri	Ends posteriorly at confluence of sinuses (usually empties into right transverse sinus)
Inferior sagittal	Free inferior margin of falx cerebri	Ends posteriorly at anterior margin of tentorium cerebelli, where it joins straight sinus Increases in size posteriorly
Straight	Junction of falx cerebri and tentorium cerebelli	Ends posteriorly at confluence of sinuses (usually empties into left transverse sinus)
Occipital	Attached margin of falx cerebelli	Ends in confluence of sinuses
Confluence of sinuses	Internal occipital protuberance	Junction of superior sagittal, straight, occipital, and transverse sinuses
Paired sinuses		
Transverse	Posterior margin of tentorium cerebelli	Begins at confluence of sinuses Extends anterolaterally Drains into sigmoid sinus
Sigmoid	Groove along parietal, temporal, and occipital bones	Begins at transverse sinus Extends inferomedially Ends in jugular bulb, which is continuous with internal jugular vein
Cavernous	Adjacent to body of sphenoid bone and hypophysial (pituitary) fossa	Receives venous blood from orbit and deep face Communicates with contralateral sinus via intercavernous sinuses Drains into superior and inferior petrosal sinuses
Superior petrosal	Anterior margin of tentorium cerebelli (along petrous temporal bone)	Connects cavernous and transverse sinuses
Inferior petrosal	Groove between temporal and occipital bones	Connects cavernous sinus with internal jugular vein

Clinical Note

Dural sinuses can be injured by trauma (e.g., skull fracture). Injury of a sinus may cause a **cerebral venous sinus thrombosis (CVST)**, a clot within a dural sinus. While rare, CVST may lead to serious complications that may have associated neurological deficits.

CLINICAL REASONING

This patient presents with a **lateral skull fracture**.

Epidural Hematoma

Epidural hematoma is an accumulation of blood in the **epidural space**—normally, a potential space between periosteal dura and neurocranial bones (**Fig. 7.1.4A**). An epidural hematoma may occur with skull fracture when bone fragments lacerate the vessels directly. This type of hematoma may also result when skull trauma (even without skull fracture) causes separation of periosteal dura from bone and, indirectly, tears (lacerates) meningeal vessels. The **temporoparietal region** and **middle meningeal artery** are involved most commonly. Anterior ethmoidal artery may be involved in frontal injuries,

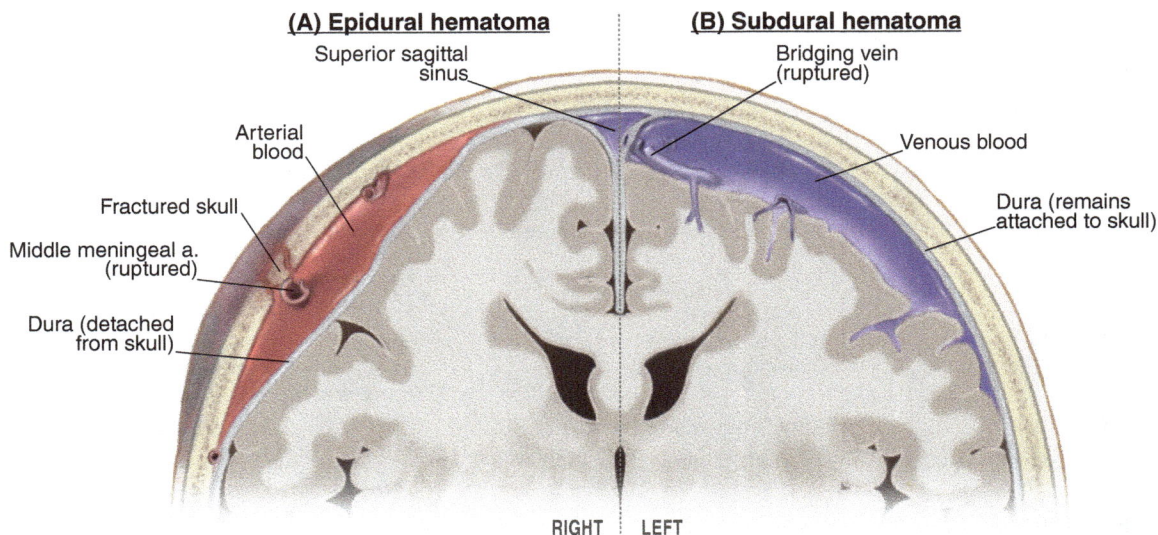

FIGURE 7.1.4 Coronal section through the skull showing epidural (**A**) and subdural hematomas (**B**).

transverse or sigmoid sinus in occipital injuries, and the superior sagittal sinus in vertex trauma.

Signs and Symptoms

- Confusion
- Dizziness
- Drowsiness or altered level of alertness
- Dilated, unreactive pupil
- Severe headache
- Nausea and/or vomiting
- Muscle weakness (usually contralateral to the injured side)

Predisposing Factors

- Head trauma
- Sex: male (4:1)

Clinical Notes

- Two-thirds of epidural hematomas involve unilateral fractures in the temporoparietal region.

- Most epidural hematomas in adult are arterial. In children, approximately half originate from venous injury.
- Mortality is greater with epidural hematoma than with subdural hematoma.

Noncontrast CT head scans may be used to image an epidural hematoma. Acute epidural hematoma usually appears as a hyperdense, biconvex (external and internal borders are convex) mass between the brain and the skull (**Fig. 7.1.5**).

Subdural Hematoma

A **subdural hematoma** is an accumulation of blood in the **subdural space** (**Fig. 7.1.4B**). The subdural space is a potential space between the meningeal layer of dura and the arachnoid mater (membrane). Subdural hematomas occur most commonly secondary to high-speed head trauma, either involving direct impact of the skull or forceful head acceleration and/or deceleration. This results in vascular damage:

- A bridging vein, which drains blood from cerebral cortex into a dural venous sinus, may

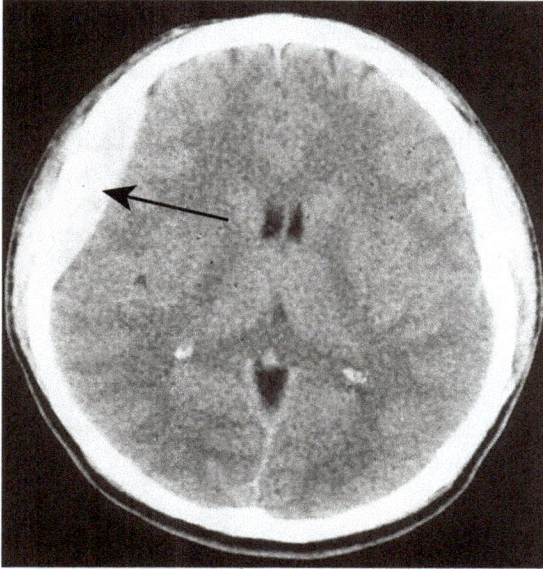

FIGURE 7.1.5 Noncontrast CT scan showing a right epidural hematoma (arrow). *Source: Fig. 378-4 in Harrison's Online.*

be torn. These veins may be "sheared" when the brain moves relative to the fixed dura mater and dural venous sinuses.

- A cortical vessel, vein or artery, can be damaged by direct injury or laceration when the surface of the brain impacts the overlying meninges and skull. Formation of a hematoma may result from a relatively minor injury and may develop slowly.

- Less common causes include anticoagulant therapy, ruptured intracranial aneurysm, and intracranial tumors.

Clinical Notes

- Subdural hematoma is the most common type of intracranial hematoma following head trauma.

- In older adults, decreased brain volume stretches cerebral bridging veins and the dural venous sinuses become more fixed to the skull. Therefore, these small veins may tear even without trauma. This produces a chronic, slowly expanding subdural hematoma with delayed signs and symptoms.

- Subdural hematomas may be characterized by their size and location, and classified on the basis of the amount of time elapsed since the precipitating event (if any):
  - **Acute subdural hematoma** occurs within 72 hours of injury.
  - **Subacute subdural hematoma** occurs 3–7 days after injury.
  - **Chronic subdural hematoma** occurs weeks-to-months after trauma.

Signs

Clinical signs are may be nonspecific, nonlocalized, or absent, and may be stable or rapidly progressive.

- Headache
- Confusion
- Depressed level of consciousness; 50% of patients present comatose

Predisposing Factors

- Sex: male (3:1)
- Age: >60 years
- Head injury
- Alcoholism
- Hypertension
- Anticoagulant medication (e.g., aspirin and warfarin)
- Coagulation disorders (e.g., hemophilia)
- Diabetes
- Postsurgical (e.g., craniotomy and CSF shunting)

Clinical Note

CT head scans provide the most conclusive evidence for subdural hematoma. These scans often show a hyperdense, crescent-shaped (external border convex internal border concave) collection of blood, which rarely crosses the falx cerebri or tentorium cerebelli (**Fig. 7.1.6**).

Aneurysm Circumscribed dilation of an artery, in direct communication lumen
Comatose Profound unconsciousness

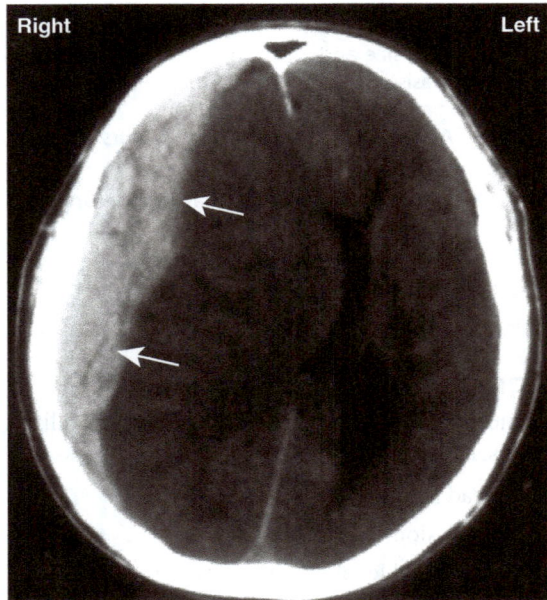

FIGURE 7.1.6 Noncontrast CT scan showing a right subdural hematoma (arrows). *Source:* Fig. 10-6A in *Clinical Neurology*, 7e.

DIAGNOSIS

The patient presentation, history, physical examination, and imaging studies support a diagnosis of **lateral skull fracture with an epidural hematoma**.

Epidural Hematoma

An epidural hematoma occurs when there is intracranial damage to meningeal arteries as a result of head trauma. Most commonly, the trauma is in the temporoparietal region (pterion) and involves the middle meningeal artery. Blood from the lacerated vessels collects in the potential space between the periosteal dura mater and the skull.

- Patients usually suffer brief loss of consciousness immediately after the trauma. This is followed by a period of lucidity for hours (commonly) to days (rarely).

- Rapid expansion of the hematoma causes increased intracranial pressure. This can cause changes in consciousness and lead to neurological deficits.

- With an expanding hematoma, the brain may "shift" within the skull. In some cases, a portion(s) of the brain may be forced into an adjacent cranial compartment (brain herniation). Serious signs and symptoms related to brain herniation include secondary loss of consciousness, hypertension, and bradycardia.

- **Epidural hematoma was confirmed in this patient** with CT imaging that revealed a characteristic biconvex hemorrhage in the temporoparietal region.

Clinical Note

Decompression of the hematoma is a surgical emergency.

Subdural Hematoma

A subdural hematoma also represents blood that has entered a potential space. In this case, the blood is found in the subdural space between the dura mater and the arachnoid mater.

- The source of bleeding in a subdural hematoma is often bridging veins and cerebral veins that pass across the subarachnoid space into the superior sagittal sinus.

- Patients may have large, chronic hematomas with minimal neurologic deficit.

- They may occur with or without a history of trauma.

- Drowsiness, inattentiveness, and incoherence are frequent signs and symptoms.

- CT scan without contrast usually shows a crescent-shaped mass subjacent to the skull.

The signs and symptoms, and radiography in this patient are not consistent with a subdural hematoma.

Patient Presentation

A 38-year-old male is admitted to the emergency department with a headache and frequent nosebleeds. He complains that his nose is painful when touched.

Relevant Clinical Findings

History

The patient reports experiencing frequent nosebleeds over the last 3-4 weeks. He has also had difficulty sleeping since he always seems "stuffed up" and has resorted to mouth breathing. More recently, he has developed headaches and his nose has become tender to touch. He does not report any serious trauma to his face, but he bumped his nose on an open cupboard door several weeks ago. There was little swelling or bruising at the time so he did not seek medical attention. He does not report having any allergies and does not suffer from asthma.

Physical Examination

The following findings were noted on physical examination:

- Nose tender to touch
- Pale skin and mild anesthesia on apex of the nose
- "Saddle nose" deformity (viewed in lateral profile)

Rhinoscopy revealed:

- Large, red, round swelling that occludes most of right nasal cavity
- Mass is tender to palpation

Imaging Studies

- Plain film radiography was inconclusive regarding fractures related to the nasal cavity.

Clinical Problems to Consider

- Adenotonsillar disease
- Fractured nasal septum
- Nasal polyps
- Sinusitis

LEARNING OBJECTIVES

1. Describe the anatomy of the nose and paranasal sinuses.
2. Describe the anatomy of the naso- and oropharynx.
3. Explain the anatomical basis for the signs and symptoms associated with this case.

Anesthesia Loss of sensation
Rhinoscopy Procedure to inspect nasal cavity
Polyp A small growth that protrudes from a mucous membrane. Polyps that are attached by a slender stalk are termed pedunculated

RELEVANT ANATOMY

Nose

The **nose** is comprised of the **external nose** and the **nasal cavity**.

External Nose

The **external nose** is a pyramidal projection on the face. Bone, cartilage, and fibro-areolar tissues determine its shape (**Table 7.2.1**).

Branches of the ophthalmic (CN V1) and maxillary (CN V2) nerves supply the skin of the nose. The **nares** (nostrils), the **anterior nasal apertures**, open into the right and left nasal cavities.

TABLE 7.2.1	Skeleton of the external nose.
Bone	**Cartilage**
Nasal bone	Lateral cartilages
Frontal processes of maxillae	Alar cartilages
Nasal processes of frontal bone	Septal cartilage

Nasal Cavity

The **nasal cavity** is the first part of the respiratory tract and is divided by the **nasal septum**. The cavities are continuous posteriorly with the nasopharynx through the **choanae (posterior nasal aperture)**. The nasal cavities are wedge-shaped, with a large inferior base and narrow superior apex. Each nasal cavity consists of three regions (**Table 7.2.2**).

The medial wall (septum) of each nasal cavity is formed by the nasal septum covered with mucous membrane (**Fig. 7.2.1A**). This mucosa is continuous with that of the nasal cavity and nasopharynx. The septum consists of three parts (**Fig. 7.2.1B**):

1. Anterior: **septal nasal cartilage**
2. Superior: **perpendicular plate of the ethmoid bone**
3. Postero-inferior: **vomer**

TABLE 7.2.2	Regions of the nasal cavity.
Region	**Description**
Vestibule	Small dilated space just inside nares Lined by skin that contains hair follicles Transitions to respiratory mucous membrane
Olfactory	Small superior part of nasal cavity Lined by olfactory epithelium (contains olfactory receptors)
Respiratory	Larger central portion of nasal cavity Lined by highly vascular mucous membrane, with, respiratory epithelium

The **nasal crests** of the **maxilla and palatine bones** complete the inferior part of the nasal septum (**Fig. 7.2.1B**).

The lateral nasal walls are characterized by the three curved shelves of bones (nasal conchae) that project medially and inferiorly into the nasal cavity (**Fig. 7.2.2A**). The **superior and middle nasal conchae** are projections of the ethmoid bone. The **inferior nasal concha** is a distinct bone. The conchae are lined with respiratory mucous membrane and, therefore, increase the surface area to facilitate warming and humidification of inspired air.

Clinical Note

A common clinical term for nasal concha is **turbinate**.

The shape and position of the conchae creates four recesses on the lateral nasal wall (**Table 7.2.3**) into which the paranasal sinuses and nasolacrimal duct open (**Fig. 7.2.2B**).

Blood Supply

The nasal mucous membranes contain a dense vascular plexus formed by anastomosing branches of the **ophthalmic, maxillary, and facial arteries** (**Fig. 7.2.3**).

- The **anterior and posterior ethmoidal arteries**, branches of the ophthalmic, supply the superior part of the nasal septum and lateral nasal walls.

- The **sphenopalatine artery**, the terminal branch of the maxillary, supplies the posterior portion of the nasal septum and lateral nasal wall.

- Branches from the **greater palatine artery**, a branch of maxillary, supply the infero-posterior lateral nasal wall. This artery also ascends through the incisive canal on the hard palate to anastomose with vessels on the nasal septum.

- The **septal branch of the superior labial artery**, a branch of the facial artery, supplies the antero-inferior part of the septum.

(A)

- Frontal sinus
- Sphenoidal sinus
- Nasal septum (with mucosa)
- Septal cartilage
- Choana
- Nasopharynx
- Soft palate
- Hard palate
- Oral cavity
- Incisive canal

(B)

- Perpendicular plate of ethmoid bone
- Frontal bone
- Sphenoid bone
- Nasal bone
- Septal cartilage
- Vomer
- Nasal crest of palatine bone
- Palatine bone
- Anterior nasal spine
- Nasal crest of maxilla
- Maxilla

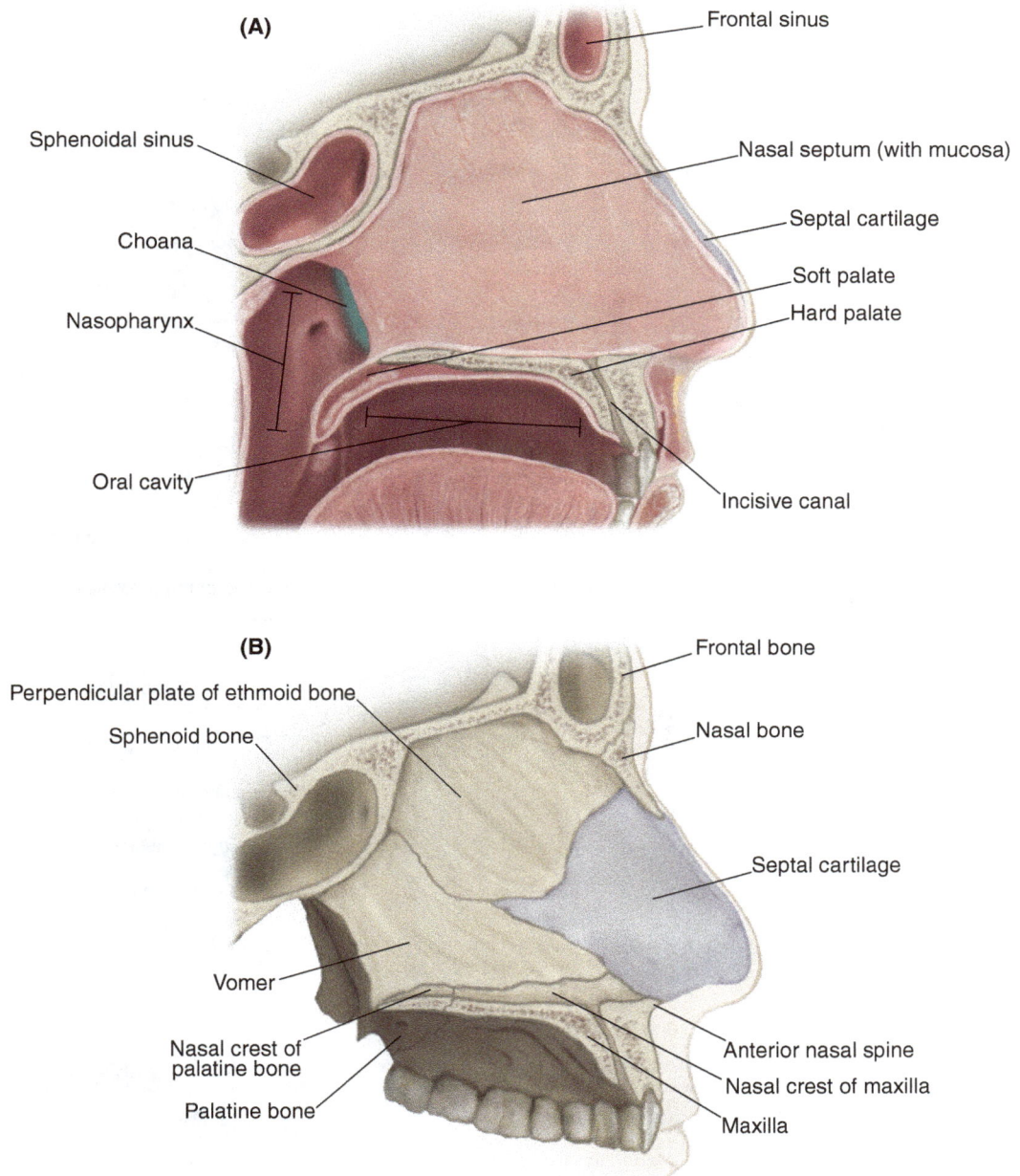

FIGURE 7.2.1 (**A**) Midsagittal view showing the nasal septum (with mucosa), nasopharynx, and oral cavity. (**B**) Midsagittal view showing skeleton of the nasal septum.

(A)

(B)

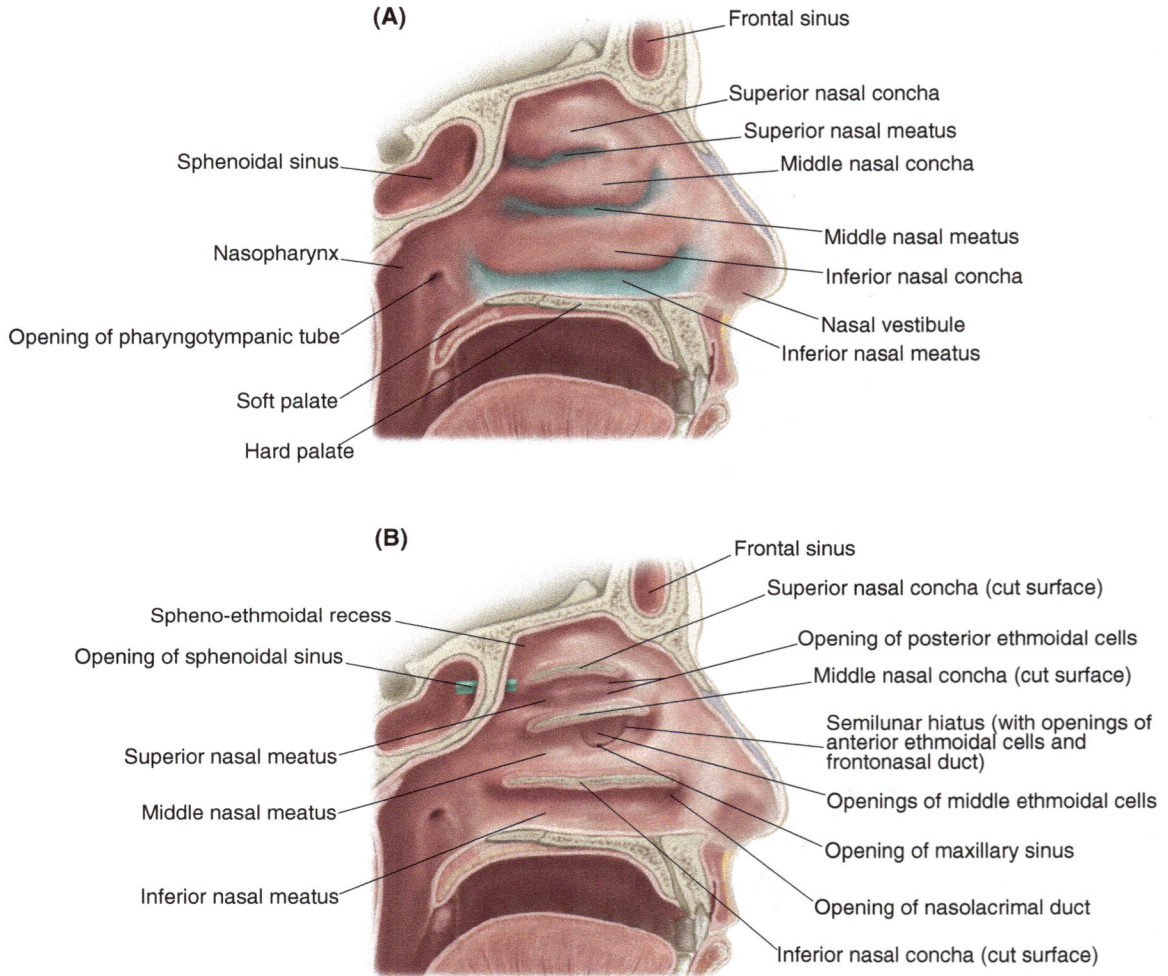

FIGURE 7.2.2 (**A**) Lateral nasal wall (nasal conchae intact). (**B**) Lateral nasal wall (conchae removed) showing the openings of the paranasal sinuses and nasolacrimal duct.

TABLE 7.2.3	Location and openings of chambers of nasal cavity.	
Name	**Location**	**Opening**
Inferior nasal meatus	Inferior to inferior nasal concha	▪ Nasolacrimal duct
Middle nasal meatus	Inferior to middle nasal concha	▪ Maxillary sinus ▪ Frontonasal duct ▪ Anterior ethmoidal air cells ▪ Middle ethmoidal air cells
Superior nasal meatus	Inferior to superior nasal concha	▪ Posterior ethmoidal air cells
Spheno-ethmoidal recess	Superior to superior nasal concha	▪ Sphenoidal sinus

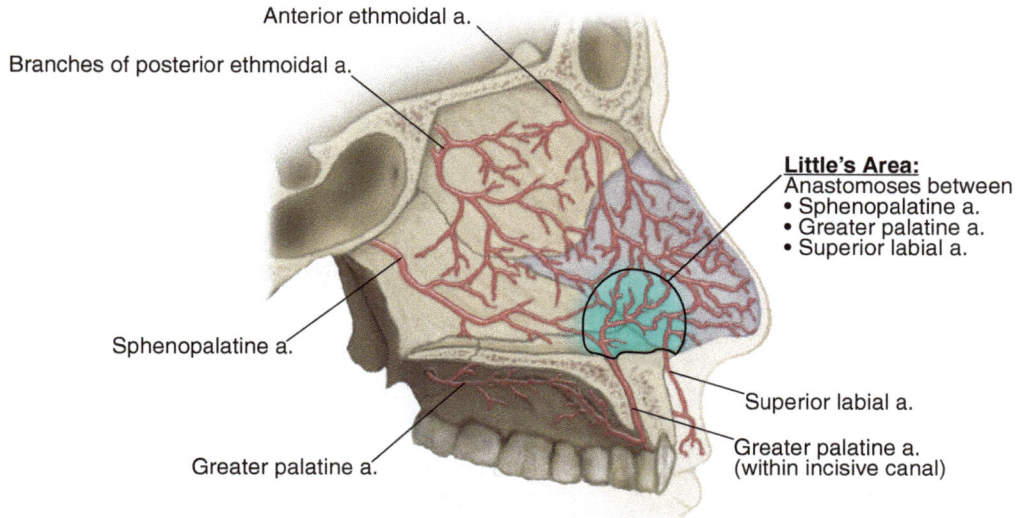

FIGURE 7.2.3 Blood supply to nasal septum.

Clinical Note

The **Kiesselbach plexus**, or **Little's area**, represents a region on the antero-inferior one-third of the nasal septum where three arteries converge: sphenopalatine, greater palatine, and superior labial. Ninety percent of nosebleeds (epistaxis) occur in Little's area. This area is exposed to the drying effect of inspired air that predisposes it to "crack" and bleed.

Nerve Supply

The mucous membrane lining the nasal cavity is supplied with multiple nerves with various origins and functions. They include the:

- **Olfactory nerve** (CN I): olfaction (smell)

- Branches of the **trigeminal nerve** (CN V1 and CN V2): general sensation

- **Parasympathetic fibers** (branches of **facial nerve** (CN VII): innervation of mucous glands

- **Sympathetic fibers** are vasoconstrictive.

The **olfactory nerves** are composed of the axons of olfactory neurons located in the nasal mucosa of the superior part of the nasal cavity. Bundles of these axons pass through the **cribriform plate of the ethmoid bone** to enter the anterior cranial fossa and synapse in the **olfactory bulb**.

The **anterior and posterior ethmoidal nerves** are branches of the **ophthalmic nerve (CN V1)**. They receive sensory information from the anterior and superior parts of the nasal cavity. The **maxillary nerve (CN V2)** gives rise to numerous, small **posterior medial** and **posterior lateral nasal** branches that supply the septum and posterior nasal wall, respectively. The **nasopalatine nerve** is also a branch of the maxillary and supplies the inferior nasal septum. It also passes through the incisive canal to supply mucous membrane on the anterior hard palate.

Preganglionic parasympathetic fibers in the **greater petrosal nerve**, a branch of the facial nerve, synapse in the pterygopalatine ganglion. **Postganglionic parasympathetic fibers** are distributed with branches of the maxillary nerve to mucous glands in the nasal cavity.

Preganglionic sympathetic fibers arise from neurons in the lateral horn of the upper thoracic spinal cord. These fibers enter the sympathetic trunk and ascend to synapse in the superior cervical ganglion. **Postganglionic sympathetic fibers** follow the internal carotid artery (internal carotid plexus) into the

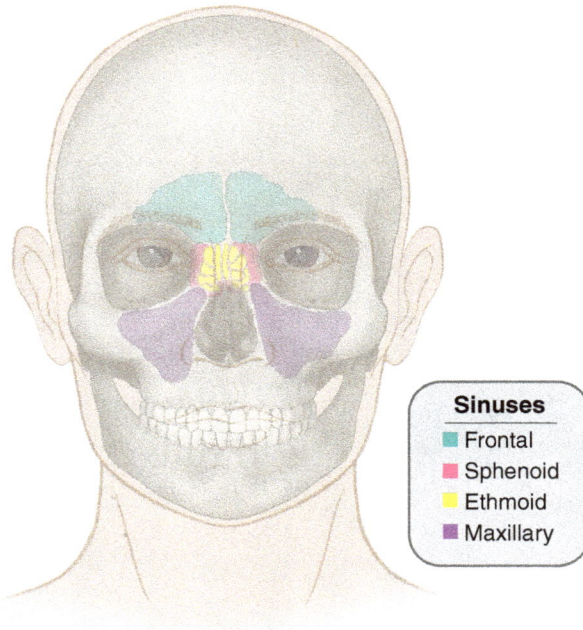

FIGURE 7.2.4 Anterior view of the paranasal sinuses.

cranial cavity. They leave the internal carotid plexus as the **deep petrosal nerve** and join the greater petrosal nerve. Like the parasympathetic fibers, sympathetic fibers are distributed along the branches of the maxillary nerve.

Paranasal Sinuses

The four **paranasal sinuses** (air spaces) develop as diverticula of the nasal cavity (**Fig. 7.2.4**). Each is named for the bone in which it is found (**Table 7.2.4**). All are lined with respiratory mucous membrane, which is continuous with the nasal cavity, and are innervated by branches of the trigeminal nerve (CN V). Each sinus is continuous with the nasal cavity through openings on the lateral nasal wall (**Fig. 7.2.2B**).

Pharynx

The pharynx is a semicircular, fibromuscular tube that opens anteriorly into the nasal and oral cavities, and lies posterior to the larynx. It is continuous inferiorly with the esophagus. Its mucous membrane is continuous with the nasal cavity, oral cavity, or larynx. The pharynx is subdivided into three regions: **nasopharynx**, **oropharynx** and **laryngopharynx**.

Nasopharynx

The nasopharynx is continuous with the nasal cavities at the choanae. The roof of the nasopharynx is formed by the base of the skull (sphenoid and occipital bones). Just deep to the mucous membrane of the roof of the nasopharynx is a collection of lymphoid tissue called the **pharyngeal tonsil** (adenoids) (**Fig. 7.2.5**).

The **auditory (pharyngotympanic) tube** is the most obvious feature on the lateral wall of the nasopharynx. It connects the tympanic cavity (middle ear) with the nasopharynx and allows for the equalization of pressure on both sides of the tympanic membrane. The auditory tube is composed of a laterally directed **bony part** and a medially directed **cartilaginous part**. The **tubal elevation** (torus tubarius) is the visible end of the cartilaginous part in the nasopharynx.

TABLE 7.2.4	Features of the paranasal sinuses.	
Name	**Location**	**Features**
Frontal sinus	Frontal bone	Triangular Drains through frontonasal duct into hiatus semilunaris in middle nasal meatus
Maxillary sinus	Maxilla (body)	Largest paranasal sinus ▪ Medial wall directed toward lateral nasal wall ▪ Roof forms floor of orbit ▪ Floor related to roots of maxillary molars and premolars Opens into center of hiatus semilunaris in middle nasal meatus
Sphenoidal sinus	Sphenoid (body): inferior to pituitary fossa, medial to cavernous sinuses	Opens into spheno-ethmoidal recess
Ethmoidal air cells	Ethmoid bone Superolateral to nasal cavity, along medial wall of orbit	8–12 small sinuses (air cells) ▪ **Anterior ethmoidal cells** open into hiatus semilunaris in middle nasal meatus ▪ **Middle ethmoidal cells** open onto ethmoidal bulla in middle nasal meatus ▪ **Posterior ethmoidal cells** open into lateral nasal wall in superior nasal meatus

Oropharynx

The oropharynx lies posterior to the oral cavity, inferior to the level of the soft palate, and superior to the epiglottis. The posterior one-third of the tongue is located in the oropharynx. Two paired folds of mucous membrane, or arches, mark the transition from the oral cavity to the oropharynx.

1. The **palatoglossal folds** are more anterior and are formed by the palatoglossus muscles and overlying mucous membrane.

2. The **palatopharyngeal folds** are more posterior and are formed by the palatopharyngeus muscles and overlying mucous membrane.

The **palatine tonsils** are collections of submucosal lymphoid tissue that lie in the tonsillar beds between the palatoglossal and palatopharyngeal folds. The **lingual tonsil** is submucosal lymphoid tissue on the posterior one-third of the tongue.

Clinical Notes

Waldeyer ring (pharyngeal tonsillar ring) describes a "ring" of lymphoid tissue in the walls of the naso- and oropharynx. The ring consists of (from superior to inferior):

- Pharyngeal tonsil (adenoids)
- Tubal tonsils (near the opening of the pharyngotympanic tube)
- Palatine tonsils
- Lingual tonsil

Waldeyer ring is located at the proximal part of the respiratory and alimentary tracts and considered mucosa-associated lymphoid tissue (MALT). These tonsils would be the first site encountered by inhaled or ingested microorganisms and, therefore, are considered the first line of defense against pathogens.

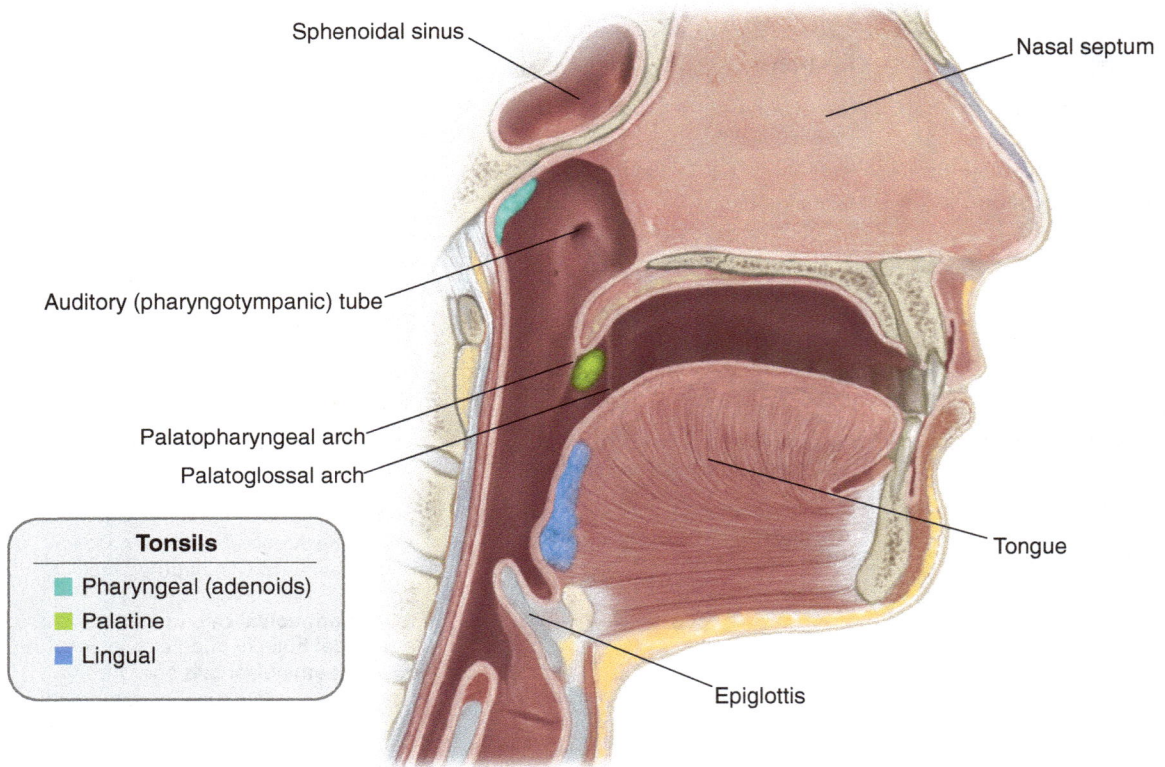

FIGURE 7.2.5 Midsagittal view showing the tonsils.

CLINICAL REASONING

The patient presentation, history, physical examination, and laboratory tests lead to the conclusion that the patient has an **irregularity in the nasal cavity or nasopharynx**.

Adenotonsillar Disease

Adenoiditis is the inflammation of the **pharyngeal tonsil**, usually caused by a viral or bacterial infection. Adenoiditis may present with cold-like symptoms; however, the symptoms often persist for 10 or more days. Severe or recurring adenoiditis may require surgical removal of the adenoids (adenotonsillectomy). Tonsillitis results when the **pala-tine tonsils** become overwhelmed by a bacterial or viral infection, which causes local inflammation and swelling. It is also very common to have concurrent inflammation of both sets of tonsils, which is known as **adenotonsillitis** or **adenotonsillar disease**.

Signs and Symptoms

- Temperature: >38°C (100.4°F)
- Tender anterior cervical lymphadenopathy
- Pus-like discharge from nose
- Pharyngotonsillar exudate
- Swollen, white-purple tonsil, often extending into the nasopharynx and/or oropharynx

---

Adenoiditis Inflammation of the pharyngeal tonsils (adenoids)
Tonsillitis Inflammation of the palatine tonsils

Lymphadenopathy Disease of the lymph nodes; used synonymously to mean swollen or enlarged lymph nodes
Exudate Fluid released from tissue due to inflammation or injury

- Sore throat with odynophagia
- Lack of cough and hoarseness

Predisposing Factors

- Age: most common in children (5–15 years)
- Family history of tonsillitis

Fractured Nasal Septum

The nose is predisposed to traumatic injury due to its central position and anterior projection on the face. Nasal fracture is the most common type of facial fracture, and most nasal fractures involve the septum. The osseous and/or cartilaginous parts of the septum may be fractured, with or without involvement of the nasal bone. These fractures usually result from isolated injuries; however, they may be associated with other head, face, and neck injuries (e.g., traumatic brain injury, fractures of the cervical spine, naso-orbital-ethmoid complex, or zygomatic-maxillary complex). They are usually simple or comminuted fractures, but seldom compound.

Clinical Notes

- Nasal fractures are often unrecognized and untreated at the time of injury.
- Confirming the presence of a septal hematoma following a septal fracture is important to avoid compressive damage to the local tissue. Septal hematoma results from bleeding within the subperichondrial/subperiosteal plane of the septum. If untreated, fibrosis of the septal cartilage may occur. Subsequent necrosis and perforation will lead to loss of structural and septal collapse. This results in a characteristic **saddle nose deformity (Fig. 7.2.6)**. A hematoma may be suspected if there is excessive septal edema and severe localized tenderness on examination.

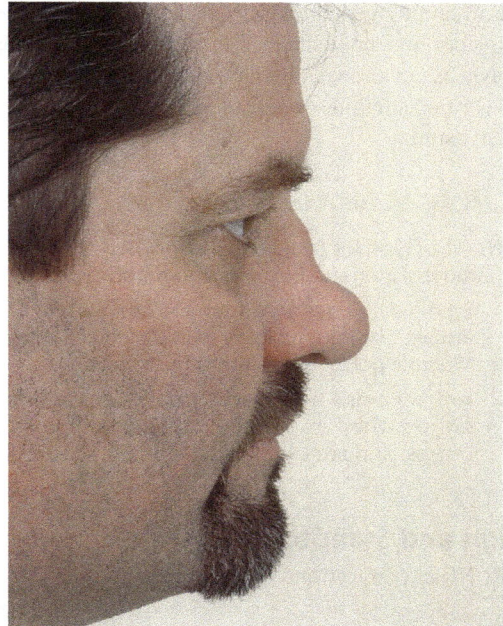

FIGURE 7.2.6 Saddle nose deformity. *Source*: Fig. 1-11B in *The Atlas of Emergency Medicine*.

Sign and Symptoms

- Epistaxis
- Infra-orbital ecchymosis
- Difficulty breathing through the nose
- Misshapen appearance of nose
- Local pain
- Nasal edema

Predisposing Factors

- History of trauma

Nasal Polyps

Nasal polyps are mucous membrane-covered, pedunculated masses that arise primarily from the

---

Odynophagia Pain on swallowing
Simple fracture A fracture in which the skin or mucous membrane is intact (synonym: closed fracture)
Comminuted fracture A fracture with three or more pieces
Necrosis Pathologic death of cells, tissues, or organs
Compound fracture A fracture in which bone fragment penetrates the skin or mucous membrane (synonym: open fracture)

Hematoma Localized extravasation of blood, usually clotted
Edema Swelling due to abnormal accumulation of fluid in subcutaneous tissue
Epistaxis Bleeding from the nose
Ecchymosis Hemorrhage into skin (bruising)

nasal cavity and paranasal sinuses. They are freely movable and nontender. The pathogenesis of nasal polyps is not known, but they are thought to be caused by chronic allergies or are associated with nonallergic adult asthma.

Clinical Notes

Nasal polyps are frequently associated with paranasal sinuses. Two more common types are:

1. **Ethmoidal** polyps arise from the ethmoidal sinuses. They are typically multiple and bilateral. Multiple polyps are referred to as polyposis.

2. **Antrochoanal** polyps arise from the maxillary sinuses. They are less common than ethmoidal polyps, and are usually single and unilateral.

Sign and Symptoms

- Nasal congestion
- Sinusitis
- Anosmia
- Headache
- Epistaxis

Predisposing Factors

- Chronic sinusitis
- Asthma
- Aspirin intolerance
- Cystic fibrosis
- Kartagerner syndrome (immotile ciliary syndrome)
- Churg–Strauss syndrome (autoimmune vasculitis)
- Nasal mastocytosis (excessive mast cells)
- Exposure to chromium

Sinusitis

Sinusitis (rhinosinusitis) is an inflammation of the mucous membrane of the paranasal sinuses. Acute sinusitis typically follows a viral infection of the upper

respiratory tract. The associated mucosal inflammation blocks the openings of one or more paranasal sinus. This causes stasis of sinus secretions and ciliary dysfunction, which promotes bacterial proliferation and acute sinus inflammation.

Clinical Notes

Sinusitis is classified by the frequency and duration of the inflammation.

- **Acute** sinusitis is a new infection that may last up to 4 weeks.
- **Recurrent acute** sinusitis refers to four or more separate episodes of acute sinusitis that occur within 1 year.
- **Subacute** sinusitis is an infection that lasts 4-12 weeks, and represents a transition between acute and chronic infection.
- **Chronic** sinusitis occurs when signs and symptoms last for more than 12 weeks.

Acute sinusitis is the most common form, but all have similar signs and symptoms and are, therefore, difficult to distinguish.

Sign and Symptoms

- Facial congestion, pain, or pressure
- Nasal drainage
- Nasal obstruction
- Epistaxis
- Fever
- Cough
- Dental pain
- Fatigue
- Purulence (assessed by nasal endoscopy)

Predisposing Factors

- Upper respiratory infection
- Allergies
- Dental infection

Sinusitis Inflammation of one or more paranasal sinus
Anosmia Loss of smell
Stasis Reduction or cessation (stagnation) of flow of a body fluid

Purulent Containing, discharging, or causing the production of pus

- Previous sinusitis
- Nasal polyps
- Cystic fibrosis
- Kartagerner syndrome (immotile ciliary syndrome)
- Environmental factors (e.g., dust, pollutants, and second-hand smoke)
- Structural abnormalities of nasal cavity (e.g., deviated septum) or auditory tube

DIAGNOSIS

The patient presentation, history, and physical examination support a diagnosis of a **fracture of the nasal septum**.

Fractured Nasal Septum

The patient suffered a direct blow to the dorsum of his nose on the edge of the kitchen cupboard door. At the time, there was only local swelling so he did not seek medical attention, and the septal fracture was undetected. Subsequently, a hematoma developed at the fracture site, which interfered with the blood supply to the nasal septum and caused necrosis of the nasal septal cartilage.

- Unilateral hematoma on the nasal septum was seen on the rhinoscopic examination. The swelling was large enough to almost completely block the flow of air through the right nasal cavity, resulting in the patient feeling "stuffed up" or congested. The mucosa associated with the hematoma, like other areas of nasal mucosa, is exposed to the drying effect of inspired air, making it a likely site for epistaxis.
- Tissue necrosis in the region of the hematoma caused local pain and tenderness.
- The saddle nose deformity is pathognomonic for nasal septal cartilage necrosis.

Pathognomonic Characteristic symptom that points unmistakably to a specific disease

Adenotonsillar Disease

The **pharyngeal tonsillar ring (Waldeyer ring)** is located around the posterior apertures of the nasal and oral cavities. Inflammation of any part of this lymphoid tissue may result in impaired airflow in the upper respiratory tract. Chronic inflammation of the pharyngeal tonsils (adenoiditis) will result in edema and vascular congestion in the superior aspect of the nasopharynx. The adenoids may become large enough to block the choanae on one or both sides of the nasal cavity. This may cause the patient to have problems with nose breathing.

- This patient did not have fever, cervical lymphadenopathy, or evidence of inflamed pharyngeal or palatine tonsils.

Nasal Polyps

Nasal polyps commonly start near the ethmoidal sinuses located superolateral to the nasal cavity. They expand into the nasal cavity. Large polyps can block the sinuses, or one or both sides of the nasal cavity.

- The swelling in the nasal cavity of this patient was midline and attached to the nasal septum. It was not movable and was clearly painful, excluding nasal polyps as a diagnosis.

Sinusitis

Paranasal sinuses are diverticula of the nasal cavity. Their mucosal secretions, therefore, drain into the nasal cavity. Sinusitis interferes with sinus drainage and mucus may collect in one or more paranasal sinus. This may make it difficult for patients to breath through their nose. The sinus congestion may be perceived as facial congestion, pain, or headache.

- Sinusitis may be a contributing factor to frequent epistaxis because it causes mucosal irritation and friable vasculature.
- Acute sinusitis is most often caused by the common cold, allergies, or fungal infections. This patient did not have a recent upper respiratory infection, does not have a fever, and has no allergies. This rules out sinusitis as a likely diagnosis.

Patient Presentation

A 52-year-old male is seen in the neurology clinic with complaints of severe, intermittent stabbing pains on the right side of the face.

Relevant Clinical Findings

History

The patient reports that pain "attacks" on his face started about 6 months ago and have recently increased in frequency. The pain occurs several times a day, but only lasts for a few seconds each time. He has been unable to shave because touching his right cheek and chin triggers excruciating pain. Sometimes, drinking or eating will initiate the pain, and while outdoors on windy days, the attacks seemed to occur more frequently. The patient has lost weight recently and saw a dentist who confirmed he has no dental problems.

Physical Examination

The following findings were noted on neurologic examination:

- A pain "attack" was prompted each time his right cheek or chin was touched.
- The pain was associated with a facial tic.

Imaging Studies

- Computed tomographic scans with contrast revealed no characteristic abnormalities, no focal area of hyperintensity, or space-filling lesions in the brain.

Clinical Problems to Consider

- Temporal arteritis
- Temporomandibular joint (TMJ) syndrome
- Trigeminal neuralgia

LEARNING OBJECTIVES

1. Describe the anatomy of the infratemporal fossa and its contents.
2. Describe the anatomy of the TMJ.
3. Explain the anatomical basis for the signs and symptoms associated with this case.

RELEVANT ANATOMY

Infratemporal Fossa

The **infratemporal fossa**, also known as the deep face, is an irregularly shaped space located infe-rior and medial to the zygomatic arch. It is continuous with the temporal fossa superiorly and opens to the neck postero-inferiorly. The boundaries of the infratemporal fossa are described in **Table 7.3.1**.

The contents of the infratemporal fossa include:

- The **medial and lateral pterygoid muscles**, two muscles of mastication.
- The **maxillary artery**, one of the two terminal branches of external carotid artery.
- Branches of maxillary artery
- The **pterygoid venous plexus**, a network of veins of deep face.

Tic Habitual, repeated contraction of specific muscles (e.g., facial tic); it may be suppressed voluntarily

Neuralgia Severe, throbbing, or stabbing pain along the course or distribution of a nerve

TABLE 7.3.1	Boundaries and openings of the infratemporal fossa.
Boundary	**Description**
Anterior	Posterior surface of maxilla
Superior (roof)	Inferior surface of greater wing of sphenoid • Foramen ovale—transmits mandibular nerve (CN V3), lesser petrosal nerve (branch of CN IX), and accessory meningeal artery • Foramen spinosum—transmits middle meningeal artery • Petrotympanic fissure—transmits chorda tympani (branch of CN VII)
Medial	Lateral pterygoid plate of sphenoid bone
Lateral	Ramus of mandible • Mandibular foramen (opening to the mandibular canal)—transmits inferior alveolar nerve, artery, and vein

- The **mandibular nerve (CN V3)**, the third division of trigeminal nerve (CN V).
- Branches of mandibular nerve.
- The **chorda tympani**, a branch of facial nerve (CN VII).
- The **lesser petrosal nerve**, a branch of glossopharyngeal nerve (CN IX).
- The **otic ganglion**, a parasympathetic ganglion associated with innervation of the parotid gland.
- The **sphenomandibular ligament**, an extracapsular ligament of the TMJ.

Muscles of Mastication

The mandible is the only movable bone of the skull. Mandibular movement occurs during mastication (chewing) and speaking. Four paired muscles comprise the **muscles of mastication** (**Fig. 7.3.1**). They are responsible for elevation, protraction (protrusion), and retraction of the mandible (**Table 7.3.2**). The muscles of mastication are derived embryologically from the first pharyngeal arch and are innervated by the **mandibular nerve (CN V3)**. Depression of the mandible (opening of the mouth) against

TABLE 7.3.2	Muscles of mastication.		
Muscle	**Origin**	**Insertion**	**Action on mandible**
Temporalis	• Temporal fossa between inferior and superior temporal lines • Temporalis fascia	• Coronoid process of mandible • Anterior margin of mandibular ramus	Elevation Retraction (posterior fibers)
Masseter 　**Superficial part** 　**Deep part**	• Zygomatic arch and maxillary process of zygomatic bone • Medial aspect of zygomatic arch	Lateral surface of mandibular ramus	Elevation Protraction
Medial pterygoid 　**Superficial** 　**Deep**	• Tuberosity of maxilla • Pyramidal process of palatine bone • Medial surface of lateral pterygoid plate	Medial surface of mandibular ramus	Elevation Protraction
Lateral pterygoid 　**Superior head** 　**Inferior head**	• Infratemporal surface of sphenoid bone • Lateral surface of lateral pterygoid plate	• Articular disc and capsule of TMJ mandible • Neck of mandible (in pterygoid fovea)	Protraction

(A)

Temporalis m.

Masseter m.
• Deep part

• Superficial part

(B)

Lateral pterygoid m.
   • Superior head
      • Inferior head

Medial pterygoid m.
   • Deep head

   • Superficial head

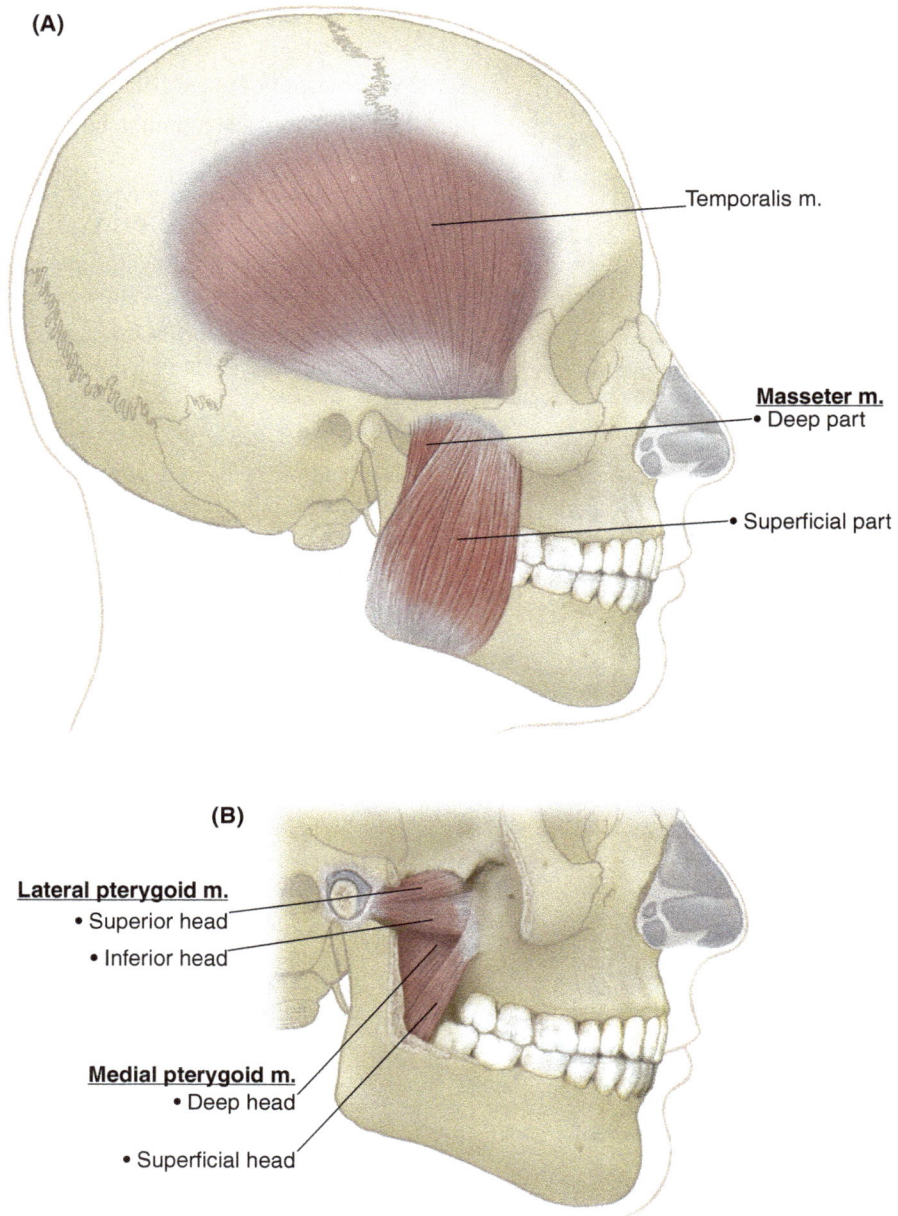

FIGURE 7.3.1 Lateral view of the head showing the superficial (A) and deep (B) muscles of mastication.

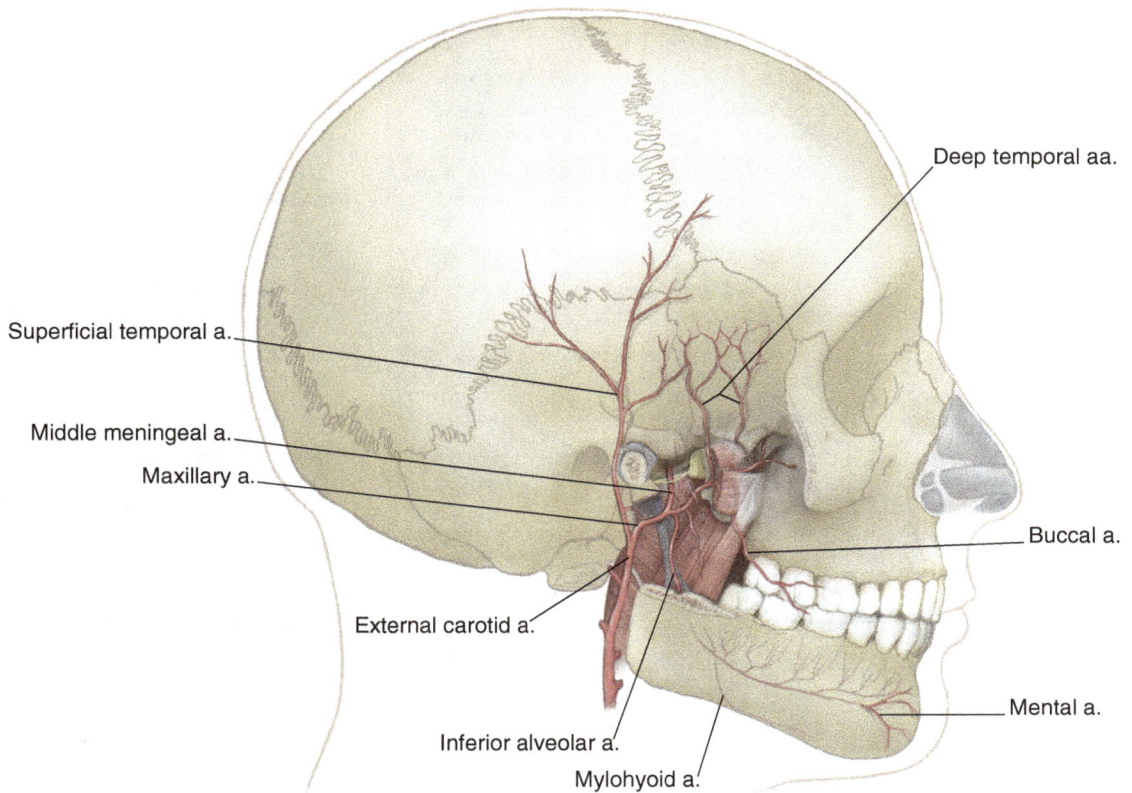

FIGURE 7.3.2 Lateral view of the infratemporal fossa showing the maxillary artery and its branches.

resistance requires muscles associated with the hyoid bone (supra- and infrahyoid muscles).

Maxillary Artery

The maxillary artery (**Fig. 7.3.2**) is the largest branch of the external carotid artery and provides the primary blood supply for the cranial dura mater, structures in the infratemporal fossa, nasal cavity, walls of the oral cavity and nasopharynx, the maxillary and mandibular teeth, and portions of the face. It branches from the external carotid within the parotid gland and passes medial to the neck of the mandible to enter the infratemporal fossa. Here, it gives rise to approximately two-thirds of its branches. It courses obliquely through the infratemporal fossa lateral (but frequently medial) to the lateral pterygoid muscle to enter the pterygopalatine fossa. From this fossa, it sends branches to the oral cavity, nasal cavity, nasopharynx, and skin of the face (**Table 7.3.3**).

Mandibular Nerve

The **mandibular nerve (CN V3)** (**Fig. 7.3.3**) is the largest of the three divisions of trigeminal nerve (CN V). It is a mixed nerve, with both sensory and motor components.

- The **sensory part** begins at the trigeminal (semilunar, Gasserian) ganglion in the middle cranial fossa. It carries general sensory information from the majority of the oral cavity mucosa, including the anterior part of the tongue, and the skin of the chin, anterior ear auricle, and temple.

- The smaller **motor part** supplies the muscles of mastication and other muscles derived from the embryonic first pharyngeal arch.

The two parts of the nerve join as they exit the middle cranial fossa to traverse the **foramen ovale** and enter the infratemporal fossa. All branches of the

TABLE 7.3.3	Branches of the maxillary artery in the infratemporal fossa.	
Branch	**Course**	**Distribution**
Middle meningeal	Passes between roots of auriculotemporal nerve and through foramen spinosum to enter cranial cavity	▪ Primary blood supply to cranial dura mater
Inferior alveolar	Passes through mandibular foramen to enter mandibular canal	▪ Mandibular teeth and gingivae
Mental	Exits mandibular canal via mental foramen	▪ Chin and lower lip
Deep auricular (2)	Ascends in parotid gland and pierces wall of external acoustic meatus	▪ External acoustic meatus
Anterior tympanic	Ascends with deep auricular branches to enter tympanic cavity	▪ Deep surface of tympanic membrane
Accessory meningeal	Enters cranial cavity through foramen ovale	▪ Cranial dura mater
Deep temporal (2)	On deep surface of temporalis muscle	▪ Temporalis muscle
Masseteric	Passes through mandibular notch	▪ Masseter muscle
Buccal	Courses obliquely from infratemporal fossa to pierce buccinator muscle	▪ Skin, muscle, and oral mucosa of cheek
Pterygoid	Small branches pass directly to pterygoid muscles	▪ Medial pterygoid muscle ▪ Lateral pterygoid muscle

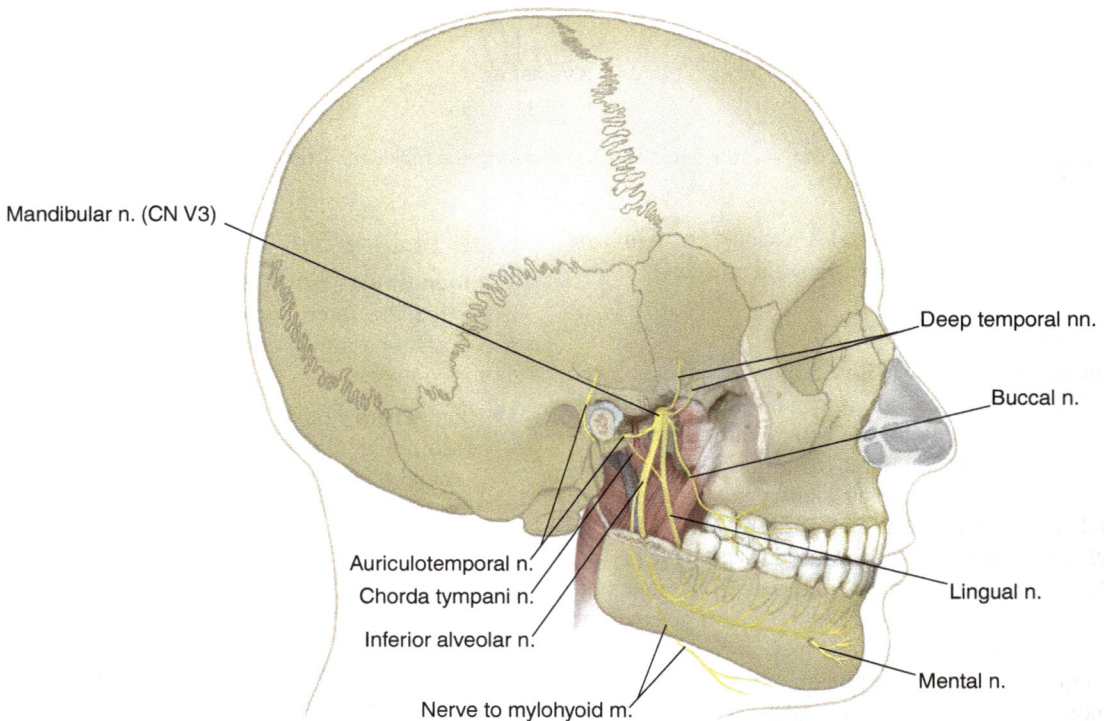

FIGURE 7.3.3 Lateral view of infratemporal fossa showing branches of the mandibular nerve. (Not all motor branches are illustrated).

TABLE 7.3.4	Branches of mandibular nerve.	
Nerve	**Component**	**Distribution**
Meningeal	Sensory	▪ Dura mater in middle cranial fossa
Nerve to medial pterygoid	Motor	▪ Medial pterygoid muscle ▪ Tensor veli palatini muscle ▪ Tensor tympani muscle
Buccal	Sensory	▪ Skin over anterior part of buccinator ▪ Buccal mucous membrane ▪ Posterior buccal gingiva
Masseteric	Motor	▪ Masseter muscle
Deep temporal (2)	Motor	▪ Temporalis muscle
Nerve to lateral pterygoid	Motor	▪ Lateral pterygoid muscle
Auriculotemporal	Sensory	▪ Skin of anterior ear and temple ▪ External acoustic meatus ▪ Temporomandibular joint
Lingual	Sensory	▪ Anterior two-thirds of tongue ▪ Mandibular lingual gingiva ▪ Oral mucosa, on floor of oral cavity
Inferior alveolar 　Nerve to mylohyoid 　Mental	 Motor Sensory	▪ Mandibular teeth and associated ▪ Mucosa and skin of lower lip and chin ▪ Mylohyoid muscle ▪ Anterior belly of digastric muscle

mandibular nerve are formed in the infratemporal fossa (**Table 7.3.4**).

Chorda Tympani and Lesser Petrosal Nerve

Chorda tympani branches from the facial nerve (CN VII) in the petrous part of the temporal bone and passes through the tympanic cavity (middle ear). Here, it is separated from the internal surface of the tympanic membrane by the handle of the malleus. It leaves the middle ear via the petrotympanic fissure and enters the infratemporal fossa where it joins the lingual nerve (branch of CN V3). Chorda tympani has two components:

1. Taste information from the **anterior two-thirds of the tongue**.

2. **Preganglionic parasympathetic fibers** to the submandibular ganglion. **Postganglionic parasympathetic fibers** from this ganglion are secretomotor to the **sublingual and submandibular salivary glands**.

The **lesser petrosal nerve** is a branch of the glossopharyngeal nerve (CN IX). It carries **preganglionic parasympathetic fibers** through foramen ovale into the infratemporal fossa, where they synapse in the **otic ganglion**. **Postganglionic parasympathetic fibers** join the auriculotemporal nerve (branch of CN V3) and are secretomotor to the **parotid salivary gland**.

Temporomandibular Joint

The **TMJ** is formed by the mandibular condyle (head), with the mandibular fossa and articular tubercle of the temporal bone (**Fig. 7.3.4**). It is a synovial joint that allows for opening and closing of the mouth, as well as other complex movements involved with mastication. A biconcave **articular disc** is interposed between the mandibular condyle and articular

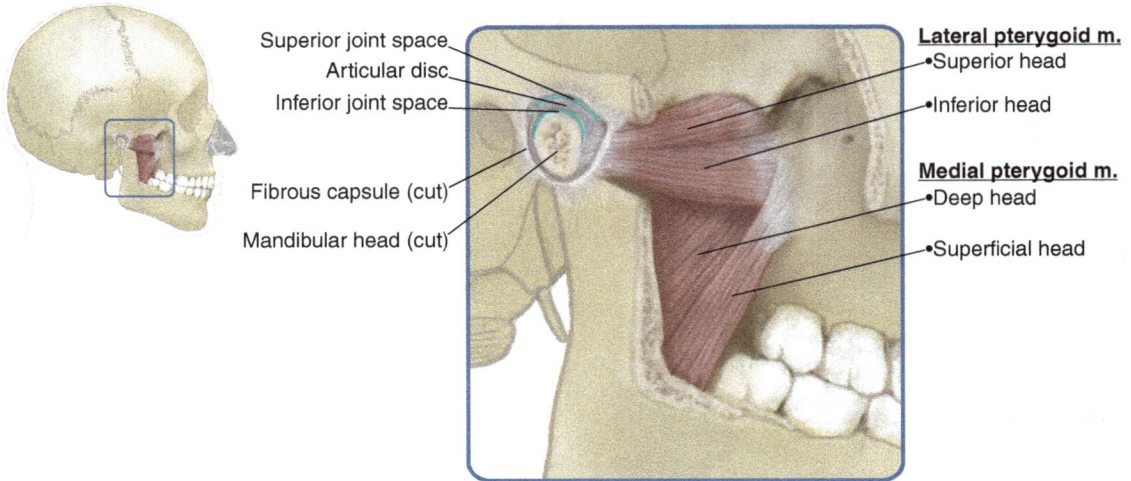

FIGURE 7.3.4 Lateral view of the temporomandibular joint with capsule open to show articular disc.

surfaces of the temporal bone. The disc divides the joint into two compartments:

1. The **lower joint space** is located between the mandibular condyle and disc. It allows for hinge-like movements that occur during depression and elevation of the mandible. During this movement, the condyle rotates on the disc.

2. The **upper joint space** is located between the disc and the temporal bone. It is designed to accommodate **translation**. Translation is the sliding movement of the mandibular condyle between the mandibular fossa and the articular tubercle. Anterior translation produces protraction (protrusion) of the mandible while translation in the posterior direction produces mandibular retraction. During translation, the condyle and disc slide as a unit.

The TMJ is surrounded by a fibrous capsule that is lined by a synovial membrane. The joint is supported by three extracapsular ligaments:

1. The **lateral temporomandibular ligament** is lateral and connects the articular tubercle with the posterolateral part of the mandibular neck.

2. The **sphenomandibular ligament** lies medial and courses between the spine on the sphenoid bone and the lingula on the medial aspect of the ramus of the mandible.

3. The **stylomandibular ligament** is also medial and extends from the styloid process (temporal bone) to the angle of the mandible.

Blood Supply and Innervation

Branches of the **maxillary and superficial temporal arteries** provide the blood supply to the TMJ. Innervation of the joint is provided primarily by the **auriculotemporal nerve**.

CLINICAL REASONING

This patient presents with **severe intermittent facial pain**.

Temporal Arteritis

Temporal (giant cell) arteritis is an inflammatory disease of blood vessels (vasculitis) that supply the head. It involves branches of the carotid arteries, particularly those that supply the temporal region. Temporal arteritis most commonly affects the deep temporal arteries. Inflammation produces swelling of the pain-sensitive vascular wall, which results in local tenderness and headache. Thrombosis may

---

Vasculitis Inflammation of blood vessels

Thrombus A fixed mass of platelets and/or fibrin (clot) that partially or totally occludes a blood vessel or heart chamber. An embolism is a mobile clot in the cardiovascular system

occur in severe cases. Although temporal arteritis is idiopathic, it has been associated with severe infections and the use of high doses of antibiotics.

Sign and Symptoms

- Fever
- Headache
- Scalp sensitivity and tenderness
- Jaw claudication
- Malaise
- Weakness and fatigue
- Muscle aches
- Vision impairment (e.g., diplopia)
- Weight loss (more than 5% of total body weight)

Clinical Notes

- Temporal arteritis is closely associated with **polymyalgia rheumatica (PMR)**, which is characterized by stiffness, aching, and pain in muscles of the neck, shoulders, lower back, hips, and thighs. Ten to twenty percent of patients who initially present with PMR develop temporal arteritis. Temporal arteritis and PMR may represent a clinical spectrum of a common disease process.

- The inflammation may affect the ophthalmic artery, resulting in ischemic optic neuropathy that may lead to blurred vision or sudden blindness. Loss of vision in both eyes may occur abruptly.

Predisposing Factors

- Age: >50 years
- Sex: female (2:1)
- Family history

Trigeminal Neuralgia

Trigeminal neuralgia (commonly called **tic douloureux**; also known as **prosopalgia** or Fothergill disease) is a neuropathic disorder characterized by unilateral episodes of intense pain on the face. Painful episodes are typically initiated by stimuli at specific areas on the face called "trigger zones." Patients describe the pain as "excruciating" with an intensity that causes a tic in which the patient grimaces or winces.

The pathophysiology of trigeminal neuralgia may be related to focal, mechanical compression of the trigeminal nerve close to the brain stem, perhaps caused by an adjacent artery that compresses, irritates, or damages the nerve. In some cases, trigeminal neuralgia has been associated with multiple sclerosis, cerebellopontine angle tumor, or other expanding brain stem lesions that impact the trigeminal nerve. In most cases, however, the condition is idiopathic.

Signs and Symptoms

- Transient episodes of sudden, excruciating, unilateral facial pain; it is described as an "electric shock" that gives way to a burning sensation that may last for seconds or minutes.

- Commonly arises near the mouth and extends toward the ear, eye, or nose.

- Pain may be triggered by touch (including drafts), movements of the head or face, including those related to eating and drinking.

Predisposing Factors

- Age: >40 years
- Sex: female:male (3:2)
- Young adults (20–40 years of age) with multiple sclerosis

Clinical Notes

- Pain-free intervals may last minutes-to-weeks, but permanent remission is rare.

- Pain is confined mainly to areas supplied by the maxillary (CN V2) and/or mandibular (CN V3) nerves (**Fig. 7.3.5**). Involvement of the ophthalmic nerve (CN V1), or bilateral disease, occurs in less than 5% of cases.

Idiopathic A condition that arises spontaneously or whose cause is obscure or unknown
Claudication Muscle pain caused by vascular insufficiency

Malaise Feeling of general body weakness or discomfort, often marking the onset of an illness

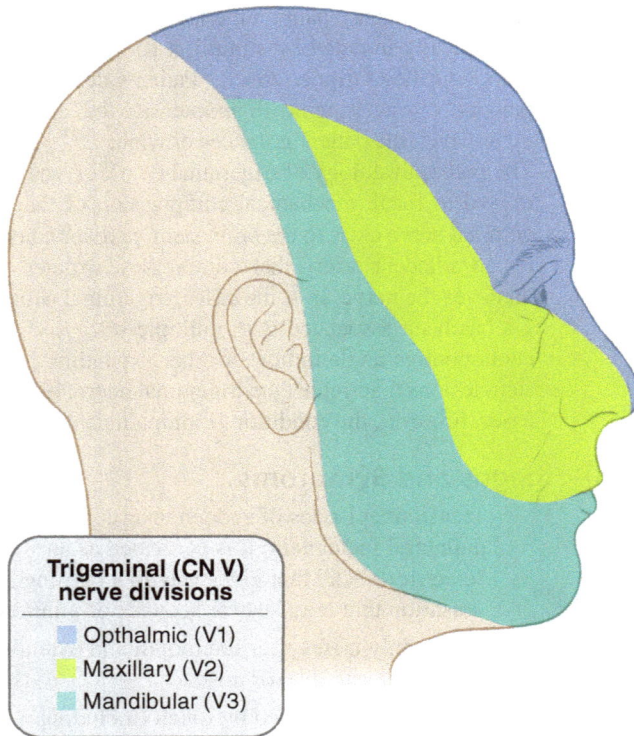

Trigeminal (CN V) nerve divisions

- Opthalmic (V1)
- Maxillary (V2)
- Mandibular (V3)

FIGURE 7.3.5 Lateral view of face showing pattern of sensory innervation by branches of trigeminal nerve (CN V).

Temporomandibular Joint (TMJ) Syndrome

TMJ syndrome refers to acute or chronic pain of the temporomandibular joint. The primary causes are muscular hyperfunction, as in the case of bruxism, and conditions that result in displacement of the articular disc during mandibular movement. Like other synovial joints, the TMJ is susceptible to ankylosis, arthritis, trauma, dislocation, developmental abnormalities, neoplasia, and reactive lesions.

TMJ dysfunction causes pain in the joint and the surrounding region. Signs and symptoms of TMJ disorder usually involve more than one joint component: muscle, nerve, tendon, ligament, bone, and/or connective tissue.

Signs and Symptoms

- Difficulty biting and/or chewing
- Clicking, popping, or grating sounds when opening or closing the mouth
- Jaw pain or tenderness
- Dull, aching facial pain
- Earache (particularly in the morning)
- Tinnitus or hearing loss
- Headache (particularly in the morning)
- Neck and shoulder pain
- Dizziness

Predisposing Factors

- Age: 30–50 years

Bruxism Grinding or clenching of teeth
Ankylosis A disease condition that results in stiffening or joint fixation

Arthritis Joint inflammation
Tinnitus Sensation of ringing or other noises in the ears

- Sex: female (3:1)
- Bruxism
- Habitual chewing of ice or gum
- Misaligned teeth or bite
- Jaw or facial deformity
- Arthritic disease
- Synovitis
- History of jaw or facial injury

DIAGNOSIS

The patient presentation, history, and physical examination support a diagnosis of **trigeminal neuralgia**.

Trigeminal Neuralgia

Trigeminal neuralgia is characterized by agonizing unilateral facial pain. It usually involves the maxillary (CN V2) and/or mandibular (CN V3) nerves.

- The presence of a trigger point is the cardinal characteristic of the neuralgia. The trigger points for this patient, the cheek and chin, are in the sensory distribution of the **mandibular nerve (CN V3)**.
- The intensity of the pain leads to a facial tic.
- The sensory or motor components of the trigeminal nerve are intact (i.e., sensation is possible, though painful, and the muscles of mastication are functional).

Temporal Arteritis

This is an inflammatory disease of temporal arteries. It is a common cause of headache in elderly people.

- Patients develop an intense headache, often with sharp, stabbing pain caused by blood flow through the inflamed vessels. The headaches may be throbbing or nonthrobbing.
- The pain is usually unilateral and localized most often to the site of the affected temporal arteries.
- Jaw claudication and ischemic nodules on the scalp with ulceration of the overlying skin have been described in severe cases.

The location and stimuli that trigger the pain in this patient is not consistent with temporal arteritis.

Temporomandibular Joint (TMJ) Syndrome

Dysfunction of the TMJ can result in craniofacial pain.

- Degenerative changes in the TMJ are associated pain that may radiate to the temple and face.
- The diagnosis of TMJ syndrome would be supported by findings of tenderness over the joint, crepitus with opening and closing the mouth, and limitation of jaw movements.

The pain in this patient is not related to the TMJ.

Synovitis Inflammation of synovial membranes

Ischemia Local anemia due to vascular obstruction

| **Facial Nerve Palsy**

Patient Presentation

A 24-year-old graduate student noticed while shaving this morning that he was unable to move the left side of his face. He worried he was having a stroke, so he sought immediate medical help in the emergency department of a local hospital.

Relevant Clinical Findings

History

The patient states that he is generally in good health, but he had flu-like symptoms the week before the onset of his facial paralysis. The patient did not complain of generalized muscle weakness in his limbs.

Physical Examination

The following findings were noted on neurologic examination:

- Oriented to time, date, and place, and able to follow simple verbal instructions
- Could not wrinkle forehead on the left side
- Could not close left eye tightly
- Could not show teeth or purse lips on left side
- Dysgeusia
- Hyperacusis
- Uvula was midline on palatal elevation
- Tongue protruded in the midline
- Vision was normal; no orbital or eye pain
- Sensation of face, scalp, and neck was intact

Imaging Studies

- Computed tomographic (CT) scans with contrast revealed no characteristic abnormalities, no focal area of hyperintensity, or space-filling lesions in the brain or cranial cavity.
- Gadolinium-enhanced MRI demonstrated diffuse linear enhancement of the left facial nerve.

Clinical Problems to Consider

- Acoustic neuroma
- Facial nerve (Bell) palsy
- Parotid tumor
- Stroke (involving facial motor cortex)

LEARNING OBJECTIVES

1. Describe the anatomy of the facial (CN VII) and vestibulocochlear (CN VIII) nerves.
2. Describe the anatomy of the muscles of facial expression.
3. Describe the anatomy of the salivary glands.
4. Describe the blood supply to the brain.
5. Explain the anatomical basis for the signs and symptoms associated with this case.

RELEVANT ANATOMY

Facial Nerve (CN VII)

The facial nerve is a mixed cranial nerve with four functional components:

1. The **motor** component supplies the muscles of facial expression and other muscles that develop from the embryonic second pharyngeal arch.

Dysgeusia Altered sense of taste
Hyperacusis Increased sensitivity to sound

2. The **general sensory** component innervates parts of the auricle and external acoustic meatus.

3. The **special sensory** component carries **taste** information from the anterior two-thirds of the tongue.

4. The facial nerve also provides **preganglionic parasympathetic** fibers involved with secretion from the lacrimal gland, nasal, oral, and nasopharyngeal mucous glands, and the submandibular and sublingual salivary glands.

Course of the Facial Nerve

The facial nerve has as two roots that arise from the lateral surface of the pons, at the cerebellopontine junction.

1. The **motor root** arises from the **facial motor nucleus** in the pons and forms the larger part of the facial nerve.

2. The sensory and parasympathetic components of the facial nerve form the smaller **intermediate nerve** (nervus intermedius).

The roots of facial nerve lie lateral to the **vestibulocochlear nerve** (CN VIII), as the two cranial nerves enter the **internal acoustic meatus** in the **petrous portion of the temporal bone** (**Fig. 7.4.1**). Within the petrous temporal bone, course of the facial nerve is the shape of an inverted "L," initially directed laterally and then inferiorly. After it passes lateral to the bony labyrinth of the inner ear, the nerve turns acutely in an inferior direction. This bend in the nerve is called the **genu**. A cluster of sensory cell bodies forms the **geniculate ganglion** at the genu.

After the genu, the nerve enters the **facial canal**, which courses in the medial wall of the tympanic cavity (middle ear). The facial nerve emerges from the facial canal at the **stylomastoid foramen**. The nerve courses anteriorly into the parotid gland.

In the parotid gland, the nerve forms the **parotid plexus**. Five major branches of the plexus emerge from margins of the parotid gland and supply the **muscles of facial expression** (**Table 7.4.1**). Although it passes through the parotid gland, the facial nerve does not innervate this gland.

Central Connections of the Facial Motor Nucleus

The neurons in the facial motor nucleus (i.e., lower motor neurons) are spatially organized: those that supply muscles of the superior part of the face are segregated from those that supply the inferior part of the face (**Fig. 7.4.2**).

All neurons in this nucleus receive direct projections from neurons in the motor cortex (i.e., upper motor neurons). The pattern of input from cortical neurons varies within the facial motor nucleus (**Fig. 7.4.2**):

■ Motor neurons that supply muscles of the **superior part of the face** receive input from ipsilateral and contralateral motor cortices (i.e., from both sides of the brain).

■ Motor neurons that supply muscles of the **inferior part of the face** receive input only from the contralateral motor cortex.

Clinical Notes

Paralysis of the muscles of facial expression may result following a lesion in the facial motor cortex or its projections, in the facial motor nucleus, or in the facial nerve itself.

■ When the facial motor nucleus (*lower motor neurons*) or the facial nerve is damaged, muscles of the entire face on the ipsilateral (same) side will be paralyzed (**Fig. 7.4.2**A).

■ Damage to the facial motor cortex (*upper motor neurons*) will affect muscles in the inferior part of the contralateral face (**Fig. 7.4.2**B). Muscles in the superior part of the contralateral face will not be paralyzed because input from the ipsilateral motor cortex is not affected by this lesion.

Vestibulocochlear Nerve (CN VIII)

The vestibulocochlear nerve is a special sensory cranial nerve with two components that have distinct functions:

1. The **vestibular portion** of CN VIII is concerned with *equilibrium*. It is composed of the central processes of bipolar neurons whose cell bodies are located in the **vestibular ganglion**.

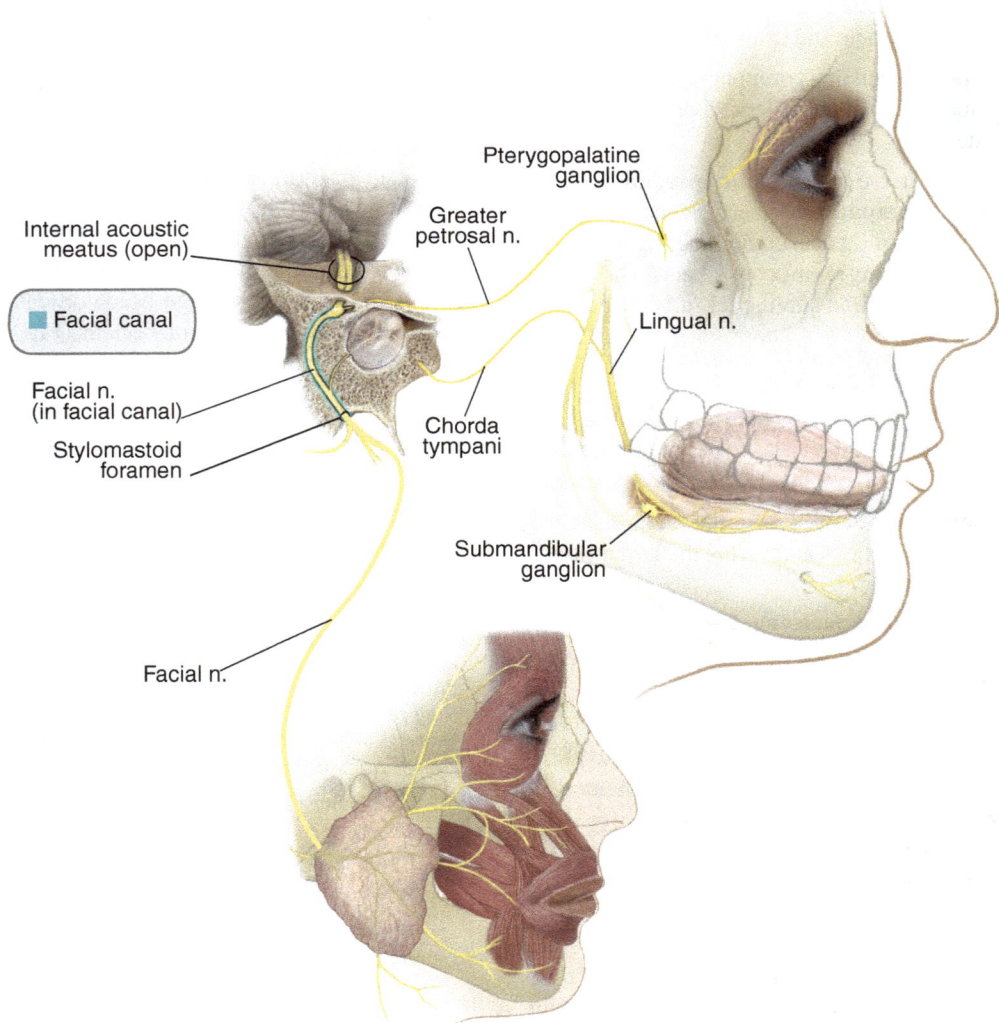

FIGURE 7.4.1 Lateral view showing the course and branches of the facial nerve.

This ganglion is located at the base of the semicircular canals of the inner ear, within the petrous portion of the temporal bone. The peripheral processes of this nerve extend to the maculae of the utricle and saccule, and the ampullae of the semicircular ducts.

2. The **cochlear portion** of CN VIII transmits *auditory information*. This portion of the vestibulocochlear nerve is formed by the central processes of bipolar neurons whose cell bodies

are located in the **spiral (cochlear) ganglion**. The spiral ganglion is located in the modiolus of the cochlea. The peripheral processes extend to the spiral organ (of Corti).

Course of the Vestibulochlear Nerve

The central processes of the primary sensory neurons in the vestibular and spiral ganglia form the vestibular and cochlear nerves, respectively. These nerves, jointly, form the vestibulocochlear nerve in the inter-

TABLE 7.4.1	Course and distribution of facial nerve.		
Branch[a]	Course	Distribution	Functional component(s)
Greater petrosal nerve	▪ Branches at the genu ▪ Courses anteriorly to middle cranial fossa to enter pterygoid canal	▪ Preganglionic parasympathetic fibers synapse in pterygopalatine ganglion	▪ **Parasympathetic (visceral efferent)**
Nerve to stapedius	▪ Branches in the facial canal ▪ Enters tympanic cavity	▪ Stapedius muscle (dampens vibration of ossicles)	▪ **Motor (skeletal muscles from second pharyngeal arch)**
Chorda tympani	▪ Branches in the facial canal ▪ Passes through tympanic cavity ▪ Enters infratemporal fossa ▪ Joins lingual nerve (branch of CN V3)	▪ Taste fibers from anterior ⅔ of tongue ▪ Preganglionic parasympathetic fibers synapse in submandibular ganglion	▪ **Special sensory** ▪ **Parasympathetic (visceral efferent)**
Posterior auricular	▪ Branches distal to stylomastoid foramen ▪ Courses posterior to external ear	▪ Small auricular muscles ▪ Occipital belly of occipitofrontalis muscle ▪ Sensory from skin of external ear	▪ **Motor (skeletal muscles from second pharyngeal arch)** ▪ **General sensory**
Nerves to posterior belly of digastric and stylohyoid muscles	▪ Branches distal to stylomastoid foramen ▪ Directly enters muscles	▪ Posterior belly of digastric muscle ▪ Stylohyoid muscle	▪ **Motor (skeletal muscles from second pharyngeal arch)**
Terminal branches 1. **Temporal** 2. **Zygomatic** 3. **Buccal** 4. **Marginal mandibular** 5. **Cervical**	▪ Branch from parotid plexus in parotid gland ▪ Distribute on face and neck	▪ Muscles of facial expression	▪ **Motor (skeletal muscles from second pharyngeal arch)**

[a]Proximal-to-distal.

nal acoustic meatus, where they travel with the facial nerve (CN VII). The vestibulocochlear nerve enters the medulla at its junction with the pons.

Muscles of Facial Expression

The muscles of facial expression (**Table 7.4.2**) are derived from the embryonic **second pharyngeal arch** and are innervated by the **facial nerve**. Most of these muscles originate in the subcutaneous tissue of the face; a few have bony origins. They all insert on the subcutaneous tissues or dermis of the face, neck, or scalp (**Fig. 7.4.3**). These muscles act primarily as sphincters or dilators of the facial openings (orbit, nose, and mouth).

Salivary Glands

There are three pairs of major salivary glands: **parotid**, **submandibular**, and **sublingual** (**Table 7.4.3**). These are compound tubulo-acinar exocrine glands that synthesize and secrete components of saliva.

Blood Supply and Innervation

The salivary glands receive arterial blood from the **facial or lingual** branches of the **external carotid**

(A) Facial nerve lesion
(Bell's palsy)

(B) Supranuclear lesion

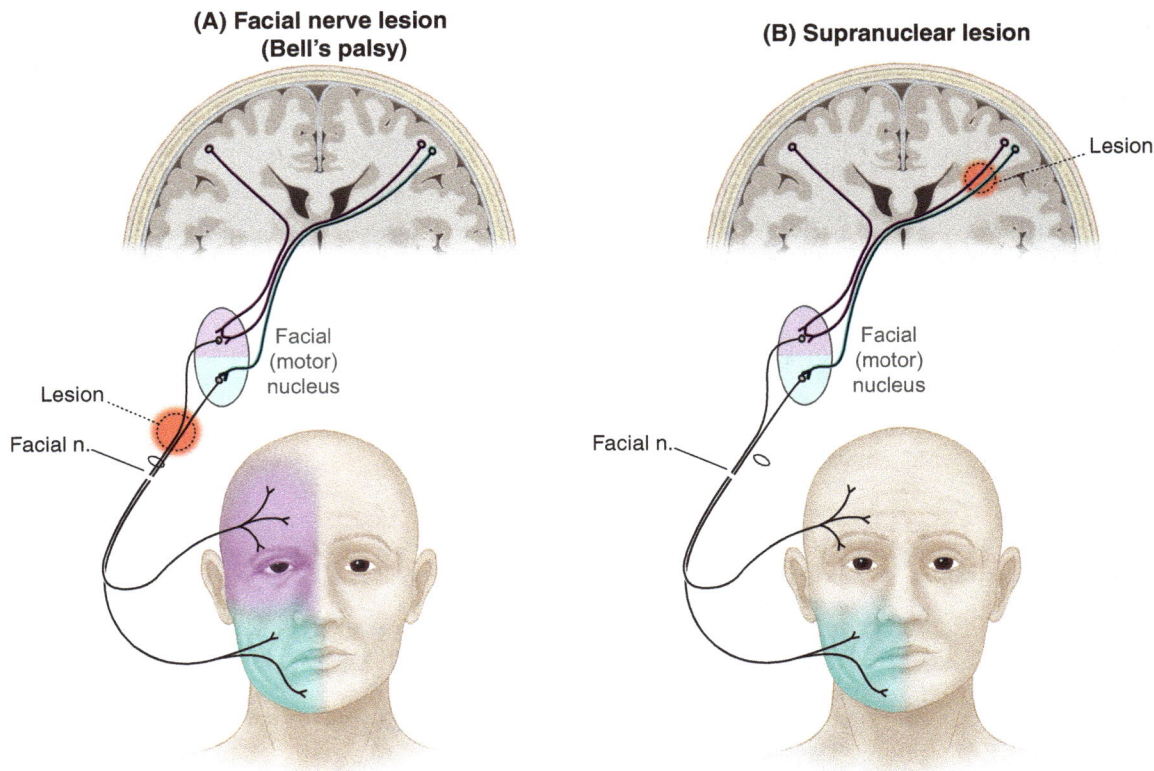

FIGURE 7.4.2 Central connections of the facial motor nucleus. Cortical (upper motor neuron) or brainstem (lower motor neruron) lesions have different clinical presentation.

artery (**ECA**), or directly from the ECA (**Table 7.4.4**). Venous blood drains into both the **internal and external jugular veins**.

The salivary glands receive both sympathetic and parasympathetic innervation. **Sympathetic post-ganglionic fibers** arise from cervical sympathetic ganglia. They form a plexus along the ECA and its branches that supply the glands. Sympathetic fibers supply these blood vessels; they may also be secretomotor to the glands.

Parasympathetic innervation to these glands involves components of the facial nerve (CN VII) or glossopharyngeal nerve (CN IX). Parasympathetic fibers (secretomotor) stimulate production of components of saliva, including some digestive enzymes.

Blood Supply to the Brain

The brain receives arterial blood from two primary sources:

1. The anterior circulation is from **internal carotid arteries**. They enter the cranial cavity and terminate as the **anterior and middle cerebral arteries.**

2. The posterior (vertebrobasilar) circulation is from the **vertebral arteries**, which join on the ventral aspect of the medulla to form the **basilar artery**.

The cerebral hemispheres receive blood from the anterior and posterior circulations. The brain stem is supplied by only the posterior circulation. The anterior and posterior circulations anastomose to form a vascular ring, **cerebral arterial circle (Willis)**, on the ventral aspect of the brain (see Fig. 6.4.2).

TABLE 7.4.2	Muscles of facial expression.	
Muscle	**Facial nerve branch**	**Action(s)**
Occipitofrontalis		
Frontal belly	Temporal	▪ Raises eyebrows ▪ Wrinkles forehead and protracts scalp
Occipital belly	Posterior auricular	▪ Retracts scalp
Orbicularis oculi	Temporal and zygomatic	▪ Closes eye tightly (e.g., winking)
Orbicularis oris	Buccal and marginal mandibular	▪ Protrudes lips (e.g., kissing)
Levator labii superioris		▪ Retract upper lip (e.g., showing upper teeth
Zygomaticus major	Zygomatic	▪ Elevates upper lip (e.g., smiling)
Zygomaticus minor	Zygomatic	▪ Elevates upper lip (e.g., smiling)
Levator anguli oris	Zygomatic	▪ Widens oral fissure (e.g., grinning)
Buccinator	Buccal	▪ Resist distension of cheek (e.g., whistling) ▪ Presses cheek against teeth to help keep food in oral cavity proper
Depressor labii inferioris	Marginal mandibular	▪ Retract lower lip (e.g., showing lower teeth or pouting)
Depressor anguli oris	Marginal mandibular	▪ Depresses corner of mouth (e.g., frowning)
Platysma	Cervical	▪ Helps depress mandible ▪ Tenses skin of chin and neck

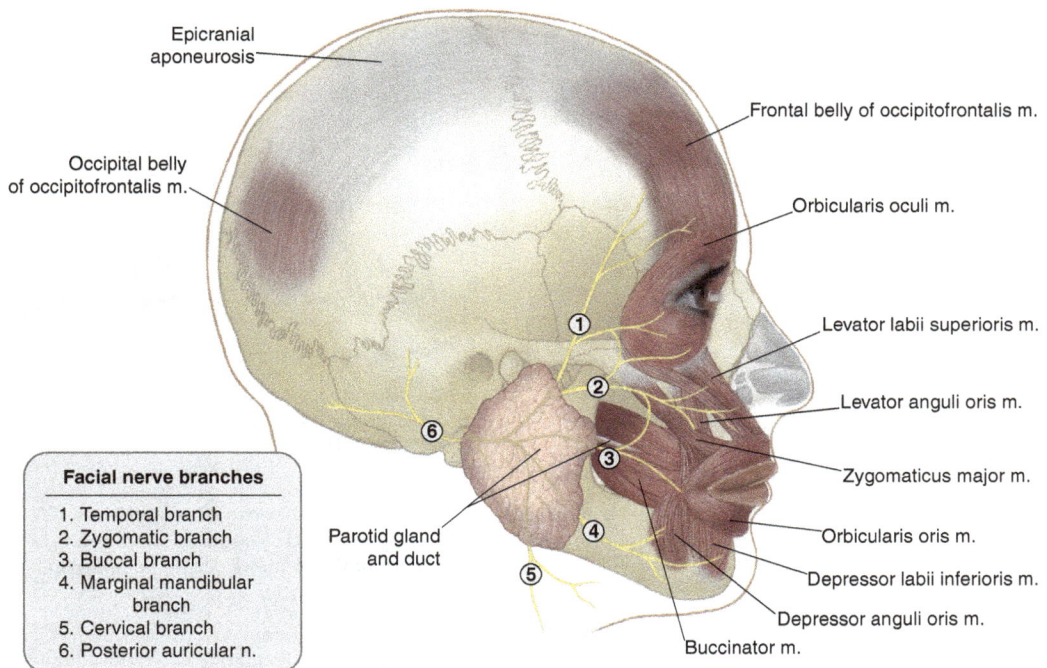

FIGURE 7.4.3 Lateral view of face showing the muscles of facial expression and terminal branches of the facial nerve (CN VII).

TABLE 7.4.3   Major salivary glands.

Salivary gland	Description
Parotid	Largest salivary gland Lateral and posterior to mandibular ramus Parotid plexus (facial nerve) embedded within the gland Drains into parotid duct at anterior margin; duct pierces buccinator and opens into oral vestibule (adjacent to second maxillary molar) at parotid papilla
Submandibular	1. Superficial part: larger part of gland; in digastric triangle of neck, attached to inferior mandibular border; covered by platysma and deep fascia 2. Deep part: between mylohyoid muscle (inferolateral) and hyoglossus and styloglossus muscles (medially) Both parts drain into the submandibular duct that opens at the sublingual papilla (adjacent to base of lingual frenulum)
Sublingual	In sublingual fossa (floor of oral cavity); deep to mucous membrane, between the mandible and genioglossus muscle Multiple ducts open directly into sublingual fossa

CLINICAL REASONING

This patient presents with signs and symptoms of **unilateral facial** paralysis, dysgeusia, and hyperacusis.

Acoustic Neuroma

An **acoustic neuroma** (vestibular schwannoma) is a benign, intracranial tumor of Schwann (myelin-forming) cells of the **vestibulocochlear nerve (CN VIII)**. These tumors may grow very large and compress adjacent brain stem structures and cranial nerves (e.g., facial nerve) (**Fig. 7.4.4**).

Paralysis Loss of muscle function, particularly related to voluntary movement

Clinical Notes

Acoustic neuromas may develop from unknown causes or may be part of von Recklinghausen neurofibromatosis. There are two forms of this neuroma:

1. **Neurofibromatosis type I** is a schwannoma that sporadically involves the facial nerve (CN VII), but may involve other cranial nerves. These tumors affect individuals in adulthood.

2. **Neurofibromatosis type II** is a schwannoma involving the entire extent of the vestibulocochlear nerve. These typically present before the age of 21 and show strong autosomal dominant inheritance.

TABLE 7.4.4   Blood supply and innervation of the major salivary glands.

Gland	Arterial supply	Parasympathetic innervation
Parotid	Direct branches of ECA	**Preganglionic fibers:** Lesser petrosal nerve (CN IX) **Parasympathetic ganglion:** Otic **Postganglionic fibers:** Travel with auriculotemporal nerve (branch of CN V3)
Sublingual	Sublingual branch of lingual artery	**Preganglionic fibers:** Chorda tympani (CN VII) **Parasympathetic ganglion:** Submandibular **Postganglionic fibers:** Travel with lingual nerve (branch of CN V3)
Submandibular	Glandular branches of facial and lingual arteries	**Preganglionic fibers:** Chorda tympani (CN VII) **Parasympathetic ganglion:** Submandibular **Postganglionic fibers:** Supply gland directly

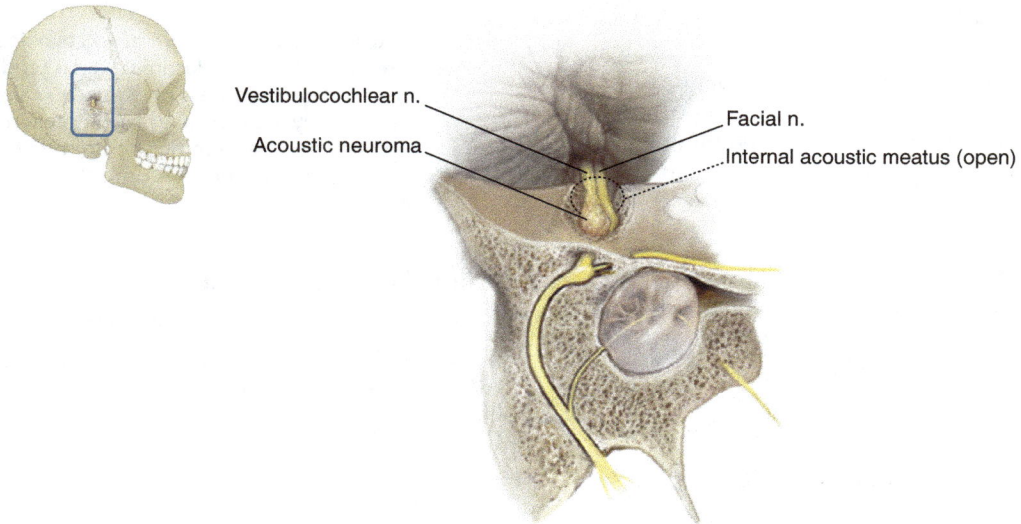

FIGURE 7.4.4 Lateral view of petrous portion of temporal bone (cut) showing path of facial nerve (CN VII) and adjacent acoustic neuroma on vestibulocochlear nerve (CN VIII) in internal acoustic meatus (cut).

Signs and Symptoms

- Vertigo
- Hearing loss
- Tinnitus
- Dizziness
- Facial or aural paresthesia or pain
- Facial muscle weakness
- Sleepiness

Increased intracranial pressure associated with these tumors may cause symptoms such as headache, vomiting, and altered consciousness.

Predisposing Factors

- Genetic risk of neurofibromatosis II

Suspected, but unconfirmed, risk factors include exposure to persistent or loud noises, head and neck exposure to low-dose radiation in children, and excessive cell telephone use.

Facial Nerve (Bell) Palsy

Bell palsy is the most common form of acute unilateral facial paralysis. This palsy exhibits a range of facial muscle deficits, from mild dysfunction (slight weakness) to total paralysis. Its onset is usually abrupt, and otalgia may precede paralysis by several days. Bell palsy typically resolves within 6–8 weeks. The precise cause(s) of Bell palsy remains unclear. It is thought that an inflammation of the facial nerve occurs in the facial canal. Nerve inflammation leads to edema, compression, and possible damage, which results in altered function. The inflammation associated with Bell palsy is consistent with local viral or bacterial infection.

The functional components of facial nerve branch at different points along its course (**Fig. 7.4.1** and

---

Vertigo Sensation that the environment is moving (spinning or "whirling") when there is not actual movement
Tinnitus Sensation of ringing or other noises in the ears
Paresthesia Numbness and tingling

Palsy Paralysis (loss of voluntary motor function; may be caused by disease or injury) or paresis (partial or incomplete paralysis)
Otalgia Ear pain

Table 7.4.1). Therefore, the site of facial nerve compression determines the sign and symptoms. For example, lesion at the stylomastoid foramen affects the muscles of facial expression. In contrast, lesion in the facial canal before the origin of chorda tympani would impact taste, salivary gland secretion, as well as muscles of facial expression.

Sign and Symptoms

- Sudden onset (hours) of unilateral facial paralysis; symptoms may develop over longer periods (up to 48 hours)
- Otalgia
- Epiphora
- Decreased secretion of lacrimal fluid
- Hyperacusis
- Dysgeusia
- Ocular pain
- Blurred vision

Clinical Notes

Maintenance of the corneal surface depends on several factors:

- Lacrimation (innervation of the lacrimal gland involves the greater petrosal nerve).
- A normal anatomical relationship between the eyelids and the ocular surface (involves orbicularis oculi, as well as the intrinsic eyelid structure).
- Blinking, which is necessary for normal lacrimal flow and drainage (regulated by orbicularis oculi and levator palpebrae superioris muscles).

Bell palsy frequently affects the corneal surface because the facial nerve controls lacrimal gland secretion and orbicularis oculi. This results in corneal drying (exposure keratitis), ocular pain, and visual deficit (blurred vision). This outcome may result in two ways:

1. **Lacrimal secretion is *reduced* and blinking is impaired.** In this case, there is insufficient production lacrimal fluid to coat the ocular surface.

2. **Lacrimal secretion is *normal* and blinking is impaired.** Sufficient lacrimal fluid is produced, but not enough stays on the ocular surface. The paralyzed orbicularis oculi prevents the eyelids from regulating the fluid flow and lacrimal fluid overflows onto the face.

Predisposing Factors

- Age: adults (slightly higher >60 years)
- Viral infection (e.g., herpes simplex, herpes zoster, and Epstein–Barr)
- Diabetes
- Hypertension
- Pregnancy (3.3 times higher risk)
- Lyme disease (spirochete *Borrelia burgdorferi*)
- Family history

Parotid Tumors

A parotid neoplasm can be benign or malignant. **Benign parotid masses** occur most commonly in the inferior portion of the gland. Physical examination usually reveals a single, firm, mobile, nontender mass. The two most common benign tumors are a pleomorphic adenoma (mixed parotid tumor) and Warthin tumor (a papillary cystadenoma lymphomatosum).

A **malignant parotid tumor** is suggested by:

- Pain and tenderness over the posterior cheek
- Rapid growth of the mass
- Fixation of the mass to the skin or underlying tissues
- Paralysis of a branch of the facial nerve.

Most malignancies arise from epithelial components of the gland, but they may also develop in components of the parotid duct.

Epiphora Overflow of tears onto the face due to inadequate drainage

Lacrimation Secretion of lacrimal (tear) fluid; sometimes refers to excess secretion

Hypertension Abnormal increase in arterial pressure

Signs and Symptoms

- Pain and/or tenderness in the posterior cheek
- Facial muscle weakness or paralysis
- Trismus
- Dysphagia
- Otalgia
- Paresthesia in maxillary (CNV2) and mandibular (CNV3) nerve distributions

Predisposing Factors

- History of prior cutaneous lesion
- History of squamous cell carcinoma
- History of malignant melanoma
- History of malignant fibrous histiocytoma

Stroke Involving Facial Motor Cortex

Stroke, also known as a **cerebrovascular accident (CVA)**, can be described as any disease process that disrupts blood flow to the brain. Stroke can be classified into two major categories:

1. **Ischemic stroke** accounts for the majority of strokes. This results when there is decreased blood flow to a part of the brain. The reduced blood flow may be due to a thrombus, an embolism, or hypoperfusion. Strokes deprive the brain tissue adjacent to the involved vessel(s) of oxygen and nutrients, which results in a compromise to function.

2. **Hemorrhagic stroke** occurs when there is a rupture of a cerebral blood vessel and leakage of blood into the brain and/or subarachnoid space. The affected area loses normal function.

Stroke symptoms typically start abruptly and depend on the area of the brain affected. The more extensive the area involved in the stroke, the more brain functions will be impacted or lost. In most cases, symptoms are unilateral and will present on the contralateral side of the body.

The facial part of motor cortex (in precentral gyrus) is supplied by the middle cerebral artery. Strokes in this cortical area result in:

Signs and Symptoms

- Contralateral hemiparesis or hemiplegia
- Contralateral facial paresis or paralysis

Predisposing Factors

- Age: >55 years
- Family history of stroke
- Cardiovascular disease
- Hypertension
- Diabetes
- Hypercholesterolemia
- Hyperlipidemia
- Over weight (body mass index >26)
- Smoking

Clinical Note

Genu VII Bell palsy results from a pontine stroke that involves the abducens motor nucleus (CN VI) and the facial motor fibers as they exit the lateral pons. This results in facial palsy with the added finding of an abducens palsy (i.e., inability to abduct the ipsilateral eye).

Trismus Jaw stiffness

Thrombus A fixed mass of platelets and/or fibrin (clot) that partially or totally occludes a blood vessel or heart chamber. An embolism is a mobile clot in the cardiovascular system

Embolus A mobile clot in the cardiovascular system, often derived from a thrombus, that is obstructive

Hypoperfusion Reduced blood flow relative to the metabolic requirements of an organ or tissue.

Hemorrhage Escape of blood from vessels:
- Petechia: <2 mm diameter
- Ecchymosis (bruise): >2 mm
- Purpura: a group of petechiae or ecchymoses
- Hematoma: hemorrhage resulting in elevation of skin or mucosa

Hypercholesterolemia Elevated serum cholesterol (total cholesterol level >200 mg/dL)

Hyperlipidemia Elevated serum cholesterol and/or triglycerides

DIAGNOSIS

The patient presentation, history, physical examination, and imaging studies support a diagnosis of **Bell palsy, a facial nerve (CN VII) lesion**.

Bell Palsy

The facial nerve has multiple functional components. Therefore, the signs and symptoms associated with a facial nerve lesion depend on the site of the lesion and the components involved. Bell palsy is always unilateral.

- This patient presented with sudden, unilateral facial paralysis. This indicates involvement of facial motor fibers.
- Dysgeusia in this patient indicates involvement of chorda tympani branch of facial nerve. This indicates involvement of special sensory component of the facial.
- Chorda tympani also carries preganglionic parasympathetic fibers (secretomotor) for the submandibular and sublingual salivary glands. This patient might have experienced dry mouth, but this was not one of his complaints.
- The patient's sensitivity to "loud noises" (hyperacusis) indicates involvement of the nerve to stapedius. Paralysis of this muscle, which normally dampens vibration of the ossicles, results in excessive stimulation of the cochlear apparatus.

The sudden onset of the signs and symptoms may point to a vascular incident. This is unlikely, however, because the signs and symptoms were restricted to the distribution of the facial nerve (CN VII). The signs and symptoms in this patient indicate that the lesion was in the facial canal, proximal to the branch points of the nerve to stapedius and chorda tympani.

Acoustic Neuroma

Acoustic neuromas are benign tumors of the vestibulocochlear nerve (CN VIII) that frequently form as the nerve enters the internal acoustic meatus. This nerve shares the internal acoustic meatus with the facial nerve.

- The most common symptoms of this disorder are vertigo, hearing loss, tinnitus, and headache. These symptoms were not seen in this patient.

- When an acoustic neuroma becomes large enough to impact the facial nerve, facial muscle weakness is usually a clinical sign. These neuromas are usually large enough to be visualized with brain imaging. The CT scan for this patient did not show a space-filling lesion associated with CN VIII.

Parotid Tumor

The parotid gland is located on the lateral aspect of the face, anterior to the ear and lateral to the mandibular ramus. The facial nerve plexus is located within the substance of the gland and is, therefore, vulnerable to injury with a parotid tumor.

- Lack of a palpable parotid mass and absence of pain or tenderness in the parotid or ear regions, are not consistent with a parotid tumor.

Stroke

Stroke involves impaired blood flow to the brain. Paresis or paralysis of facial muscles can result following a stroke in the facial part of motor cortex (upper motor neurons) or facial motor nucleus in the pons (lower motor neurons) (**Fig. 7.4.2**).

- Damage to upper motor neurons in the facial part of motor cortex will result in paresis/paralysis of facial muscles in the *inferior* part of the contralateral face. Muscles in the *superior* part of the contralateral face will be spared because the facial motor nucleus that innervates them receives bilateral cortical innervation.
- Patients with damage to lower motor neurons in the facial motor nucleus will have paresis/paralysis of facial muscles in the entire ipsilateral face (i.e., in both the superior and inferior parts of the face).

The signs and symptoms in this patient are not consistent with a stroke that affected either the facial motor cortex or the facial motor nucleus in the pons. Clinical signs associated with these lesions would be limited to motor loss (i.e., functional components of chorda tympani would not be affected).

Paresis Partial or incomplete paralysis

7.5 Blowout Fracture of Orbital Floor

Patient Presentation

A 22-year-old male is admitted to the emergency department with trauma to his left eye, including bruising, swelling, and pain. The patient complains of double vision.

Relevant Clinical Findings

History

The patient reports having been punched in the eye during a bar fight.

Physical Examination

Results of physical examination of the left eye and orbit:

- Periorbital ecchymosis and edema
- Ptosis of upper eyelid
- Enophthalmos
- Hypotropia and vertical diplopia that is worse with attempted upward or downward gaze (**Fig. 7.5.1**)
- Periorbital pain
- Hypesthesia from the maxillary region
- Visual acuity and pupillary response is comparable to the right eye

Laboratory Tests

Based on patient presentation, history, and physical examination, no laboratory tests were ordered.

Imaging Studies

- Coronal computed tomography (CT) showed a fracture of the floor of the left orbit, with prolapse of orbital contents into the maxillary sinus (**Fig. 7.5.2**).

Clinical Problems to Consider

- Blowout fracture of orbital floor
- Oculomotor palsy

LEARNING OBJECTIVES

1. Describe the anatomy of the orbit and contents.
2. Describe the anatomy of the extra-ocular muscles (**EOMs**).
3. Describe the anatomy of the eyelids.
4. Describe the autonomic innervation of the eye and orbit.
5. Explain the anatomical basis for the signs and symptoms associated with this case.

RELEVANT ANATOMY

Orbit

The orbit contains the eyeball, EOMs, cranial nerves (CN II, II, IV, V1, VI), and vasculature. These structures are invested in and supported by orbital (intraconal) fat. The maxillary (CN V2) nerve and vessels lie within the orbital floor.

Ecchymosis Hemorrhage into skin (bruising)
Ptosis Drooping eyelid
Enophthalmos Recession (sinking in) of the eye into the orbit

Hypotropia Downward (inferior) deviation of one eye relative to the other, a form of strabismus (misalignment of the eye)
Diplopia Double vision
Hypesthesia Diminished sensation

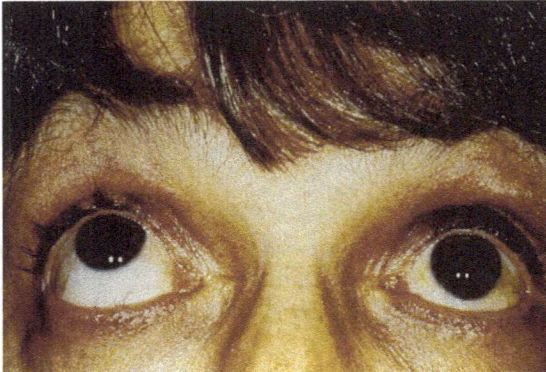

FIGURE 7.5.1 Hypotropia of the left eye with upward gaze. *Source:* Fig. 236-47 in *Tintinalli's Emergency Medicine: A Comprehensive Study Guide,* 7e.

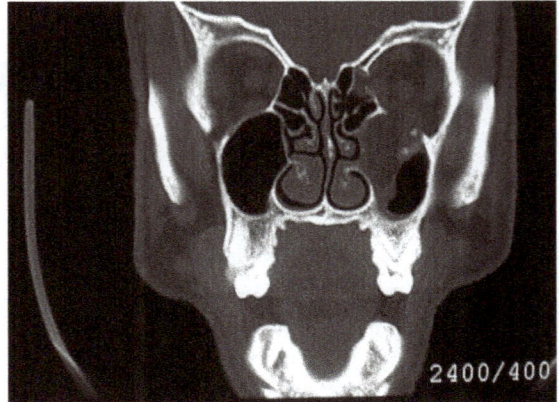

FIGURE 7.5.2 Coronal CT image showing a depressed fracture of the floor of the left orbit. *Source:* Fig. 18-19 in *Schwartz's Principles of Surgery,* 9e.

Orbital Walls and Related Structures

The orbit is a pyramidal space, with its apex posteriorly. Seven bones contribute to the orbit:

- **Apex:** sphenoid (lesser wing)
- **Superior wall (roof):** primarily frontal, with small contribution from sphenoid (lesser wing)
- **Inferior wall (floor):** primarily maxilla, with contributions from zygomatic and palatine
- **Medial wall:** primarily ethmoid, with contributions from frontal, lacrimal, and sphenoid

- **Lateral wall:** zygomatic and sphenoid (greater wing)

Nerves and vessels pass through openings in or between these bones (**Table 7.5.1**).

Several of the paranasal sinuses are adjacent to the orbit:

- The **floor** separates the orbit from the **maxillary sinus**.
- The **medial wall** separates the orbit from the **ethmoidal sinuses**.

TABLE 7.5.1	Orbital openings and associated nerves and vessels.	
Opening	**Location**	**Nerves and vessels**
Optic canal	Lesser wing of sphenoid	- Optic nerve (CN II) - Ophthalmic artery
Superior orbital fissure	Between greater and lesser wings of sphenoid	- Oculomotor nerve (CN III) - Trochlear nerve (CN IV) - Abducens nerve (CN VI) - Ophthalmic veins
Inferior orbital fissure	Between greater wing of sphenoid, maxilla, palatine, and zygomatic bones	- Infra-orbital nerve (from CN V2) - Infra-orbital vessels - Tributary of inferior ophthalmic vein
Supraorbital notch (foramen)		- Supra-orbital nerve (from CN V1) - Supra-orbital vessels
Infra-orbital canal (groove)		- Infra-orbital nerve (from CN V2) - Infra-orbital vessels
Ethmoidal foramina		- Ethmoidal nerves (from nasociliary) - Ethmoidal vessels

Extra-ocular Muscles

The normal resting position of the eye, with gaze straight-ahead fixed at infinity, is called **primary position**. Six EOMs attach to and move the eye. They can be divided into three groups:

1. **Horizontal rectus muscles:** medial rectus and lateral rectus
2. **Vertical rectus muscles:** superior rectus and inferior rectus
3. **Oblique muscles:** superior oblique and inferior oblique

A seventh muscle, levator palpebrae superioris, attaches to the upper eyelid. The innervation and actions of the EOMs are outlined in **Table 7.5.2**. A mnemonic for the nerve supply to the EOMs is: **LR$_6$ SO$_4$ AO$_3$** (LR = lateral rectus; SO = superior oblique; AO = "all others"). The oculomotor nerve also carries presynaptic parasympathetic fibers to the ciliary ganglion. Postsynaptic fibers from this ganglion supply the sphincter pupillae and ciliary muscles.

Eyelids

The eyelids protect the anterior aspect of the eye. The opening between the eyelids is known as the **palpebral fissure**.

Each eyelid contains a central **tarsal plate** (tarsus) composed of dense connective tissue that provides a structural core for the lid. The eyelids also contain **sebaceous tarsal glands** (Meibomian) and **ciliary glands** (Zeiss), as well as modified **apocrine sweat glands** (Moll).

- Sebaceous secretions contribute to the superficial lipid layer of lacrimal (tear) film. With blinking, the tear film is spread over the eye surface and the lipid component of tears serves to stabilize and reduce evaporation of the film. It also helps to constrain the fluid to the eye surface and prevent it from spilling onto the face.

- Modified sweat glands contribute to the aqueous component of tear film. The function of the glands of Moll is not understood completely, although their secretions include antibacterial agents.

The eyelids are closed and opened reflexively during blinking, but they may be closed tightly or held open voluntarily. The upper eyelid is elevated slightly when looking upward and the lower eyelid is depressed when looking downward. Overall, the upper eyelid exhibits greater movement that the lower lid.

TABLE 7.5.2	Innervation and actions of extraocular muscles.	
Muscle	**Innervation**	**Action(s) from primary position**
Horizontal rectus		
Lateral rectus	CN VI	Abduction
Medial rectus	CN III	Adduction
Vertical rectus		
Superior rectus	CN III	Primary: elevation Secondary: adduction, intorsion[a]
Inferior rectus	CN III	Primary: depression Secondary: adduction, extorsion[a]
Oblique		
Inferior oblique	CN III	Primary: intorsion[a] Secondary: depression, abduction
Superior oblique	CN IV	Primary: extorsion[a] Secondary: elevation, abduction
Levator palpebrae superioris	CN III	Elevates upper eyelid

[a]Intorsion and extorsion also referred to clinically as incyclotorsion and excyclotorsion, respectively. Some anatomy texts refer to these actions as medial and lateral rotation, which may lead to confusion because the eye can be rotated medially or laterally about either its supero-inferior (vertical) axis or its antero-posterior axis.

TABLE 7.5.3 **Eyelid muscles.**

Muscle	Innervation	Origin	Insertion	Action
Orbicularis oculi	CN VII	Medial orbital margin Medial palpebral ligament Lacrimal bone	Skin of eyelids, orbital margin, and tarsal plates (superior and inferior)	Closes eyelids
Levator palpebrae superioris	CN III	Lesser wing of sphenoid	Skin of upper eyelid	Elevates upper eyelid
Superior tarsal muscle (Müller)	Sympathetic	Levator palpebrae superioris	Superior tarsus	Elevates upper eyelid
Inferior tarsal muscle (Riolan)	Sympathetic	Inferior rectus	Inferior tarsus	Depresses lower eyelid

Skeletal and smooth muscles are involved in eyelid movements (**Table 7.5.3**). The eyelids are closed by **orbicularis oculi**, a muscle of facial expression. Palpebral and orbital parts of this muscle are distinguished:

- The **palpebral portion** (i.e., in the eyelids) is responsible for blinking.
- The **orbital portion** surrounds the orbit and is attached to facial bones. It closes the eyes tightly.

The muscles that open the eyelids are also summarized in **Table 7.5.3**. The tarsal muscles, composed of smooth muscle, regulate the width of the palpebral fissure.

The **inferior tarsal muscle** is significantly less prominent than the **superior tarsal muscle**. Since the inferior tarsal muscle originates from inferior rectus, the lower lid is depressed (i.e., pulled downward) when one looks downward, thereby moving the lower lid out of the way of the pupil.

Autonomic Innervation of the Eye and Orbit

The structures in the eye and orbit that receive autonomic innervation are outlined in **Table 7.5.4**.

Sympathetic Innervation of the Eye

A three-neuron "oculosympathetic" pathway provides sympathetic innervation to structures in the eye and orbit.

1. **First-order neurons are located in the hypothalamus.** Their axons descend through the reticular formation of the brain stem to the (C8)T1-T2 spinal cord, where they synapse on second-order neurons.

2. **Second-order neurons are located in the lateral horn of the (C8)T1-T2 spinal cord.** This group of cells is referred to as the ciliospinal center (Budge–Waller). Their axons exit the spinal cord via the T1-T2 ventral rami and follow white rami communicans to the sympathetic trunk. These axons ascend to synapse on third-order neurons.

TABLE 7.5.4 **Autonomic innervation of the eye and orbit.**

Autonomics	Ganglion	Structure	Function
Sympathetic	Superior cervical	Superior tarsal muscle Inferior tarsal muscle Dilator pupillae	Elevates upper eyelid Depresses lower eyelid Dilates pupil
Parasympathetic	Ciliary	Sphincter pupillae Ciliary muscle	Constricts pupil Regulates lens shape
	Pterygopalatine	Lacrimal gland	Secretomotor

a. Classical anatomy places the preganglionic sympathetic cell bodies in the interomedio-lateral cell columns at the T1-L2 (or L3) spinal cord. However, some clinical and neuroanatomy references include C8 as a location for preganglionic sympathetic cell bodies for the eye and orbit.

3. **Third-order neurons are located in the superior cervical ganglion.** A plexus of postganglionic axons follow the internal carotid artery, including the ophthalmic artery and its branches. In the orbit, these fibers are distributed along several routes:

a. Fibers may join, sequentially, the ophthalmic nerve (CN V1) and then its nasociliary branch. From the nasociliary, two pathways are available to distribute these axons to the **dilator pupillae**:

   i. **Long ciliary nerves** pass directly to the sclera and enter the posterior aspect of the eye.

   ii. The **nasociliary (sympathetic) root** of the ciliary ganglion carries postganglionic sympathetic axons to the ganglion and then, without synapse, into **short ciliary nerves**. These axons also pierce the sclera and enter the posterior aspect of the eye.

b. Other fibers follow the ophthalmic artery and its branches to reach the **tarsal muscles** and vasculature of the lacrimal gland.

Parasympathetic Innervation of the Eye

Preganglionic parasympathetic axons from cell bodies in the Edinger–Westphal nucleus travel in the oculomotor nerve (CN III) to reach the orbit. These fibers enter the **ciliary ganglion** through its **oculomotor (parasympathetic) root** to synapse on postganglionic parasympathetic cell bodies. Postganglionic axons reach the eye through **short ciliary nerves** and innervate the **constrictor pupillae** and **ciliary muscles**.

CLINICAL REASONING

Based on the history and physical examination, the patient suffered traumatic injury to the face that resulted in a **dysfunction of ocular movement and strabismus**.

Strabismus Misalignment of the eyes

Blowout Fracture of the Orbital Floor

Fracture of the orbital floor may affect eye movements and/or sensation from the face.

Signs and Symptoms

Sensory Deficits
- Orbital and periorbital pain
- Vertical diplopia
- Hypesthesia from maxillary region of face

Motor Deficits
- Hypotropia, exacerbated with attempted upward or downward gaze

Other Deficits
- Periorbital ecchymosis and edema
- Hyphema
- Enophthalmos

Pain, ecchymosis, edema, and compromised eye movements are external signs and symptoms related to mechanical trauma of the orbit.

Clinical Notes

The pathophysiologic mechanism by which a peri-orbital blow results in an orbital fracture is not understood completely. Two theories have been proposed:

Hydraulic Theory
- This theory suggests that the sudden blow to the peri-orbital region by a blunt object (e.g., a ball, fist, or elbow) causes a sudden increase in intra-orbital pressure. Because the orbit is an enclosed space, the force of the blow is transmitted to the orbital contents and walls. The thinnest orbital walls (i.e., floor and medial wall) are most vulnerable to fracture. Typically, the eye does not rupture.

Buckling Theory
- Another theory suggests that forces from a blow to the orbital rim are transmitted posteriorly to adjacent bones of the orbit. The thin orbital floor and medial wall are vulnerable because they are either part of (maxilla), or adjacent to (ethmoid), the margin.

Hyphema Blood in the anterior chamber of the eye
Edema Swelling due to abnormal accumulation of fluid in subcutaneous tissue

(A) Blow-out fracture of orbital floor

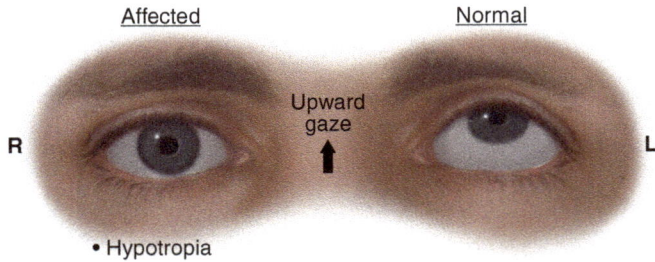

Affected   Normal

Upward gaze

R   L

• Hypotropia

(B) Oculomotor (CN III) palsy

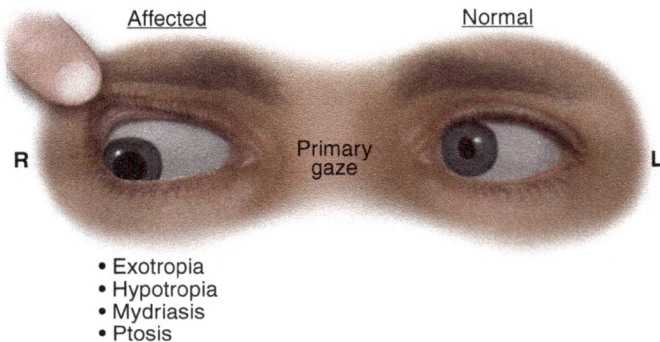

Affected   Normal

Primary gaze

R   L

• Exotropia
• Hypotropia
• Mydriasis
• Ptosis

FIGURE 7.5.3 (**A**) Blowout fracture of the floor of the right orbit. With attempted upward gaze, the affected eye is hypotropic (i.e., lower) than the normal eye. (**B**) Oculomotor (CN III) palsy showing the exotropic and hypotropic eye, ptosis of the upper eyelid (partial or full), and a mydriatic pupil. The affected eyelid is raised by the examiner.

Most fractures of the orbital floor occur in the orbital surface of the maxilla, posterior and medial to the infra-orbital groove. This region underlies the inferior rectus muscle. As a result, the muscle can be "trapped" (i.e., get caught) between bone fragments, leading to impaired and painful function of the muscle. The resulting strabismus typically presents as hypotropia because the trapped inferior rectus impairs downward movement of the eye, especially when the patient is asked to look upward. As a result, patients commonly experience vertical diplopia (**Fig. 7.5.3A**).

Ipsilateral hypesthesia from the region of the face supplied by the infra-orbital nerve is common in orbital floor fractures. This occurs because the infra-orbital nerve, which passes through the infra-orbital canal with accompanying blood vessels, is damaged by the fractured bone.

Palsy Paralysis (loss of voluntary motor function; may be caused by disease or injury) or paresis (partial or incomplete paralysis)

Clinical Note

Pupillary dysfunction and decreased visual acuity may reflect traumatic or compressive damage to the optic nerve (CN II).

Predisposing Factors

- Sex: >81% occur in males

- Age: due to the most common mechanisms of injury, mostly occur in teens and young adults

Oculomotor Palsy

Palsy of the oculomotor nerve (CN III) can affect movements of the eye and upper eyelid, and/or pupillary constriction.

Signs and Symptoms

Sensory Deficits

- Diplopia

Motor Deficits

- Ophthalmoplegia: affected eye is typically exotropic and hypotropic
- Ptosis of upper eyelid (partial or full)
- Mydriasis

Predisposing Factors

- Age: >60 years
- Atherosclerosis, especially associated with conditions such as diabetes or hypertension

Clinical Notes

Causes of oculomotor nerve palsies include:

- Expanding or ruptured aneurysms (e.g., berry aneurysm of the posterior communicating artery; internal carotid artery aneurysm in the cavernous sinus)
- Infections
- Cavernous sinus thrombosis
- Head or brain trauma
- Tumors, especially of the pituitary gland or at the base of the brain

DIAGNOSIS

The patient presentation, history, physical examination, and imaging studies support a diagnosis of **blowout fracture of the orbital floor**.

Blowout Fracture of the Orbital Floor

- The mechanism of injury (periorbital blow) and accompanying signs and symptoms (pain, ecchymosis, edema, hypesthesia) are consistent with a blowout fracture of the orbital floor. CT confirms the fracture and entrapment of orbital contents (**Fig. 7.5.2**).
- Vertical diplopia is consistent with this diagnosis. Most fractures of the orbital floor occur in the orbital surface of the maxilla. As a result, the inferior rectus muscle can be "trapped" between fracture fragments. The impaired and painful function of this muscle results is strabismus. Typically, the affected eye is hypotropic because the entrapped inferior rectus "holds" the eye in a position that directs the eye downward and the patient is unable to look upward with the affected eye.
- Ipsilateral hypesthesia from the area of distribution of the infra-orbital nerve is common because the infra-orbital nerve courses through the orbital floor and may be damaged by the fractured bone.

Oculomotor Nerve Palsy

A patient with **oculomotor nerve palsy** would also present with diplopia, but it would be distinguished by:

- Ophthalmoplegia with the diplopia. Thus, the patient would present with exotropia because the affected eye can be moved laterally due to the unopposed action of the lateral rectus (CN VI). The eye is also hypotropic (directed inferiorly or downward) due to the unopposed action of superior oblique (CN IV).
- Ptosis of the upper eyelid, which may be partial or total, is due to paralysis or paresis of levator palpebrae superioris.
- Mydriasis (i.e., dilated pupil) if the parasympathetic fibers in CN III are involved. This results from the unopposed action of the dilator pupillae, which receives sympathetic innervation.

Ophthalmoplegia Paralysis or paresis in one or more extraocular muscles

Exotropia Outward (lateral) deviation of one eye relative to the other, a form of strabismus (misalignment of the eye)

Mydriasis Excessive pupillary dilation (opposite of miosis)

Hypertropia Upward (superior) deviation of one eye relative to the other, a form of strabismus (misalignment of the eye)

Hypertension Abnormal increase in arterial pressure

Aneurysm Circumscribed dilation of an artery, in direct communication lumen

Thrombus A fixed mass of platelets and/or fibrin (clot) that partially or totally occludes a blood vessel or heart chamber. An embolism is a mobile clot in the cardiovascular system

Paralysis Loss of muscle function, particularly related to voluntary movement

Paresis Partial or incomplete paralysis

CASE **7.6** | **Horner Syndrome**

Patient Presentation

A 49-year-old female is admitted to the emergency department complaining of right upper limb pain and weakness of her right hand. She has headaches, and notes that her right eyelid droops. Her voice has become hoarse.

Relevant Clinical Findings

History

The patient has a 35-year history of smoking two packs of cigarettes a day. She has no history of previous surgery or trauma. Her ocular history is unremarkable.

Physical Examination

The following findings were noted on physical examination:

- Partial ptosis of the upper eyelid and reverse ptosis of lower eyelid on the right side
- Miosis on the right, with the anisocoria more apparent in dim illumination
- Right facial anhidrosis
- Pain along medial aspect of right upper limb
- Weak grip and impaired fine movements of the right hand
- Reduced pulses in the right upper limb (compared to the left)

Laboratory Tests

Based on patient presentation, history, and physical examination, no laboratory tests were ordered.

Imaging Studies

- Posterior–anterior chest radiographs and MRI (**Fig. 7.6.1**) showed a $6 \times 9 \times 6$ cm soft tissue mass at the apex of the right lung. The trachea was deviated to the left.

Clinical Problems to Consider

- Cavernous sinus syndrome
- Horner syndrome
- Oculomotor palsy

LEARNING OBJECTIVES

1. Describe the anatomy of the eyelids.
2. Describe the autonomic innervation of the eye and orbit.
3. Describe the anatomy of the extraocular muscles.
4. Describe the anatomy of the cavernous sinus.
5. Explain the anatomical basis for the signs and symptoms associated with this case.

RELEVANT ANATOMY

Eyelids

The eyelids protect the anterior aspect of the eye. The opening between the eyelids is known as the **palpebral fissure**.

Skeletal and smooth muscles are involved in eyelid movements (see **Table 7.5.3**). The eyelids are closed by **orbicularis oculi**, a muscle of facial expression. Muscles that open the eyelids include **levator palpebrae superioris** and the **tarsal muscles**.

Ptosis Drooping eyelid
Miosis Excessive pupillary constriction (opposite of mydriasis)

Anisocoria Unequal size of pupils
Anhidrosis Absence of sweating

(A)

(B)

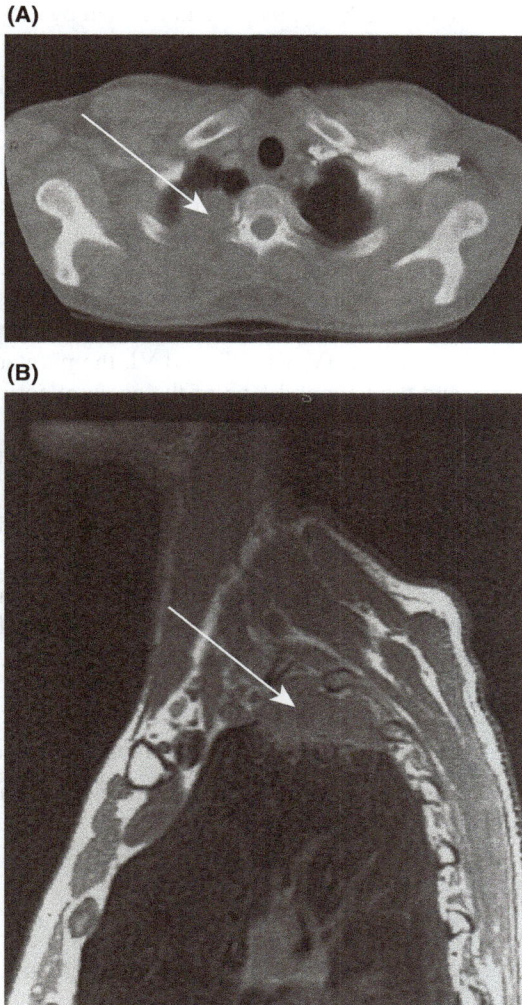

FIGURE 7.6.1 Transverse **(A)** and sagittal **(B)** MRI showing a tumor (arrows) at the apex of the right lung. The tumor invades the second rib and extends into the right apical fat. *Source*: Fig. 15-17 in *The MD Anderson Manual of Medical Oncology*, 2e.

Autonomic Innervation of the Eye and Orbit

The structures in the eye and orbit that receive autonomic innervation are outlined in **Table 7.6.1**.

Sympathetic Innervation of the Eye

A three-neuron "oculosympathetic" pathway provides sympathetic innervation to structures in the eye and orbit.

1. **First-order neurons are located in the hypothalamus.** Their axons descend through the reticular formation of the brain stem to the (C8)T1-T2 spinal cord, where they synapse on second-order neurons.

2. **Second-order neurons are located in the lateral horn of the (C8)T1-T2 spinal cord.** This group of cells is referred to as the cilio-spinal center (Budge–Waller). Their axons exit the spinal cord via the T1-T2 ventral rami and follow white rami communicans to the sympathetic trunk. These axons ascend to synapse on third-order neurons.

   a. Classically, anatomy places the preganglionic sympathetic cell bodies in the interomediolateral cell columns at the T1-L2 (or L3) spinal cord. However, some clinical and neuroanatomy references include C8 as a location for preganglionic sympathetic cell bodies for the eye and orbit.

3. **Third-order neurons are located in the superior cervical ganglion.** A plexus of postganglionic axons follow the internal carotid artery, including the ophthalmic artery and its branches. In the orbit, these fibers are distributed along several routes:

TABLE 7.6.1 Autonomic innervation of the eye and orbit.

Autonomics	Ganglion	Structure	Function
Sympathetic	Superior cervical	Superior tarsal muscle Inferior tarsal muscle Dilator pupillae	Elevates upper eyelid Depresses lower eyelid Dilates pupil
Parasympathetic	Ciliary	Sphincter pupillae Ciliary muscle	Constricts pupil Regulates lens shape
	Pterygopalatine	Lacrimal gland	Secretomotor

a. Fibers may join, sequentially, the ophthalmic nerve (CN V1) and then its nasociliary branch. From the nasociliary, two pathways are available to distribute these axons to the **dilator pupillae**:
   i. **Long ciliary nerves** pass directly to the sclera and enter the posterior aspect of the eye.
   ii. The **nasociliary (sympathetic) root** of the ciliary ganglion carries postganglionic sympathetic axons to the ganglion and then, without synapse, into **short ciliary nerves**. These axons also pierce the sclera and enter the posterior aspect of the eye.
b. Other fibers follow the ophthalmic artery and its branches to reach the **tarsal muscles** and vasculature of the lacrimal gland.

Parasympathetic Innervation of the Eye

Preganglionic parasympathetic axons from cell bodies in the Edinger–Westphal nucleus in the brainstem travel in the oculomotor nerve (CN III) to reach the orbit. These fibers enter the **ciliary ganglion** through its **oculomotor (parasympathetic) root** to synapse on postganglionic parasympathetic cell bodies. Postganglionic axons reach the eye through **short ciliary nerves** and innervate the **constrictor pupillae** and **ciliary muscles**.

Extraocular Muscles

The extraocular muscles (EOMs) were described in the preceding case (7.5). Briefly, the six EOMs attach to and move the eye can be divided into three groups:

1. **Horizontal rectus muscles:** medial rectus (MR) and lateral rectus (LR)
2. **Vertical rectus:** superior rectus and inferior rectus muscles
3. **Oblique:** superior oblique and inferior oblique muscles

Three cranial nerves innervate these muscles, as indicated by the memory device "**LR$_6$ SO$_4$ AO$_3$**":

1. LR = lateral rectus: it is innervated by the abducens nerve (CN VI)
2. SO = superior oblique: it is innervated by the trochlear nerve (CN IV)

3. AO = "all others": they are innervated by the oculomotor nerve (CN III)

The actions of these muscles are described in Table 7.5.2.

Cavernous Sinus

The cavernous sinuses are paired dural venous sinus located posterior to each orbit, on either side of the body of the sphenoid bone (**Fig. 7.6.2**). Numerous structures are related to the cavernous sinus, including the cavernous part of the internal carotid artery, cranial nerves III, IV, V1, V2, and VI, the pituitary gland, and the temporal lobe of the cerebral cortex (**Table 7.6.2**).

Tributaries of the cavernous sinus from the orbit are the ophthalmic veins.

- The **superior ophthalmic vein** passes through the superior orbital fissure.
- The **inferior ophthalmic vein** either joins the superior ophthalmic vein in the orbit or itself passes through the superior orbital fissure. The inferior ophthalmic vein also connects via the inferior orbital fissure with the pterygoid venous plexus in the infratemporal fossa.

The cavernous sinus drains into the **superior and inferior petrosal sinuses**.

CLINICAL REASONING

Based on the physical examination, the patient demonstrates **dysfunction of the pupil and eyelids**.

Cavernous Sinus Syndrome

Cavernous sinus syndrome involves damage to one or more of the structures associated with the cavernous sinus. This includes:

Signs and Symptoms

Sensory Deficits
- Periorbital pain and erythema
- Trigeminal sensory loss from the distributions of CN V1 and/or V2
- Potential visual field deficits

Erythematous Reddened skin

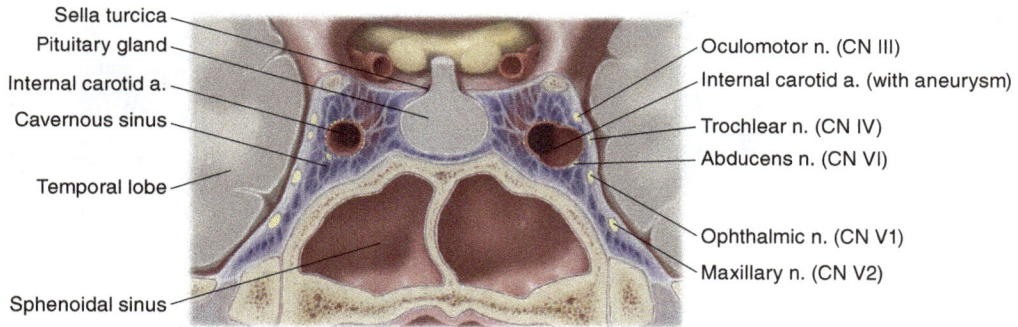

FIGURE 7.6.2 Coronal section through the cavernous sinus and sella turcica.

Motor Deficits

- Ophthalmoplegia (acute or slowly progressive)—see below
- Diplopia
- Pupillary dysfunction (miosis and/or mydriasis)

The position of the affected eye depends on which cranial nerves are involved:

- Involvement of the abducens nerve (CN VI) would result in esotropia of the affected eye (**Fig. 7.6.3A**).
- Involvement of the oculomotor nerve (CN III) would result in exotropia and hypotropia of the affected eye (see **Fig. 7.6.3D**).
- Involvement of the trochlear nerve (CN IV) would result in exotropia and hypotropia of the affected eye (not shown).

TABLE 7.6.2 | **Anatomical relationships of the cavernous sinus.**

Relationship to sinus	Structure	Notes
Medial to sinus	• Sphenoidal air sinus • Pituitary gland	
Within the sinus	• Internal carotid artery • Abducent nerve (CN VI)	Cavernous part Adjacent and lateral to ICA
In lateral wall	• Oculomotor nerve (CN III) • Trochlear nerve (CN IV) • Ophthalmic nerve (CN V1) • Maxillary nerve (CN V2)	
In inferoposterior part of sinus	• Trigeminal ganglion	
Lateral to sinus	• Temporal lobe (uncus)	

Ophthalmoplegia Paralysis or paresis in one or more extraocular muscles
Diplopia Double vision
Mydriasis Excessive pupillary dilation (opposite of miosis)
Esotropia Inward (median) deviation of one eye relative to the other, a form of strabismus (misalignment of the eye)

Exotropia Outward (lateral) deviation of one eye relative to the other, a form of strabismus (misalignment of the eye)
Hypotropia Downward (inferior) deviation of one eye relative to the other, a form of strabismus (misalignment of the eye)

(A) Cavernous Sinus Syndrome - CN VI

Affected   Normal

R   Right Gaze   L

- Estropia

(B) Cavernous Sinus Syndrome - CN VI, CN II, & CN VI

Affected   Normal

R   Right Gaze   L

- Paralyzed eye
- Mydriasis
- Ptosis

(C) Horner Syndrome

Affected   Normal

R   Primary Gaze   L

- Miosis
- Partial ptosis/reverse ptosis
- No opthalmoplegia

(D) Oculomotor (CN III) Palsy

Affected   Normal

R   Right Gaze   L

- Exotropia
- Hypotropia
- Mydriasis
- Ptosis

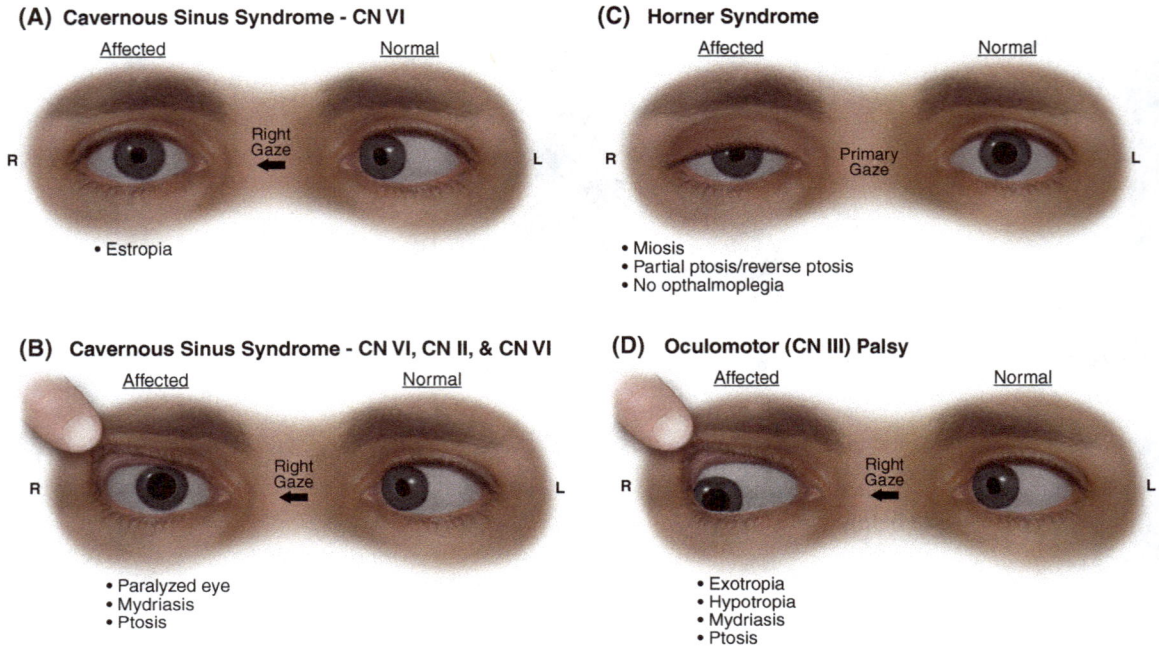

FIGURE 7.6.3 **(A)** Right cavernous sinus syndrome involving the abducens nerve (CN VI). This results in esotropia of the affected eye. **(B)** Cavernous sinus syndrome involving CN VI, as well as III and IV. The eye is paralyzed due to involvement of all cranial nerves that supply all extraocular muscles. **(C)** Horner syndrome showing miosis and partial ptosis. **(D)** Oculomotor (CN III) palsy showing the exotropic and hypotropic eye (with primary gaze), ptosis of the upper eyelid (partial or full), and a mydriatic pupil. In **B** and **D**, the affected eyelid is raised by the examiner.

- The eye would be paralyzed if all of the above nerves (CN III, IV, VI) were involved (**Fig. 7.6.3B**).

Other Deficits
- Chemosis
- Proptosis

Predisposing Factors
- Related to underlying causes of the syndrome (e.g., tumor and aneurysm)

Clinical Notes

- The most common cause of cavernous sinus syndrome is cavernous sinus thrombosis, often secondary to orbital or facial infection or sinusitis. Other etiologies include tumors (pituitary, meningioma, nasopharyngeal carcinoma), cavernous ICA aneurysm, and carotid-cavernous fistula.

- Mortality from septic cavernous sinus thrombophlebitis has decreased significantly with improved diagnosis and therapeutics.

- Pituitary tumors may also affect vision due to involvement of the optic nerve and/or chiasm.

Paralysis Loss of muscle function, particularly related to voluntary movement

Chemosis Edema of bulbar conjunctiva

Proptosis Protrusion of the eye (synonymous with exophthalmos)

Thrombophlebitis Venous inflammation with thrombus formation

Horner Syndrome

Horner syndrome (or Bernard–Horner syndrome) results from an interruption of the three-neuron oculosympathetic pathway. This pathway extends from the brain stem, into the upper spinal cord, and through the neck and base of the skull. Consequently, diseases and clinical conditions distant to the eye can present with ocular signs and symptoms.

Signs and Symptoms

Motor Deficits
- Partial ptosis of upper eyelid
- Lower eyelid often exhibits reverse ptosis
- Miosis

Other Deficits
- Facial anhidrosis
- Facial flushing

Predisposing Factors

- None with respect to race, sex, or age

Clinical Notes

- Facial anhidrosis may not be evident in the clinical setting because the controlled environment (temperature, humidity) may not favor sweat production. It may be more apparent with exercise.
- Enophthalmos is erroneously considered a clinical sign of Horner syndrome; this sign is "apparent" only. Due to the narrowed palpebral fissure, the eye may appear to be enophthalmic (i.e., to "sink inward").

Oculomotor Palsy

Palsy of the oculomotor nerve (CN III) can affect movements of the eye and upper eyelid, and/or pupillary constriction.

Signs and Symptoms

Sensory Deficit
- Diplopia

Motor Deficit
- Ophthalmoplegia: affected eye is typically exotropic and hypotropic
- Ptosis of upper eyelid (partial or full)
- Mydriasis

Predisposing Factors

- Age: >60 years
- Atherosclerosis, especially associated with conditions such as diabetes or hypertension

Clinical Notes

Causes of oculomotor nerve palsies include:

- Expanding or ruptured aneurysms (e.g., berry aneurysm of the posterior communicating artery; internal carotid artery aneurysm in the cavernous sinus)
- Infections
- Cavernous sinus thrombosis
- Head or brain trauma
- Tumors, especially of the pituitary gland or at the base of the brain

DIAGNOSIS

The patient presentation, medical history, physical examination, and imaging studies support a diagnosis of **Horner syndrome secondary to an apical lung tumor** (Pancoast). This lung tumor is located in the pulmonary apex ("top" of either right or left lung). As such, it can compress nearby structures, including: the brachiocephalic vein, subclavian artery, the phrenic, recurrent laryngeal, and vagus

---

Enophthalmos Recession (sinking in) of the eye into the orbit

Hypertension Abnormal increase in arterial pressure

Aneurysm Circumscribed dilation of an artery, in direct communication lumen

Thrombus A fixed mass of platelets and/or fibrin (clot) that partially or totally occludes a blood vessel or

heart chamber. An embolism is a mobile clot in the cardiovascular system

Palsy Paralysis (loss of voluntary motor function; may be caused by disease or injury) or paresis (partial or incomplete paralysis)

nerves, as well as the sympathetic trunk. Compression of the sympathetic trunk can result in a Horner syndrome.

Horner Syndrome

Horner syndrome is usually unilateral and affects the pupil and eyelids.

- Miosis and partial ptosis are hallmarks of Horner syndrome. Miosis (i.e., dilated pupil) results from the combination of diminished, or absent, sympathetic innervation to the *dilator* pupillae and the resultant unopposed action of the *constrictor* pupillae. Interruption of sympathetic innervation to the tarsal muscles leads to partial ptosis. Complete ptosis is unlikely because the oculomotor nerve is unaffected and, thus, levator palpebrae superioris can elevate the upper eyelid.
- Facial anhidrosis erythmotosis (flushing) may also occur, depending on the involvement of sympathetic neurons that supply facial sweat glands and vasculature (T2-T4 spinal levels).
- EOMs do not receive sympathetic innervation. Therefore, ophthalmoplegia and diplopia are not part of the syndrome.

Oculomotor Palsy

Oculomotor palsy may be distinguished from Horner syndrome on the basis of eye movements and pupillary responses.

- A patient with oculomotor nerve palsy would present with ophthalmoplegia and diplopia. Thus, the patient would present with exotropia because the affected eye can be moved later-

ally due to the unopposed action of the lateral rectus (CN VI). The eye is also hypotropic (directed inferiorly or downward) due to the unopposed action of superior oblique (CN IV).

- Ptosis of the upper eyelid, which may be partial or total, is due to paralysis or paresis of levator palpebrae superioris.
- If the parasympathetic fibers that travel in the oculomotor nerve are involved, the patient would exhibit mydriasis (i.e., dilated pupil). This results from the unopposed action of the dilator pupillae, which receives sympathetic innervation.

Cavernous Sinus Syndrome

Patients with cavernous sinus syndrome may present with ophthalmoplegia, diplopia, proptosis, Horner syndrome, and/or trigeminal sensory loss.

- A patient with cavernous sinus syndrome would likely present with exotropia due to involvement of the lateral rectus muscle (CN VI).
- Chemosis and proptosis are related to impaired venous drainage of the orbit.
- Miosis may be observed if the carotid sympathetic plexus is involved.
- An oculomotor and/or trochlear palsy, as well as trigeminal sensory loss, may also occur if the cranial nerves in the lateral wall of the sinus (i.e., CN III, IV, V1, V2) are involved.

Palsy Paralysis (loss of voluntary motor function; may be caused by disease or injury) or paresis (partial or incomplete paralysis)

REVIEW QUESTIONS

1. A 25-year-old male undergoes emergency surgery to repair an internal carotid thrombosis following a gunshot injury in the right carotid triangle. Following the procedure, the patient exhibits miosis (pupillary constriction) and partial ptosis (drooping) of the right upper and lower eyelids. Which structure is injured?

   A. C4 spinal cord
   B. Ciliary ganglion

   C. Frontal nerve
   D. Internal carotid plexus
   E. Oculomotor nerve (CN III)

2. A 44-year-old man is admitted to the emergency department with multiple facial contusions after a fight. Physical examination fails to detect any visual defects (e.g., loss of vision or double vision). However, there is a loss of sensation from the cornea. Which nerve is most likely injured?

A. Frontal
B. Infra-orbital
C. Nasociliary
D. Oculomotor (CN III)
E. Optic (CN II)

3. A 30-year-old female presents in the ophthalmology clinic complaining of double vision (diplopia). The extra-ocular muscles are assessed using the "H-test." Upon left gaze (i.e., patient looking to their left), the right (adducted) eye "drifts" upward into elevation. Lesion of which nerve on the affected side would cause the observed eye movement defect?

A. Abducent nerve (CN VI)
B. Inferior branch of the oculomotor nerve (CN III)
C. Superior branch of the oculomotor nerve (CN III)
D. Trochlear nerve (CN IV)
E. None of the above

4. A 23-year-old male is referred to the ophthalmology clinic following an automobile accident in which he sustained trauma to the right side of his head. He complains that his right eye is dry, and he feels a burning sensation from that eye. Lateral radiographs reveal a fracture of the zygomatic bone, which suggests the possibility that secretomotor fibers to the lacrimal gland were interrupted. Where do the cell bodies for these secretomotor fibers reside?

A. Ciliary ganglion
B. Otic ganglion
C. Pterygopalatine ganglion
D. Submandibular ganglion
E. Superior cervical ganglion

5. Injury to which of the following cranial nerves is most likely to result in hyperacusis (increased sensitivity to sound)?

A. Accessory (CN XI)
B. Facial (CN VII)
C. Glossopharyngeal (CN IX)
D. Hypoglossal (XII)
E. Vagus (X)

6. A tumor is discovered in the cranial cavity of a female patient. Her primary symptoms are decreased sensation from the anterior two-thirds of the tongue, mandibular teeth and gums, and skin of the chin cheek and temples. Which nerve is most likely damaged by the tumor?

A. Chorda tympani
B. Facial (CN VII)
C. Glossopharyngeal (CN IX)

D. Mandibular (CN V3)
E. Maxillary (CN V2)

7. A women suffering from severe headaches and neck stiffness is seen by her physician. Neurologic evaluation reveals that her uvula deviates to the right upon elevation of the palate. Which nerve is most likely affected by the tumor?

A. Left glossopharyngeal
B. Left vagus
C. Right glossopharyngeal
D. Right hypoglossal
E. Right vagus

8. A male patient is seen in the family practice clinic with a complaint of hearing loss in his left ear. Physical examination reveals that he also has loss of taste and significant muscle weakness on the left side of his face. The physician orders a head CT, which shows an intracranial tumor. The tumor is most likely located at which opening?

A. Foramen ovale
B. Foramen rotundum
C. Internal acoustic meatus
D. Jugular foramen
E. Stylomastoid foramen

9. A 12-year-old girl complaining of ear pain is seen in the pediatric clinic. During the physical examination, it is noted that the patient has a low-grade fever, and her oropharynx is inflamed. Which structure provides a pathway for the spread of infection from the pharynx to the middle ear?

A. Auditory (pharyngotympanic) tube
B. Choanae
C. External acoustic meatus
D. Internal acoustic meatus
E. Pharyngeal recess

10. A 10-year-old boy is kicked in the left temporal region during an intramural soccer game. He complained of a headache, but since he did not lose consciousness, he went home. Later in the day he vomited and became groggy. His parents immediately took him to the emergency department where a CT scan revealed a biconvex hyperdense collection of blood subjacent to the pterion. What is the source of the blood?

A. Facial artery
B. Middle meningeal artery
C. Posterior auricular artery
D. Superficial temporal artery
E. Superior cerebral vein(s)

11. A patient presents in the clinic with a complaint of changes in sensation on the "end" of his tongue. Careful physical examination reveals a loss of general sensation from the anterior two-thirds of his tongue on the right side, but taste and salivation are intact. A CT scan indicates the presence of a tumor in the right infratemporal fossa. Which is the most affected by the tumor?

   ` A. Chorda tympani
   B. Glossopharyngeal nerve
   C. Inferior alveolar nerve
   D. Lesser petrosal nerve
   E. Lingual nerve, proximal to the junction with the chorda tympani

12. A 36-year-old female patient is seen in the dental clinic complaining of pain in her right maxillary molars. Dental radiography does not show abnormalities of her teeth. The dentist refers the patient to an ENT physician because he believes her pain is most likely related to:

   A. Adenoiditis
   B. Mandibular dental abscess
   C. Maxillary sinusitis
   D. Orbital cellulitis
   E. Sphenoidal sinusitis

13. A patient suffered head trauma in an automobile accident. Radiographic images indicate a fracture of the sphenoid bone, which suggests the possibility of damage to structures that pass through foramen ovale. If this is the case, which muscle is impaired?

   A. Buccinator
   B. Lateral pterygoid
   C. Levator palpebrae superioris
   D. Posterior belly of digastric
   E. Stapedius

14. A 26-year-old patient complains of persistent nosebleeds. He is prescribed a nasal steroid to help with seasonal allergies, but the nosebleeds continue. His ENT physician is preparing to cauterize the source of the bleeding, which appears to be a vessel on the postero-inferior aspect of the nasal septum. Which artery is the mostly likely source of the bleeding?

   A. Anterior ethmoidal
   B. Greater palatine
   C. Lateral nasal branch of infra-orbital
   D. Sphenopalatine
   E. Superior labial branch of facial

Back

Epidural Analgesia

Patient Presentation

A 26-year-old woman in labor is admitted to the hospital birthing center. She requests an "epidural" for pain management.

Relevant Clinical Findings

History

The patient has no history of chronic illness, and has had appropriate prenatal care. The pregnancy is full term.

Physical Examination

Noteworthy vital signs of the fetus:

- Weight: 7.5 lb
- Pulse: 145 bpm (adult resting rate: 60–100 bpm)

Results of physical examination:

- Uterine cervix dilated to 4 cm
- Contractions 3 minutes apart, each lasting 40-60 seconds

Clinical Problems to Consider

- Lumbar epidural analgesia
- Spinal analgesia (block)
- Pudendal nerve block

Clinical Notes

There are three stages of childbirth:

1. **Stage 1** is characterized by the onset of uterine contractions and ends when the uterine cervix is dilated completely (10 cm). This stage can last up to 20 hours and is divided into three phases:
   i. **Latent phase:** cervix dilated 1–4 cm
   ii. **Active phase:** cervix dilated 4–8 cm
   iii. **Transition phase:** cervix dilated 8–10 cm

   During this stage, contractions typically increase in frequency, duration, and intensity. It is important to distinguish labor contractions from Braxton–Hicks ("false labor") contractions, which are irregular, do not increase in frequency, and may change with body position and activity.

2. **Stage 2** begins when the cervix is dilated completely and ends with delivery. Delivery may last from 20 minutes to 2 hours.

3. **Stage 3** involves delivery of the placenta and lasts between 5 and 30 minutes.

LEARNING OBJECTIVES

1. Describe the anatomy of the lumbosacral region of the vertebral column and the hip bone.
2. Describe the anatomy of the lumbosacral region of the spinal cord and spinal nerves.
3. Describe the anatomy of the spinal meninges and associated spaces.
4. Describe the sensory innervation from the uterus, vagina, and female perineum.
5. Explain the anatomical basis for the signs and symptoms associated with this case.

Analgesia Loss of painful sensation

RELEVANT ANATOMY

Lumbosacral Region of the Vertebral Column and Hip Bone

The five **lumbar vertebrae** and the **sacrum** form the skeleton of the inferior vertebral region (**Fig. 8.1.1**). The lumbar spinous processes are connected by **supraspinous** and **interspinous ligaments**. Adjacent laminae are connected by **ligamenta flava** (singular, *ligamentum flavum*). On the sacrum, the **posterior median crest** represents the spinous processes and extends inferiorly to the **sacral hiatus**.

The hip bone is formed by the fusion of the **ischium, ilium, and pubic bones**. The ischium has two prominent projections that are relevant to obstetrics and gynecology (**Fig. 8.1.2**):

- The **ischial tuberosity** is a roughened area for attachment of muscles of the pelvic floor (urogenital diaphragm) and lower limb. It is subcutaneous when the hip is flexed.

- The **ischial spine** is a sharp, medially directed process. Structures attaching to it are the sacrospinous ligament, the tendinous arch, and the coccygeus and superior gemellus muscles.

The **supracristal plane** (also known as the **intercristal line**) of the **ilium** joins the highest points of the iliac crests (**Fig. 8.1.3**). In radiographic images, this line crosses the vertebral column at the L4 spinous process.

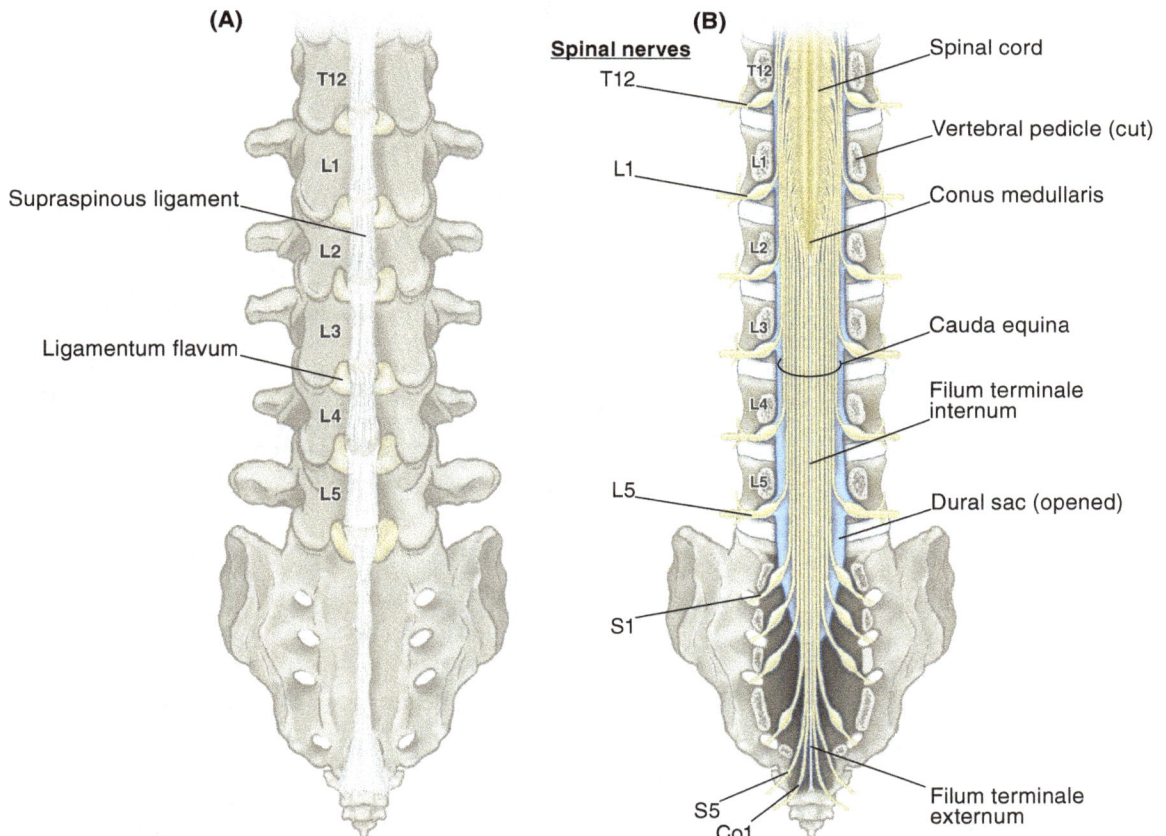

FIGURE 8.1.1 Posterior view of the lumbosacral region showing (**A**) the intact vertebrae with supraspinous ligament and ligamentum flavum, and (**B**) the opened vertebral canal and contents.

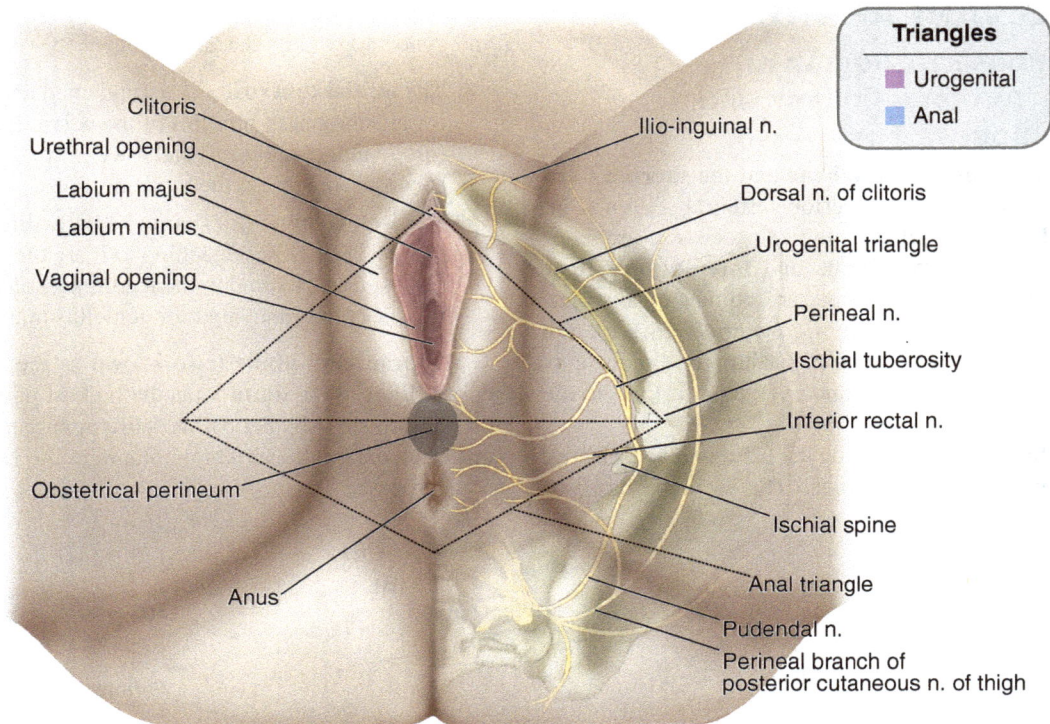

FIGURE 8.1.2 Inferior view of the female perineum and bony pelvis.

Lumbosacral Spinal Cord and Spinal Nerves

The conical inferior end of the adult **spinal cord** is referred to as the **medullary cone** (*conus medullaris*). The tip of the medullary cone usually lies at the level of the L2 vertebra (**Fig. 8.1.1B**), although it may terminate as high as T12 or as low as L3. Inferior to the medullary cone, the rootlets and roots of the L2-Co1 spinal nerves, and the spinal nerves of S1-Co1, form the **cauda equina**.

Clinical Note

To avoid spinal cord damage, a needle used for lumbar epidural or spinal procedures should be introduced inferior to the medullary cone. The cone typically lies at the L2 vertebral level and, thus, a needle introduced at the L4 vertebral level is considered safe. The commonly accepted landmark for the **L4 vertebral level** is the **supracristal plane**. Recent research shows, however, that the palpated **supracristal plane (Fig. 8.1.3)**

actually crosses the midline one vertebral level higher, at the **L3 spinous process. Radiologically**, the supracristal plane does indeed cross the L4 spinous process. The reason for this discrepancy is due to subcutaneous tissue over the iliac crests and this should be considered when establishing a landmark for performing a lumbar epidural or spinal procedures.

Spinal Meninges and Associated Spaces

The spinal meninges include (internal-to-external):

- The **pia mater** (Latin, *delicate mother*) is a thin, delicate membrane that adheres intimately to the spinal cord, spinal nerve roots, spinal nerves, and their associated blood vessels. Inferior to the spinal cord, the pia mater forms the **filum terminale (internum)**.

Palpitation Forcible or irregular pulsation of the heart that is perceptible to the patient

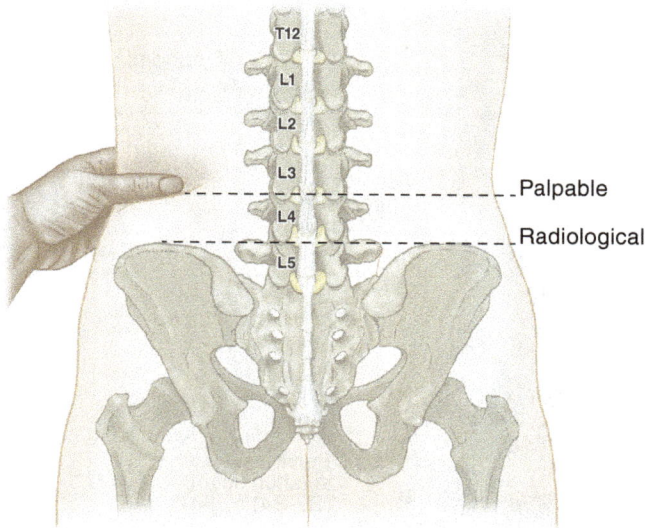

FIGURE 8.1.3 Identification of the supracristal plane by palpation and radiography.

- The **arachnoid mater** is a delicate vascular membrane. The **subarachnoid space** lies between the pia and arachnoid mater and contains **cerebrospinal fluid (CSF)**. This space is traversed by rootlets, roots, spinal nerves, and associated vasculature. Inferior to the spinal cord, the subarachnoid space is enlarged to form the **lumbar cistern**, which contains the cauda equine and filum terminale internum.

- The **dura mater** (Latin, *tough mother*) forms the thick, outer meningeal layer. Superiorly, the dura is fused to the rim of foramen magnum. In the lumbosacral region, dura forms a sac that ends at the level of the S2 vertebral body (**Fig. 8.1.1B**). The dura continues inferior to this level as the **filum terminale externum**, a thin sleeve around the filum terminale internum. The filum terminale passes through the sacral hiatus to attach on the coccyx.

A potential space, the **spinal epidural (extradural) space** extends from foramen magnum to the sacral hiatus and lies between the dura and bones that form the vertebral canal. Spinal nerve roots, covered by dura, cross the epidural space to reach intervertebral or sacral foramina. The epidural space also contains the **internal vertebral venous plexus** and **extradural fat**.

Sensory Innervation of the Uterus, Vagina, and Female Perineum

Visceral and somatic pathways are involved in the sensory innervation of the pelvis and perineum (**Fig. 8.1.4**). Visceral afferent innervation of the uterus and superior vagina follows sympathetic or parasympathetic pathways as determined by the **pelvic pain line** (**Table 8.1.1**). The pelvic pain line is determined by the relationship between an organ and the peritoneum:

- Pelvic organs, or portions of organs, that are **not in contact with peritoneum** are "below" the pelvic pain line, and their afferents follow parasympathetic pathways.

The female perineum receives somatic innervation (**Fig. 8.1.2** and **Table 8.1.2**) from nerves of the lumbar plexus (anterior vulva) and sacral plexus (posterior vulva and anal region). The pudendal nerve (S2-S4) supplies most of the perineum and is considered its primary nerve supply.

Vulva Female external genitalia

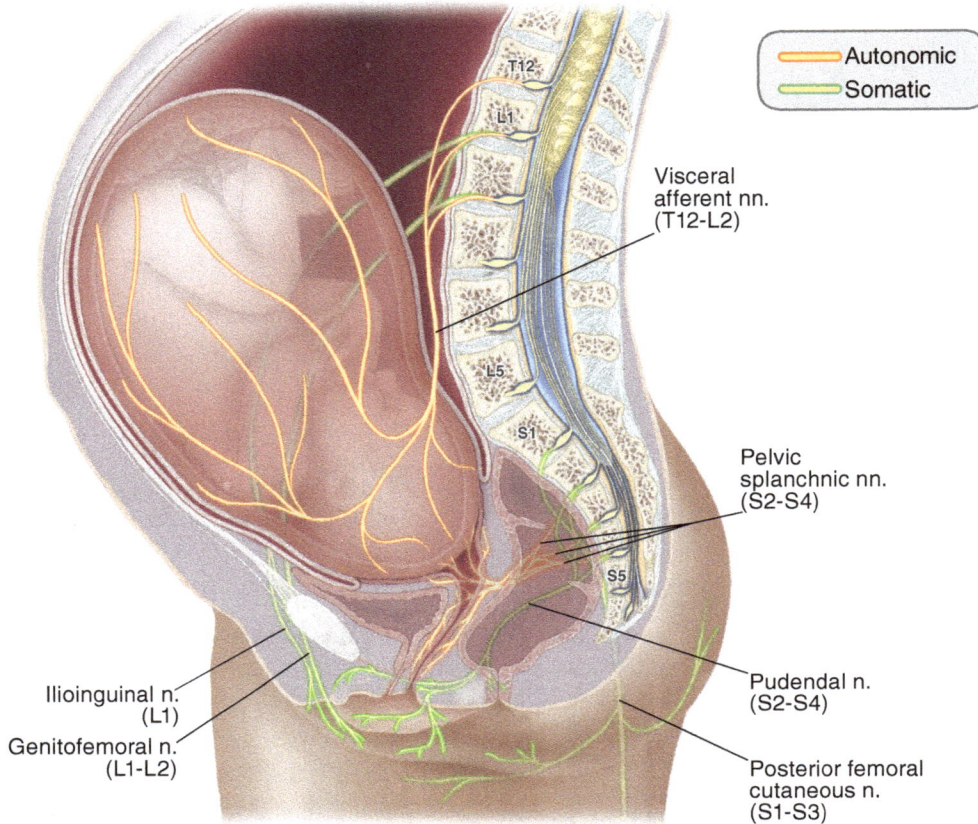

FIGURE 8.1.4 Medial view of the innervation of the pelvis and perineum. Pelvic organs, or portions of organs, that are in contact with peritoneum are "above" the pelvic pain line, and their afferents follow sympathetic pathways.

Clinical Note

The perineum is defined anatomically as a diamond-shaped region between the pubic symphysis and coccyx, bounded laterally by the ischial tuberosities. This region includes the anus and external genitalia. The obstetrical perineum is defined as the region between the posterior vaginal wall and the anus; it overlies the perineal body.

TABLE 8.1.1   Relationship of female reproductive organs to pelvic pain line.

Organ	Peritoneal covering	Pelvic pain line	Nerve pathway
Uterus—fundus and body	Yes	Above	Sympathetic fibers via T12-L2 spinal ganglia
Uterus—cervix Vagina—superior ¾	No	Below	Parasympathetic fibers via S2-S4 spinal ganglia
Vagina—inferior ¼	No	Below	Somatic afferents innervation via deep perineal branch of pudendal nerve

TABLE 8.1.2	Somatic innervation of the female perineum.		
Nerve	**Spinal level(s)**	**Branch**	**Region**
Ilioinguinal	L1	Anterior labial	▪ Mons pubis ▪ Anterior labium majus
Genitofemoral	L1-L2	Genital	▪ Anterior labium majus
Posterior cutaneous nerve of thigh	S2	Perineal	▪ Labium majus ▪ Perianal skin
Pudendal	S2-S4	Perineal (superficial branch)	▪ Posterior labium majus ▪ Labium minus ▪ Vaginal orifice ▪ Perineal muscles
		Perineal (deep branch) Dorsal nerve of clitoris	▪ Sensation from clitoris

CLINICAL REASONING

Based on the physical examination, the patient arrived at the hospital in **stage 1 of childbirth**. Dilation of her uterine cervix (4 cm) and the duration and timing of her contractions indicated that she was entering the **active phase**.

Regional analgesia during labor and delivery is commonly achieved by injection of anesthetic agents into the lumbar epidural space, into the lumbar cistern of the subarachnoid space, or along the course of peripheral nerves that supply the perineum (pudendal and/or ilioinguinal).

Epidural Analgesia

Epidural analgesia (also called an epidural "block") during labor involves the perfusion of a local anesthetic agent (e.g., lidocaine or bupivacaine) and an opioid analgesic agent (e.g., morphine or fentanyl) into the lumbar epidural space (**Fig. 8.1.5A**). Anesthetic and analgesic agents diffuse across the dura where they affect primarily the spinal nerve roots, spinal nerves and, to a lesser extent, the spinal cord. Epidural analgesia for labor and delivery is usually applied during phase 1 of stage 1 (i.e., cervical dilation 4–5 cm).

Procedure

▪ The patient is in a sitting position or lying on their side, with their vertebral column flexed (**Fig. 8.1.6**). In either position, adjacent laminae move apart and ligamenta flava are stretched. This widens the space for insertion of the introducer needle and catheter, usually between L4 and L5 vertebrae.

▪ Antiseptic solution (e.g., Betadine) is applied to the area around the injection site and local anesthetic (1% lidocaine) injected into both subcutaneous and deeper tissue.

▪ For a lumbar epidural, an introducer needle is inserted through the skin and advanced through the ligamentum flavum. Entry into the epidural space is confirmed by a loss of resistance to a syringe plunger attached to the needle. A catheter is threaded through the needle into the epidural space, the needle is removed, and the catheter remains in place to allow periodic injections or continuous infusion.

Epidural injection typically results in analgesia or anesthesia below the umbilicus. The number of nerve roots affected depends on the amount of anesthetic/analgesic agent(s) injected and the vertebral level at which they are introduced. Typically, the level is 3–4 spinal segments higher than the point of insertion.

▪ *This procedure is similar to a lumbar puncture except that the dura mater is not penetrated and anesthetic is infused rather than CSF being withdrawn.*

▪ For **caudal epidural**, a catheter is inserted into the sacral hiatus and anesthetic is infiltrated

---

Anesthesia Loss of sensation

(A)

Supraspinous ligament
Interspinous ligament
Catheter
Epidural needle
L4
L5
Dura mater
Epidural space
Subarachnoid space

(B)

Spinal needle
Epidural needle
L4
L5

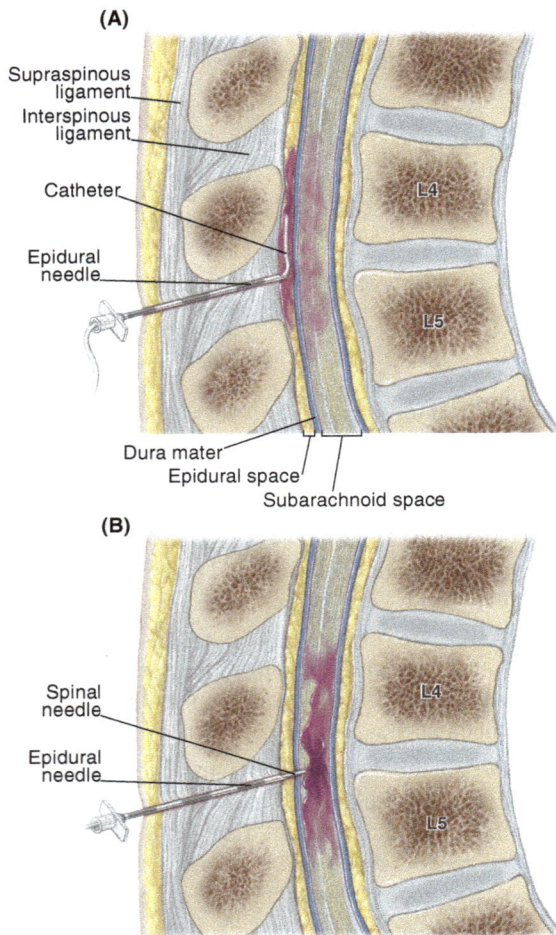

FIGURE 8.1.5 Midsagittal view of the lumbar spinal canal and contents. (**A**) Epidural analgesia and (**B**) spinal analgesia.

ligamentum flavum, dura mater, and arachnoid mater to reach the subarachnoid space and CSF. Anesthetic and analgesic agents bathe the spinal cord and nerve roots directly. As a result, spinal analgesia has more rapid onset than epidural analgesia. This results in anesthesia of structures inferior to the waist, including the perineum, pelvic floor, and birth canal. Motor supply to the uterus may be affected, as well as sensory and motor innervation of the lower limb.

Pudendal Nerve Block

Local anesthetic administered along the peripheral course of the nerves that supply the vulva can be used to provide pain relief during delivery. **Pudendal nerve block** anesthetizes the posterior vulva (S2-S4), while an **ilioinguinal nerve block** anesthetizes the anterior vulva (L1-L2). Pain associated

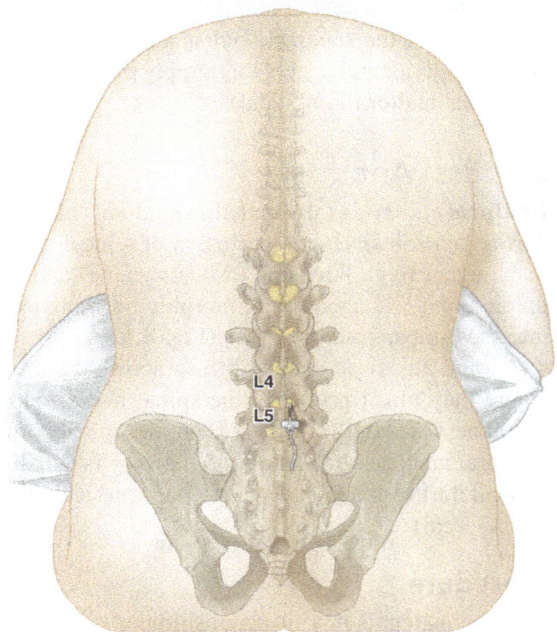

around the caudal end of the dural sac. This infiltrates anesthetic around the **sacral spinal nerves** and affects the entire birth canal (it does not typically affect lumbar spinal nerves).

Spinal Analgesia

Spinal analgesia (spinal block) refers to the administration of anesthetic into the subarachnoid space (**Fig. 8.1.5B**). For this procedure, a needle is inserted at the L3-4 vertebral level through the

FIGURE 8.1.6 Patient positioned for administration of a lumbar epidural or spinal block.

with uterine contractions and cervical dilation is not affected by a pudendal block since this sensation is conducted along sympathetic (T10-L2) and parasympathetic (S2-S4) fibers, respectively (**Table 8.1.1**).

Procedure

- Palpate the ischial spine transvaginally with the patient in the lithotomy position (**Fig. 8.1.7**).

- The needle is passed through the lateral vaginal wall, and guided toward the ischial spine until it touches the sacrospinous ligament.

- The tissue in this region is infiltrated with anesthetic.

- The needle is then advanced through the sacrospinous ligament (the area of the

pudendal nerve) and anesthetic is again administered.

CLINICAL SUMMARY

Epidural and Spinal Analgesia

- Epidural and/or spinal analgesia are the preferred methods for regional analgesia during labor and delivery because it is possible with one procedure to provide prolonged pain relief through continuous, controlled infusion of anesthetic.

Pudendal Nerve Block

- Pudendal nerve block is an alternative in some circumstances, for example, during stage 2 of labor (after full cervical dilation and during delivery) or repair of episiotomy.

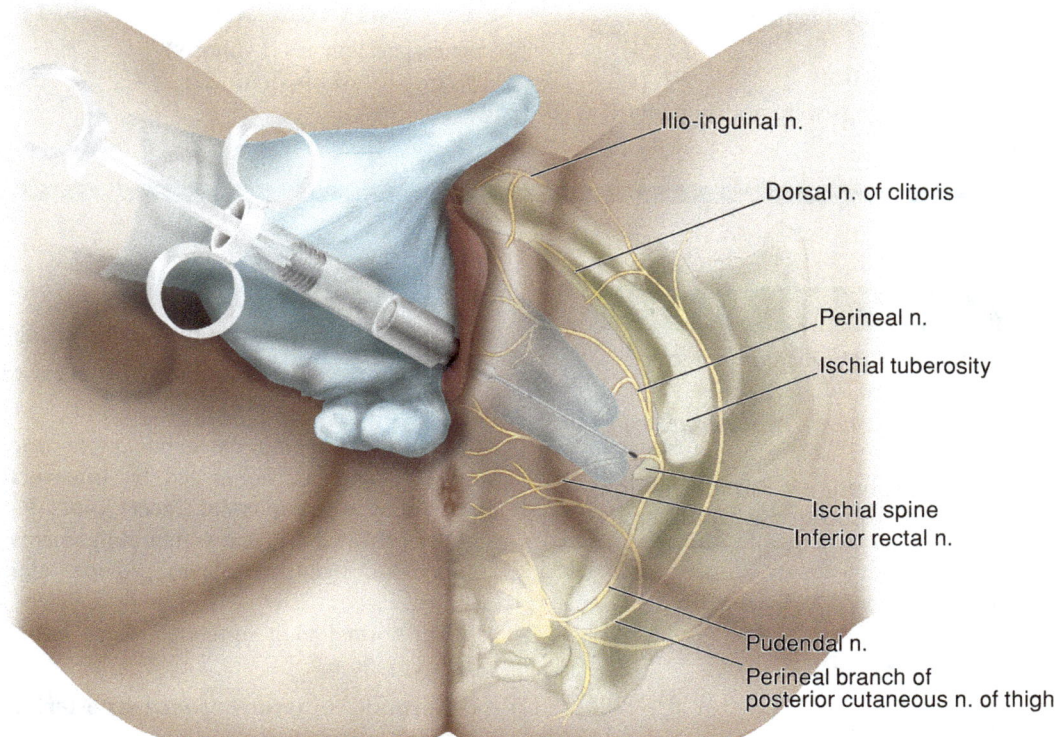

Ilio-inguinal n.

Dorsal n. of clitoris

Perineal n.

Ischial tuberosity

Ischial spine
Inferior rectal n.

Pudendal n.

Perineal branch of posterior cutaneous n. of thigh

FIGURE 8.1.7 Administration of a pudendal block.

Patient Presentation

A 13-year-old girl visits her pediatrician for an annual physical examination. Her mother expresses concern that her daughter's right shoulder is slightly higher than the left, that the right shoulder blade sticks out, and that the spine is twisted slightly to the right.

Relevant Clinical Findings

History

The patient complains that she is self-conscious about the appearance of her body, and that she has difficulty participating in school sports.

Physical Examination

The following findings were noted on physical examination:

- The medial border of the right scapula protrudes.
- Right lateral curvature in thoracic region with vertebral column flexed (Adams forward bending test).
- The patient does not describe any pain associated with her condition.

Imaging Studies

- Lateral and anterior–posterior radiographs show an abnormal lateral spine curvature.

Clinical Problems to Consider

- Hyperkyphosis (kyphosis)
- Hyperlordosis (lordosis)
- Scoliosis

LEARNING OBJECTIVES

1. Describe the vertebral column and its postural muscles.
2. Describe the relevant surface anatomy of the back.
3. Explain the anatomical basis for the signs and symptoms associated with this case.

RELEVANT ANATOMY

Vertebral Column

The vertebral column is composed of 31 vertebrae (**Fig. 8.2.1**):

- 7 Cervical
- 12 Thoracic
- 5 Lumbar
- 5 Sacral (fused to form the sacrum in adult)
- 2–4 Coccygeal (fused to form the coccyx in adult)

Adjacent vertebrae in the cervical, thoracic, and lumbar regions are connected by **intervertebral discs** and by **zygapophysial (facet) joints**. Postural muscles (**Table 8.2.1**) and vertebral ligaments (e.g., longitudinal, ligamenta flava) provide additional stability.

The normal adult vertebral column has characteristic curvatures:

- **Lordotic curvatures (*concave* posteriorly)** are present in the cervical and lumbar regions.
- **Kyphotic curvatures (*convex* posteriorly)** are present in the thoracic and sacral regions.

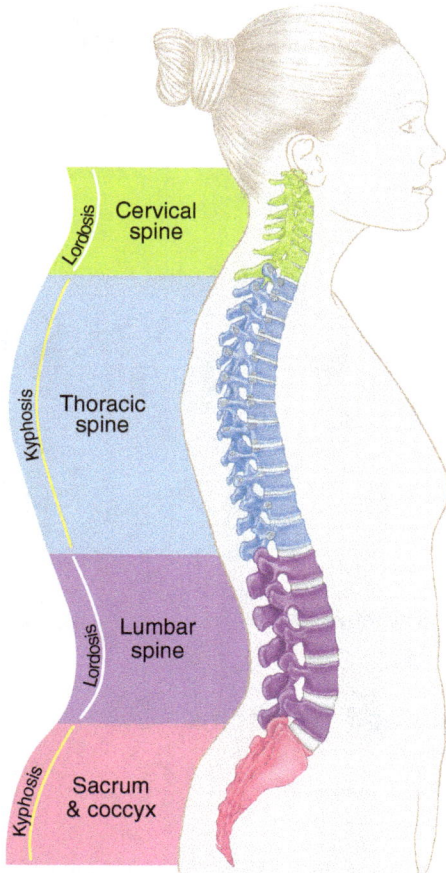

FIGURE 8.2.1 Lateral view of the vertebral column and its normal curvatures.

These curvatures play an important role in maintaining posture and contribute to the flexibility and shock-absorbing properties of the vertebral column.

Clinical Notes

Pathological conditions that exaggerate one or more of the normal spinal curvatures can occur in the cervical, thoracic, and/or lumbar regions.

- **Hyperkyphosis**, sometimes referred to simply as **kyphosis**, is an exaggeration of the normal kyphotic curvature in the thoracic region.
- **Hyperlordosis**, sometimes referred to simply as **lordosis**, is an exaggeration of the normal lordotic curvature in the cervical and/or lumbar regions.
- **Scoliosis** is an abnormal lateral deviation of the vertebral column. It is frequently accompanied by changes in kyphotic and/or lordotic curves.

Postural Muscles

Posture and movement of the back and neck involves intrinsic muscles of the back, as well as muscles anterior to the vertebral column (**Table 8.2.1**).

Surface Anatomy of the Back

Physical examination of the back includes an assessment of vertebral column alignment. The following structures are usually palpable:

- External occipital protuberance
- Spinous process of C7 vertebra (referred to as vertebral prominens)
- Thoracic spinous processes
- Scapulae (spine, medial border, inferior angle)
- Iliac crests
- Sacrum
- Paravertebral muscles

TABLE 8.2.1	Muscles that act on the vertebral column.
Location	**Muscles**
Deep back	▪ Splenius ▪ Erector spinae (iliocostalis, longissimus, spinalis) ▪ Deep intrinsic muscles (e.g., semispinalis, multifidus, rotatores)
Anterior neck	▪ Longus capitis ▪ Longus colli
Lateral neck	▪ Scalene muscles ▪ Levator scapulae
Posterior abdominal wall	▪ Psoas ▪ Quadratus lumborum
Anterior abdominal wall	▪ Rectus abdominis ▪ Abdominal oblique muscles

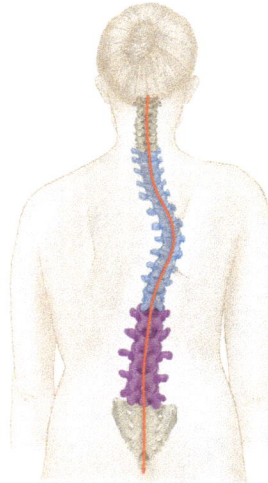

(A) Hyperkyphosis **(B) Hyperlordosis** **(C) Scoliosis** **(D) Scoliosis (spine flexed)**

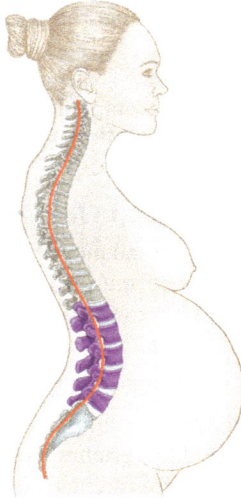

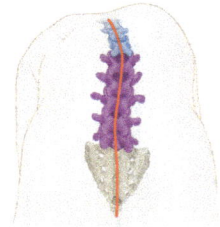

FIGURE 8.2.2 Exaggerated curvatures of the vertebral column. (**A**) Hyperkyphosis, (**B**) hyperlordosis, (**C**) scoliosis, and (**D**) scoliosis, with back flexed. The lateral deviation of the vertebral column characteristic of scoliosis and associated asymmetry of the trunk are more obvious when the patient bends forward.

Clinical Note

Some surface features of the back (e.g., spinous processes) can be better appreciated when the patient flexes their vertebral column (Adams forward bending test). In many cases, this will make abnormal curvatures more obvious. Pathological changes in the vertebral column may affect the thoracic cavity and/or scapular function.

CLINICAL REASONING

This patient presents with signs and symptoms indicating a clinical condition of an abnormal curvature of the vertebral column.

Hyperkyphosis (*Kyphosis*)

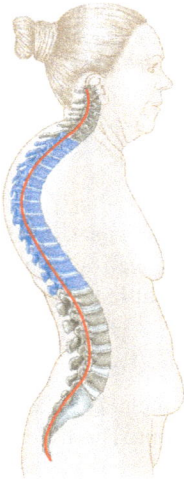

Hyperkyphosis is an exaggerated posterior convexity in the thoracic region of the vertebral column (**Fig. 8.2.2A**).

Signs and Symptoms

Sensory
- Mild back pain

Other
- Postural "hunchback" (rounded back)
- Fatigue
- Back stiffness or tenderness
- Dyspnea (in severe cases)

Predisposing Factors

In Adults
- Degenerative spine diseases (arthritis, disc degeneration, spondylolisthesis)
- Fractures caused by osteoporosis (especially in thoracic vertebrae)
- Infection of thoracic vertebrae (e.g., tuberculosis)
- Trauma

Dyspnea Difficulty breathing and shortness of breath

Spondylolisthesis A condition in which a vertebra "slips" out of alignment with adjacent vertebrae

In Adolescents

- Girls with poor posture
- Boys, 10–15 years old, at risk for adolescent kyphosis (Scheuermann disease)
- Connective tissue disorders (e.g., Marfan syndrome)

Clinical Note

Mild hyperkyphosis can be treated conservatively unless the deformity is debilitating or causes pain. Congenital hyperkyphosis is typically treated surgically at an early age.

Hyperlordosis (*Lordosis*)

Hyperlordosis is an exaggerated posterior concavity in the cervical or lumbar regions of the vertebral column (**Fig. 8.2.2B**).

Signs and Symptoms

Sensory

- Lower back pain

Other

- Postural "swayback"

Predisposing Factors

- Obesity
- Pregnancy, especially later stages
- Tight low back muscles or weak anterior abdominal muscles ("beer-gut")
- Previous back surgery
- Degenerative spine diseases (arthritis, disk degeneration, spondylolisthesis)
- Osteoporosis (wedge fractures in thoracic vertebrae)
- Kyphosis
- Hip joint disease (hip fixed in flexed position)

Scoliosis

Scoliosis is described classically as an abnormal lateral deviation of the vertebral column. In reality, however, it may also involve rotational distortions of the vertebral column and/or changes in the kyphotic and/or lordotic curvatures. Thus, these conditions are referred to as rotoscoliosis and kyphoscoliosis, respectively (**Fig. 8.2.2C, D**).

Scoliosis may involve the thoracic region (**Fig. 8.2.3A**), the lumbar region (**Fig. 8.2.3B**), or both

(A) Thoracic   **(B) Lumbar**   **(C) Thoracolumbar**

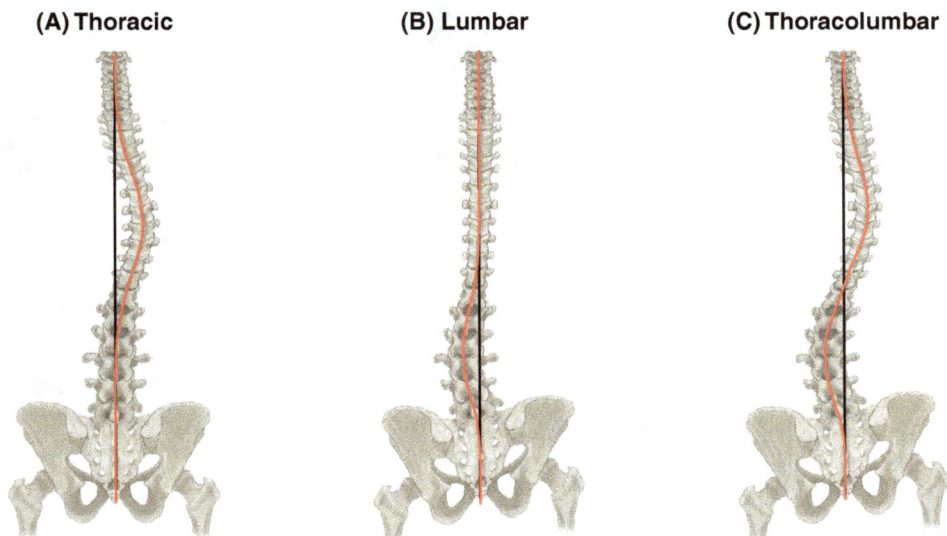

FIGURE 8.2.3 Posterior view of the vertebral column showing different curvature patterns associated with scoliosis, which are differentiated according to the vertebral region(s) affected. (**A**) Thoracic, (**B**) lumbar, (**C**) thoracolumbar (double curvature).

(A)

(B)

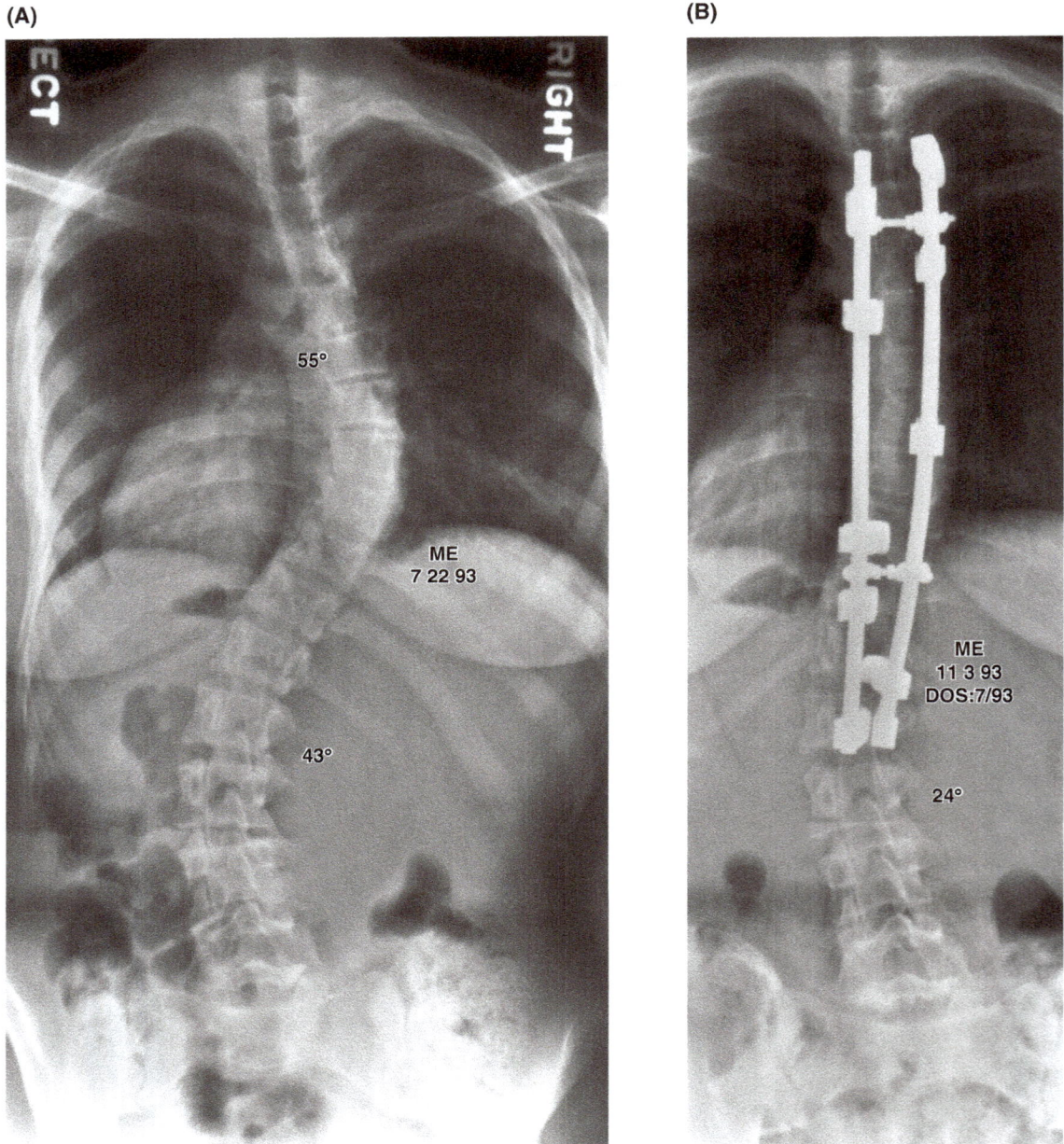

FIGURE 8.2.4 Radiographs of a patient with adolescent idiopathic scoliosis.
(**A**) Before surgery, showing moderate-to-severe thoracic (55°) and lumbar (43°)
curvatures, and (**B**) after surgery, showing a degree of correction with instrumentation.
Source: Fig. 5-21 in *Current Diagnosis & Treatment in Orthopedics,* 4e.

(**Fig. 8.2.3C**). This results in single curvatures (C shaped) or double curvatures (S shaped).

Clinical Notes

- Scoliosis is typically classified on the degree of lateral deviation from normal of the thoracic curvature: mild (5–15°), moderate (20–45°), severe (≥50°).
- Curvature pattern and location are important factors in determining treatment. Double S-shaped curvature is more likely to progress than do single C-shaped curvature, and thoracic curvatures progress more often than those located in cervical or lumbar regions.

Signs and Symptoms

- One shoulder higher than the other
- One scapula (medial border and inferior angle) protrudes posteriorly
- Abnormal gait and posture (leaning slightly to one side)
- Postural "hunchback"

The radiographic appearance of a thoracolumbar (double) curvature is shown in **Figure 8.2.4**.

Predisposing Factors

- Age: incidence increases just before puberty
- Sex: adolescent girls
- Family history

Clinical Note

Idiopathic **scoliosis** is the most common form and is classified according to the patient's age at diagnosis.

Infantile idiopathic scoliosis

- Diagnosed at <3 years of age
- More common in boys of European descent
- Usually a single thoracic deviation toward the **left**
- Significant rate of spontaneous resolution (as high as 90%)
- Prevalent treatment is nonoperative, using an orthosis.

Juvenile idiopathic scoliosis

- Recognized in patients 3–10 years of age
- More common in girls
- Usually a single thoracic deviation toward the **right**

Adolescent idiopathic scoliosis is the most common type of scoliosis (**Fig. 8.2.4**).

- Recognized in children >10 years of age
- More common in girls
- Usually a thoracic deviation toward the **right**

Treatment for idiopathic scoliosis depends on severity of the curvature(s) and growth status:

- **Observation** is used for curvatures <20°.
- **Orthosis** is used to prevent curves from getting worse, particularly if the child is still growing, and has a spinal curvature between 25° and 45°.
- **Operative intervention.** Surgery may be recommended if the child is still growing and has a curvature >45°. At skeletal maturity, surgery may be recommended for scoliotic curvatures >50–55°.

DIAGNOSIS

A diagnosis of **adolescent idiopathic scoliosis** is evident from the abnormal lateral spinal curvature.

Idiopathic A condition that arises spontaneously or whose cause is obscure or unknown

Orthosis Correction of limbs or spine using braces or other similar devices for alignment or support

Questions 1–3 refer to the following clinical case.

A 16-year-old female is admitted to the emergency depart-ment with fever, neck stiffness, and a change in mental status. Her mother reports that she had complained of a headache and had become progressively tired and con-fused over the past 48 hours. Bacterial meningitis is sus-pected and a lumbar puncture is performed to assess cere-brospinal fluid.

1. Which structure would *not* be pierced when performing this procedure?

    A. Arachnoid mater
    B. Dura mater
    C. Ligamentum flavum
    D. Pia mater
    E. All of the above would be pierced.

2. Which is the palpable bony landmark may be used to identify a safe vertebral "level" to perform the lumbar puncture?

    A. Coccyx
    B. Rib 12
    C. Spinous process of L2
    D. Supracristal plane
    E. Transverse process of L4

3. Lumbar puncture is facilitated by the fact that the needle is inserted into the region of the vertebral canal occupied by the cauda equina. Which state-ment best describes the cauda equina?

    A. It is composed of lumbar spinal nerves.
    B. It is composed of lower lumbar and sacral dorsal and ventral rami.
    C. It represents the caudal extension of the spinal cord.
    D. It is composed of lower lumbar and sacral dorsal and ventral roots.
    E. It is composed of sacral spinal nerves.

4. A 30-year-old male complains to his primary care physician that he has lost sensation from the skin on the back of his head. Neurological testing revealed sensory innervation for the rest of the head was intact. Which nerve was injured?

    A. C2 dorsal ramus
    B. C2 dorsal root
    C. C2 spinal nerve
    D. C2 ventral ramus
    E. C2 ventral root

5. During labor, anesthetic may be injected through the sacral hiatus into the inferior portion of the sacral vertebral canal. Which meningeal space is bathed by the anesthetic?

    A. Epidural
    B. Subarachnoid
    C. Subdural
    D. Subpial

CHAPTER 9

Upper Limb

Patient Presentation

A 15-year-old male comes to the emergency department complaining of severe pain over his right collarbone. He uses his left hand to hold his right arm close to and across the front of his body.

Relevant Clinical Findings

History

While playing baseball, the patient reports colliding with the pitcher when running to first base. The impact knocked him off his feet and he fell on his right shoulder. His mother, and others nearby, reported hearing a loud snap. When he stood up, he was slouched and complained of extreme pain over his right clavicle.

Physical Examination

The following findings were noted on physical examination:

- Prominence, ecchymosis, and tenderness over the right midclavicular region
- Diffuse swelling over the right clavicle, shoulder, and arm
- Right upper limb depressed (i.e., lower than left)
- Mild numbness over right upper limb
- Right radial pulse weaker than left
- Lung sounds normal

Imaging Studies

- Radiography of right shoulder revealed a midclavicular fracture (**Fig. 9.1.1**).

Clinical Problems to Consider

- Anterior shoulder dislocation
- Clavicle fracture
- Shoulder separation

LEARNING OBJECTIVES

1. Describe the anatomy of the shoulder.
2. Describe the structures that stabilize and move the shoulder.
3. Explain the anatomical basis for the signs and symptoms associated with this case.

RELEVANT ANATOMY

Shoulder and Arm

The **shoulder** includes the clavicle, scapula, and proximal humerus, with their associated ligaments, muscles, and tendons (**Fig. 9.1.2**). The **shoulder girdle** is formed by the clavicle and scapula and is anchored to the axial skeleton at the sternoclavicular (SC) joint.

Clavicle

The clavicle is an *f*-shaped bone that acts as a strut to keep the shoulder joint and arm away from the body wall so that the upper limb can move freely. It contributes to two joints:

1. Medially, the **SC joint** represents the sole articulation between the upper limb and axial skeleton. Injuries to the shoulder rarely involve the SC joint. It is stabilized by the several ligaments: **anterior and posterior SC**, **costoclavicular**, and **interclavicular**.

Ecchymosis Hemorrhage into skin (bruising)

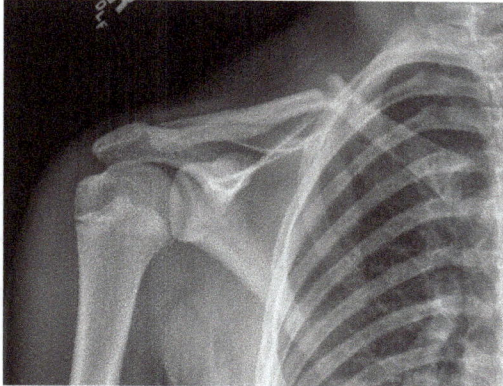

FIGURE 9.1.1 Anteroposterior radiograph of the right shoulder showing a midclavicular fracture.

2. Laterally, the **acromioclavicular (AC) joint** is surrounded by a relatively weak fibrous capsule, making it more vulnerable to injury. The joint is stabilized by ligaments: **AC** (along the *superior* aspect of the capsule) and **coracoclavicular (CC)** (*trapezoid* and *conoid* parts).

The **subclavian vessels** and **brachial plexus** pass between the clavicle and rib 1, a passage known as (the **cervico-axillary canal**), to enter the upper limb (**Fig. 9.1.3**).

Clinical Note

Proximity of the subclavian vessels, brachial plexus, and cervical pleura to the clavicle makes it essential to conduct neurovascular and lung exams with suspected clavicle fracture.

Muscles that act on the clavicle include the **sternocleidomastoid, trapezius, subclavius**, and **pectoralis minor** (the latter acts indirectly via its attachment to the coracoid process and CC ligaments).

Scapula

The scapula is located on the posterolateral thoracic wall. It has no direct articulation with the axial skeleton and is held in place by its articulation with the clavicle at the AC joint and by numerous muscles. Three scapular features contribute to the shoulder:

1. The **glenoid cavity** forms the glenohumeral joint with the humeral head.

2. The **acromion** is the lateral end of the scapular spine. It lies superior to the glenohumeral joint and contributes to the AC joint.

3. The **coracoid process** is a prominent anterolateral projection of the scapula located medial to the glenoid cavity. This process serves as an attachment for the CC ligament and several muscles: coracobrachialis, short head of biceps brachii, and pectoralis minor (**Fig. 9.1.2**).

Humerus

Anatomical features of the proximal humerus that contribute to the shoulder include:

- The **humeral head** forms the glenohumeral joint with the glenoid cavity of the scapula. The head is large relative to the size of the cavity.

- The narrow **anatomical neck** separates the head from the humeral tubercles (greater and lesser).

- The **greater tubercle** serves as attachment for three rotator cuff muscles: supraspinatus, infraspinatus, and teres minor.

- The **lesser tubercle** serves as an attachment for subscapularis (a rotator cuff muscle).

- The **intertubercular sulcus** (bicipital groove) is a deep channel between the tubercles of the humerus. The tendon of the long head of biceps brachii lies in this channel before entering the glenohumeral joint capsule and attaching to the supraglenoid tubercle of the scapula.

Glenohumeral Joint

The glenohumeral (shoulder) joint is formed by the **humeral head** and the **glenoid cavity** (fossa). The relative mismatch in size between these elements (often described as a "golf ball on a tee") accounts for the great mobility of the joint, as well as for its inherent instability. Numerous ligaments and muscles stabilize this joint (see **Fig. 9.1.2**).

- The **glenoid labrum** is a ring of fibrocartilage on the rim of the glenoid cavity. It deepens the cavity slightly (similar to the rim on a dinner plate).

(A)

Anatomical neck of humerus
Surgical neck of humerus

1. Head of humerus
2. Greater tubercle of humerus
3. Lesser tubercle of humerus
4. Acromion
5. Coracoid process
6. Clavicle
7. Glenoid cavity

(B)

1. Fibrous capsule of glenohumeral joint
2. Coracohumeral ligament
3. Coraco-acromial ligament
4. Fibrous capsule of acromioclavicular joint
5. Coracoclavicular ligament
   a. Conoid ligament
   b. Trapezoid ligament
6. Tendon of long head of biceps brachii (cut)
7. Tendon of subscapularis (cut)
8. Tendon of supraspinatus (cut)
9. Costoclavicular ligament
10. Sternoclavicular ligament

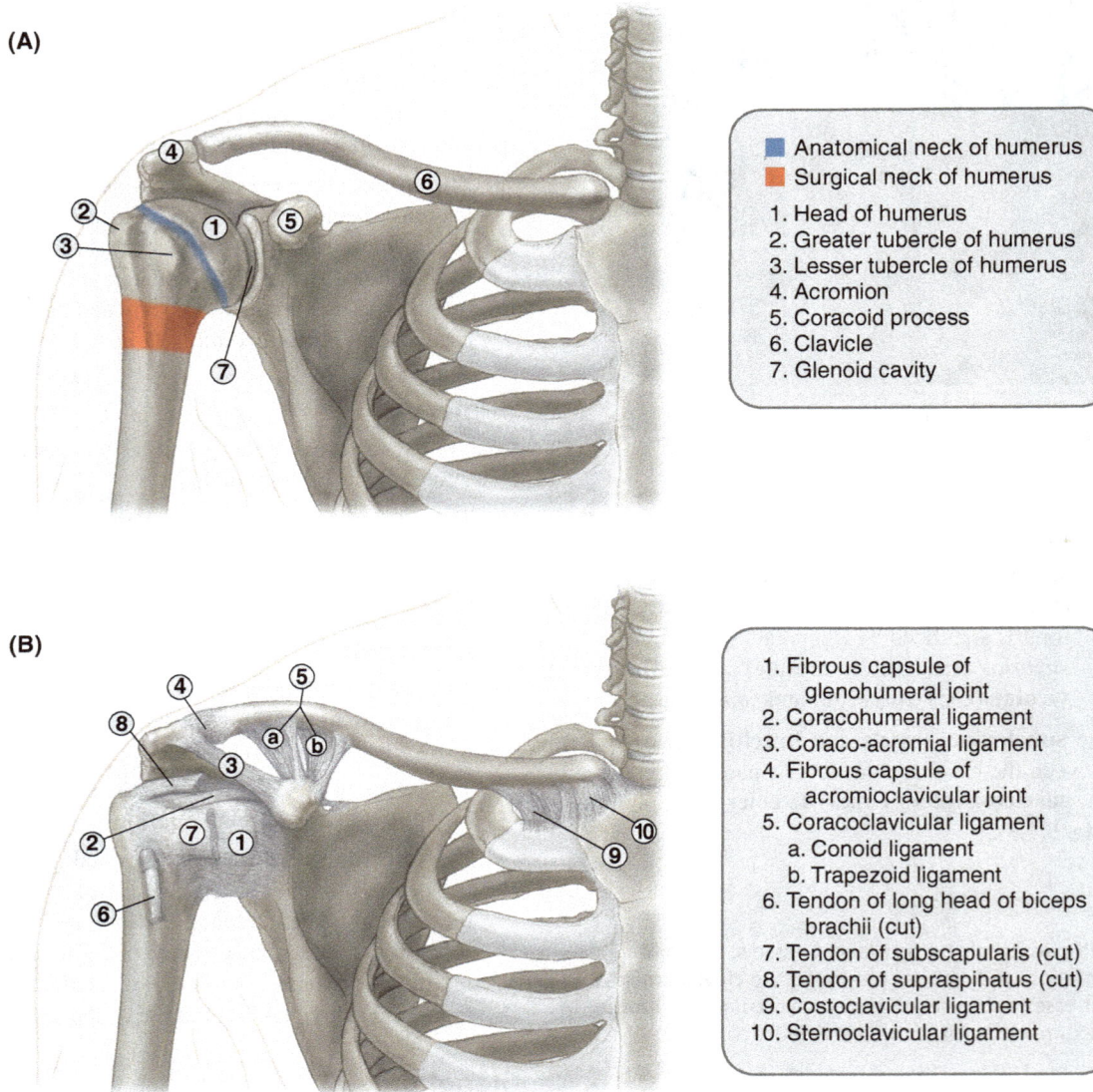

FIGURE 9.1.2 Anterior view of the right shoulder showing the skeleton **(A)** and ligaments **(B)**.

- The **fibrous capsule** encloses the joint cavity. It extends from the glenoid neck to the anatomical neck of the humerus.

- **Glenohumeral ligaments** (*superior, middle, inferior*) are thickenings of the anterior aspect of the capsule; they are apparent only on its internal surface.

- The **coracohumeral ligament** extends from the base of the coracoid process to the greater tubercle.

- The **coraco-acromial ligament**, with its bony attachments, forms the **coracoacromial arch**. This arch lies superior to the humeral head and limits its superior displacement.

(A)

Anterior scalene m.

Middle scalene m.

1. Subclavian a.
2. Subclavian v.
3. Axillary a
4. Axillary v.
5. Superior thoracic a.
6. Thoraco-acromial a.
    a. Clavicular branch
    b. Acromial branch
    c. Deltoid branch
    d. Pectoral branch
7. Lateral thoracic a.
8. Subscapular a.
    e. Circumflex scapular a.
    f. Thoracodorsal a.
9. Anterior circumflex humeral a.
10. Posterior circumflex humeral a.
11. Brachial a.
12. Profunda brachii a.

Pectoralis minor m.

(B)

Ventral rami

C5
C6
C7
C8
T1

1. Superior trunk
2. Middle trunk
3. Inferior trunk
4. Lateral cord
5. Posterior cord
6. Medial cord
7. Axillary n.
8. Musculocutaneous n.
9. Median n.
10. Ulnar n.
11. Medial cutaneous n. of forearm
12. Medial cutaneous n. of arm
13. Radial n.

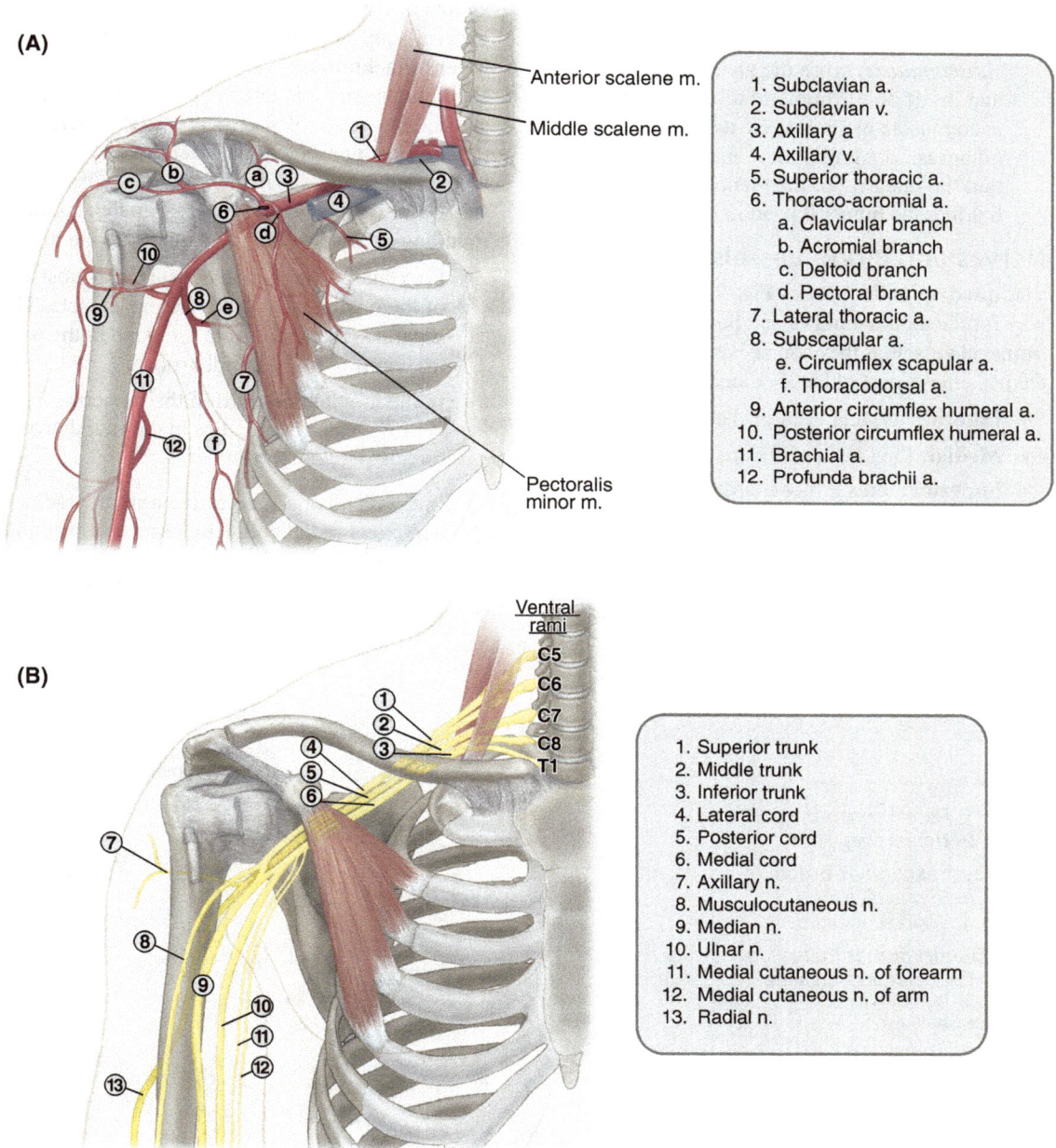

FIGURE 9.1.3 Anterior view of the right shoulder showing the subclavian vessels (**A**) and brachial plexus (**B**) and their relationship to the anterior and middle scalene muscles, pectoralis minor muscle, the clavicle, and rib 1.

- The tendons of the **rotator cuff muscles** (*supraspinatus, infraspinatus, teres major, subscapularis*) cross the glenohumeral joint and insert on the humeral tubercles. While each muscle has a discrete action on the humerus, as a group these muscles help maintain the integrity of the glenohumeral joint by holding the humeral head in the glenoid cavity.

Nerves and Blood Vessels

The **quadrangular space** (Fig. 9.1.4) is a passageway for the **axillary nerve** and **posterior circumflex humeral** vessels as they course between the axilla and the posterior aspect of the shoulder. Its borders are:

- **Lateral:** surgical neck of the humerus
- **Medial:** long head of triceps brachii muscle
- **Superior:** teres minor muscle
- **Inferior:** teres major muscle

The **radial nerve** and **profunda brachii artery** (deep artery of the arm) pass from the axilla to the posterior aspect of the humerus through the triangular **triceps hiatus** (sometimes called the triangular interval), an interval between the long head of the triceps brachii muscle and the humeral shaft (some sources consider this interval to be between the long and lateral heads of the triceps). (**Fig. 9.1.4**). The radial nerve and profunda brachii artery then enter in the radial groove of the humerus. The relationship between this neurovascular bundle and the surrounding structures changes from proximal to distal:

- Proximally, they lie on the superior-most fibers of the medial head of the triceps muscle. This separates the nerve from the bone in the proximal portion of the spiral groove.

- In the middle one-third of the humerus, the neurovascular bundle lies in the radial groove directly against the humerus.

- In the distal one-third of the arm, the radial nerve and radial collateral branch of profunda brachii pierce the lateral intermuscular septum. They pass to the anterior compartment as they cross the elbow joint.

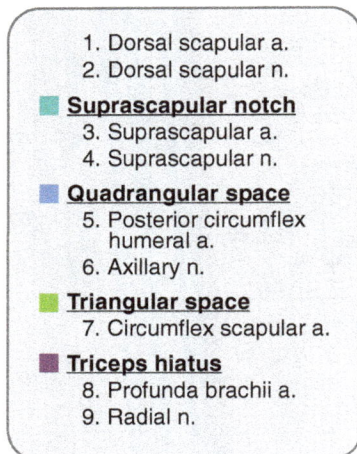

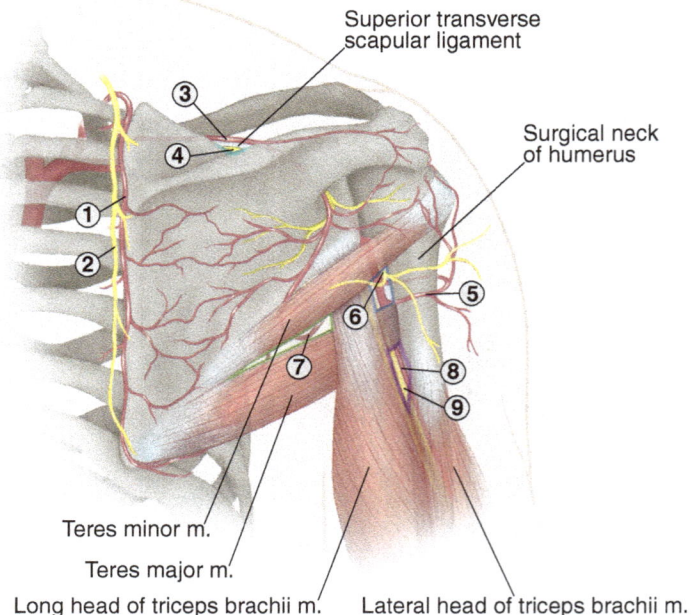

1. Dorsal scapular a.
2. Dorsal scapular n.

Suprascapular notch
3. Suprascapular a.
4. Suprascapular n.

Quadrangular space
5. Posterior circumflex humeral a.
6. Axillary n.

Triangular space
7. Circumflex scapular a.

Triceps hiatus
8. Profunda brachii a.
9. Radial n.

Superior transverse scapular ligament

Surgical neck of humerus

Teres minor m.
Teres major m.
Long head of triceps brachii m.   Lateral head of triceps brachii m.

FIGURE 9.1.4 Posterior view showing the neurovascular tracts of the right shoulder and arm.

Clinical Notes

- The position of the axillary nerve in the quadrangular space places it at risk with fracture of the surgical neck of the humerus or shoulder dislocation. Injury of the axillary nerve can lead to weakness of the deltoid muscle (manifest as an inability to fully abduct the arm) and diminished sensation over the deltoid region.

- The close proximity of the radial nerve to the posterior aspect of the humerus places it at risk with humeral shaft fractures, especially distally where the nerve lies directly against the bone.

CLINICAL REASONING

This patient presents with signs and symptoms of traumatic shoulder injury.

Anterior Shoulder Dislocation

Disruption of the glenohumeral joint, with the humeral head subluxated from the glenoid cavity, is commonly referred to as a shoulder dislocation. The joint is inherently unstable because of the size disparity between the humeral head and the glenoid cavity ("golf ball on the tee").

Signs and Symptoms

Sensory Deficits

- Severe shoulder pain
- Diminished sensation over deltoid region

Motor Deficits

- Restricted range of movement
- Arm held in slight abduction with external rotation
- Disruption of contour over deltoid and humeral head; acromion may be more prominent
- Weakness of the deltoid muscle

Subluxation Partial dislocation of a joint (*luxation* is a complete dislocation)

Predisposing Factors

- **Sex and age:** most common in males 20–30 years of age and women >60 years of age
- **Activities:** most common when arm is in abduction, extension, and external rotation and vulnerable to trauma

Clinical Note

95–98% of shoulder dislocations are anterior.

With anterior shoulder dislocation, the anterior portion of the glenohumeral joint capsule may be ruptured or detached. This often results from the forceful abduction, extension, and lateral rotation of the arm. For example:

- When throwing a ball overhand
- When the upper limb is hit while throwing a football

In these circumstances, the capsule is stretched tightly across the anterior aspect of the joint, and the forces on the upper limb displace the humeral head anteriorly and inferiorly from the glenoid cavity (**Fig. 9.1.5**). The anteroinferior part of the glenoid labrum may also tear (Bankart lesion) because it is attached to glenohumeral ligaments.

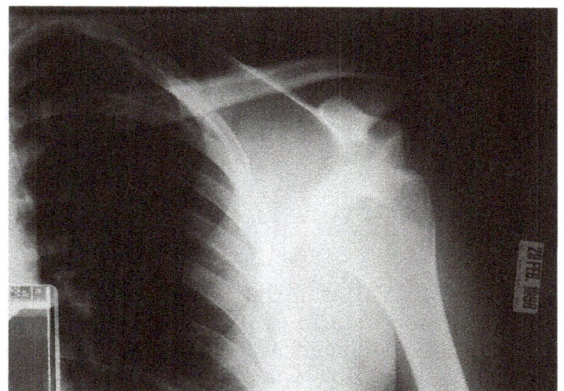

FIGURE 9.1.5 Radiograph showing shoulder dislocation, with the humeral head subluxated anterior and inferior to the glenoid cavity. *Source:* Fig. 11.5 in *The Atlas of Emergency Medicine*, 3e.

In older adults, the combination of a weakened joint capsule and supporting structures with a susceptibility to falls increases the incidence of shoulder dislocation.

Clavicle Fracture

Signs and Symptoms

Sensory Deficit

- Localized pain over clavicle (commonly in the middle ⅓), exacerbated by arm movement

Motor Deficit

- Shoulder depressed with arm medially rotated and adducted

Other Deficits

- Localized swelling and ecchymosis over clavicle
- Skin may be "tented" or penetrated by fracture
- Possible decreased peripheral perfusion, and/or brachial plexus injury, pneumothorax, hemothorax

Predisposing Factors

- Sex: more common in males, usually between 13 and 20 years of age; incidence increases in the seventh decade

Fractures of the clavicle typically result from force applied to the lateral end of the clavicle, either directly (e.g., direct impact to the shoulder from a fall or motor vehicle accident) or indirectly (e.g., falling on an outstretched hand). The patient usually recalls hearing a snapping sound at the time of injury. Because the clavicle is subcutaneous, the fracture edges may be visible, and localized ecchymosis and swelling may be present.

Clinical Notes

- Most (70–80%) fractures of the clavicle occur via a direct trauma to the shoulder.
- The clavicle fractures most commonly in its middle third, medial to the CC ligament, where the bone is narrowest and lacks supporting ligaments.

- Since the clavicle is the only osseous connection between the upper limb and the trunk, the majority of force transmitted to the trunk from the upper limb passes along the clavicle to the SC joint. The anatomy of the clavicle (i.e., size, shape, and position) makes it the "weakest link" in the chain.

Physical examination often reveals displacement of the fractured portions of the clavicle (**Fig. 9.1.6**):

- The **proximal (sternal) part** is displaced superiorly (relative to the distal fragment) due to the upward pull of the sternocleidomastoid muscle, whose clavicular head inserts on the medial third of the clavicle.

- The **distal (acromial) part** is displaced inferiorly because the weight of the arm combined with contraction of the latissimus dorsi and pectoralis major muscles pull the arm downward. The trapezius muscle is inadequate to counterbalance these forces.

To maximize comfort and minimize pain, patients typically present with the affected limb internally rotated and adducted, and supported by the uninjured extremity.

Clinical Note

The proximity of the clavicle to the subclavian vessels, brachial plexus, and pleural cavities makes it important to assess the upper limb for vascular perfusion, sensation, muscle strength, as well as breath sounds to rule out pneumothorax and/or hemothorax.

Shoulder Separation

Disruption of the AC joint is commonly referred to as shoulder separation.

Signs and Symptoms

Sensory Deficit

- Focal tenderness and swelling

Pneumothorax Air or gas in pleural cavity

Hemothorax Blood in pleural cavity

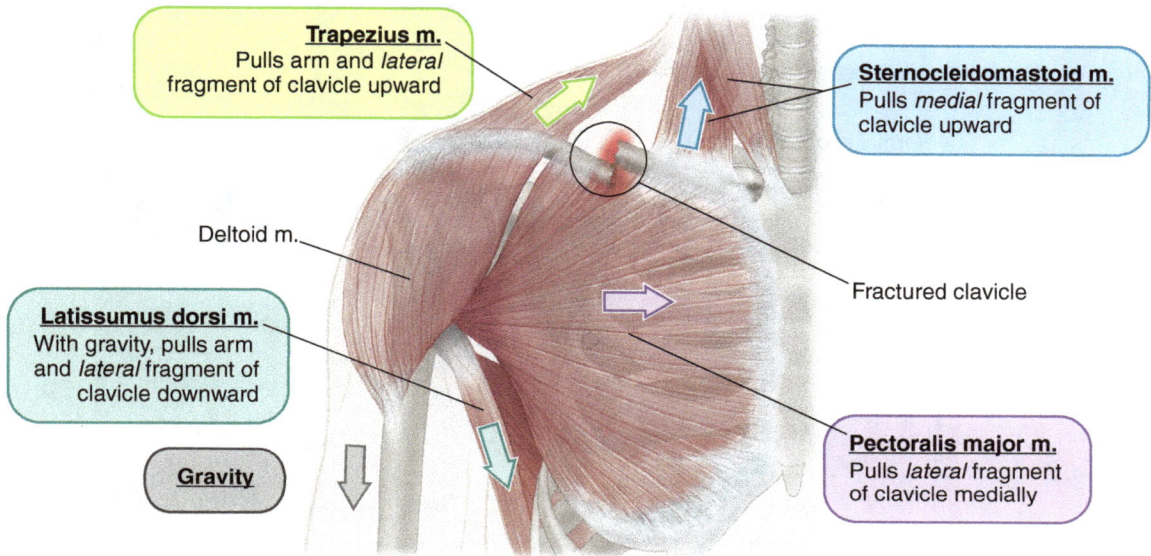

FIGURE 9.1.6 Muscle forces acting on the shoulder after midclavicular fracture.

Motor Deficit

- Restricted range of shoulder movement

Other Deficit

- With more severe injury, palpable deformity of the AC joint

Clinical Note

- The most common cause of shoulder separation is direct trauma to the shoulder, usually from a fall. AC joint injuries are classified according to the integrity of the AC and CC ligaments and the relative positions of the clavicle and acromion (**Table 9.1.1**).
- With severe separation (Grades IV-VI), the scapula moves inferiorly due to the weight of the arm and creates a "shelf" over the shoulder (**Fig. 9.1.7**).

Predisposing Factors

- **Sex and age:** more common in males, usually in the third decade

DIAGNOSIS

The patient presentation, medical history, physical examination, and imaging studies support a diagnosis of **midclavicular fracture**.

Midclavicular Fracture

Signs and symptoms of midclavicle fractures are related to disruption of its function as a strut between the manubrium of the sternum (axial skeleton) and the glenohumeral joint (appendicular skeleton), and

TABLE 9.1.1	Grades of shoulder separation.
Grade	**Structures disrupted or torn**
Grade I	AC and CC ligaments sprained (stretched or partially torn)
Grade II	AC ligament disrupted (completely torn), CC ligament sprained
Grade III	AC and CC ligaments disrupted
Grade IV–VI	AC and CC ligaments disrupted; clavicle displaced from acromion at increasing distances and/or directions

Torn coracoclavicular ligament

Torn fibrous capsule of acromioclavicular joint

Normal shoulder contour

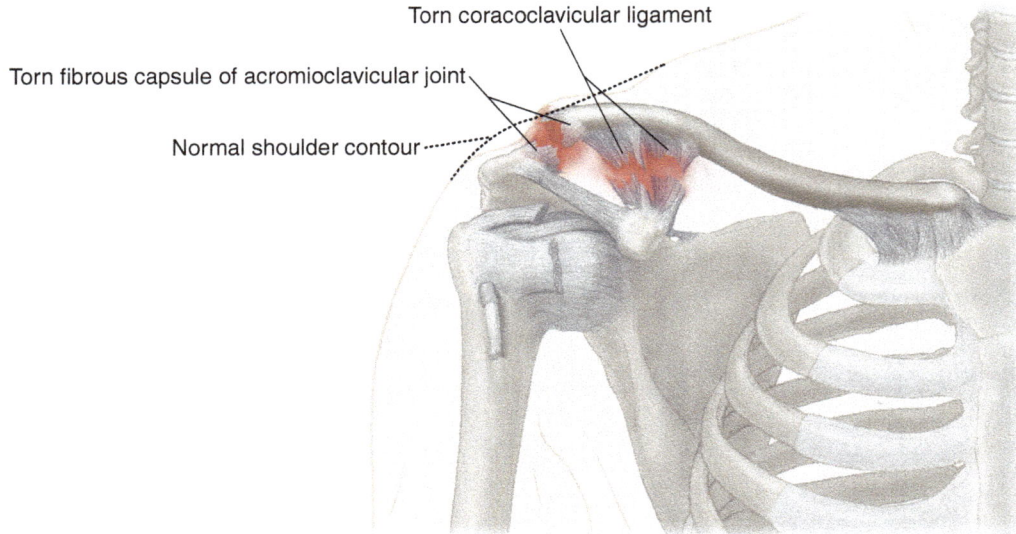

FIGURE 9.1.7 Grade IV shoulder separation.

to possible injury of anatomically related structures (e.g., subclavian artery and brachial plexus).

- The obvious deformity over the middle one-third of the clavicle and the displacement of the proximal and distal fragments is diagnostic of this type of fracture.

Shoulder Dislocation

Shoulder dislocation involves anatomical disruption of the glenohumeral joint. This may include its stabilizing ligaments or the rotator cuff muscles.

- If the patient had a shoulder dislocation, it would be characterized by shoulder pain and restricted range of movement.

- The arm would be abducted slightly and externally rotated, and there would be deltoid weakness and diminished sensation from skin over the deltoid.

- The rounded contour of the deltoid is lost and the humeral head is palpable because the bone is subluxated from the glenoid cavity.

Shoulder Separation

Shoulder separation involves anatomical disruption of the stabilizing ligaments of the AC joint.

- Shoulder separation is characterized by pain and tenderness over the SC joint and limited range of movement.

- In severe cases, a "shelf" may be apparent over the shoulder where the SC joint has been disrupted, with the weight of the limb causing the acromion to be displaced inferior to the distal end of the clavicle.

Patient Presentation

A 63-year-old white female is admitted to the emergency department with acute pain and swelling in her distal left forearm after falling onto her outstretched hand.

Relevant Clinical Findings

History

The patient is postmenopausal. Prior bone mineral density measurement of the forearm and femoral neck indicated osteoporosis.

Physical Examination

The following findings were noted on physical examination of the left upper limb:

- Deformity of the distal forearm, with posterior displacement of the distal radius and wrist
- Acute tenderness and swelling of the distal forearm and wrist

- Paresthesia over the area of median nerve distribution in the hand
- Weakness of thumb opposition
- Radial and ulnar pulses are normal

Imaging Studies

- Posterior–anterior and lateral radiographs of the left distal forearm and hand (**Fig. 9.2.1**) revealed a fracture of the distal radius, with posterior displacement of the distal fracture fragment.
- MRI showed posterior displacement of the median nerve.

Clinical Problems to Consider

- Colles fracture
- Scaphoid fracture
- Smith fracture

LEARNING OBJECTIVES

1. Describe the structure and relationships of the distal forearm and wrist.
2. Describe the nerve supply of the distal forearm and hand.
3. Explain the anatomical basis for the signs and symptoms associated with this case.

RELEVANT ANATOMY

Distal Forearm and Wrist

Distal Forearm

The distal ends of the radius and ulna are located in the distal forearm (**Fig. 9.2.2**). The **radius** has an expanded distal end, which includes a styloid process. The **brachioradialis muscle** is inserted on this process. The distal end of the **ulna** is narrowed, but also includes a prominent styloid process. The distal

---

Postmenopausal A period in female life following menopause

Paresthesia Numbness and tingling

(A)

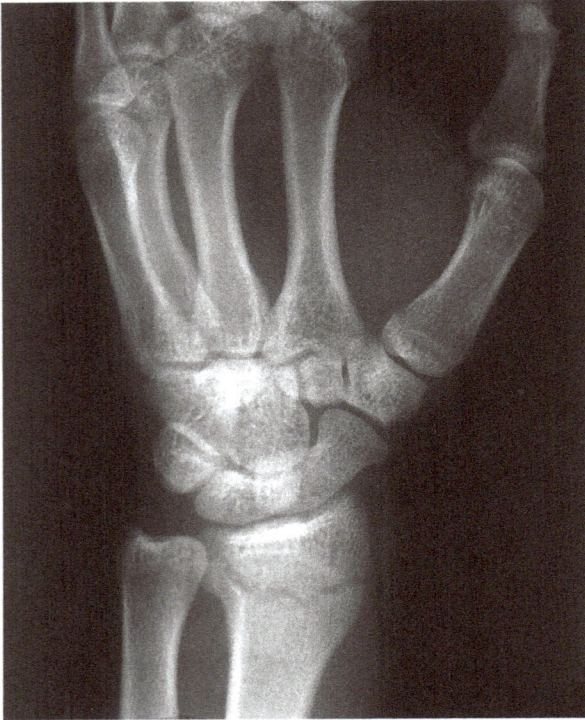

(B)

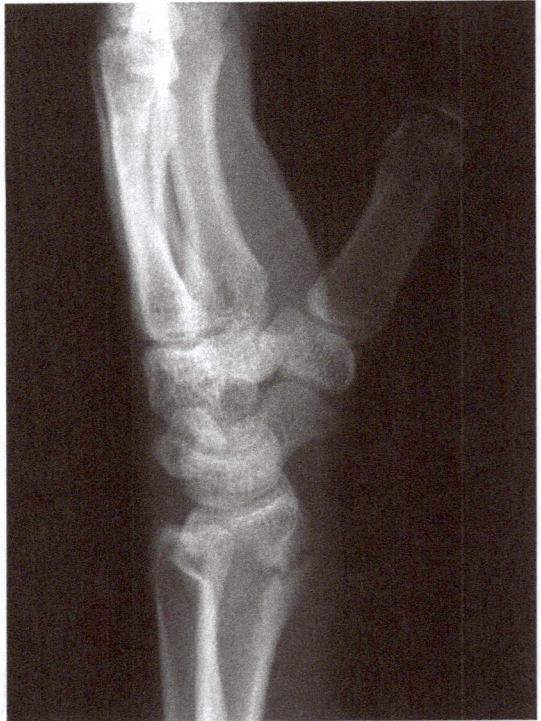

FIGURE 9.2.1 Posterior–anterior and lateral and radiographs of the left distal forearm and hand showing a distal fracture of the radius. *Source:* Fig. 266-17. *Colles' fracture.* Posteroanterior (A) and lateral (B) views of the ... *Tintinalli's Emergency Medicine,* Chapter 266.

ends of these bones articulate at the **distal radioulnar joint**, a synovial joint that permits the radius to move across the ulna during pronation/supination movements.

Wrist

The **wrist** (carpus) is the most proximal part of the hand. It includes eight **carpal bones** that are arrayed in two rows. From lateral-to-medial, these are:

1. **Proximal row:** scaphoid–lunate–triquetrum–pisiform
2. **Distal row:** trapezium–trapezoid–capitate–hamate

The **radiocarpal (wrist) joint** is formed by the closely apposed surfaces of the distal radius and scaphoid bones. To a lesser extent, the lunate also contributes to this articulation. The distal end of the ulna is separated from the lunate and triquetrum by an articular disc.

Carpal Tunnel

The **carpal tunnel** is an osseofibrous passageway formed by the concave anterior surfaces of the carpal bones and the flexor retinaculum, also known as the transverse carpal ligament (**Fig. 9.2.3**). The flexor retinaculum holds the contents of the carpal tunnel in place during wrist flexion, preventing "bowstringing" of the tendons. The **long digital flexor tendons** and the **median nerve** enter the

(A) Anterior view

(B) Posterior view

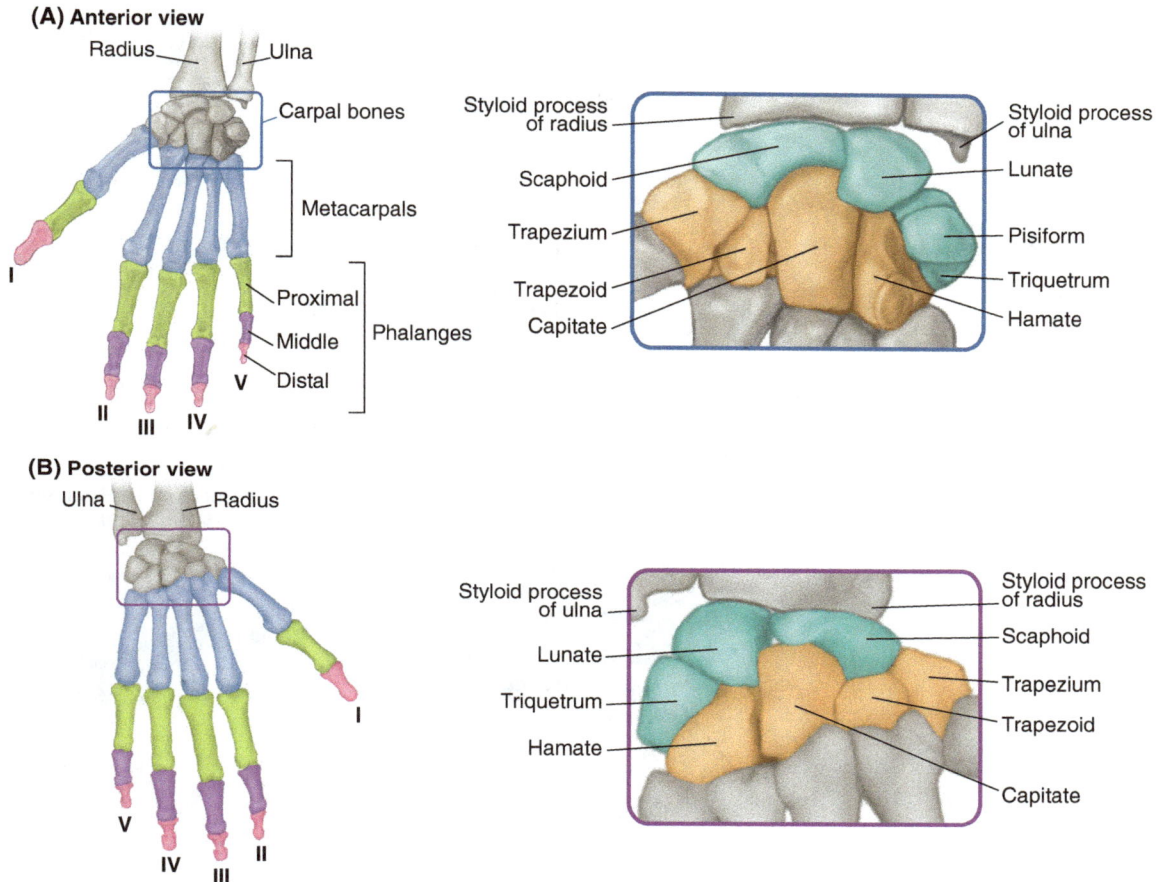

FIGURE 9.2.2 Anterior (**A**) and posterior (**B**) views of distal forearm and hand.

hand via the carpal tunnel. The median nerve lies immediately deep to the flexor retinaculum. Distal to the carpal tunnel, the nerve supplies the thenar muscles (recurrent branch), the lateral two lumbrical muscles, and the lateral palmar and digital skin.

Posterior and Lateral Aspects of the Wrist and Hand

Tendons on the posterior and lateral aspects of the wrist are subcutaneous and their positions are maintained by the **extensor retinaculum**. The **anatomical snuffbox** is a shallow depression on the lateral wrist, bounded by three tendons that attach to the thumb (**Fig. 9.2.4**):

Anterior

1. Abductor pollicis longus
2. Extensor pollicis brevis

Posterior

3. Extensor pollicis longus

The snuffbox is most obvious when the thumb is fully extended. Branches of the superficial radial nerve supply the overlying skin. Palpable structures in the snuffbox include:

- Radial artery
- Scaphoid and trapezium in the floor

(A)

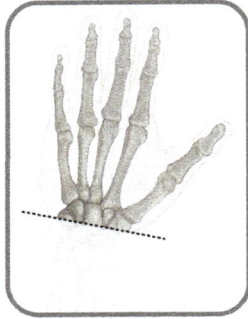

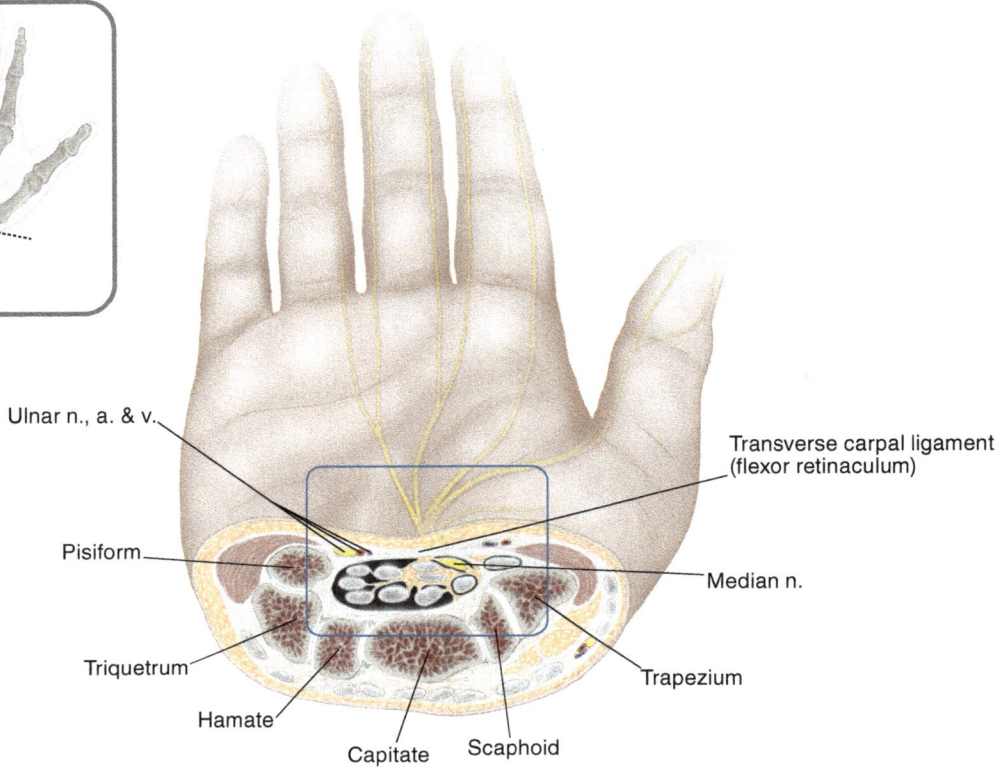

Ulnar n., a. & v.

Transverse carpal ligament (flexor retinaculum)

Pisiform

Median n.

Triquetrum

Trapezium

Hamate

Capitate   Scaphoid

(B)

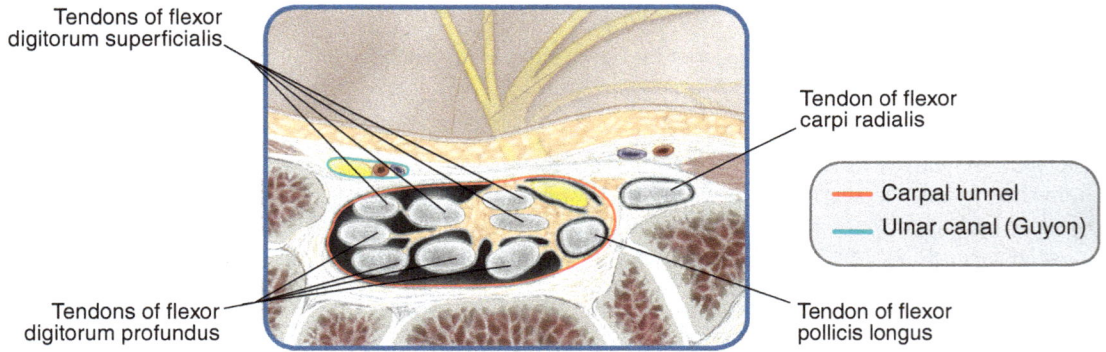

Tendons of flexor digitorum superficialis

Tendon of flexor carpi radialis

Carpal tunnel
Ulnar canal (Guyon)

Tendons of flexor digitorum profundus

Tendon of flexor pollicis longus

FIGURE 9.2.3 (**A**) Carpal tunnel. (**B**) Cross section of carpal tunnel (box in A is shown enlarged in B).

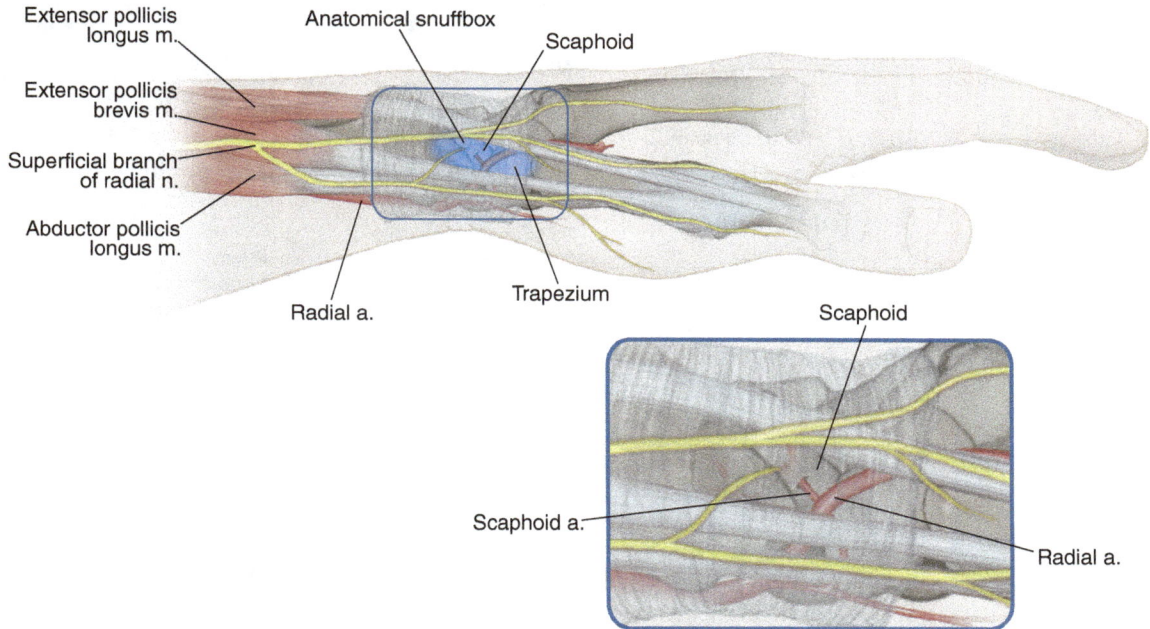

FIGURE 9.2.4 Anatomical snuffbox.

- Radial styloid process proximally
- Base of the first metacarpal distally

Median Nerve in the Forearm and Hand

The median nerve supplies structures in the anterior forearm and palmar hand (**Table 9.2.1**). At the elbow, it crosses the cubital fossa and then passes between the two heads of **pronator teres**. The nerve courses distally between **flexor digitorum superficialis** and **flexor digitorum profundus**. It gives off the **anterior interosseous nerve** to supply deep anterior forearm muscles. The **palmar cutaneous branch** arises in the distal forearm and enters the hand superficial to the carpal tunnel to supply the proximal lateral palm. The median nerve lies superficial and lateral within the carpal tunnel (**Fig. 9.2.3**). As the nerve enters the palm, it gives off a **recurrent branch** to supply the thenar muscles; other motor branches dis-

tribute to the first and second lumbricals. **Palmar digital** nerves are sensory branches that supply the distal lateral palm and the lateral 3 fingers, including the dorsum distal to the distal interphalangeal joint.

CLINICAL REASONING

This patient presents with signs and symptoms indicating a **fracture of the distal radius and/or wrist**.

Colles Fracture

Colles fracture involves the distal radius and usually results from a fall onto the outstretched, that is, extended hand (FOOSH). Described by Dr. Abraham Colles in 1814, it is defined as a displaced fracture of the distal two centimeters of the radius (**Fig. 9.2.5A**). The median nerve is commonly compressed by a FOOSH, and may be further compromised by

TABLE 9.2.1 **Distribution of the median and ulnar nerves.**

Spinal levels	Sensory	Motor
Median nerve C(5)6-T1	**Palmar cutaneous branch** • Proximal lateral palm	**Forearm** • Flexor–pronator muscles
	Palmar digital branches • Distal lateral palm • Lateral 3½ fingers	*Except flexor carpi ulnaris and ulnar part of flexor digitorum profundus*
		Hand • Recurrent branch: thenar muscles • Palmar branches: lumbricals 1 and 2
Ulnar nerve C8-T1	**Palmar cutaneous branch** • Proximal medial palm	**Forearm** • Flexor carpi ulnaris • Ulnar part of flexor digitorum profundus
	Dorsal branch • Medial dorsum of hand	**Hand** • Deep branch: intrinsic hand muscles
	Superficial branch • Distal medial palm • Medial 1½ fingers	Hypothenar muscles Lumbricals 3 and 4 All interossei Adductor pollicis Deep part of flexor pollicis brevis

subsequent swelling leading to acute carpal tunnel syndrome.

Signs and Symptoms

Sensory Deficits
- Pain and swelling of the wrist
- May include sensory changes (e.g., paresthesia, hypesthesia) from distribution of median nerve in the hand

Motor Deficit
- Dysfunction of intrinsic hand muscles (thenar muscles and the first two lumbricals) innervated by the median nerve

Other Deficits
- "Dinner fork" deformity of the distal forearm, that is, the distal fracture fragment displaced posteriorly
- Involvement of the radial or ulnar arteries may impair vascular supply to the hand

Predisposing Factors

- Age: simple FOOSH in adults >50 years of age more likely to result in fracture
- Sex: higher incidence in women due to osteoporosis

- Sports: activities with high likelihood of forward falls (e.g., skiing, rollerblading, skateboarding, and running)
- Osteoporosis

Clinical Notes

Distal radius fractures occur in two broad groups of patients:

1. More than 50 years of age, due to low-impact trauma to osteoporotic bone

2. Young adults, due to high-impact trauma (e.g., sports)

Scaphoid Fracture

The scaphoid is the most frequently fractured carpal bone (**Fig. 9.2.5B**). Approximately 60% of scaphoid fractures occur in the middle ("waist") of the bone. These fractures may result from a FOOSH or from a direct blow to the palm.

Signs and Symptoms

Sensory Deficit
- Pain and swelling along lateral aspect of wrist, with maximal tenderness in anatomical snuffbox

(A) Colles fracture ("Dinner Fork" deformity)

Radius

(B) Smith's fracture ("Garden Spade" deformity)

Radius

(C) Scaphoid fracture

Scaphoid   Trapezium

Radial a.   Branch to
            scaphoid

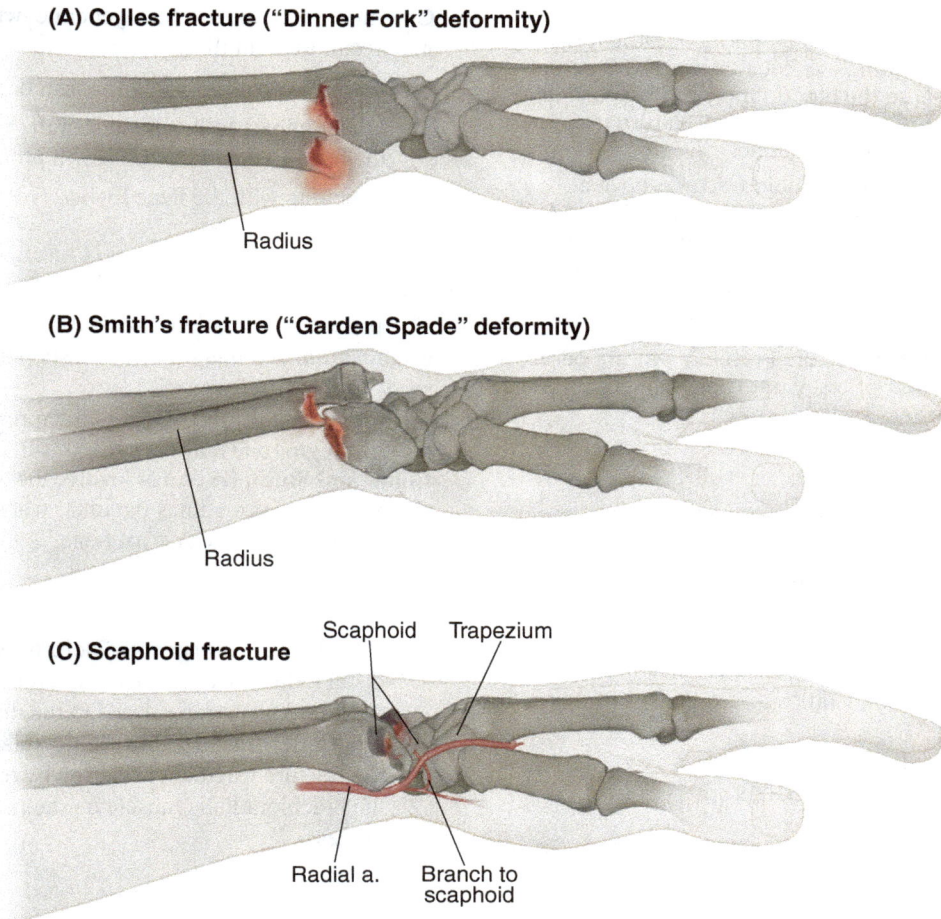

FIGURE 9.2.5 Fractures of the distal forearm and wrist. (**A**) Colles fracture, (**B**) Smith fracture, and (**C**) scaphoid fracture.

Motor Deficit

- Abduction (radial deviation) or hyperextension of wrist may elicit pain

Predisposing Factors

- Age: 20–30 years of age, but may occur at any age
- Sex: more common in males
- Most often results from a sports injury or motor vehicle accident

The likely fracture mechanism is that wrist extension with FOOSH causes the scaphoid to be com-pressed between the radial styloid process and adjacent carpal bones.

Clinical Notes

- Initial radiographs may fail to reveal a scaphoid fracture, which is sometimes mis-diagnosed as a severe wrist sprain. The fracture may become obvious in radiographs 1–2 weeks later as the proximal fragment

---

Sprain Excessive stretching of a ligament; may include partial tearing

has reduced radio-opacity because it is undergoing avascular necrosis. This occurs because the blood supply to the proximal fragment depends on vasculature from the distal end of the bone.

- Misaligned scaphoid fractures have a nonunion rate of 60–90%, and this leads in almost all cases to avascular necrosis of the proximal fragment. In this case, a vascularized bone graft (i.e., with its own blood supply) is an effective treatment.

- Scaphoid fracture in children is rare because it does not ossify until approximately 6 years of age.

Smith Fracture

A Smith fracture, also known as a reverse Colles fracture, is a flexion fracture of the distal radius. It usually results from a FOOSH but with the wrist in flexion. Though relatively uncommon compared to Colles fractures, it also results in a displaced fracture of the distal two centimeters of the radius (**Fig. 9.2.5C**).

Signs and Symptoms

Sensory Deficit
- Pain and swelling around wrist

Other Deficit
- "Garden spade" deformity of the distal forearm, that is, volar displacement of distal fracture fragments

Predisposing Factors

- Age: >50 years
- Sex: higher incidence in women due to osteoporosis

Volar Related to the palm of the hand or sole of the foot

Common mechanisms that generate wrist flexion may result in a Smith fracture:

- Falling onto a supinated forearm or hand
- Direct impact to the dorsum of the wrist, with the hand in flexion and forearm pronated
- Punching with the flexed wrist

DIAGNOSIS

The patient presentation, medical history, physical examination, and imaging studies support a diagnosis of **Colles fracture**.

Colles, Smith, and scaphoid fractures all involve the wrist region and may occur as a result of **FOOSH**. **Colles and Smith fractures** involve the distal end of the radius, and sometimes the ulna, whereas **scaphoid fracture** involves a carpal bone.

Colles Fracture

- The obvious "dinner fork" deformity of the distal radius, with the distal fracture fragment displaced posteriorly, leads to the diagnosis of a Colles fracture. Posterior displacement of the fracture fragment is caused by traction of the brachioradialis muscle on the radial styloid process.

Smith Fracture

- Volar displacement of the fracture fragment ("garden spade") would indicate a Smith fracture.

Scaphoid Fracture

- A scaphoid fracture is distinguished by maximal point of tenderness in the anatomical snuffbox.
- The likely fracture mechanism is that wrist extension with FOOSH causes the scaphoid to be compressed between the radial styloid process and adjacent carpal bones.

Upper Brachial Plexus (Erb) Palsy

Patient Presentation

A 35-year-old multiparous Latina was in labor for 20 hours and, after a difficult vaginal delivery, gave birth to a 9.9 lb baby girl. Postpartum evaluation of the baby revealed a limp right upper limb.

Clinical Note

The average newborn is approximately 51 cm in length and weighs 5.9–8.8 lb.

Relevant Clinical Findings

History

The mother had two previous pregnancies that produced large babies. During this delivery, the baby had shoulder dystocia and forceps was used (**Fig. 9.3.1**). Bruising and swelling of the right upper extremity was observed.

Physical Examination

Noteworthy vital signs:

- Respiratory rate: 19 cycles/min (Normal adult: 14–18 cycles/min; women slightly higher)

The following findings were noted during the neonatal examination.

- Flail right upper extremity, with arm adducted and medially rotated, elbow extended, forearm pronated, and wrist flexed (**Fig. 9.3.2**)

- Biceps tendon and Moro reflexes absent on the affected side
- Hypesthesia from the lateral aspect of the affected limb

Clinical Notes

The **Moro reflex** is a "startle" reflex that is present at birth; it diminishes in the third or fourth month, and usually disappears by the sixth month.

- From a partially reclined position, the supported head and trunk are allowed to "fall" to approximately 30°. The infant will abduct and extend the arms, forearms supinated, the fingers abducted, and the thumbs and index fingers flexed. The infant may exhibit a "startled" look. At the end of the reflex, the arms will be adducted and the elbows flexed, drawing the upper limbs across the body. The infant will then relax.

  Unilateral absence of the Moro reflex in a newborn suggests the possibility of a clavicle fracture, shoulder dislocation, or upper brachial plexus (Erb) palsy.

Clinical Problems to Consider

- Upper brachial plexus (Erb) palsy
- Lower brachial plexus (Klumpke) palsy

Dystocia Difficult birth
Flail A joint with impaired or absent function
Hypesthesia Diminished sensation

Palsy Paralysis (loss of voluntary motor function; may be caused by disease or injury) or paresis (partial or incomplete paralysis)

1. Describe the anatomy of the brachial plexus.
2. Describe the sensory and motor innervation of the upper limb.
3. Explain the anatomical basis for the signs and symptoms associated with this case.

RELEVANT ANATOMY

Brachial Plexus

The brachial plexus is formed by ventral rami of the C5-T1 spinal nerves. It extends through the neck and axilla to supply the upper limb. It also receives post-ganglionic sympathetic fibers from the middle and inferior cervical ganglia that supply sweat glands, arrector pili muscles, and blood vessels.

The brachial plexus can be divided into **supraclavicular** and **infraclavicular parts** (**Fig 9.3.3**).

The supraclavicular part includes:

- **Roots:** C5-T1 ventral rami
- **Trunks:** superior, middle, and inferior
- **Supraclavicular branches**

The infraclavicular part includes:

- **Divisions:** anterior and posterior
- **Cords:** lateral, medial, and posterior
- **Infraclavicular branches**
- **Terminal branches**

The supraclavicular, infraclavicular, and terminal branches of the plexus innervate the upper limb (**Table 9.3.1**).

In general, nerves derived from C5-C7 innervate muscles in the proximal part of the upper limb (i.e., shoulder and arm), whereas those that are derived from C8-T1 supply muscles in the distal part (i.e., forearm and hand).

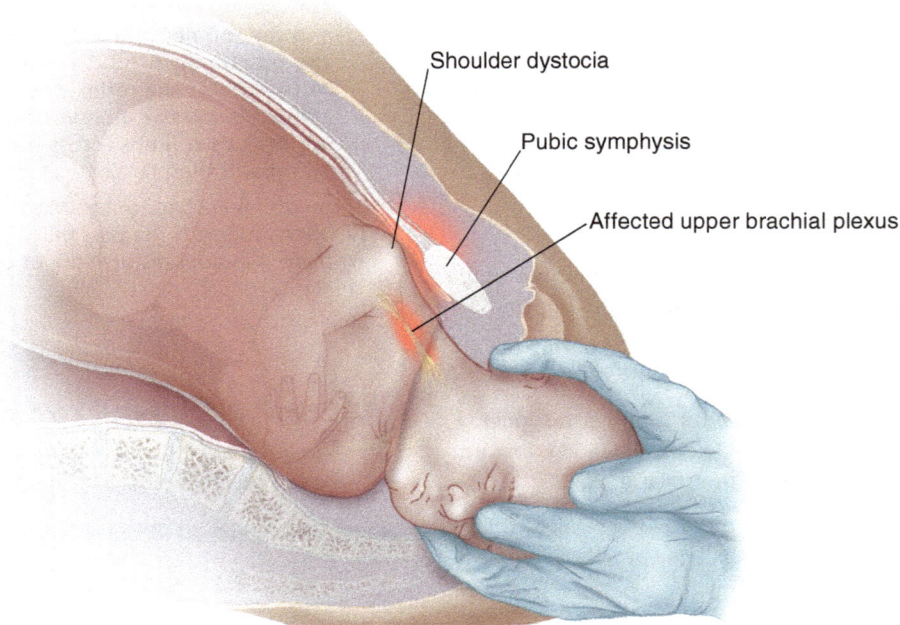

FIGURE 9.3.1 Shoulder dystocia during delivery. The right shoulder is impeded in the birth canal by the pubic symphysis. This can result in obstetric palsy of the upper brachial plexus.

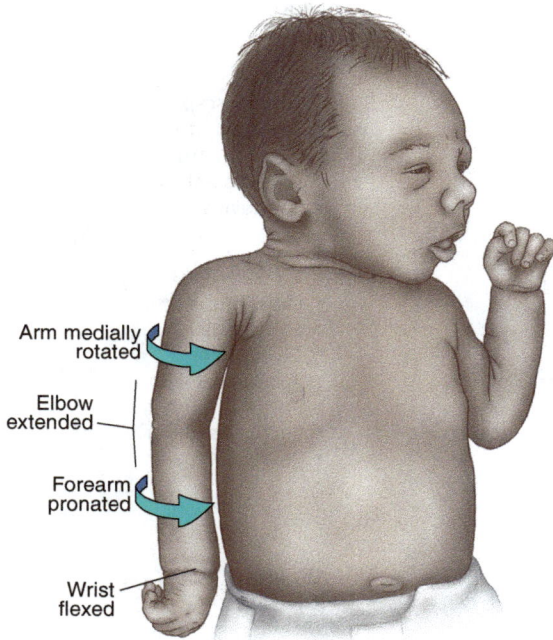

Arm medially rotated

Elbow extended

Forearm pronated

Wrist flexed

FIGURE 9.3.2 Infant with flail upper limb.

CLINICAL REASONING

This patient presents with signs and symptoms of traumatic brachial plexus injury. When the injury results from childbirth, this is also called **obstetric brachial plexus palsy** (**OBPP**).

The two forms of OBPP are related to different parts of the proximal brachial plexus:

1. **Upper brachial plexus palsy** (Erb or Erb–Duchenne) involves the **C5 and C6 roots** (sometimes C7).

2. **Lower brachial plexus palsy** (Klumpke) involves the **C8 and T1 roots**.

The relationship between the spinal levels that contribute to specific parts of the brachial plexus and its motor and sensory distribution is foundational to understanding the signs and symptoms associated with OBPP (**Table 9.3.2**).

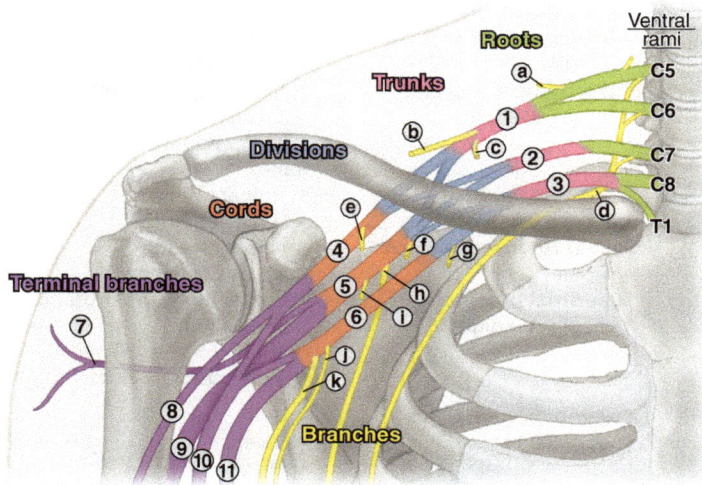

Trunks
1. Superior
2. Middle
3. Inferior

Supraclavicular branches
a. Dorsal scapular n.
b. Suprascapular n.
c. Subclavian n.
d. Long thoracic n.

Cords
4. Lateral
5. Posterior
6. Medial

Infraclavicular branches
e. Lateral pectoral n.
f. Upper subscapular n.
g. Medial pectoral n.
h. Thoracodorsal n.
i. Lower subscapular n.
j. Medial cutaneous n. of arm
k. Medial cutaneous n. of forearm

Terminal branches
7. Axillary n.
8. Musculocutaneous n.
9. Radial n.
10. Median n.
11. Ulnar n.

FIGURE 9.3.3 Anterior view of brachial plexus and its relationship to the clavicle.

TABLE 9.3.1 Spinal levels and distribution of brachial plexus.

Nerve	Spinal level(s)[a]	Sensory	Motor
Supraclavicular branches			
Dorsal scapular	(C4) **C5**		▪ Rhomboids ▪ Levator scapulae
Suprascapular	(C4) **C5-C6**	▪ Shoulder joint	▪ Supraspinatus ▪ Infraspinatus
Long thoracic	**C5-C6**, C7		▪ Serratus anterior
Subclavian	(C4) C5-**C6**		▪ Subclavius
Infraclavicular branches			
Upper subscapular	C5		▪ Subscapularis
Thoracodorsal	C6, **C7**, C8		▪ Latissimus dorsi
Lower subscapular	C6		▪ Subscapularis ▪ Teres major
Lateral pectoral	C5, **C6**, C7	▪ Shoulder joint	▪ Pectoralis major ▪ Pectoralis minor
Medial pectoral	C8-T1		▪ Pectoralis minor ▪ Pectoralis major
Medial cutaneous of arm	C8-T1	▪ Medial arm—distal ⅓	
Medial cutaneous of forearm	C8-T1	▪ Medial forearm	
Terminal branches			
Axillary	**C5-C6**	▪ Skin over deltoid ▪ Shoulder joint	**Shoulder** ▪ Deltoid ▪ Teres minor
Musculocutaneous	C5-C7	**Lateral cutaneous nerve of forearm** ▪ Lateral forearm	**Arm** ▪ Coracobrachialis ▪ Biceps brachii ▪ Brachialis
Median	(C5) C6-T1	**Palmar cutaneous branch** ▪ Lateral palm **Palmar digital branches** ▪ Palmar aspects of lateral 3½ digits ▪ Distal dorsal aspect of lateral 3⅓ digits	**Anterior forearm** ▪ Pronators ▪ Flexor carpi radialis ▪ Flexor digitorum superficialis ▪ Flexor digitorum profundus (lateral ½) ▪ Palmaris longus **Hand** ▪ Thenar muscles ▪ Lumbricals 1–2
Ulnar	C8-T1	**Palmar cutaneous branch** ▪ Medial palm **Superficial branch** ▪ Medial 1½ digits	**Anterior forearm** ▪ Flexor carpi ulnaris ▪ Flexor digitorum profundus (medial ½) **Hand (deep branch)** ▪ Most intrinsic muscles
Radial	C5-T1	Posterior arm, forearm **Superficial branch** ▪ Lateral dorsal hand	**Arm** ▪ Triceps ▪ Anconeus ▪ (Brachialis) **Posterior forearm** ▪ All extensor muscles ▪ Supinator ▪ Brachioradialis ▪ Snuffbox muscles

[a]**Boldface** indicates the primary level(s) of contribution.

TABLE 9.3.2	Cervical nerve roots and potential motor and sensory deficits.	
Spinal level	**Motor deficit(s)**	**Sensory deficit(s)**
C4	▪ Shoulder elevation (trapezius, levator scapulae, rhomboids) ▪ Diaphragm	▪ Top of shoulder ▪ Diaphragmatic pleura and peritoneum
C5	▪ Shoulder abduction (deltoid, rotator cuff)	▪ Skin over deltoid
C6	▪ Shoulder abduction (deltoid, rotator cuff) ▪ Elbow flexion (biceps brachii, brachialis)	▪ Distal lateral forearm ▪ Thenar region
C7	▪ Elbow extension (triceps) ▪ Finger extension (extensor digitorum)	▪ Lateral palm ▪ Index and middle fingers
C8	▪ Thumb opposition (opponens pollicis)	▪ Medial distal forearm and palm ▪ Digits 4–5
T1	▪ Thumb adduction (adductor pollicis)	▪ Medial arm ▪ Proximal medial forearm

Upper Brachial Plexus (Erb) Palsy

Injury to the upper brachial plexus involves its **C5 and C6 roots**, the **superior trunk**, and their branches (**Table 9.3.1**). The C7 root, and its associated middle trunk, may be involved. This type of injury affects muscles and cutaneous areas supplied by these spinal cord levels (**Table 9.3.2**).

Signs and Symptoms

Sensory Deficit
- Lateral aspect of upper limb

Motor Deficit
- Loss of arm abduction and lateral rotation
- Loss of elbow flexion
- Loss of forearm supination
- Weakness of wrist extension

An upper brachial plexus palsy typically affects **intrinsic shoulder muscles** (deltoid, rotator cuff), **elbow flexors** (brachialis, biceps brachii, brachioradialis), **forearm supinators** (biceps brachii and supinator), and one strong **wrist extensor** (extensor carpi radialis longus) (**Table 9.3.3**). The unopposed actions of muscles result in the characteristic "waiter's tip" position of the limb (**Fig. 9.3.2**):

- Arm adducted and internally rotated
- Elbow extended
- Forearm pronated
- Wrist flexed

Predisposing Factors

Baby
- Macrosomia
- Shoulder dystocia

Birth Mother
- Gestational diabetes
- Small pelvis

Clinical Notes

- Many cases of OBPP are associated with traction used during birth, most commonly with **shoulder dystocia** (i.e., the shoulder is impeded in the birth canal by the pubic symphysis). Traction applied to assist the delivery may cause excessive lateral flexion of the neck away from the involved side. This widens the angle between the head and shoulder and stretches the roots of the upper brachial plexus.

- A similar traction injury can result from a hard fall that forcefully and suddenly increases the angle between the shoulder and neck (e.g., when an individual is thrown to the road from a motorcycle).

Macrosomia A newborn with excessive birth weight.

TABLE 9.3.3	Muscles affected by an upper brachial plexus injury.

Muscle	Nerve	Spinal levels[a]	Action(s) impaired
Deltoid	Axillary	**C5**, C6	Arm abduction and medial/lateral rotation
Supraspinatus	Suprascapular	**C5**, C6	Arm abduction
Infraspinatus	Suprascapular	**C5**, C6	Arm lateral rotation
Teres minor	Axillary	**C5**, C6	Arm lateral rotation
Subscapularis	Subscapular	**C5**, C6, C7	Arm medial rotation
Teres major	Lower subscapular	**C5**, C6, C7	Arm medial rotation and extension
Serratus anterior	Long thoracic	**C5**, C6, C7	Scapular retraction and upward rotation
Pectoralis major (clavicular head)	Lateral pectoral	**C5**, C6	Arm flexion and medial rotation
Coracobrachialis	Musculocutaneous	**C5**, C6, C7	Arm flexion
Biceps brachii	Musculocutaneous	**C5**, C6	Elbow flexion and forearm supination
Brachialis	Musculocutaneous	C5, **C6**, C7	Elbow flexion
Brachioradialis	Radial	**C5**, C6	Elbow flexion
Supinator	Radial (posterior interosseous)	**C6**, C7	Arm supination
Extensor carpi radialis longus	Radial	C5, **C6**	Wrist extension and abduction (radial deviation)

[a]**Boldface** indicates the primary level(s) of contribution.

- Impaired respiration with upper brachial plexus injury may indicate injury to the **phrenic nerve** because of the contributions of C4 or C5 to this nerve.

Lower Brachial Plexus (Klumpke) Palsy

A Klumpke palsy involves the C8 and T1 roots of the plexus and, thus, the inferior trunk and its branches (**Table 9.3.4**). This type of injury typically results in hand paralysis. It affects muscles in the forearm and hand that are supplied by nerves that contain primarily C8 and T1 fibers, that is, ulnar, median, and radial.

Signs and Symptoms

Sensory Deficits

- Medial aspect of upper limb (not including axillary skin)

Motor Deficits

- Loss of fine finger movements
- Weakened forearm pronation
- Weakened wrist and finger extension
- Weakened wrist and finger flexion

Other Deficits

- Horner syndrome may be associated

The unopposed actions of the affected muscles result in an upper limb position characterized primarily by a "claw hand" (**Fig. 9.3.4**).

Predisposing Factors

Baby

- Macrosomia
- Shoulder dystocia
- Vertex presentation

Paralysis Loss of muscle function, particularly related to voluntary movement

TABLE 9.3.4	Muscles affected by a lower brachial plexus injury.		
Muscle	**Nerve**	**Spinal levels**[a]	**Action(s) impaired**
Flexor carpi ulnaris	Ulnar	C7, **C8**	Wrist flexion and adduction (ulnar deviation)
Flexor digitorum superficialis	Median	C7, C8, T1	Wrist and finger flexion (MP and PIP joints)
Flexor digitorum profundus	Median Ulnar	**C8**, T1 C8, **T1**	Wrist and finger flexion (MP, PIP, and DIP joints)
Pronator quadratus	Median (anterior interosseous)	**C8**, T1	Pronation
Extensor digitorum	Radial	**C7**, C8	Extensor digitorum
Extensor indicis	Radial (posterior interosseous)	C7, **C8**	Extensor indicis
Thenar muscles	Median (recurrent branch)	**C8**, T1	Thumb movements
Adductor pollicis	Ulnar (deep)	C8, **T1**	Thumb adduction
Hypothenar muscles	Ulnar (deep)	C8, **T1**	Little finger movements
Dorsal interossei	Ulnar (deep)	C8, **T1**	Finger abduction
Palmar interossei	Ulnar (deep)	C8, **T1**	Finger adduction
Lumbricals	Median Ulnar (deep)	C8, **T1** C8, **T1**	Flexion of MP joint, extension of IP joints

[a]**Boldface** indicates the primary level(s) of contribution.

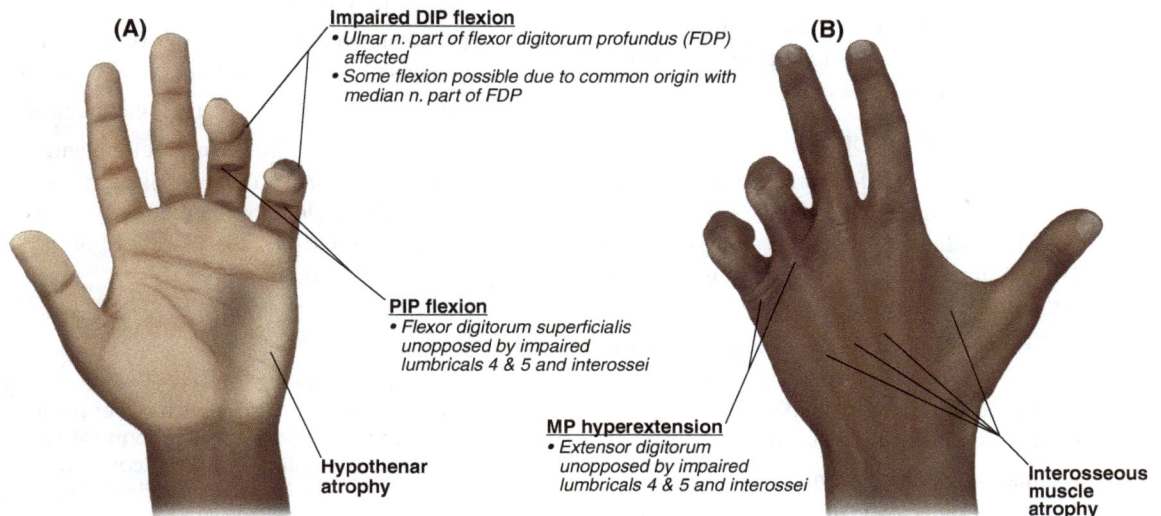

(A)

Impaired DIP flexion
• *Ulnar n. part of flexor digitorum profundus (FDP) affected*
• *Some flexion possible due to common origin with median n. part of FDP*

(B)

PIP flexion
• *Flexor digitorum superficialis unopposed by impaired lumbricals 4 & 5 and interossei*

Hypothenar atrophy

MP hyperextension
• *Extensor digitorum unopposed by impaired lumbricals 4 & 5 and interossei*

Interosseous muscle atrophy

FIGURE 9.3.4 Clinical signs of lower brachial plexus injury are related principally to ulnar nerve damage. They include "claw hand" (**A:** palmar view) and wasting of intrinsic hand muscles (**B:** dorsal view). Abbreviations: DIP, distal interphalangeal; PIP, proximal interphalangeal; MP, metacarpophalangeal.

Birth Mother
- Gestational diabetes
- Small pelvis

Damage to the T1 nerve root may also impact the cervicothoracic (stellate) ganglion or sympathetic chain. This may injure preganglionic sympathetic fibers related to the orbit and result in **Horner syndrome** (pupillary miosis, ptosis of the upper eyelid, reverse ptosis of the lower eyelid).

Clinical Notes

- Many cases of OBPP are related to traction during birth associated with **vertex presentation** (i.e., the occiput is first to enter the birth canal). Traction applied to the arm or head to assist delivery may narrow the angle between the head and arm and stretch the lower roots brachial plexus.

- A similar, non-obstetrical injury may occur when the upper limb is suddenly pulled superiorly, decreasing the angle between the shoulder and neck. This might occur, for example, when an individual grasps for support when falling.

DIAGNOSIS

The patient presentation, medical history, and physical examination support a diagnosis of **upper brachial plexus (Erb) palsy**.

Upper Brachial Plexus (Erb) Palsy

Damage to the upper part of the brachial plexus affects components that carry axons from the C5-C6(7) spinal levels. This affects dermatomes along the lateral aspect of the upper limb, as well as muscles in the shoulder, arm, and forearm.

- The primary clinical sign is "waiter's tip position," which is characterized by an adducted and internally rotated arm, extended elbow, pronated forearm, and a flexed wrist.

- Diminished sensation from the lateral aspect of the upper limb is also consistent with damage

to cutaneous nerves that contain C5-C6, and sometimes C7 axons. These include the lateral cutaneous nerve of the forearm (C6) and the median nerve and superficial branch of the radial nerve (C6-C7).

- Shoulder dystocia, the use of traction, and stretching of the neck during delivery may **increase** the angle between the shoulder and head. This predisposes the upper part of the plexus (C5-C6 and sometimes C7) to injury.

Lower Brachial Plexus (Klumpke) Palsy

Damage to the lower part of the brachial plexus affects components that carry axons from the C8-T1 spinal levels. This affects dermatomes along the medial aspect of the upper limb, as well as muscles that move the wrist and fingers.

- The primary clinical sign is "claw hand", due to the unopposed actions of long digital flexors and extensors (**Fig. 9.3.4**). With long-standing injury, wasting of intrinsic hand muscles may occur.

- A supinated forearm is due to weakened pronator quadratus.

- Diminished sensation from the medial aspect of the upper limb involves damage to cutaneous nerves that contain C8-T1 axons (i.e., medial cutaneous of the arm and forearm, and ulnar).

- It may be accompanied by a Horner syndrome.

- Shoulder dystocia, the use of traction, and stretching of the neck during delivery may **decrease** the angle between the shoulder and head. This predisposes the lower part of the plexus (C8-T1) to injury.

Clinical Notes

- Most cases of OBPP (>70%) resolve spontaneously. However, babies with upper plexus lesions tend to have a better prognosis for spontaneous resolution (>90%) compared to those with lower plexus injuries (60%).

- Depending on the degree of nerve injury, some recovery may be observed within several days. In most cases, spontaneous recovery is apparent within 1 month, although full recovery may not occur for several months.

Miosis Excessive pupillary constriction (opposite of mydriasis)
Ptosis Drooping eyelid

Patient Presentation

An 18-year-old male is admitted to the emergency department with severe trauma to his left upper limb as a result of a motor vehicle accident. There is pain and swelling, with an obvious deformity of the distal arm.

Relevant Clinical Findings

History

The patient is an otherwise healthy young adult male.

Physical Examination

The following findings were noted on physical examination:

- Swelling and pain in the left arm and proximal forearm
- Inability to extend the wrist
- Diminished sensation from skin on the dorsum of the hand, especially over the first dorsal interosseous muscle

Imaging Studies

- Anterior–posterior radiograph revealed a compound fracture of the distal one-third of the humerus.

Clinical Problems to Consider

- Axillary neuropathy with fracture of the surgical neck
- Compressive radial neuropathy
- Fracture of the humeral shaft, with radial neuropathy
- Ulnar neuropathy with fracture of the medial epicondyle

LEARNING OBJECTIVES

1. Describe the humerus and its relationships to the axillary, radial, and ulnar nerves.
2. Describe the distribution of the axillary, radial, and ulnar nerves.
3. Explain the anatomical basis for the signs and symptoms associated with this case.

RELEVANT ANATOMY

Humerus

Anatomical features of the humerus relevant to fractures include:

- The **surgical neck** is a narrowed portion of the shaft of the humerus, immediately distal to the greater and lesser tubercles.

- The **shaft** of the humerus is its long, narrowed part between the anatomical neck and the condyles.

- **Radial (*spiral*) groove** is a shallow, oblique channel on the posterior aspect of the middle one-third of the shaft. The **radial nerve** and profunda brachii artery and veins (deep artery and veins of the arm) traverse this groove from medial to lateral.

- The **medial epicondyle** is a prominent projection near the distal end of the bone. The **ulnar nerve** lies in a groove on the posterior aspect of this projection.

Neuropathy Damage to one or more nerves that results in impaired function

Axillary, Radial, and Ulnar Nerves

The **axillary nerve** (C5, C6), **radial nerve** (C5-T1), and **ulnar nerve** (C8-T1) are terminal branches of the brachial plexus (**Fig. 9.4.1**).

- The **axillary nerve** passes through the **quadrangular space**, adjacent to the surgical neck of the humerus. It supplies the deltoid and teres minor muscles, and skin over the deltoid.

- The **radial nerve** courses obliquely from medial to lateral in the radial groove. In this region, the nerve is closely associated with the bone of the humerus. Distal to the radial groove, the nerve penetrates the **lateral intermuscular septum** to enter the anterior (flexor) compartment of the arm. It passes anterior to the lateral **supracondylar ridge of the humerus** and then crosses the elbow joint. In the forearm, it divides into **superficial (cutaneous)** and **deep (motor) branches**. The deep branch passes through the supinator muscle and becomes the **posterior interosseous nerve**. The radial nerve supplies all extensor muscles of the upper limb, as well as supinator and brachioradialis. Its sensory distribution includes the posterior arm, lateral forearm, and lateral dorsal hand.

- The **ulnar nerve** courses along the medial arm. It pierces the medial intermuscular septum to reach the posterior aspect of the medial epicondyle. It enters the forearm via the cubital tunnel (see Fig. 9.5.4). It supplies flexor carpi ulnaris, medial part of flexor digitorum profundus, and most intrinsic muscles of the hand. Its sensory distribution is restricted to the medial aspect of the hand.

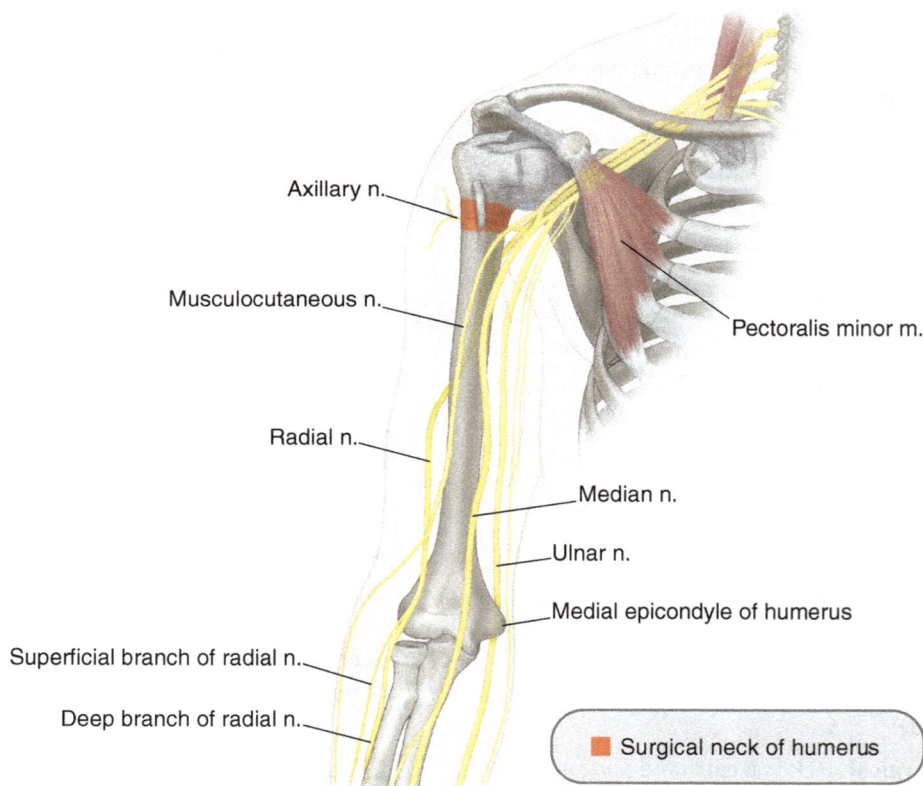

FIGURE 9.4.1 Nerves in the arm and proximal forearm.

CLINICAL REASONING

This patient presents with signs and symptoms consistent with peripheral neuropathy.

Axillary Neuropathy Following Surgical Neck Fracture

The proximity of the contents of the quadrangular space may result in axillary neuropathy with a displaced fracture of the surgical neck of the humerus.

Signs and Symptoms

Sensory Deficit

- Skin over the lateral deltoid

Motor Deficit

- Arm abduction beyond 30°

Predisposing Factors

- Age and sex: more common in middle aged and elderly women
- Osteoporosis

Abduction of the arm (0–90°) occurs at the glenohumeral joint. Beyond horizontal, movement of the arm depends on scapular rotation.

- Abduction is initiated by supraspinatus (0–30°).
- Deltoid is the prime mover for arm abduction beyond 30°.
- Scapular rotation involves serratus anterior and trapezius.

With a displaced surgical neck fracture and axillary nerve injury, the abduction of the arm may be initiated (supraspinatus) but not carried beyond 30° due to the compromise of the nerve supply to the deltoid. Sensation from skin on the lateral aspect of the deltoid would be diminished because it is also supplied by the axillary nerve. Sensation from skin over the proximal deltoid would not be impaired because it is supplied by supraclavicular nerves (C3-C4).

Compressive Radial Neuropathy

Injury to the radial nerve in the axilla can result from compression against the humeral shaft. This can occur, for example, in individuals who use improperly fitted crutches ("crutch palsy") or when an individual falls asleep with their arm over the back of a chair, especially when intoxicated ("Saturday night palsy").

Signs and Symptoms

Sensory Deficit

- Radial side of dorsum of the hand (superficial branch of radial nerve), especially between metacarpals 1 and 2 (i.e., over first dorsal interosseous muscle)

Motor Deficit

- Elbow extension (triceps brachii)
- "Wrist-drop" (extensor carpi and long digital extensors)
- Extension of metacarpophalangeal (MP) joints (long digital extensors)
- Extension of interphalangeal (IP) joints (long digital extensors). Some extension is spared (lumbricals and interossei, innervated by median and ulnar nerves).
- Elbow flexion (brachioradialis)
- Supination (supinator)

Distal Humeral Fracture with Radial Neuropathy

Fracture of the humeral shaft may be associated with radial nerve injury.

Signs and Symptoms

Sensory Deficit

- Similar to compressive radial neuropathy

Motor Deficit

- Similar to compressive radial neuropathy, but little or no effect on elbow extension because triceps brachii innervated in proximal arm

Predisposing Factors

- Age and sex: more common in men <35 years of age; women >50 years of age or older also at risk (e.g., from falls)

Palsy Paralysis (loss of voluntary motor function; may be caused by disease or injury) or paresis (partial or incomplete paralysis)

(A)

Impaired DIP flexion
- Ulnar n. part of flexor digitorum profundus (FPD) affected
- Some flexion possible due to common origin with median n. part of FDP

(B)

PIP flexion
- Flexor digitorum superficialis unopposed by impaired lumbricals 4 & 5 and interossei

MP hyperextension
- Extensor digitorum unopposed by impaired lumbricals 4 & 5 and interossei

Hypothenar atrophy

Interosseous muscle atrophy

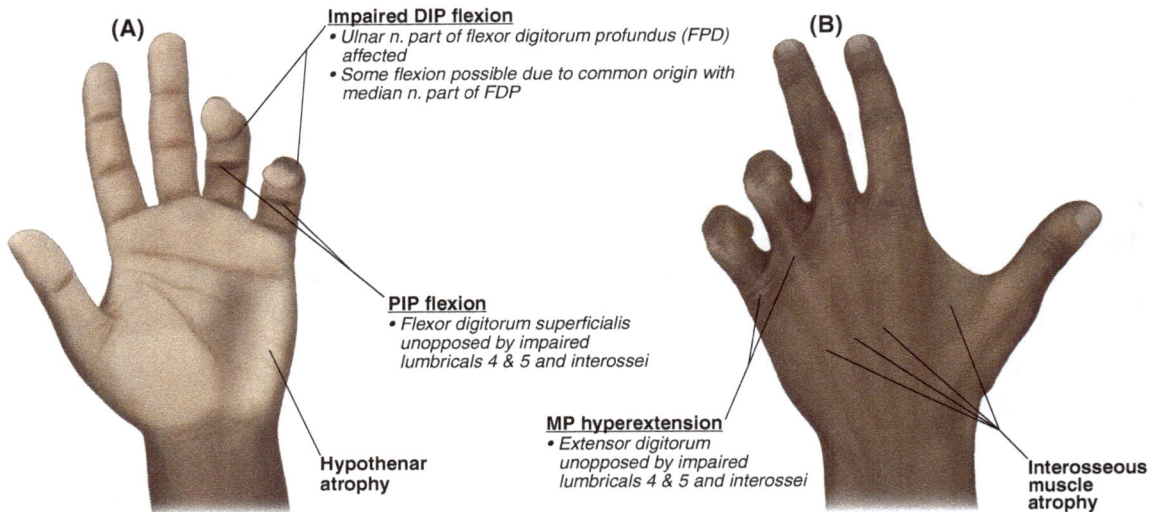

FIGURE 9.4.2 Wasting of intrinsic hand muscles and claw hand may result from chronic ulnar nerve damage. (**A**) palmar view, (**B**) dorsal view. Abbreviations: DIP, distal interphalangeal; PIP, proximal interphalangeal; MP, metacarpophalangeal.

Clinical Note

Fractures of the humeral shaft are usually caused by trauma. The incidence of associated radial nerve palsy is approximately 12%.

Ulnar Neuropathy with Medial Epicondyle Fracture

Fracture of the medial epicondyle may injure the ulnar nerve.

Signs and Symptoms

Sensory Deficits
- Medial aspect of palm
- Medial 1½ fingers

Motor Deficits
- Wrist adduction (flexor carpi ulnaris)
- Wrist flexion (ulnar part of flexor digitorum profundus)
- Flexion of digits 4 and 5 (ulnar part of flexor digitorum profundus)
- Abduction and adduction of digits (interossei and adductor pollicis)

- Extension of IP joints of digits 4 and 5 (interossei and medial two lumbricals)

Motor impairment may be tested by asking the patient to hold a piece of paper between the thumb and index finger (adductor pollicis). Chronic ulnar neuropathy may result in atrophy of intrinsic hand muscles (excluding thenar muscles) and "claw hand" (**Fig. 9.4.2**).

Clinical Notes

"Claw hand" deformity results from an imbalance between extrinsic and intrinsic hand muscles. The deformity, most apparent when the patient is asked to extend their fingers, affects the fourth and fifth fingers. Muscles affected by ulnar neuropathy depend on the lesion location:

- **High ulnar nerve lesion** (e.g., at the medial epicondyle) involves flexor carpi ulnaris, flexor digitorum profundus (ulnar part), and intrinsic hand muscles. The weakened

---

Atrophy Wasting of tissues, organs, parts of the body, or the entire body

MP flexion by the denervated medial lumbricals combined with the unopposed actions of extensor digitorum and extensor digiti minimi to these fingers tends to draw these MP joints into hyperextension. Denervation of the ulnar part of flexor digitorum profundus, however, also weakens IP flexion of these fingers. As a result, their IP joints may be extended, again due to unopposed actions of long extensors.

- **Low ulnar nerve lesion** (e.g., at the wrist) involves only intrinsic hand muscles. The sparing of flexor digitorum profundus results in the continued flexion of the IP joint.

Predisposing Factors

- Age: more common in children and adolescents

DIAGNOSIS

The patient presentation, medical history, physical examination, and imaging studies support a diagnosis of a **humeral fracture with radial neuropathy**.

Humeral Fracture with Radial Neuropathy

Humeral shaft fracture may injure the radial nerve as it passes along the radial (spiral) groove. Injury to this nerve may lead to sensory deficits on the dorsomedial aspect of the hand. It also affects extensor muscles distal to the injury, as well as the supinator muscle.

- The humeral fracture was confirmed by, eliminating compressive radial neuropathy as a possible diagnosis.
- The radial neuropathy was diagnosed by weakness in wrist extension ("wrist drop") and sensory deficit from the distribution of the superficial radial nerve. Elbow extension was normal because the motor innervation to the triceps muscle branched from the radial nerve proximal to the fracture.

Axillary Neuropathy

Fracture of the surgical neck of the humerus may injure the axillary nerve as it passes through the quadrangular space. The most notable effect of the injury is on the ability to abduct the arm.

- A patient with axillary neuropathy would present with deltoid weakness and sensory loss in the shoulder region.

Ulnar Neuropathy

Injury of the ulnar nerve affects dermatomes on the anteromedial aspect of the hand as well as the motor innervation of intrinsic hand muscles.

Patient Presentation

A 30-year-old white female visits the orthopedic clinic complaining of severe pain in her right wrist and hand. She reports that her right hand is weak.

Relevant Clinical Findings

History

The patient is a medical illustrator who makes extensive use of a computerized drawing tablet. Her right hand is dominant. She reports that her wrist and hand pain was exacerbated during and after her first pregnancy.

Physical Examination

The following findings were noted on physical examination of the right hand:

- Numbness, paresthesia, and pain along the lateral palm and lateral 3½ fingers
- Mild atrophy of the thenar muscle group, with a weak grip compared to the left hand
- Positive Tinel sign and Phalen wrist flexion maneuver (**Fig. 9.5.1**)

Clinical Problems to Consider

- Carpal tunnel syndrome
- Cubital tunnel syndrome
- Pronator syndrome

LEARNING OBJECTIVES

1. Describe the anatomy of the cubital fossa, and the carpal and cubital tunnels.
2. Describe the anatomy of the median and ulnar nerves in the forearm and hand.
3. Describe the distribution of the median and ulnar nerves.
4. Explain the anatomical basis for the signs and symptoms associated with this case.

RELEVANT ANATOMY

Cubital Fossa

The cubital fossa is a shallow depression on the anterior aspect of the elbow (**Fig. 9.5.2**). Its boundaries are summarized in **Table 9.5.1**.

Important nerves and vessels that cross the elbow region are related to the cubital fossa.

Within the Fossa

- Terminal part of the **brachial artery,** its **radial and ulnar** branches, and accompanying paired, **deep veins**
- **Median nerve**, medial to the brachial artery
- **Radial nerve** in the lateral aspect of the fossa, covered by brachioradialis
- **Biceps brachii tendon**

Superficial to the Fossa

- **Median cubital vein**
- Skin and subcutaneous tissue

Clinical Note

The median cubital vein is used commonly for venipuncture because of its superficial position, its relatively large size, and the ability to easily position the vessel. In this procedure, care must be exercised to direct the needle superficially into the vein. The bicipital aponeurosis, a tendinous sheet that reinforces the roof of the cubital fossa, separates the vein from the underlying brachial artery and median nerve.

Atrophy Wasting of tissues, organs, parts of the body, or the entire body

(A) Tinel sign

Median n.

(B) Phalen maneuver

■ Sensory distribution of median n.
after passing through carpal tunnel

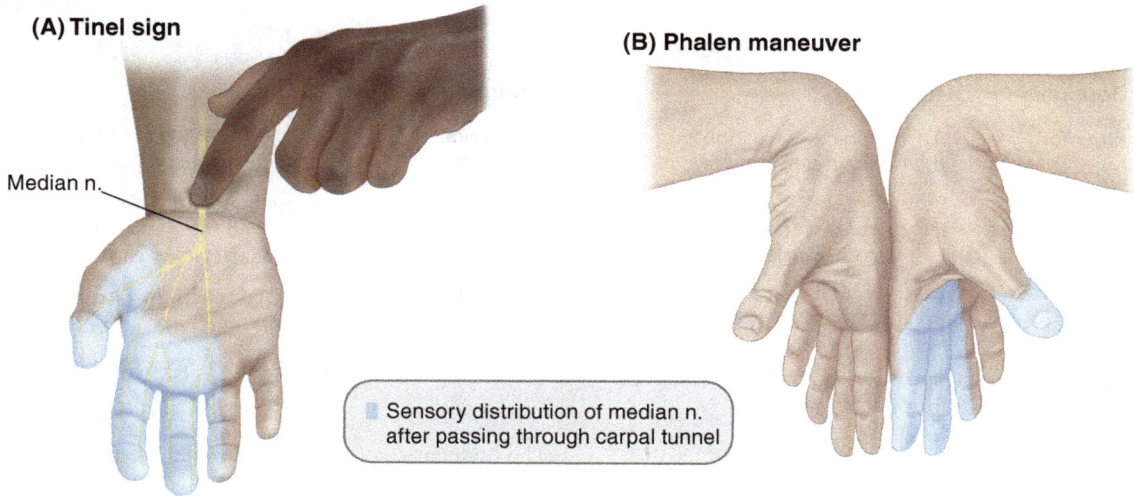

FIGURE 9.5.1 Provocative tests for median nerve function in the hand. (**A**) A positive **Tinel sign** refers to distally radiating pain and/or paresthesia elicited by percussing a superficial peripheral nerve, in this case the median. (**B**) The **Phalen maneuver** is performed by apposing the wrists in 90° of flexion. Paresthesia in the hand within 60 seconds is considered a positive test.

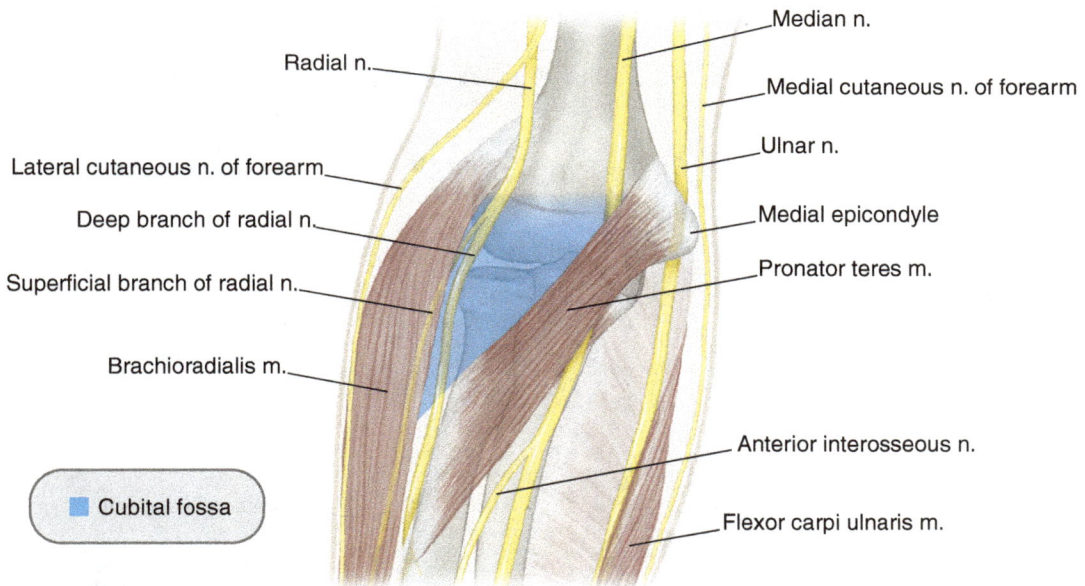

Radial n.

Lateral cutaneous n. of forearm

Deep branch of radial n.

Superficial branch of radial n.

Brachioradialis m.

■ Cubital fossa

Median n.

Medial cutaneous n. of forearm

Ulnar n.

Medial epicondyle

Pronator teres m.

Anterior interosseous n.

Flexor carpi ulnaris m.

FIGURE 9.5.2 Anterior view showing the cubital fossa, its borders, and nerves that cross the elbow. The median and radial nerves pass through this fossa.

Paresthesia Numbness and tingling

TABLE 9.5.1	Anatomical boundaries of the cubital fossa.
Boundary	**Structure(s)**
Superior	▪ An imaginary line connecting the humeral epicondyles
Medial	▪ Pronator teres ▪ Flexor muscles that attach to medial epicondyle
Lateral	▪ Brachioradialis ▪ Extensor muscles that attach to lateral epicondyle
Roof	▪ Brachial and antebrachial fasciae ▪ Bicipital aponeurosis
Floor	▪ Brachialis muscle ▪ Supinator muscle

TABLE 9.5.2	Boundaries of the cubital tunnel.
Boundary	**Structure(s)**
Anterior	▪ Medial epicondyle of humerus ▪ Ulnar collateral ligament
Posterior	▪ Arcuate ligament (Osborne)—a transverse band between the two heads of flexor carpi ulnaris
Medial	▪ Humeral head of flexor carpi ulnaris
Lateral	▪ Ulnar head of flexor carpi ulnaris

The boundaries of the cubital tunnel are reviewed in **Table 9.5.2.**

Cubital Tunnel

The **cubital tunnel** is a passageway for the ulnar nerve as it passes the elbow to enter the forearm. This tunnel lies posterior to the medial epicondyle of the humerus (**Fig. 9.5.3**).

Carpal Tunnel and Its Contents

The **carpal tunnel** (**Fig. 9.5.4**) is an osseofibrous passage formed by the concave, anterior surfaces of the carpal bones, and the **transverse carpal ligament** (**flexor retinaculum**). This nearly square ligament

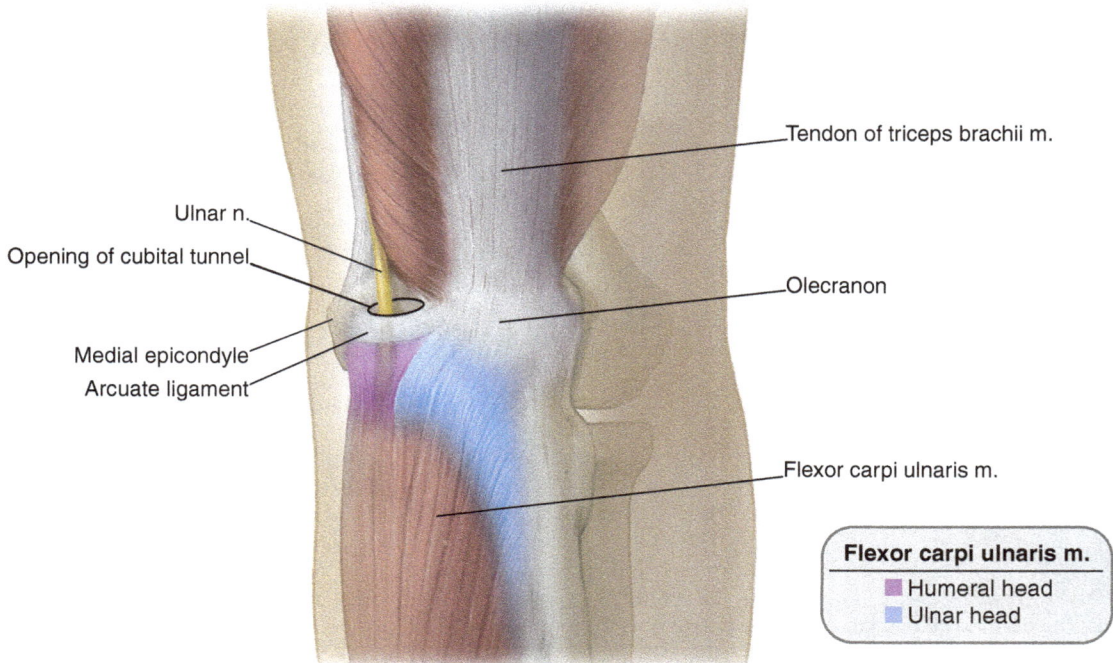

Flexor carpi ulnaris m.
■ Humeral head
■ Ulnar head

FIGURE 9.5.3 Posterior view of the elbow showing the ulnar nerve and cubital tunnel.

Carpal tunnel
Ulnar canal (Guyon)

Ulnar n., a. and v.

Transverse carpal ligament
(flexor retinaculum)

Pisiform

Median n.

Triquetrum

Hamate

Trapezium

Capitate   Scaphoid

Tendons of flexor
digitorum superficialis

Tendon of flexor
carpi radialis

Tendons of flexor
digitorum profundus

Tendon of flexor
pollicis longus

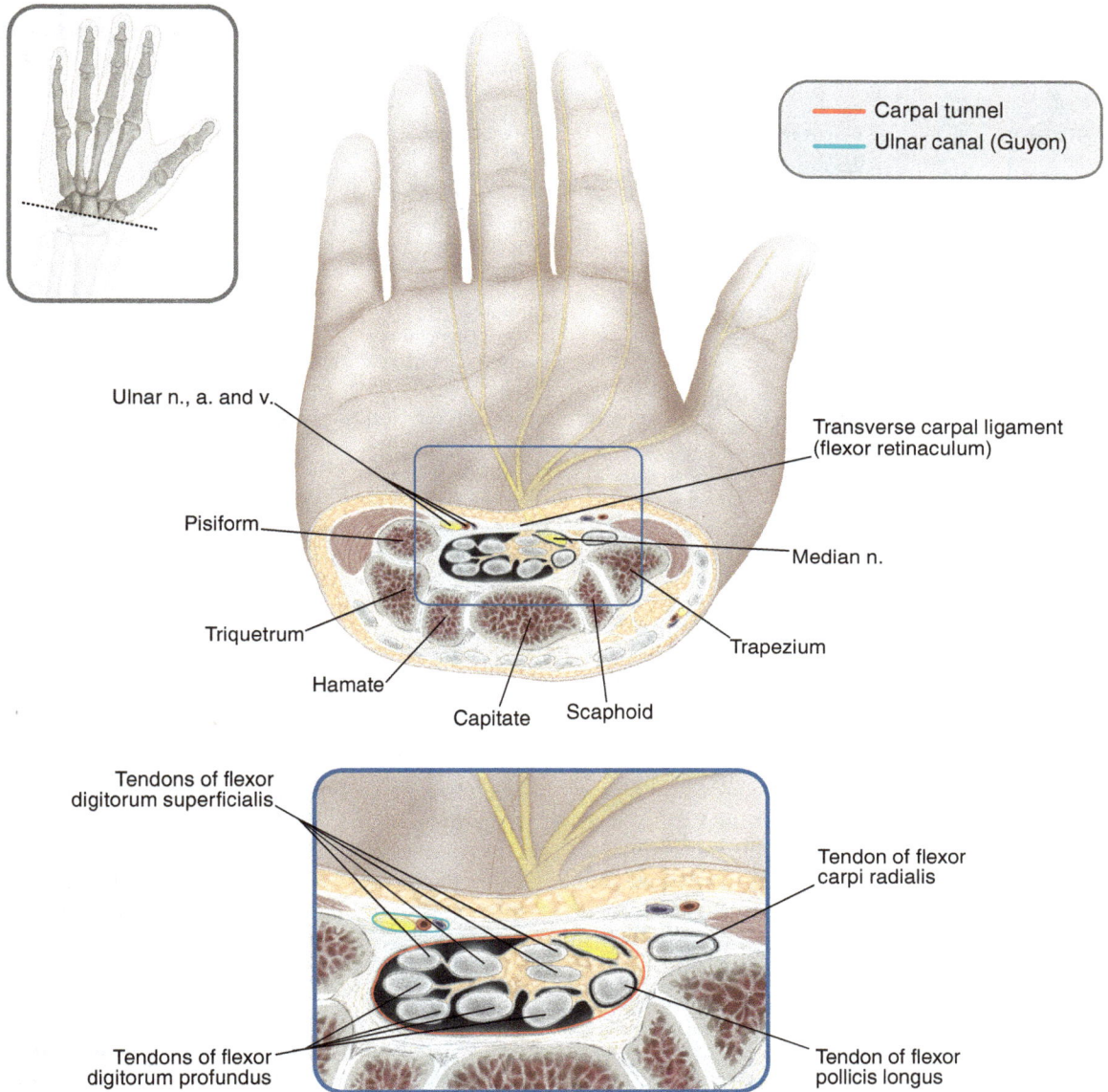

FIGURE 9.5.4 View of the carpal tunnel and its contents.

(2.5–3.0 cm on a side) is attached laterally to the scaphoid and trapezium, and medially to the pisiform and hamulus of the hamate. The ligament also serves as an attachment for the palmaris longus, flexor carpi ulnaris, and several thenar and hypothenar muscles.

The **median nerve** and long **digital flexor tendons** and their synovial sheaths pass through this tunnel. The median nerve lies immediately deep to the transverse carpal ligament. The ligament holds the contents of the carpal tunnel in place during wrist flexion, preventing "bowstringing" of the tendons.

TABLE 9.5.3	Distribution of the median and ulnar nerves.		
Spinal levels	**Sensory**	**Motor**
Median nerve C(5)6-T1	**Palmar cutaneous branch** ▪ Proximal lateral palm **Palmar digital branches** ▪ Distal lateral palm ▪ Lateral 3½ fingers	**Forearm** ▪ Flexor–pronator muscles *Except flexor carpi ulnaris and ulnar part of flexor digitorum profundus* **Hand** ▪ Recurrent branch: thenar muscles ▪ Palmar branches: lumbricals 1 and 2
Ulnar nerve C8-T1	**Palmar cutaneous branch** ▪ Proximal medial palm **Dorsal branch** ▪ Medial dorsum of hand **Superficial branch** ▪ Distal medial palm ▪ Medial 1½ fingers	**Forearm** ▪ Flexor carpi ulnaris ▪ Ulnar part of flexor digitorum profundus **Hand** ▪ Deep branch: intrinsic hand muscles Hypothenar muscles Lumbricals 3 and 4 All interossei Adductor pollicis Deep part of flexor pollicis brevis

Median Nerve in the Forearm and Hand

The median nerve supplies structures in the anterior forearm and palmar hand (**Table 9.5.3**). At the elbow, it crosses the cubital fossa and then passes between the two heads of **pronator teres** (**Fig. 9.5.2**). The nerve courses distally between **flexor digitorum superficialis** and **flexor digitorum profundus**. It gives off the **anterior interosseous nerve** to supply deep anterior forearm muscles. The **palmar cutaneous branch** arises in the distal forearm and enters the hand superficial to the carpal tunnel to supply the proximal lateral palm (see **Fig. 9.5.9**).

The median nerve lies superficial and lateral within the carpal tunnel (**Fig. 9.5.4**). As the nerve enters the palm, it gives off a **recurrent branch** to supply the thenar muscles; other motor branches distribute to the first and second lumbricals. **Palmar digital nerves** are sensory branches that supply the distal lateral palm and the lateral 3 fingers (**Figs. 9.5.4** and **9.5.5**), including the dorsum distal to the distal interphalangeal joint (**Fig. 9.5.6**).

Ulnar Nerve in the Forearm and Hand

The ulnar nerve supplies structures in the anterior forearm and hand (**Table 9.5.3**). The nerve passes through the **cubital tunnel** and courses distally between flexor carpi ulnaris and flexor digitorum profundus. Near the wrist, it gives off **palmar** and **dorsal cutaneous branches**:

▪ The **dorsal branch** is directed to the dorsal medial side of the hand (**Fig. 9.5.6**).

▪ The **palmar branch** passes superficial to the transverse carpal ligament (flexor retinaculum) and distributes to the proximal medial palm (**Fig. 9.5.9**).

At the wrist, the ulnar nerve and vessels pass through the **ulnar canal** (Guyon), a superficial channel in the transverse carpal ligament. This canal lies immediately lateral to the pisiform and hamulus of the hamate. Distal to the canal, the ulnar nerve divides into a **superficial (sensory) branch** and a **deep (motor) branch** (**Fig. 9.5.5**). The superficial branch supplies the distal medial palm and the

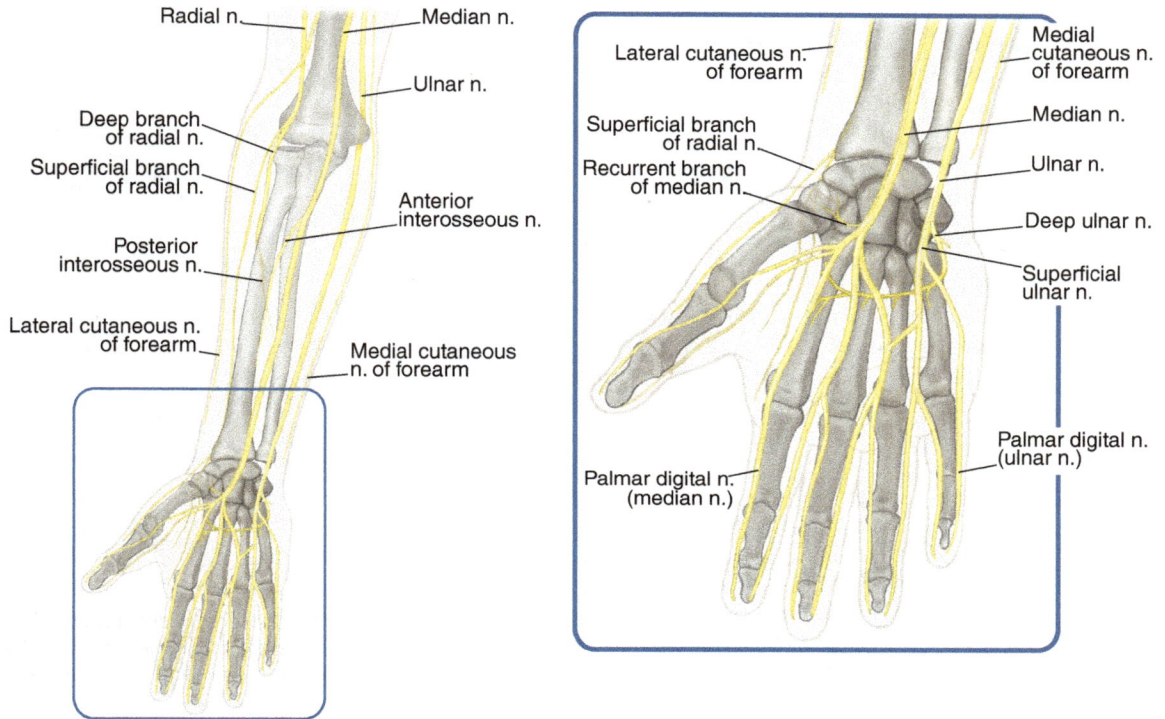

FIGURE 9.5.5 Anterior view of the median, radial, and ulnar nerves in the forearm and hand.

medial 2 fingers. The deep branch passes between the hypothenar muscles to reach the level of the metacarpals, where it courses laterally with the deep palmar arterial arch. It supplies most of the intrinsic hand muscles (**Table 9.5.3**).

CLINICAL REASONING

This patient presents with signs and symptoms indicating a peripheral neuropathy of the upper limb.

Carpal Tunnel Syndrome

Carpal tunnel syndrome is a peripheral mononeuropathy caused by compression of the median nerve in the carpal tunnel.

Neuropathy Damage to one or more nerves that results in impaired function

Signs and Symptoms

Sensory Deficits
- Distal, lateral palmar aspect of hand
- Lateral 3½ fingers
- Nocturnal pain common

Motor Deficits
- Reduced grip strength and impaired thumb movements (thenar muscles: abductor pollicis brevis, flexor pollicis brevis, opponens pollicis)

Predisposing Factors

- Age: 30–60 years
- Sex: more common in women
- Pregnancy and childbirth: mild and intermittent carpal tunnel syndrome is associated with pregnancy. Signs and symptoms typically diminish after delivery, although they may persist or increase afterward.

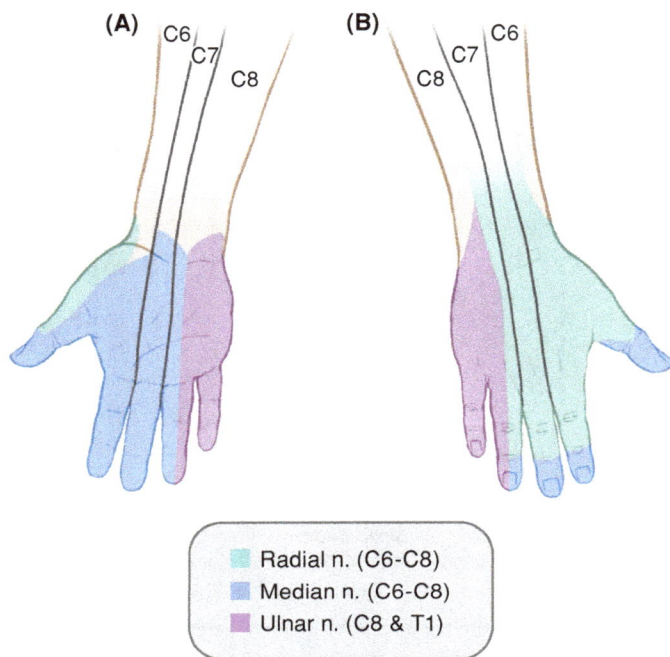

(A) C6 C7 C8

(B) C8 C7 C6

Radial n. (C6-C8)
Median n. (C6-C8)
Ulnar n. (C8 & T1)

FIGURE 9.5.6 Sensory distribution of the median, ulnar, and radial nerves in the hand. (**A**) Palmar view and (**B**) dorsal view.

- Occupation: repetitive tasks using hands and wrists (e.g., computer users, typists, meat and fish packing workers, and musicians) increase risk

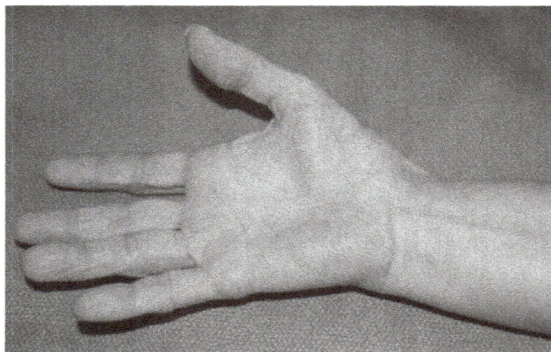

FIGURE 9.5.7 Atrophy of thenar muscles due to long-standing carpal tunnel syndrome. *Source:* Fig. 6.4 in *Current Rheumatology Diagnosis & Treatment*, 2e

Clinical Notes

- Sensory deficits may be more severe with hand use and at night (depending on the position of the wrist).
- Tinel sign and Phalen maneuver may be positive (**Fig. 9.5.1**).
- Thenar muscle atrophy becomes apparent with long-term carpal tunnel syndrome (**Fig. 9.5.7**).
- Carpal tunnel syndrome is the most common compressive peripheral mononeuropathy.

Cubital Tunnel Syndrome

Cubital tunnel syndrome is a peripheral mononeuropathy caused by compression of the ulnar nerve as it passes between the two heads of flexor carpi ulnaris, posterior to the medial epicondyle of humerus (**Fig. 9.5.3**).

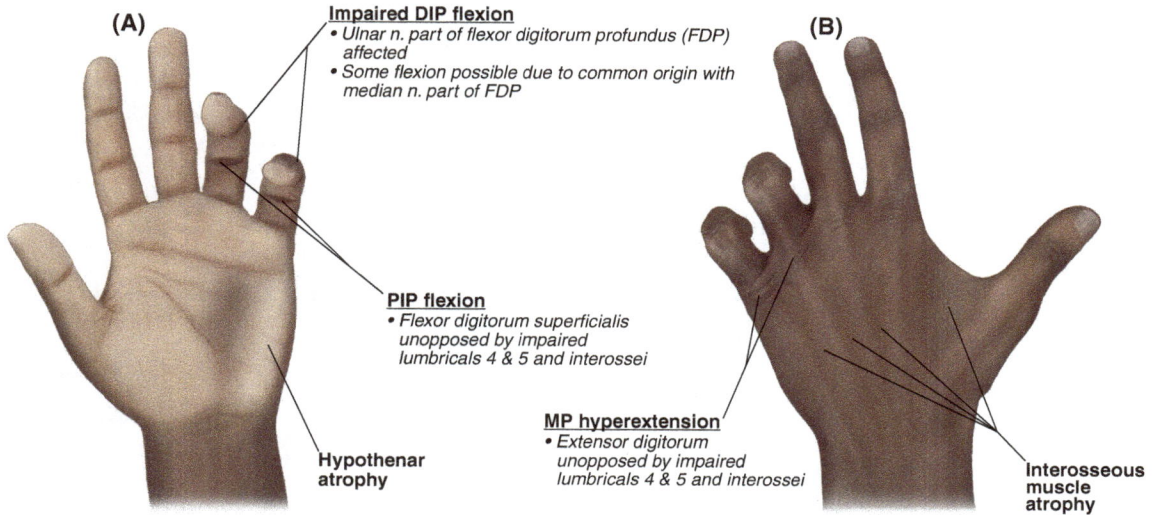

Impaired DIP flexion
- *Ulnar n. part of flexor digitorum profundus (FDP) affected*
- *Some flexion possible due to common origin with median n. part of FDP*

PIP flexion
- *Flexor digitorum superficialis unopposed by impaired lumbricals 4 & 5 and interossei*

Hypothenar atrophy

MP hyperextension
- *Extensor digitorum unopposed by impaired lumbricals 4 & 5 and interossei*

Interosseous muscle atrophy

FIGURE 9.5.8 Palmar (**A**) and dorsal (**B**) views of the hand with severe ulnar nerve lesion. Abbreviations: DIP, distal interphalangeal; PIP, proximal interphalangeal; MP, metacarpophalangeal.

Signs and Symptoms

Sensory Deficits
- Medial aspect of hand (palmar and/or dorsal)
- Medial 1½ fingers

Motor Deficits
- Wrist adduction (flexor carpi ulnaris)
- Wrist flexion (flexor carpi ulnaris and ulnar part of flexor digitorum profundus)
- Flexion of digits 4 and 5 (ulnar part of flexor digitorum profundus)
- Abduction and adduction of digits (interossei and adductor pollicis)
- Extension of interphalangeal joints of digits 4 and 5 (interossei and medial two lumbricals)

Predisposing Factors

- Sex: more common in men
- Prior elbow fracture or dislocation, bone spurs or arthritis
- Swelling or cysts near the elbow joint
- Activities that require a repetitive or prolonged elbow flexion and extension

Clinical Notes

- Positive Tinel sign may be present in up to 24% of individuals who are asymptomatic for cubital tunnel syndrome.
- Positive **elbow flexion test** elicits paresthesia (performed by flexing patient's elbow (>90°) with forearm supinated and wrist extended)
- Chronic ulnar neuropathy typically results in the appearance of a "claw hand" (**Fig. 9.5.8**).
- Cubital tunnel syndrome is the second most common compressive peripheral mononeuropathy after carpal tunnel syndrome.

Pronator Syndrome

Pronator syndrome is an overuse injury of the forearm that results in neuropathy of the median nerve due to its compression as it passes between the humeral and ulnar heads of pronator teres (see **Fig. 9.5.2**).

Signs and Symptoms

Sensory Deficits
- Pain and paresthesia from anterior forearm, exacerbated by applied pressure to pronator teres

Paresthesia Numbness and tingling

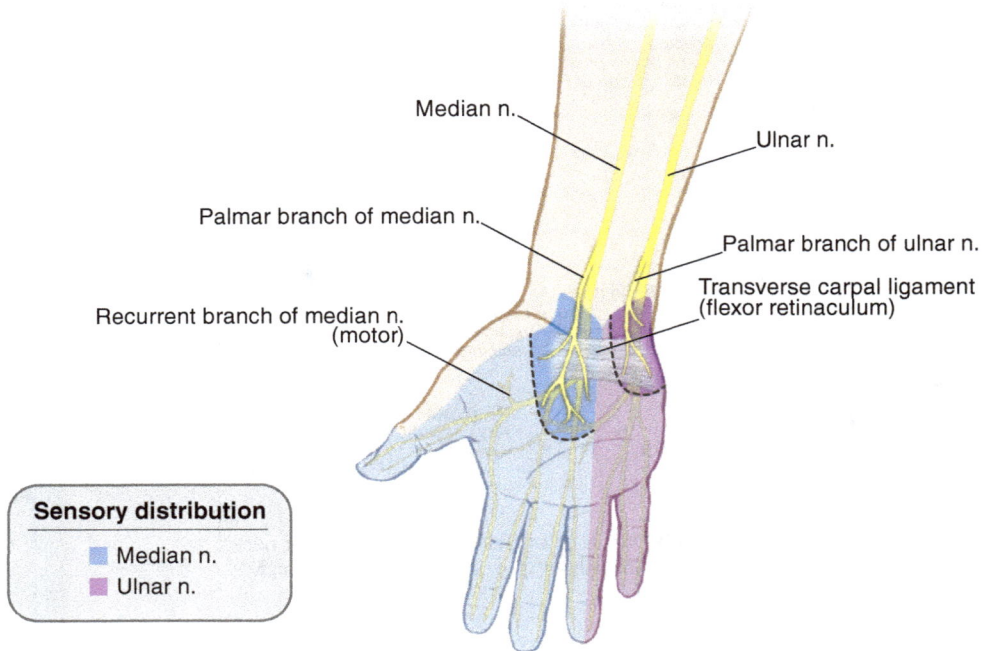

FIGURE 9.5.9 Anterior view showing the sensory distribution of median and ulnar nerves to the palm.

- Lateral aspect of palm, including proximal part
- Lateral 3½ fingers
- Nocturnal pain uncommon

Motor Deficits
- Reduced grip strength and impaired thumb movements (thenar muscles: abductor pollicis brevis, flexor pollicis brevis, opponens pollicis)

Predisposing Factors
- Sports and work activities that involve forceful, repetitive forearm pronation, for example, weight lifting, baseball, racquet sports, carpenters, and assembly line workers.

Clinical Notes

- A Tinel sign may be positive.
- The Phalen maneuver does not provoke symptoms.

DIAGNOSIS

The patient presentation, medical history, physical examination, and imaging studies support a diagnosis of peripheral neuropathy consistent with **carpal tunnel syndrome**.

Carpal Tunnel Syndrome

Carpal tunnel syndrome affects structures supplied by the median nerve distal to the carpal tunnel.

- Sensory deficits are caused by compression of the median nerve in the carpal tunnel. These include paresthesia, hypesthesia, and/or anesthesia from the *distal* lateral palm, and the lateral 3½ fingers (**Fig. 9.5.9**). Nocturnal pain is common.
- There is atrophy of the thenar eminence due to involvement of the recurrent median nerve, which branches from the median nerve after it exits the carpal tunnel. This results in the loss of thumb strength.

Anesthesia Loss of sensation

TABLE 9.5.4	Signs and symptoms associated with carpal tunnel and pronator syndromes.	
Sign/Symptom	**Carpal tunnel syndrome**	**Pronator syndrome**
Sensory deficit on palm	▪ Distal lateral	▪ Proximal and distal
Regional pain	▪ Hand ▪ Nocturnal pain common	▪ Forearm ▪ Nocturnal pain uncommon
Pain with muscular activity	▪ Aggravated by wrist movements ▪ Associated with compression of median nerve in carpal tunnel	▪ Aggravated by elbow movements ▪ Contraction of pronator teres or flexor digitorum superficialis
Tinel sign	▪ Positive at wrist over carpal tunnel	▪ Positive in forearm over pronator teres
Phalen maneuver	▪ Positive	▪ Negative

Pronator Syndrome

In pronator syndrome, the median nerve is compressed between the humeral and ulnar heads of pronator teres. As a result, it can affect structures in the forearm and hand that are supplied by the median nerve.

- Patients typically complain of pain and tenderness along the *anterior forearm*. Skin over the *proximal* lateral palm is also involved because the palmar cutaneous branch is affected (**Fig. 9.5.9**). Unlike carpal tunnel syndrome, nocturnal pain is uncommon.

- Motor deficits are related to the anterior interosseous branch of the median nerve. This includes weakness in the ability to pinch the thumb (flexor pollicis longus) and index finger (flexor digitorum profundus to the index finger).

There are notable differences in the sensory and motor deficits associated with carpal tunnel and pronator syndromes (**Table 9.5.4**). In both cases, thumb movements are affected and thenar wasting occurs.

Cubital Tunnel Syndrome

Cubital tunnel syndrome is a compression neuropathy of the ulnar nerve posterior to the elbow. This nerve injury, as well as any proximal to the cubital tunnel, affects all targets of the ulnar nerve.

- Sensory deficits include the medial aspect of the hand (**Fig. 9.5.9**).

- Motor deficits are related to the intrinsic hand muscles. This results in the loss of fine motor function of the fingers.

R E V I E W   Q U E S T I O N S

Questions 1–2 refer to the following clinical case.

1. A 54-year-old male is brought to the emergency department after a motor vehicle accident. Physical examination reveals some sensory loss from the upper limb and radiologic imaging confirms herniation of the C5-C6 intervertebral disc. Which nerve carries sensory fibers from the affected dermatome?

   A. Deep branch of ulnar
   B. Lateral cutaneous nerve of forearm
   C. Long thoracic
   D. Medial cutaneous nerve of forearm
   E. None of the above

2. Which actions would also be likely impaired by this injury?

   A. Elbow flexion and thumb opposition
   B. Shoulder abduction and wrist adduction
   C. Shoulder abduction and elbow flexion
   D. Shoulder abduction and shoulder elevation

3. A 44-year-old woman is diagnosed with breast cancer. Following mastectomy and dissection of axillary lymph nodes, the patient is unable to raise her arm over her head. Which nerve is most likely damaged?

   A. Long thoracic
   B. Pectoral branch of thoraco-acromial
   C. Suprascapular

D. Thoracodorsal

E. None of the above

4. Which clinical sign would *not* be associated with injury of the ulnar nerve due to fracture of the medial epicondyle?

A. Hypesthesia from proximal part of the medial half of the palm

B. Hypesthesia from the middle finger

C. Hypesthesia from the ring finger

D. Weakened flexion of digits 4 and 5

E. Weakened flexion of the wrist

5. A 35-year-old male patient visits the orthopedics clinic complaining of weakness of his upper limb. Physical examination suggests involvement of the posterior cord of the brachial plexus. Which action would *not* be affected?

A. Elbow extension

B. Elbow flexion

C. Pronation

D. Shoulder abduction

E. Wrist extension

6. A 52-year-old male undergoes rotator cuff surgery to repair the supraspinatus muscle. During the procedure, the suprascapular artery is identified. Which other vessels contribute to scapular collateral circulation.

A. Circumflex scapular and posterior circumflex humeral

B. Dorsal scapular and circumflex scapular

C. Thoracodorsal and dorsal scapular

D. Lateral thoracic and circumflex scapular

7. During anatomy lab dissection, a medical student accidentally receives a superficial laceration on the anterior aspect of his wrist. Physical examination in the emergency department reveals no loss of function of intrinsic hand muscles, but the skin on lateral aspect of his palm is numb. Which nerve has been severed?

A. Lateral cutaneous nerve of the forearm

B. Palmar branch of median

C. Palmar branch of radial

D. Recurrent branch of median

E. Superficial branch of ulnar

8. Intravenous lines are often placed in subcutaneous veins on the posterolateral aspect of the hand? Which nerve would be stimulated in this location?

A. Deep branch of radial

B. Deep branch of ulnar

C. Lateral cutaneous nerve of forearm

D. Superficial branch of radial

E. Superficial branch of ulnar

9. A 12-year-old boy is brought to the emergency department after a collision while playing baseball. He complains of extreme pain over his right clavicle, which is obviously deformed. Radiologic imaging confirms a right midclavicular fracture. Which vein is especially vulnerable to injury as a result of the fracture?

A. Subclavian

B. Brachiocephalic

C. External jugular

D. Internal jugular

E. Brachial

10. A 33-year-old female comes to the orthopedics clinic complaining of weakness in her right (dominant) hand. During physical examination, you ask her to hold a piece of paper tightly between adjacent surfaces of the index and middle fingers. You find, however, that it is easy, compared with her left hand, to pull the paper from between her fingers. Which muscles are involved?

A. First dorsal and first palmar interosseous

B. First dorsal and second palmar interosseous

C. First lumbrical and second dorsal interosseous

D. First palmar and second dorsal interosseous

E. None of the above

11. You are asked to obtain an arterial blood sample from the radial artery. Prior to the procedure, you apply the Allen test to assess the patency of the ulnar artery, which provides the major contribution to the superficial palmar arterial arch. Which statement about the palmar arterial arches is *correct*?

A. The deep arch is located immediately deep to the palmar aponeurosis.

B. The deep arch lies on the distal row of carpal bones.

C. The deep branch of the radial artery passes through the thenar muscles and contributes to the superficial arch.

D. The superficial arch is formed mainly by the deep branch of the ulnar artery.

E. The superficial arch lies distal to the deep arch.

12. While rollerblading, a 25-year-old male falls on his outstretched right wrist. Physical examination in the emergency department reveals severe pain upon palpation of the anatomical "snuffbox" and radiological studies confirm a fracture. Which bone is most likely fractured?

A. Head of first metacarpal

B. Lunate

C. Scaphoid

D. Trapezoid

E. Styloid process of ulna

CHAPTER 10

Lower Limb

10.1 | Piriformis Syndrome

Patient Presentation

A 32-year-old female complains of a nagging tenderness in the center of her right buttock and episodes of numbness and tingling in the posterior thigh and calf regions on this same side.

Relevant Clinical Findings

History

The patient relates that she stays physically fit by running and playing hockey on a local semiprofessional women's team. Until recently, she participated in most local charity runs. The presenting complaint has greatly reduced her training and participation. She believes that the muscles of her right lower limb are weaker than those of the left.

Physical Examination

The following findings were noted on physical examination:

- Dysesthesia elicited in right gluteal region with deep pressure or medial rotation and adduction of hip.
- Dysesthesia not present with pressure to ischial tuberosity or posterior thigh.
- Deep gluteal pressure caused paresthesia in right posterior thigh.
- Crossed straight-leg raise (SLR) test was negative.
- Side-to-side strength comparison of gluteal and posterior thigh muscles was inconclusive.

Imaging Studies

- MRI of the lumbar region was normal.

Clinical Problems to Consider

- Chronic hamstring tendinitis
- Herniated lumbar intervertebral (IV) disc
- Piriformis syndrome

LEARNING OBJECTIVES

1. Describe the structure and relationships of the gluteal region and posterior thigh.
2. Describe the greater sciatic foramen and list structures that pass through it.
3. Explain the anatomical basis for the signs and symptoms associated with this case.

RELEVANT ANATOMY

Gluteal Region

The tissue plane on the deep side of **gluteus maximus** contains nerves and vessels that distribute to the **lower limb** and **perineum**. Most of these structures enter the gluteal region from the true pelvis by passing through the **greater sciatic foramen** (**Fig. 10.1.1**). The boundaries of this foramen are rigid and include:

Dysesthesia Pain

Paresthesia Numbness and tingling

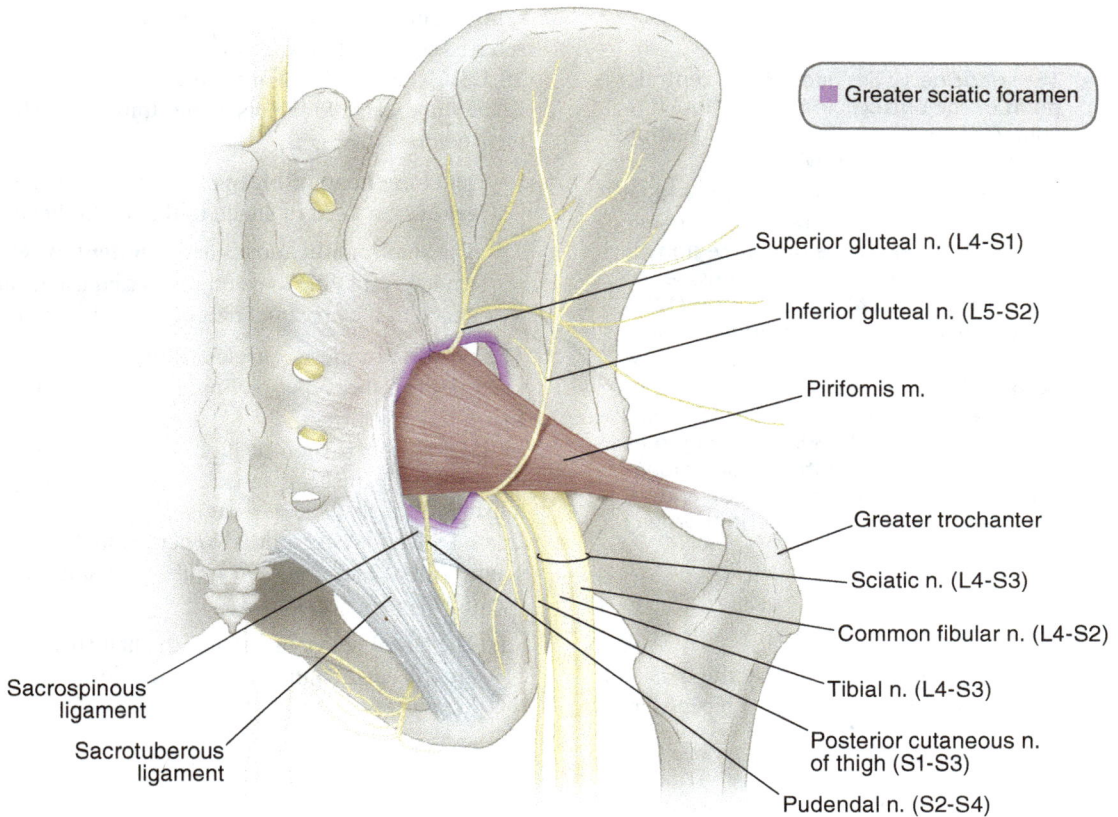

FIGURE 10.1.1 Posterior view of the gluteal region (gluteal muscles removed) showing structures passing through the greater sciatic foramen.

- **Greater sciatic notch** of the ilium
- **Sacrotuberous ligament**
- **Sacrospinous ligament**

In addition to nerves and blood vessels, this foramen also transmits the belly of the **piriformis** muscle as it passes from the pelvic surface of the sacrum to the greater trochanter of the femur. The position of piriformis in the deep gluteal region serves as a landmark for identifying other structures that accompany it through the foramen.

Superior to Piriformis

- **Superior gluteal nerve** (L4-S1) and vessels

Inferior to Piriformis

- **Inferior gluteal nerve** (L5-S2) and vessels
- **Sciatic nerve** (L4-S3)
- **Posterior cutaneous nerve of the thigh** (S1-S3)
- **Pudendal nerve** (S2-S4) and **internal pudendal vessels**

The gluteal nerves and vessels distribute to the gluteal muscles. The **sciatic nerve** and **posterior cutaneous nerve of the thigh** descend through the gluteal region and enter the posterior thigh at the lower border of the gluteus maximus. The **pudendal nerve** enters the **lesser sciatic foramen** to distribute to the perineum.

Clinical Notes

- The sciatic nerve lies deep to the central point of the buttock. In this position, it is vulnerable during intramuscular injections administered by the novice.

- The sciatic nerve is composed of the **tibial** (L4-S3) and **common fibular** (L4-S2) nerves, which usually enter the gluteal region bound in a common connective tissue sheath (**Fig. 10.1.2A**). In some cases (12%), the two nerves enter the gluteal region separately: the tibial nerve at the inferior border of piriformis and the common fibular nerve pierces piriformis (**Fig. 10.1.2B**). In another variation (0.5%), the common fibular nerve enters the gluteal region along the superior border of the muscle and the tibial nerve along its inferior border (**Fig. 10.1.2C**).

Posterior Thigh

The posterior compartment of the thigh is dominated by three muscles (long head of biceps femoris, semitendinosus, and semimembranosus) that are referred to collectively as the hamstring group (**Fig. 10.1.3**). The proximal attachment for each is the ischial tuberosity. In the distal one-third of the thigh, their tendons diverge:

- The **long head of biceps femoris** passes **lateral** to the knee to attach to the fibular head.

- The **semimembranosus** and **semitendinosus** tendons cross the knee on its **medial** aspect and attach to the proximal part of the tibial shaft.

The position of these muscles as they approach the knee defines the superior borders of the popliteal fossa.

In the proximal posterior thigh, the sciatic nerve is deep to the hamstring muscles. As the hamstring muscles diverge, the nerve enters the popliteal fossa. In a majority of cases, the two nerves that form the sciatic (tibial and common fibular) separate in the popliteal fossa (**Fig. 10.1.3**):

1. The **tibial nerve** innervates the hamstring muscles and continues near the midline across the posterior aspect of the knee.

(A)

(B)

(C)

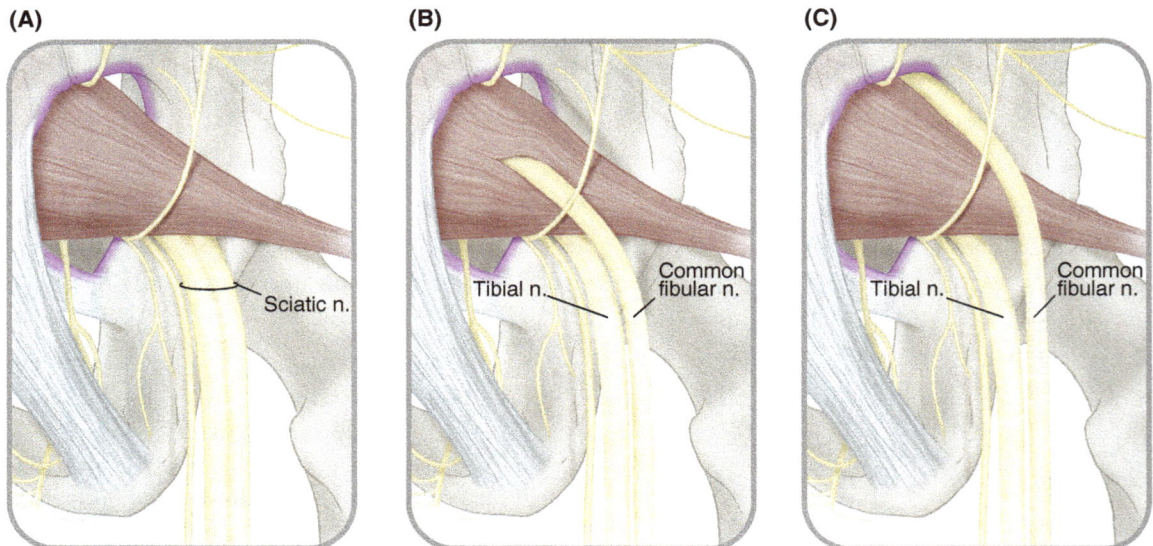

Sciatic n.

Tibial n.

Common fibular n.

Tibial n.

Common fibular n.

FIGURE 10.1.2 Posterior view showing the relationship between the piriformis muscle and the sciatic nerve. (**A**) The tibial and common fibular nerves are bound in a common connective tissue sheath and pass inferior to piriformis. (**B**) The common fibular nerve passes through piriformis. (**C**) The common fibular nerve passes superior to piriformis.

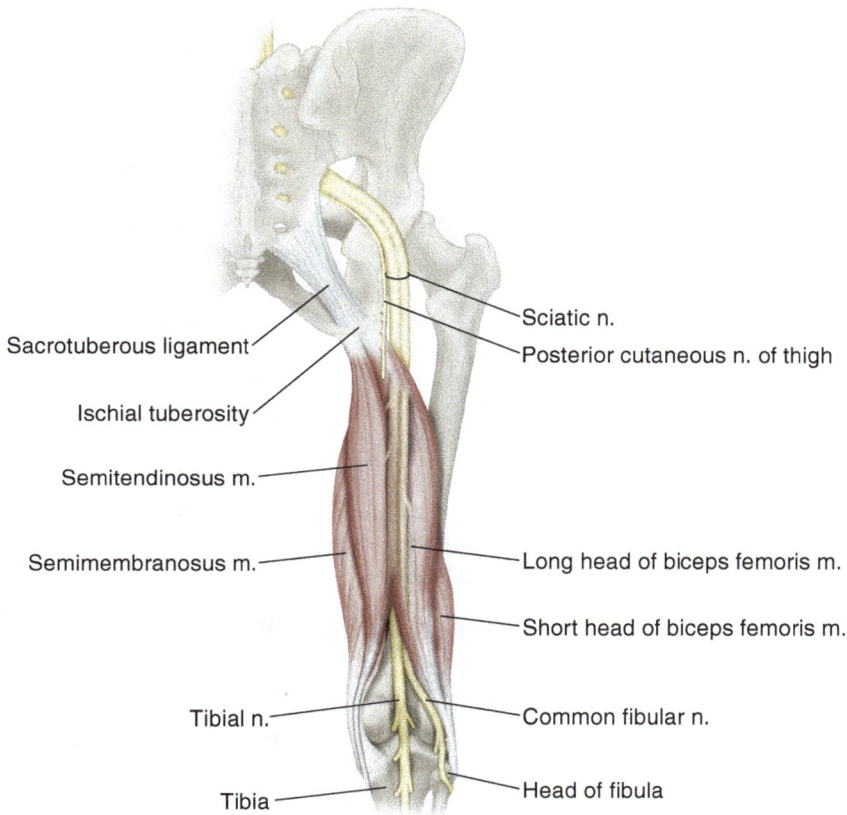

Sacrotuberous ligament

Ischial tuberosity

Semitendinosus m.

Semimembranosus m.

Sciatic n.

Posterior cutaneous n. of thigh

Long head of biceps femoris m.

Short head of biceps femoris m.

Tibial n.

Tibia

Common fibular n.

Head of fibula

FIGURE 10.1.3 Posterior compartment of the thigh.

2. The **common fibular nerve** innervates the short head of biceps femoris and parallels the tendon of the long head of the biceps (the short head of biceps femoris muscle is located in the posterior compartment of the thigh but is not part of the hamstring group).

Sensory Innervation of the Posterior Thigh and Leg

The **posterior cutaneous nerve of the thigh** (S1-S3) provides sensory innervation for the skin of the posterior thigh (**Fig. 10.1.4**). This nerve courses deep to the **fascia lata** and only its terminal branches lie in the superficial fascia. Commonly, this nerve supplies the skin as far distal as the mid-calf and distributes to a larger area of skin than any other cutaneous nerve.

Intervertebral Discs

IV discs occupy the spaces between vertebral bodies from C2 to S1. Collectively, the IV discs account for approximately 25% of the length of the spinal column. These symphysis-type joints are thickest in the lumbar region. Each disc is composed of two parts:

1. An outer **annulus fibrosus** formed by concentric rings of collagen and fibrocartilage. The anulus is thinnest posteriorly. It is tightly bound to adjacent vertebral bodies and is important for the integrity of the spinal column. The outer lamellae of fibrocartilage are vascularized.

2. An inner, eccentrically placed (i.e., off-center) core of gelatinous tissue is called the **nucleus pulposus** (**Fig. 10.1.5**). The more-fluid consistency (85% water) of the nucleus pulposus gives it an effective shock-absorbing

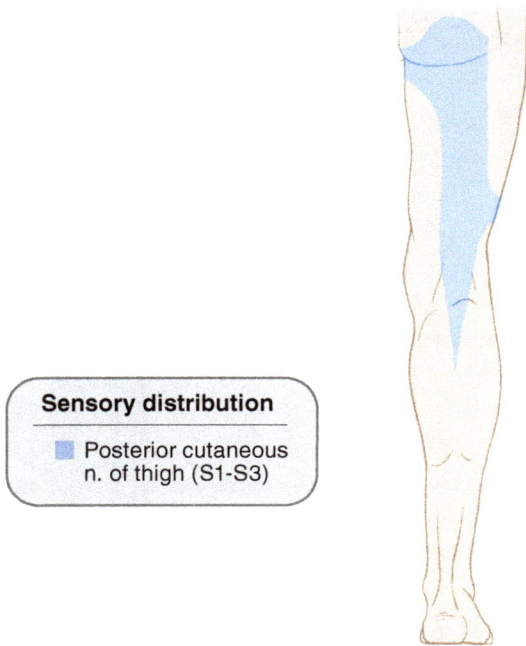

Sensory distribution

■ Posterior cutaneous n. of thigh (S1-S3)

FIGURE 10.1.4 Sensory innervation of the posterior thigh and leg.

function. The central portion of the disc is the largest area of avascular tissue in the body.

Clinical Note

The anulus fibrosus thickens with age and develops cracks and cavities. These areas of weakness in the anulus may predispose the disc to herniation of the nucleus pulposus.

CLINICAL REASONING

Based on the history, physical examination, and radiological imaging, the patient has symptoms of **sciatica**.

Chronic Hamstring Tendinitis

This is an inflammation of the tendon of the hamstring muscles on, or near, their attachment to the ischial tuberosity.

Signs and Symptoms

Sensory Deficits

■ Pain and tenderness over ischial tuberosity

■ Pain during stretching or contraction of hamstring muscles

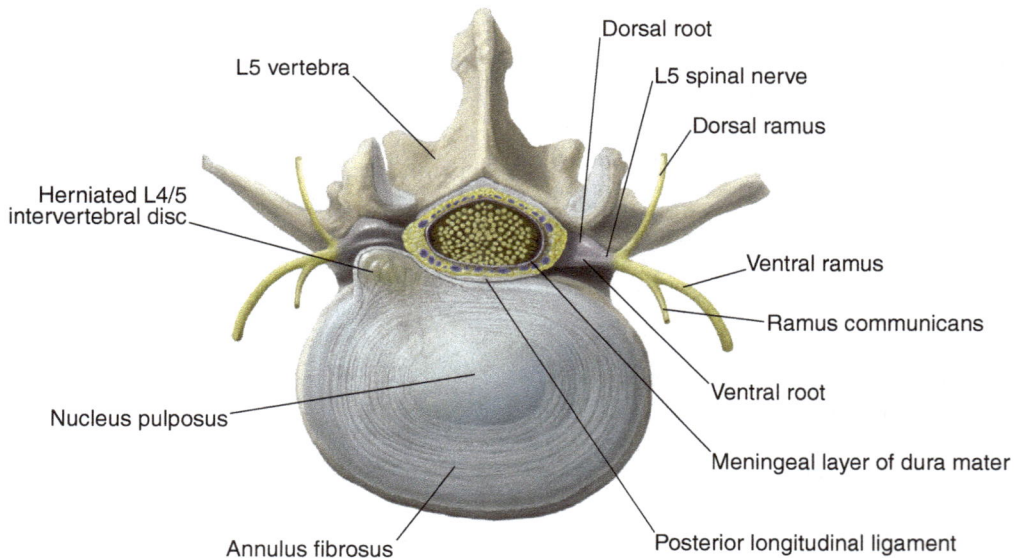

FIGURE 10.1.5 Herniated lumber intervertebral disc.

Predisposing Factors

- Overuse of hamstring muscles, usually through excessive training, distance running, jumping, or kicking
- Inadequate warm-up prior to strenuous activity

Herniated Lumbar Intervertebral Disc

This involves a herniation of the nucleus pulposus through a ruptured anulus fibrosus of an IV disc (**Fig. 10.1.5**). Most lumbar disc herniations are posterolateral, typically involving the L4-L5 and/or L5-S1 discs. The herniated tissue may put pressure on a spinal nerve in its IV foramen. A posterolateral herniation of the L4-L5 disc will affect the L5 spinal nerve; herniation of the L5-S1 disc will affect the S1 spinal nerve.

Signs and Symptoms

Sensory Deficits

- Constant pain that radiates unilaterally from the lower back into the gluteal region, posterior thigh, leg, and/or foot
- Electric shock-like pain, burning, and/or tingling in posterior lower limb

Motor Deficit

- Muscle weakness, cramping, or spasm in posterior thigh, leg, and/or foot

Predisposing Factors

- Age: peak incidence at 40 years
- Posture
- Obesity
- Improper lifting techniques

This condition results from a weakened or ruptured anulus fibrosis of a lumbar IV disc. This allows the gelatinous nucleus pulposus to protrude into surrounding tissues. Most herniations occur in a pos-

terolateral direction because of the thinner anulus in this region. The herniated disc is displaced laterally because of the posterior longitudinal ligament, which reinforces the anulus in the midline.

The herniated tissue may contact a spinal nerve on the affected side, compressing and/or chemically irritating the nerve. This will manifest as compromised impulse conduction, producing symptoms such as muscle weakness, pain, numbness, and tingling in areas of distribution for that spinal nerve. Compression to the L5 or S1 nerve by a herniated disc will manifest symptoms in the branches of the gluteal and sciatic nerves.

Clinical Note

IV discs may herniate in other regions of the spinal column, especially the lower cervical portion. Nevertheless, 90% of herniated discs occur at the L4-L5 or L5-S1 levels. Spinal nerves L4 and L5 contribute to most of the nerves that form the lumbosacral plexus (L4-S4).

Pain in the lower back and/or lower limb during the crossed SLR test is almost pathognomonic for a herniated lumbar disc. In this test, the patient is supine and the examiner flexes the *contralateral* thigh (with the knee in full extension and the ankle in the neutral position) in a 35-70° arc. As the thigh is flexed, increased traction is placed on the sciatic nerve. This will be transmitted to the dural sac surrounding the cauda equina. At 70° of thigh flexion, the traction is sufficient to slightly displace on the dura mater and the L4 and L5 spinal nerve roots on the opposite (affected) side. If the roots and surrounding dura mater are compressed by a herniated disc, pain will be elicited.

Piriformis Syndrome

Hypertrophy or spasm of the piriformis muscle due to overuse may compress the sciatic nerve in the greater sciatic foramen (**Fig. 10.1.6**).

Pathognomonic Characteristic symptom that points unmistakably to a specific disease

Hypertrophy Increase in size of a tissue or organ due to increased cell size, that is, without increased cell number (antonym: hyperplasia)

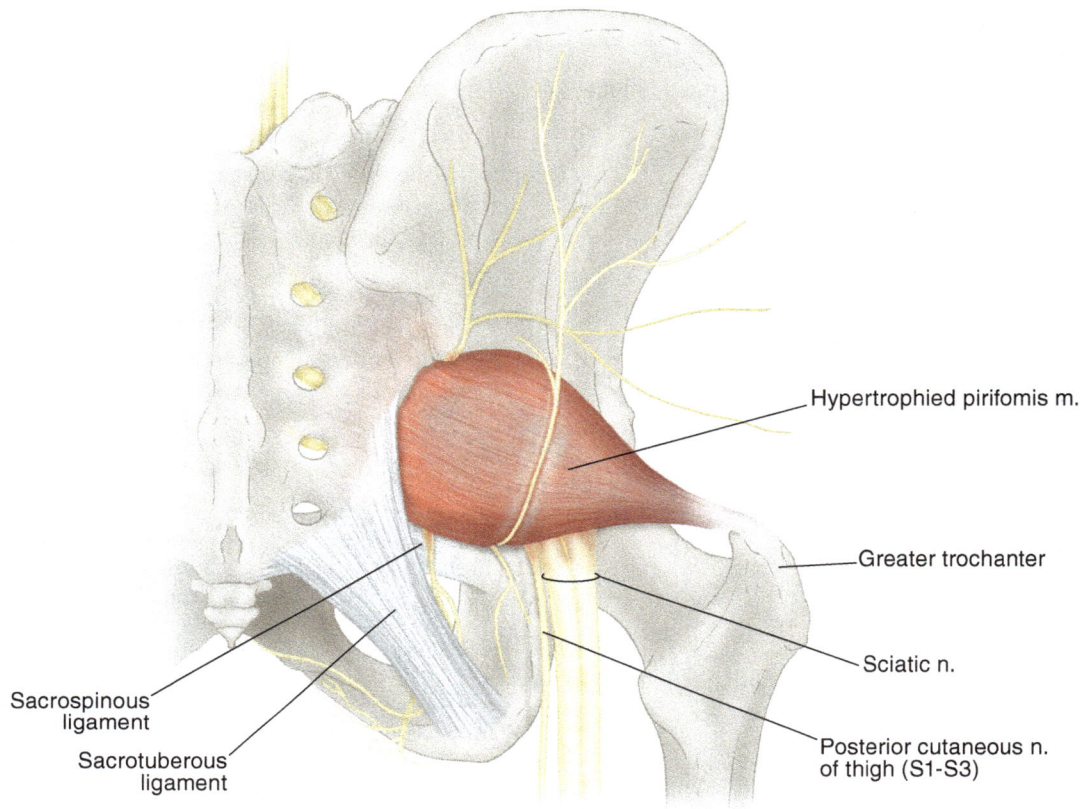

Hypertrophied pirifomis m.

Greater trochanter

Sciatic n.

Posterior cutaneous n. of thigh (S1-S3)

Sacrospinous ligament

Sacrotuberous ligament

FIGURE 10.1.6 Posterior view of gluteal region (gluteal muscles removed) showing hypertrophied piriformis muscle and compression of surrounding nerves.

Signs and Symptoms

Sensory Deficits

- Pain in gluteal region
- Pain radiating into posterior thigh and/or calf

Motor Deficit

- Decreased range of motion in hip, especially lateral rotation

Predisposing Factors

- Sex: 6:1 female
- Overuse of piriformis muscle (distance runners, hockey players)
- Common fibular nerve passing through piriformis muscle (**Fig. 10.1.2B**)

Piriformis syndrome is caused by hypertrophy, spasm, or scarring of the muscle, all brought about by overuse. Activity that involves repetitive, external rotation of the hip joint will contribute to overuse of piriformis. Piriformis muscle, together with other small muscles of the deep gluteal region, is important in holding the head of the femur within the acetabulum. Distance running, for example, will require piriformis to be active for long periods of time not only in movements of the hip joint but also its stabilization.

Clinical Note

The signs and symptoms for this condition often mimic those of a herniated L4-L5 or L5-S1 IV disc. Indeed, until ruled out through the SLR test or by imaging, a herniated disc is usually suspected.

DIAGNOSIS

The patient presentation, medical history, physical examination, and imaging studies support a diagnosis of **piriformis syndrome**.

Piriformis Syndrome

Activity that involves repetitive, external rotation of the hip joint will contribute to overuse of piriformis. The underlying mechanism for this syndrome is compression of components of the sciatic nerve as it passes through the greater sciatic foramen with the piriformis muscle. Hypertrophy, spasm, or trauma of the muscle can compress the nerve as it passes near, or through, the muscle in the foramen.

- The patient's history (hockey player and runner) suggests the potential for overuse of piriformis muscle.
- Pain in the gluteal region and posterior thigh with deep gluteal pressure suggests irritation of the piriformis muscle and/or the sciatic nerve.
- Further support is gained by eliciting pain by stretching the piriformis muscle (medial rotation and adduction of the hip joint).
- Pain may be referred to the S1-S2 dermatomes (posterior thigh and posterior leg).

Hamstring Tendinitis

Diagnosis of this condition is usually straightforward based on the presenting factors. Pain in the hamstring muscle group or over the ischial tuberosity during hip flexion is nearly pathognomonic for injury to the posterior thigh muscles. This may include tendinitis, tearing of muscle tissue, microvascular hemorrhage into muscle, or scarring from earlier injuries. Usually, there is no sensory loss.

- Hamstring tendinitis can be ruled out since there is an absence of pain when pressure is applied to the ischial tuberosity and to muscles of the posterior thigh (hamstrings).

Herniated Lumbar Intervertebral Disc

Herniated lumbar discs usually involve the L4-L5 or L5-S1 vertebral levels. They are typically the result of a failure of the anulus fibrosis to retain the nucleus pulposus, the more liquid core of the disc. The nucleus pulposus may distend the annulus, or it may rupture through it. The anulus is weakest posterolaterally and most hernias occur at this position. The herniated tissue applies pressure to the spinal nerve in the IV foramen, producing characteristic symptoms of radiating pain through the back and lower limb, and lower limb muscle weakness.

- MRI shows a normal lower lumbar spine and the negative crossed SLR test makes a herniated lumbar IV disc unlikely in this patient.

10.2 Femoral Neck Fracture

Patient Presentation

An 83-year-old female complains of severe pain in her left hip following a fall the previous day.

Relevant Clinical Findings

History

The patient relates that she was taking her wheeled garbage can to the curbside when she lost her balance and fell after stepping on an uneven area in her concrete driveway. She is height-weight proportionate and has had annual physical examinations the past 30 years. Bone density evaluations during the past 15 years show a progressive loss of cortical bone.

Physical Examination

The following findings were noted on physical examination:

- Limited range of motion in left hip joint (compared to right hip).
- Weight bearing on, or any movement of, left hip produces pain.
- Left lower limb externally rotated and appears "shorter" than the right.
- Small contusion and abrasion on left knee.

Imaging Studies

- Radiographs of the left hip revealed intra-capsular bone fragments.

Clinical Problems to Consider

- Acetabular fracture
- Femoral neck fracture
- Intertrochanteric fracture

LEARNING OBJECTIVES

1. Describe the structure and relationships of the hip joint.
2. Describe the structures that maintain hip joint integrity.
3. Describe the vasculature and nerve supply of the hip joint.
4. Explain the anatomical basis for the signs and symptoms associated with this case.

RELEVANT ANATOMY

Hip Joint

The **hip joint** is a ball-and-socket joint formed by the **head of the femur** and the **acetabulum** of the

hip bone (os coxae). The femur is the longest bone in the body and its proximal end presents a **head**, **neck**, and **shaft**. The proximal shaft has two projections for muscle attachment, the **greater and lesser trochanters**. The **intertrochanteric line** (anteriorly) and **intertrochanteric crest** (posteriorly) connect the trochanters.

The globular head is smooth and covered by articular cartilage except at its **fovea**, a shallow depression for the **ligament of the head of the femur**. The short, thick neck joins the head to the shaft and is set at an angle (of inclination) that varies with age and sex. The socket of the hip joint is the **acetabulum**, a deep depression on the lateral aspect of the hip bone that

---

Contusion Mechanical injury beneath skin that results in subcutaneous hemorrhage (i.e., a bruise)

accommodates the head of the femur. Until the late teen years, the hip bone is represented by three separate bones: **ilium**, **ischium**, and **pubis**. These fuse in the acetabulum, with each bone contributing approximately one-third of the acetabulum.

- The interior of the acetabulum has a horseshoe-shaped, **articular surface**. The articular surface is absent inferiorly.

- An **acetabular notch** allows vessels and nerves to enter the joint.

- A **nonarticular surface** provides attachment for the ligament of the head of the femur.

- The **brim of the acetabulum** has a thick fibrocartilaginous **labrum** that deepens the socket and is important in retaining the femoral head within the acetabulum.

Supporting Structures of the Hip Joint

The hip joint is surrounded by a thick fibrous capsule that encloses the joint and neck of the femur. Anteriorly the capsule attaches to the intertrochanteric line (**Fig. 10.2.1A**). Posteriorly, the capsule does not reach the intertrochanteric crest so the lateral part of the femoral neck is extracapsular (**Fig. 10.2.1B**).

The capsule is thickened by three ligaments:

1. Iliofemoral
2. Pubofemoral
3. Ischiofemoral

The spiral course of these ligaments provides the mechanism for their two primary functions:

1. They act as check ligaments for hip movements.
2. They become "tightly wound," especially during hip extension, to hold the femoral head in the acetabulum.

A deep group of small, short muscles collectively are important in pulling the femoral head deep into the acetabulum:

- Piriformis
- Obturator internus
- Obturator externus
- Superior gemellus
- Inferior gemellus
- Quadratus femoris

These muscles extend between the pelvic bones and the femur and lie in contact with, or near, the posterior

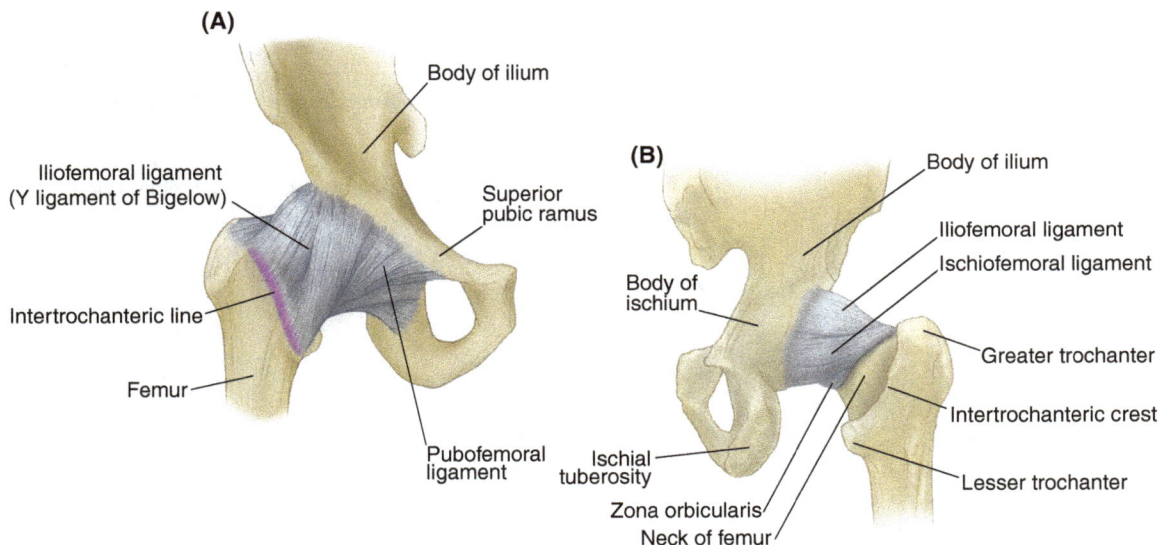

FIGURE 10.2.1 Fibrous capsule and ligaments of the hip joint. (**A**) Anterior view and (**B**) posterior view.

and inferior aspects of the fibrous capsule. They act primarily as lateral rotators of the hip joint.

Blood Supply to the Hip Joint

The bones of the hip joint receive blood from the **medial and lateral circumflex femoral** and the **obturator** arteries (**Fig. 10.2.2**). Both circumflex femoral arteries contribute to an **extracapsular vascular ring** located on the femoral neck, near its union with the shaft. From this vascular ring, small **retinacular arteries** pierce the joint capsule to enter the neck and head of the femur. In addition, an **acetabular artery**, derived from either the obturator or medial circumflex femoral artery, passes through the acetabular notch to enter the hip joint. This artery courses along the ligament of the head of the femur and supplies a small variable area of the femoral head.

Sensory Innervation of the Hip Joint

The capsule of the hip joint receives its sensory nerve supply from several sources:

Anterior Capsule
- Obturator
- Femoral

Posterior Capsule
- Nerve to quadratus femoris
- Superior gluteal
- Sciatic nerve

CLINICAL REASONING

This elderly patient presents with signs and symptoms indicating a clinical condition of hip injury following a fall at home.

Acetabular Fracture

This involves a fracture of the acetabulum of the hip joint (**Fig. 10.2.3A**).

Signs and Symptoms

Sensory Deficit
- Hip pain on weight bearing

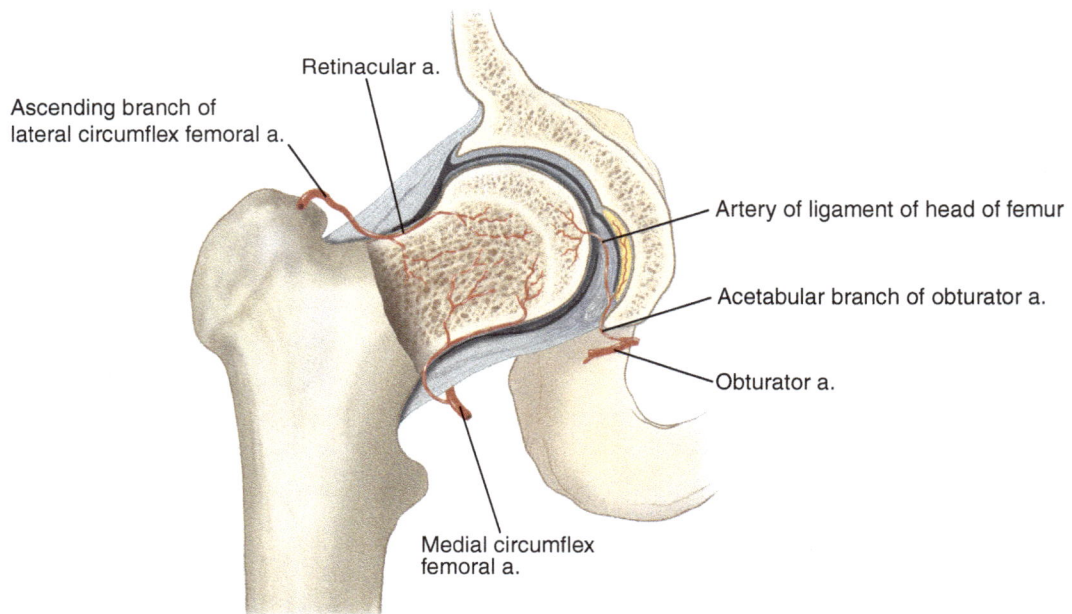

FIGURE 10.2.2 Anterior view showing the blood supply to the hip joint. The hip joint is shown in a coronal section.

(A)

(B)

(C)

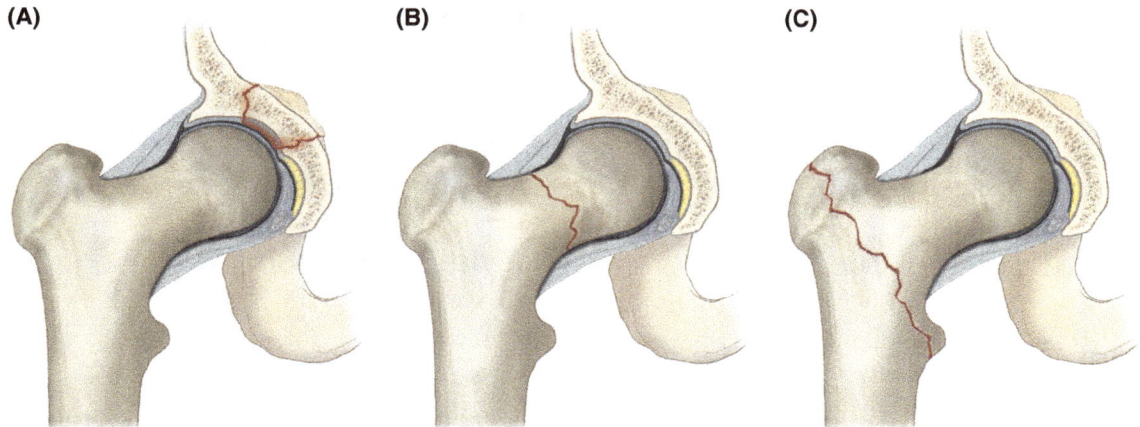

FIGURE 10.2.3 Anterior view of the hip illustrating different types of fracture. **(A)** Acetabular, **(B)** femoral neck, and **(C)** intertrochanteric. The acetabulum is shown in a coronal section.

Other Deficit

- Contusions in knee or hip area

Predisposing Factors

- Age: young adults

Fractures of this type are classified by the location of the fracture in the socket. A primary concern is that bone fragments or uneven distribution of pressure will disrupt the articular cartilage and lead to degenerative disease of the joint. Repair is sometimes made more difficult because the acetabular fracture may damage genitourinary structures in the pelvis.

Acetabular fractures before middle age are most often associated by severe, high-energy trauma (motorized vehicle accident). Typically, force is transmitted from the foot or knee along the femoral shaft, neck, and head to the acetabulum. The compressive force of the femoral head to the acetabulum causes the acetabulum to fracture. In this population, 50% of cases are accompanied by trauma to internal organs. The patient presentation is often very similar for acetabular fracture and femoral neck fracture. Distinction can be made with radiographic evidence.

Clinical Note

Acetabular fractures in the elderly occur most often in females with a history of osteoporosis. If the articular cartilage is not damaged, surgery may not be indicated and the fracture is allowed to heal, even if the alignment is not ideal.

Femoral Neck Fracture

This involves a fracture of the neck of the femur (**Fig. 10.2.3B**) and is often referred to as a "hip fracture."

Signs and Symptoms

Sensory Deficit

- Hip pain on weight bearing

Motor Deficit

- External rotation of femur

Other Deficit

- Contusions in knee or hip area

Predisposing Factors

- Age: >65 years (75% of hip fractures occur in patients >75 years)

- Sex: female (approximately 2:1)
- Osteoporosis

These fractures are broadly categorized as undisplaced and displaced. In undisplaced fractures, bone alignment is not disrupted. Displaced fractures are characterized by:

- Comminuted fracture
- Disruption of retinacular vessels

Retinacular vessels provide the primary blood supply to the femoral neck and head. These arteries are very small and anastomoses are not well established. Their disruption makes bone fragments vulnerable to avascular necrosis (AVN) and, therefore, nonunion during the healing process. The femoral head is also vulnerable to AVN in displaced fractures, since most of its blood supply is from arteries on the posterior surface of the neck. The artery in the ligament of the head of the femur may not be sufficient to support its viability; furthermore, it is absent in 20% of the population. In addition, these fractures can be problematic in the healing process because the periosteum of the femoral neck is thin and, thus, poorly supports bone healing through callus formation.

Clinical Note

AVN is reported to occur in as many as 80% of displaced femoral neck fractures and 11% of undisplaced fractures. There is unresolved controversy whether early or late surgical fixation of displaced femoral fractures is more effective in reducing AVN.

Intertrochanteric Fracture

This involves a fracture that connects the greater and lesser trochanters of the femur, usually along the intertrochanteric line.

Comminuted fracture A fracture with three or more
  pieces
Necrosis Pathologic death of cells, tissues, or organs

Signs and Symptoms
Sensory Deficit
- Hip pain on weight bearing

Other Deficit
- Contusions in knee or hip area

Predisposing Factors
- Age: >65 years
- Sex: female
- Osteoporosis

The patient presentation is often very similar for intertrochanteric fracture and femoral neck fracture. Distinction can be made with radiographic evidence. In general, these fractures are considered to be extracapsular even though, anteriorly, the fibrous joint capsule attaches along the intertrochanteric line.

Often with intertrochanteric fracture, the hip joint will be in a position of extension and lateral rotation. Due to the presence of ample cortical bone and its associated blood supply, fractures in this part of the femur usually heal without complication.

Clinical Note

Intertrochanteric fractures are treated by the surgical procedure known as open reduction internal fixation (ORIF) that involves realigning the fractured elements during surgery and reattaching the fragments with hardware (screws and plates).

DIAGNOSIS

The patient presentation, medical history, physical examination, and imaging studies support a diagnosis of **comminuted fracture of the femoral neck**.

Hip fractures associated with falls in the elderly, especially females with a history of osteoporosis, typically involve the femoral neck, femoral trochanters, or the acetabulum. Radiographic imaging is necessary to distinguish among these.

- Radiographs in this patient showed a comminuted fracture of the **femoral neck**.

- An **intertrochanteric fracture** would be extracapsular.

- An **acetabular fracture** involves the hip bone rather than the femur.

An important predisposing factor is the degree of osteoporosis in the femoral neck and trochanters, and the acetabulum.

Clinical Note

There is a growing belief that fractures of the neck of the femur *cause the patient to fall* rather than the opposite, that is, the patient falls, resulting in a fracture. In this case, the patient with osteoporosis missteps on an uneven surface and loses her balance, resulting in a fall. When the fracture occurred cannot be determined, however, the possibility that the fracture was caused by the altered joint biomechanics when she misstepped cannot be ruled out.

Patient Presentation

A 20-year-old female college lacrosse player presents in the emergency department with right knee pain.

Relevant Clinical Findings

History

The patient describes an incident during a practice session the previous evening when she made a sudden turn to avoid a teammate. The patient had her right foot planted when she heard a "pop" sound and immediately felt pain in her knee. Overnight, the knee area became swollen with persistent pain, especially when trying to stand on her right foot. She feels that the knee is unstable when bearing weight.

Physical Examination

The following findings were noted on physical examination:

- Swelling of soft tissues have made the diameter of the right knee significantly greater than the left.
- Pain is increased with attempted torsion and sliding movements of the knee joint.
- Lachman and anterior drawer tests were positive.
- Posterior drawer, McMurray, valgus MCL, and tibial sag tests were negative.

Imaging Studies

- Magnetic resonance imaging (MRI) of the right knee revealed trauma to the anterior cruciate ligament (ACL).

Clinical Problems to Consider

- ACL rupture
- ACL strain
- Posterior cruciate ligament (PCL) rupture
- Medial collateral ligament (MCL) tear

LEARNING OBJECTIVES

1. Describe the anatomy of the knee joint including the ligaments that support it.
2. Describe the role of each knee ligament in movements of the joint.
3. Explain the anatomical basis for the signs and symptoms associated with this case.

RELEVANT ANATOMY

The weight-bearing knee joint is vulnerable to trauma due to the relationship of the two primary bones involved. The distal end of the **femur**, with its rounded condyles, rests on the relatively flat **tibial plateau**. Thus, bone structure adds little stability to the joint. Soft tissues stabilize the joint: ligaments, muscles and their tendons, and fascia—the most important among these being the ligaments.

Ligaments of the Knee Joint

Typical of synovial joints, the knee has a **fibrous capsule**. Knee ligaments are classified as extracapsular and intracapsular (**Fig. 10.3.1**). The two major **extracapsular ligaments** that are important in holding the femur and tibia in contact are:

1. Medial (tibial) collateral ligament
2. Lateral (fibular) collateral ligament

These ligaments extend from the respective femoral condyle to the medial condyle of the tibia or the head of the fibula.

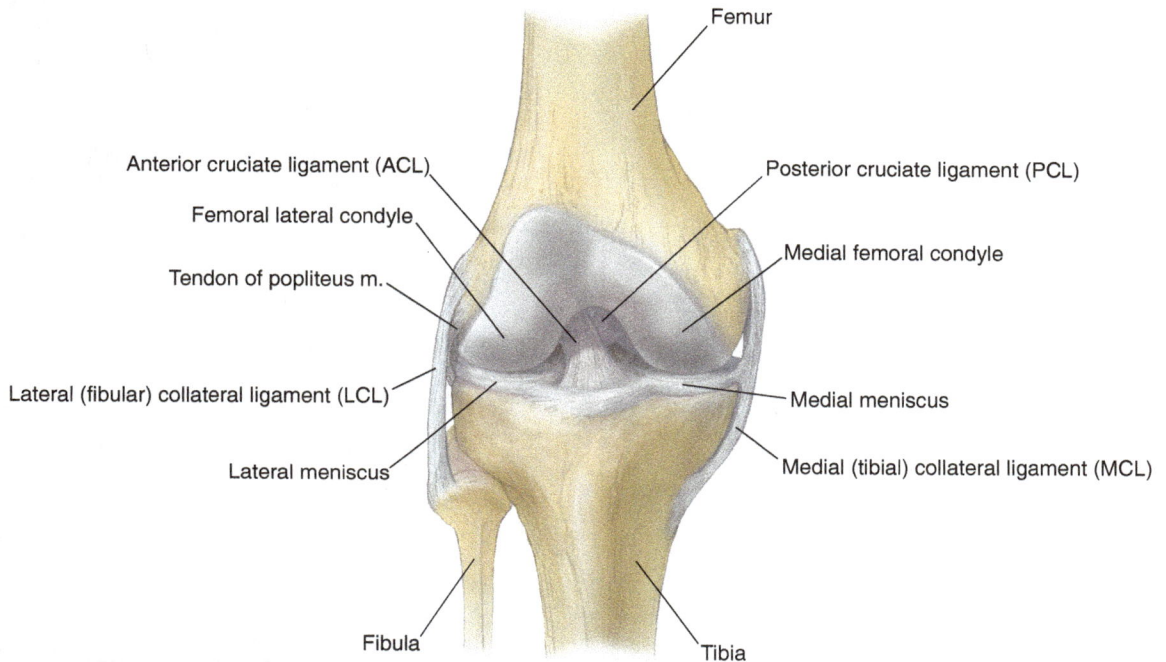

FIGURE 10.3.1 Anterior view of the knee showing its ligament and menisci.

Two **intracapsular ligaments** have a major role in knee joint integrity:

1. **ACL**
2. **PCL**

The ACL and PCL extend from respective points of the intercondylar eminence of the tibia. These ligaments cross one another so that the ACL attaches to the medial side of the lateral condyle of the femur and the PCL attaches to the medial side of the medial condyle of the femur. In the fully extended knee, the ACL is taut. In contrast, the PCL is taut in the flexed knee. The PCL is the strongest of the knee ligaments in that it can sustain greater stress before injury.

While not ligaments, the fibrocartilaginous **medial and lateral menisci** are classically included with descriptions of intracapsular knee ligaments. These "C-shaped" plates rest on the respective articular surfaces of the tibial plateau. The limbs of the "C" are attached to the intercondylar eminence. The medial meniscus is less mobile due to the additional attachment of its lateral margin to the MCL. The lateral meniscus is separated from the LCL by the intracapsular part of the tendon of popliteus muscle. The functions of the ligaments and menisci of the knee are outlined in **Table 10.3.1**.

Clinical Notes

The following tests are commonly used to assess for knee injury:

Lachman Test for ACL Injury

For this test, the patient is supine and the knee is flexed to 30°. One of the examiner's hands supports the tibia with the thumb against the tibial tuberosity. The other hand should be placed on the anterior thigh, just above the knee. The examiner pulls the tibia anteriorly with the femur fixed by the other hand. A soft or "mushy" end point of the anterior translation of the tibia is positive for an **ACL injury**. A hard or fixed end point is considered a negative result.

Tibial Sag Test for PCL Injury

During this test, the patient is supine with the hip and knee flexed to 90°. In this position, the weight of the tibia will "sag" posteriorly without the support of the PCL.

Drawer Test for ACL or PCL Injury

The patient is supine and the hip is at 45° of flexion and the knee is flexed to 90°. The examiner places one hand around the proximal tibia, near the joint line. The digits should wrap around the tibia and contact the hamstring tendons to assure that these muscles are relaxed. The tibia is then drawn anteriorly and pushed posteriorly (as in opening and closing a drawer) to assess excessive translation between the tibia and femur. Excessive **anterior sliding is indicative of ACL injury** and excessive **posterior displacement indicates PCL injury**.

Valgus Test for MCL

The valgus MCL test involves placing the knee in 30° flexion. With one hand on the thigh and the other grasping the heel, the heel laterally displaced (valgus) to test the integrity of the MCL. Increased lateral displacement of the tibia and a widening of the knee joint line on the medial side supports the diagnosis of MCL damage.

McMurray Test for Meniscal Injury

During this test, the patient is supine and the thigh flexed to 90°. The knee is tightly flexed (heel against buttocks) and the examiner grasps the knee with one hand so that the thumb is over one meniscus and fingers are over the other. With the other hand grasping the heel, the examiner laterally rotates the knee and extends the knee toward 90° of flexion. This same procedure is followed with medial rotation of the knee. A palpable or audible click during either procedure is a strong indication of **injury to the meniscus** on the side of rotation.

It is essential in all of these tests to evaluate the uninjured limb to establish a baseline for comparison.

CLINICAL REASONING

This patient presents with signs and symptoms indicating a clinical condition of knee pain associated with a sporting event accident.

TABLE 10.3.1	Structure and function of ligaments and menisci of the knee.
Structure	**Function(s)**
MCL and LCL	▪ Stabilize respective sides of knee joint during its full range of motion
ACL	▪ Prevents posterior displacement ("sliding") of distal femur on tibial plateau ▪ Resists rotation at knee joint when foot is planted ▪ Resists hyperextension of knee joint
PCL	▪ Prevents anterior displacement ("sliding") of distal femur on tibial plateau ▪ Resists hyperflexion of knee joint ▪ Major support for knee when weight-bearing limb is flexed (e.g., walking downhill)
Menisci	▪ Deepen the articular surface for the femoral condyles ▪ Deaden shock to knee joint during locomotion ("shock absorbers")

ACL Rupture

Rupture of the ACL (**Fig. 10.3.2**) is a common injury in athletes participating in court and field competitions. It often occurs in the absence of contact with another player. The classic scenario is the athlete who changes direction quickly, planting one foot firmly with the knee slightly flexed. The torsion produced by the change in direction, with the knee and foot in this position, places increased stress on the ACL.

Signs and Symptoms

Sensory Deficit
- Immediate knee pain and disability

Other Deficit
- Audible "pop" sound at the time of injury
- Sense of knee instability
- Swelling and hematoma
- Positive Lachman and anterior drawer tests

Hematoma Localized extravasation of blood, usually clotted

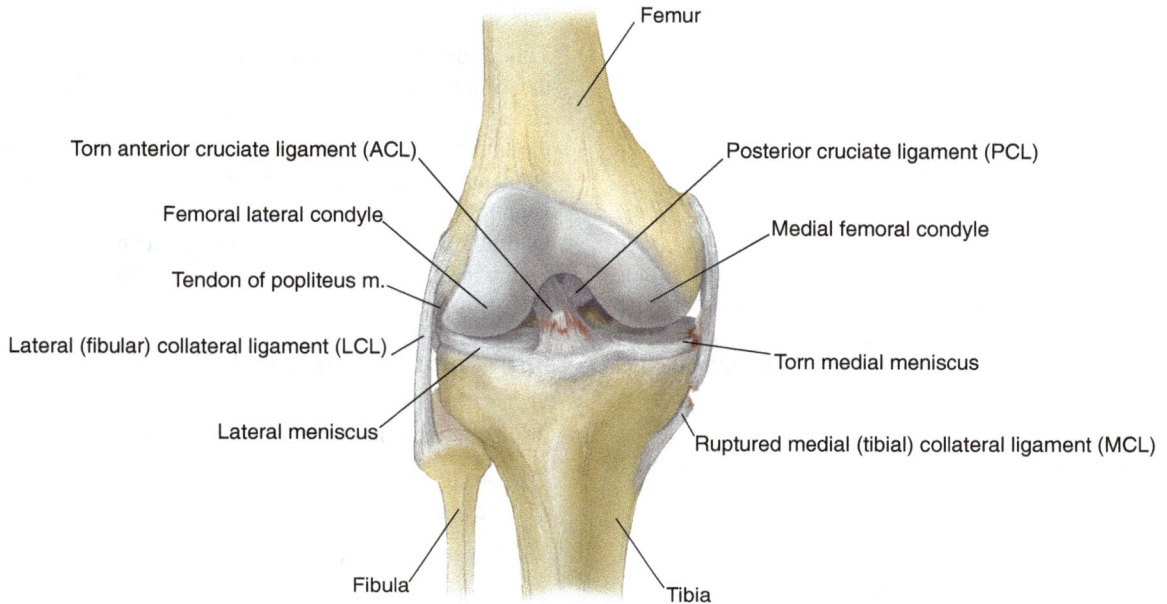

FIGURE 10.3.2 Anterior view of the knee showing the "unhappy triad," which involves injury to the ACL, MCL, and medial meniscus.

Predisposing Factors

- Sex: 10:1 female
- Participant in court/athletic field competition (e.g., soccer, football, downhill skiing, and volleyball)
- Improper conditioning and training

Some of the collagen bundles that make up the ligament are more vulnerable to trauma than others, so rupture may be only partial. The ligament may also avulse its bony attachments, most commonly from the tibia.

Clinical Notes

ACL trauma is more prevalent in female athletes. The evidence is that, compared to males, females have:

- Less strength in thigh muscles, putting more stress on ligaments for knee support
- A wider pelvis and greater angle of inclination (see case 10.2)
- Cyclic hormone changes that may affect muscle and ligament tension

Strengthening regimens for muscles acting on the knee can add sufficient stability to the knee joint to compensate for a compromised ACL.

ACL Strain

The cruciate ligaments are relatively inelastic. A diagnosis of strain to the ACL suggests that the ligament has been stretched beyond its normal limits, but the ligament has not been partially or totally separated

Signs and Symptoms

Sensory Deficit

- Knee pain at time of injury

Other Deficits

- Swelling
- Positive Lachman and anterior drawer tests

Predisposing Factors

- Sex: female
- Participant in downhill skiing or court/athletic field competition
- Improper conditioning and training

PCL Rupture

PCL rupture occurs most often during a fall with the knee flexed, with force of the impact received by the tibial tuberosity. This forces the tibia posteriorly, increasing the stress on the PCL. Normally, the PCL limits anterior displacement of distal femur on the tibial plateau. With this type of injury, the PCL is compromised and the tibial plateau is displaced with respect to the femoral condyles.

Signs and Symptoms

- Contusion over proximal tibia
- Positive posterior drawer test
- Positive tibial sag sign

Predisposing Factors

- Participant in court sports (basketball, volleyball)
- Front seat passenger in automobile accident (with passenger seated and knee flexed, the tibia is forced against the dashboard)

PCL injuries often present with minimal symptoms. Pain may be absent and injuries to the PCL may produce a range of impairments. Because the PCL is the stronger than the ACL, and the presenting symptoms with an injured PCL may be minimal, these injuries have received relatively little attention compared to those involving the ACL.

Clinical Note

A **false-positive Lachman test** may result from a damaged PCL. With PCL damage, the tibial plateau is more posterior than normal. Therefore, during the Lachman test there is more anterior movement of the tibia before it reaches its end point, leading to the erroneous conclusion (false-positive Lachman) of ACL damage. The end point in the Lachman test, however, will still be firm with a PCL injury.

MCL Tear

This injury is most often associated with trauma to the lateral knee region. The trauma is usually associated with a teammate or opponent "crashing" into the lateral leg of an athlete, especially when the foot is planted and the knee is partially flexed. In this situation, the tibia and femur are forced apart on the medial side of the knee (the bones do not separate on the lateral side) and the stress is transferred to the MCL.

Signs and Symptoms

Sensory Deficit

- Pain and tenderness to medial knee

Other Deficits

- Swelling
- Sense of instability in knee joint
- Positive valgus test

Predisposing Factors

- Participant in court or field sports

MCL injuries often occur in association with ACL damage and medial meniscus tears (**Fig. 10.3.2**). This "unhappy triad" commonly involves the medial meniscus because of its attachment to the MCL.

DIAGNOSIS

The patient presentation, medical history, physical examination, and imaging studies support a diagnosis of a **ruptured ACL**.

ACL Rupture

A sudden change in direction with one foot firmly planted and the knee of the same side in slight flexion places increased stress on the ACL.

- The audible sound described by the patient and the circumstances surrounding the accident is highly indicative of rupture of the ACL. Positive Lachman and anterior drawer tests, which assess the ACL, support this diagnosis.
- The MRI in this patient confirmed a damaged ACL.

ACL Strain

Patients with ACL strain usually do not report an audible sound at the time of the knee injury. Otherwise, the signs and symptoms are very similar to ACL rupture.

- Distinction between strain and rupture will many times rely on information derived from imaging.

PCL Rupture

PCL rupture occurs when the tibia is forced posteriorly, placing increased stress on this ligament. Falls, with the knee flexed and the tibial tuberosity making contact with a hard surface, account for many of these injuries. In other cases, a force to the anterior leg with the foot firmly planted can damage the PCL (blocking below the waist in football). Patients with PCL rupture should present with positive:

- Posterior drawer test
- Tibial sag test
- Diagnosis is often aided by the patient's history, and may be supported by imaging or arthroscopic examination.

MCL Tear

Lateral trauma to the partially flexed knee when the foot is planted ("clipping" in football) forces the tibia and femur apart on the medial side of the knee and the produces the resultant stress to the MCL. These injuries often also involve the medial meniscus since it is anchored to the MCL. Patients with MCL tears should exhibit:

- Positive valgus test
- Positive McMurray test, if the medial meniscus is involved

10.4 | Common Fibular Nerve Trauma

Patient Presentation

A 38-year-old male complains of numbness on the dorsum of his right foot and difficulty lifting the front part of the foot.

Relevant Clinical Findings

History

The patient describes an accident the previous day when he was struck by an automobile as he stepped from the curb at a busy intersection. He was knocked to the pavement, but did not believe any serious injuries were sustained and did not seek medical attention at the time. Subsequently, he noticed numbness over the dorsum of his right foot and reports having to make a "high step" to avoid dragging his toes when he walks. He reports a progressive loss of the ability to raise his toes.

Physical Examination

The following findings were noted on physical examination:

- Sensory deprivation to the entire dorsum of the right foot
- Dorsiflexion and eversion against resistance weaker on the right side (compared to left side)
- Right foot drop and high steppage during the swing phase of gait
- Tenderness, edema, and a hematoma just distal to the head of the right fibula (**Fig. 10.4.1**)
- Normal deep tendon reflexes (DTR) for the quadriceps and calcaneal tendons
- Stable knee and ankle joints

Imaging Studies

- Radiographic imaging revealed a nondisplaced fracture of the neck of the right fibula.

Clinical Problems to Consider

- Anterior compartment syndrome
- Common fibular nerve trauma

LEARNING OBJECTIVES

1. Describe the anatomy of the lateral aspect of the knee region.
2. Describe the course and distribution of the common fibular nerve.
3. Explain the anatomical basis for the signs and symptoms associated with this case.

RELEVANT ANATOMY

Lateral Knee Region

Skeletal elements that contribute to the lateral aspect of the knee region include the **lateral condyles of the femur and tibia** and the **proximal end of the fibula**. The proximal fibula is composed of the **head, neck,** and **shaft**. The fibular head articulates with the lateral condyle of the tibia; the fibula does not contribute to the knee joint. The head of the fibula provides the distal attachment point for the **tendon of biceps femoris** and the **lateral (fibular) collateral ligament** of the knee.

Common Fibular Nerve and Its Branches

The subcutaneous portion of the **common fibular nerve (L4-S2)** is palpable on the neck of the fibula

---

Hematoma Localized extravasation of blood, usually clotted

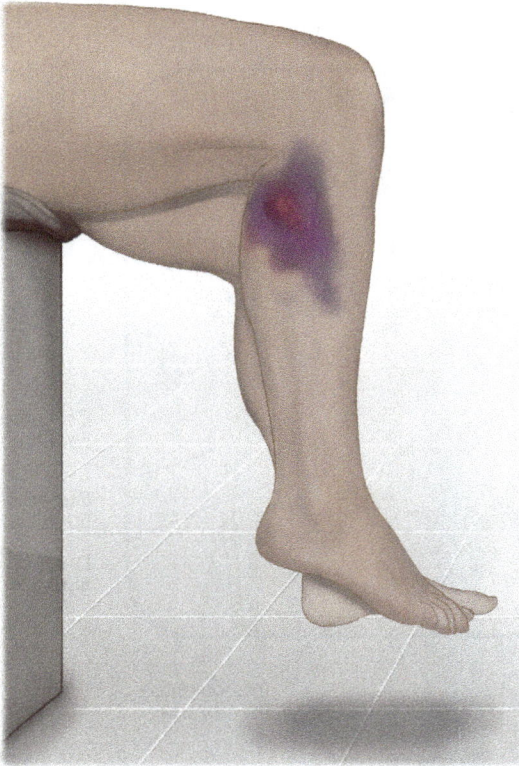

FIGURE 10.4.1 Illustration of trauma to right lateral leg and resultant foot drop.

(**Fig. 10.4.2**). This nerve is not accompanied by a major blood vessel. The common fibular nerve does not have a sensory distribution (although its terminal branches do) and innervates one muscle: the short head of biceps femoris.

On the neck of the fibula, the common fibular nerve divides into terminal branches:

- **Superficial fibular (L4-S1)**
- **Deep fibular (L4-L5)**

Both of these nerves have motor and sensory components.

The **superficial fibular nerve** supplies the muscles of the **lateral compartment of the leg**:

- **Fibularis longus**
- **Fibularis brevis**

In the distal one-third of the leg, the nerve pierces the crural fascia to supply sensory innervation from the distal anterolateral leg and most of the dorsum of the foot (**Fig. 10.4.3**).

The **deep fibular nerve** pierces the anterior intermuscular septum of the leg to enter the **anterior compartment of the leg**. It lies deep within this compartment and is joined by the **anterior tibial artery and vein**. From its position on the anterior aspect of the interosseous membrane (**Fig. 10.4.2**), it provides motor innervation to the muscles of the compartment:

- **Tibialis anterior**
- **Extensor digitorum longus**
- **Extensor hallucis longus**
- **Fibularis tertius**

The deep fibular nerve continues onto the dorsum of the foot (**Fig. 10.4.2**). Here, accompanied by the **dorsal artery of the foot** (the continuation of the anterior tibial artery), the nerve is located on the lateral side of the extensor hallucis longus tendon. On the foot, the nerve supplies **extensor hallucis brevis** and **extensor digitorum brevis**.

The deep fibular nerve supplies sensory innervation from the adjacent sides of toes 1 and 2 (**Fig. 10.4.3**).

Clinical Note

The subcutaneous position of the common fibular nerve on the fibular neck makes it vulnerable with trauma to the lateral knee region.

CLINICAL REASONING

This patient presents with signs and symptoms indicating a clinical condition of **foot drop**.

Anterior Compartment Syndrome

This serious syndrome involves the muscles, nerves, and vessels within the anterior compartment of the leg. The inelastic crural fascia, the interosseous membrane of the leg, and the tibia tightly enclose this compartment. The contents of this compartment include (see also **Fig. 10.5.1**):

- Deep fibular nerve

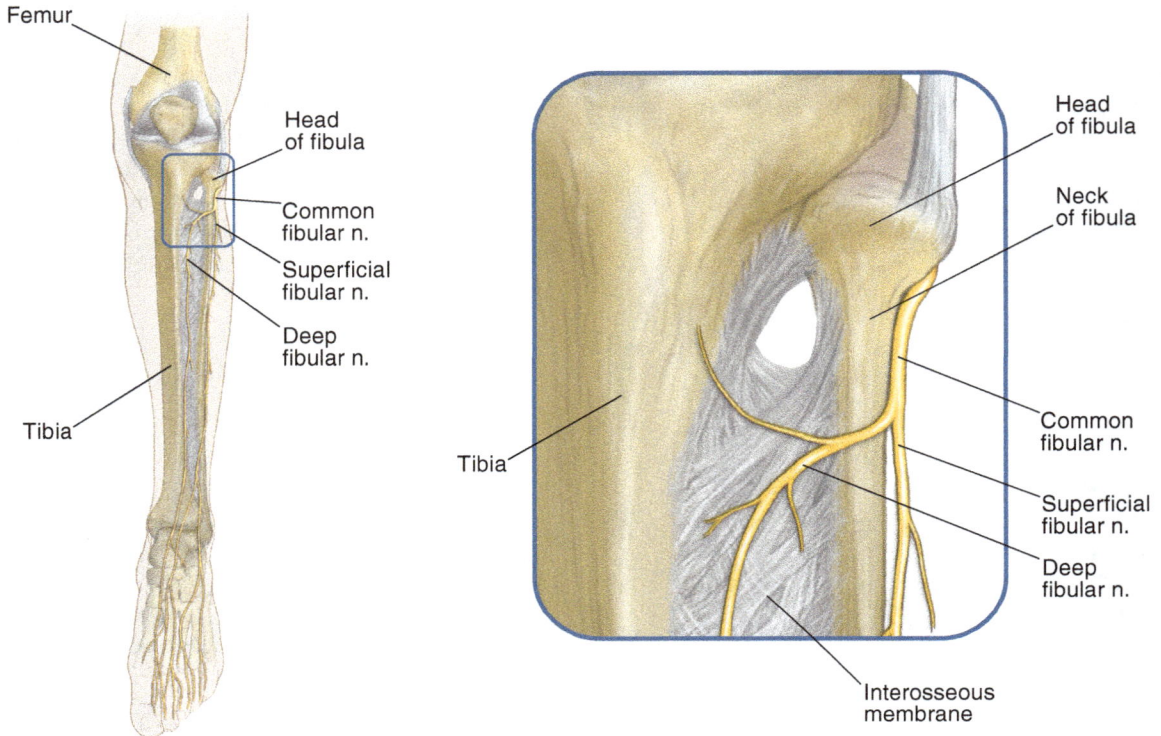

FIGURE 10.4.2 Anterior view of the common fibular nerve showing its course and branches in the leg and foot.

- Anterior tibial vessels
- Tibialis anterior muscle
- Extensor hallucis longus muscle
- Extensor digitorum longus muscle
- Fibularis tertius muscle

Excessive swelling of these muscles and the accumulation of interstitial fluid increase the pressure in the compartment. The increased pressure may result in ischemia. If not treated, structures in the compartment may become necrotic.

Signs and Symptoms

Sensory Deficits
- With acute anterior compartment syndrome, an early sign may be numbness from the adjacent

sides of the toes 1 and 2 (the cutaneous distribution of the deep fibular nerve)
- Pain in anterior leg compartment, especially during plantar flexion

Motor Deficit
- Weakness of dorsiflexion

Other Deficits
- Visible bulging of anterior compartment muscles
- Pallor over anterior compartment

Predisposing Factors

- Trauma to leg, including fractures (acute compartment syndrome)

Ischemia Local anemia due to vascular obstruction
Necrosis Pathologic death of cells, tissues, or organs

Pallor Pale skin

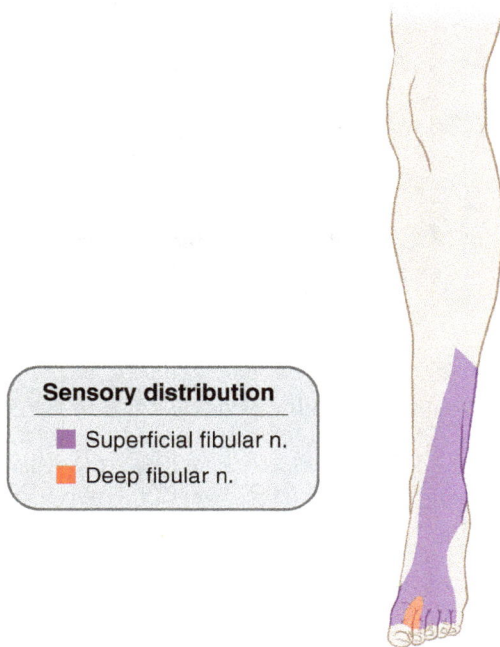

Sensory distribution

■ Superficial fibular n.
■ Deep fibular n.

FIGURE 10.4.3 Sensory distribution of the superficial and deep fibular nerves.

- Surgery to leg
- Athletic exertion, for example, running or dancing (chronic compartment syndrome)

Since this patient is presenting with signs and symptoms related to trauma, he must be evaluated for **acute anterior compartment syndrome**. If diagnosed, this is a medical emergency requiring immediate surgery (fasciotomy) to relieve increased compartment pressure.

Chronic anterior compartment syndrome is rarely a medical emergency, and the signs and symptoms tend to be transitory. Conservative treatments such as decreasing the intensity of athletic training, cross-training, running on cushioned surfaces, and using different footwear may reduce or eliminate the syndrome.

Clinical Note

It is important to distinguish chronic anterior compartment syndrome from "shin splints." While the two conditions have similar presenting complaints, shin splints are thought to be caused by overuse of muscles. Micro-tears may develop in muscles as a result of muscle fiber rupture, separation of muscle fibers from their tendon, or the detachment of the tendon from the bone. Inflammation of muscle (myositis), tendon (tendinitis), and/or periosteum (periostitis) will cause pain. Shin splints are common in individuals who run or walk long distances without proper conditioning (too much too soon).

Common Fibular Nerve Trauma

The common fibular nerve is vulnerable during even mild trauma due to its subcutaneous position on the lateral aspect of the knee region.

Signs and Symptoms

Sensory Deficit
- Distal anterior leg and dorsum of foot

Motor Deficits
- Foot drop
- Weakened eversion of foot

Predisposing Factors
- There are no predisposing factors for this injury.

It is important to establish whether trauma to the fibular neck has involved the common fibular nerve or its superficial and/or deep branches. The physical examination of this patient indicates that both the anterior and lateral leg compartments have been compromised (**Table 10.4.1**):

- Sensory deprivation from the dorsum of foot (superficial and deep fibular nerves)
- Weakened dorsiflexion and inversion with foot drop and high steppage gait (anterior compartment muscles)

Fasciotomy Incision through a fascia
Myositis Inflammation of muscle

Tendinitis Inflammation of a tendon
Periostitis Inflammation of periosteum

TABLE 10.4.1	Muscles affected by injury to superficial and deep fibular nerves.	
Muscle	**Nerve**	**Action(s) impaired**
Lateral compartment		
Fibularis longus	Superficial fibular	Plantar flexion and eversion
Fibularis brevis	Superficial fibular	Plantar flexion and eversion
Anterior compartment		
Tibialis anterior	Deep fibular	Dorsiflexion and inversion
Extensor hallucis longus	Deep fibular	Dorsiflexion and extension of toe 1
Extensor digitorum longus	Deep fibular	Dorsiflexion and extension of toes 2–5
Fibularis tertius	Deep fibular	Eversion

- Weakened eversion (lateral compartment muscles)

This indicates that the injury is to the common fibular nerve or both branches. Evidence presented in this case would indicate that the nerve was crushed during the accident because sensory and motor activity are compromised but not lost. The nondisplaced fracture reduces the chance that bone fragments have severed the nerve (**Fig. 10.4.4**).

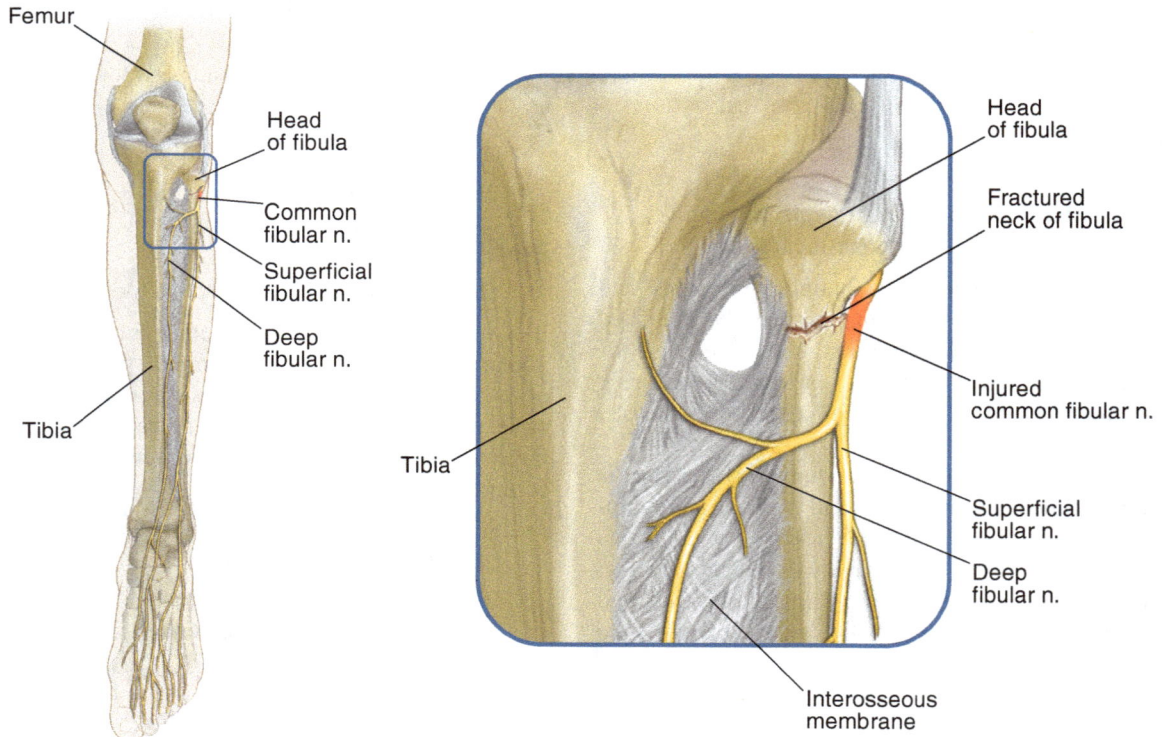

FIGURE 10.4.4 Illustration of nondisplaced fracture of the fibular neck and injured common fibular nerve.

Clinical Notes

- Isolated fracture of the fibula as a result of trauma is not common. Most often there is also damage to ligaments of the knee joint, other soft tissues, and the tibia.

- Foot drop is produced when the balance between dorsiflexors and plantar flexors is lost. In this scenario, the dorsiflexors are compromised and the unopposed and powerful plantar flexors dominate to pull the foot into plantar flexion. To compensate for the foot drop during locomotion, patients adopt a high steppage gait: flexing the hip joint to a greater extent on the affected side in order for the foot to clear the surface during the swing phase of the gait cycle.

DIAGNOSIS

The patient presentation, medical history, physical examination, and imaging studies support a diagnosis of **trauma to the common fibular nerve**.

Trauma to the Common Fibular Nerve

The subcutaneous position of the common fibular nerve on the neck of the fibula makes it vulnerable during trauma to this region. This patient's accident as a pedestrian struck by an automobile (an automobile bumper is at approximately knee level) is consistent with injury to the common fibular nerve or both of its terminal branches:

- Foot drop, weakness in muscles of eversion, and the areas of sensory loss

Acute Anterior Compartment Syndrome

The contents of the anterior compartment of the leg are tightly bound in place by the shaft of the tibia, crural fascia, and the interosseous membrane. Accumulation of fluid (edema, hemorrhage) in the closed anterior compartment can quickly lead to a condition in which compartment pressure exceeds arterial pressure, limiting vascular perfusion of the tissues. Acute anterior compartment syndrome is a medical emergency. Progressive increase in anterior compartment pressure due to muscle swelling, edema, and/or hemorrhage will compromise vascular supply, leading to ischemia and, potentially, necrosis. Acute anterior compartment syndrome can be ruled out since:

- Visible signs of swelling and pallor over the anterior leg compartment are absent.

- The patient does not complain of pain during foot movements.

Edema Swelling of skin due to abnormal accumulation of fluid in subcutaneous tissue

Hemorrhage Escape of blood from vessels:
- Petechia: <2 mm diameter
- Ecchymosis (bruise): >2 mm
- Purpura: a group of petechiae or ecchymoses
- Hematoma: hemorrhage resulting in elevation of skin or mucosa

Patient Presentation

A 24-year-old female presents in the emergency department 18 hours after being released from the hospital for treatment of a tibial fracture.

Relevant Clinical Findings

History

The patient was diagnosed with a displaced midshaft fracture of her left tibia caused during a downhill skiing accident. Following closed reduction in the fracture, she was casted and released from the hospital the following morning. She was given a prescription for pain management, but relates that the pain has progressed to the point that it is now worse than with the original injury.

She also states that the cast has progressively become tighter to the point of causing discomfort.

Physical Examination

The following findings were noted on physical examination *after removal of the cast*:

- Severe pain is elicited during dorsi- and plantar flexion.

- Weak pulse in the left dorsal artery of the foot (dorsalis pedis).
- Visible bulging of muscles of the anterior compartment of the leg.
- Skin over the anterior compartment of the leg is tense, shiny, and shows pallor.
- Hypesthesia along adjacent sides of toes 1 and 2.

Laboratory Tests

Test	Value	Reference value
Compartment Pressure Test	33	<10 mm Hg
Serum creatinine phosphokinase	2.1	0.5-0.9 mg/mL

Clinical Note

Creatinine phosphokinase (CK) is an enzyme found primarily in muscle and brain cells. Elevated serum CK may reflect muscle cell injury or death.

Clinical Problems to Consider

- Acute anterior compartment syndrome
- Deep vein thrombosis
- Necrotizing fasciitis

LEARNING OBJECTIVES

1. Describe the anatomy of the myofascial compartments of the leg.
2. Explain the anatomical basis for the signs and symptoms associated with this case.

RELEVANT ANATOMY

Myofascial Compartments of the Leg

The **crural fascia** (deep fascia of the leg) is a thick, investing layer that blends with the periosteum of the subcutaneous anterior border of the tibia (shin). Laterally, anterior and posterior **intermuscular septae**

Pallor Pale skin

extend between the crural fascia and the fibula. Three myofascial compartments—anterior, lateral, and posterior—are formed by combinations of the intermuscular septae, the fused crural fascia and tibial periosteum, and the interosseous membrane (**Fig. 10.5.1**).

Anterior Compartment of the Leg

The borders of the anterior compartment are indicated in **Table 10.5.1**.

The borders are inelastic and there is little fat and loose connective tissue in this compartment. Thus, the contents are tightly packed and there is minimal room for expansion.

Four extrinsic foot muscles are in the anterior compartment (**Fig. 10.5.1**). From medial to lateral they are:

1. Tibialis anterior
2. Extensor hallucis longus

TABLE 10.5.1	Borders of the anterior compartment of the leg.
Border	**Structure**
Anterior	Crural fascia
Posterior	Interosseous membrane
Medial	Tibia
Lateral	Anterior intermuscular septum

3. Extensor digitorum longus
4. Fibularis tertius

The blood supply for this compartment is provided by the **anterior tibial artery** (**Fig. 10.5.1**). This artery enters the compartment with concomitant veins through an opening in the proximal end of the interosseous membrane and courses toward the foot on the anterior aspect of the membrane. At the

Anterior compartment

1. Tibialis anterior m.
2. Deep fibular n. & anterior tibial vessels
3. Extensor hallucis longus m.
4. Extensor digitorum longus m.

Deep posterior compartment

5. Flexor digitorum longus m.
6. Tibialis posterior m.
7. Posterior tibial n. & vessels
8. Fibular vessels
9. Flexor hallucis longus m.

Lateral compartment

10. Fibularis brevis m.
11. Fibularis longus m.

Superficial posterior compartment

12. Tendon of plantaris m.
13. Soleus m.
14. Gastrocnemius m.

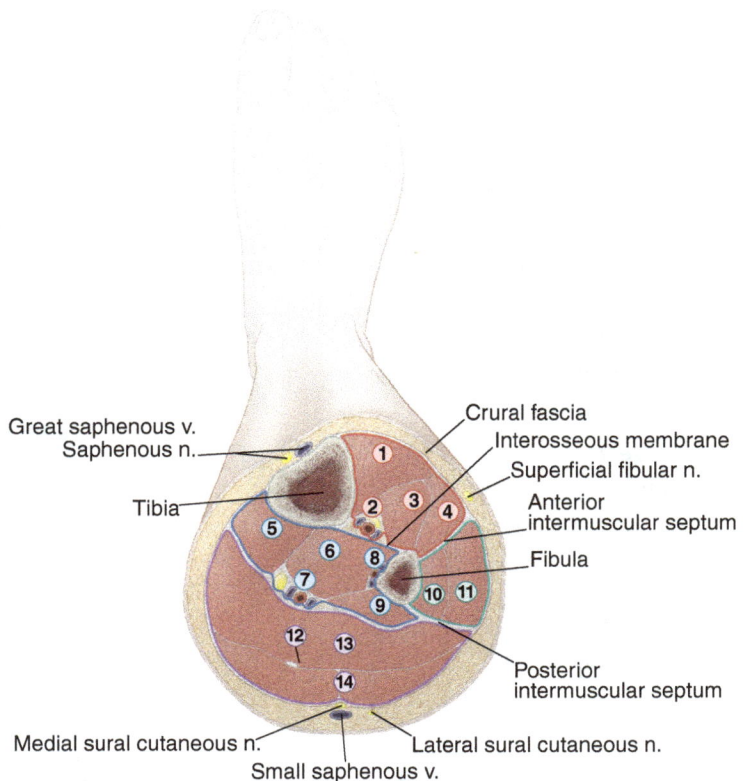

FIGURE 10.5.1 Compartments of the leg and their contents.

ankle, the names of these vessels change to dorsal artery and vein of the foot (dorsalis pedis artery and vein).

Innervation for structures in this compartment is supplied by the **deep fibular nerve** (**Fig. 10.5.1**). This terminal branch of the common fibular nerve enters the compartment proximally by piercing the anterior intermuscular septum. The nerve joins the anterior tibial artery in its course toward the foot. The tendons of the muscles and the neurovascular bundle cross the ankle deep to the extensor retinacula.

CLINICAL REASONING

This patient presents with signs and symptoms indicating a clinical condition of severe leg pain secondary to a tibial fracture.

Acute Anterior Compartment Syndrome

This medical emergency is most often associated with trauma to structures associated with the anterior compartment of the leg. The condition develops when the tissue (intercellular) pressure is greater than arterial pressure.

Signs and Symptoms

Sensory Deficits
- Increasing pain that may exceed than that of the original injury
- Severe pain during dorsi- and plantar flexion
- Loss of two-point discrimination on the skin of adjacent sides of toes 1 and 2

Other Deficits
- Skin pallor
- Compartment pressure >30 mm Hg
- Laboratory evidence of rhabdomyolysis

Predisposing Factors
- Tibial fracture
- Leg surgery
- Cast applied too tightly
- Heavy pressure for extended period (e.g., leg trapped under debris following an accident)

When tissue pressure in a myofascial compartment exceeds arterial perfusion pressure, blood vessels in the area of hypertension are compressed and cannot deliver adequate oxygen and nutrients. The resulting ischemia will trigger pain receptors. With sustained elevated tissue pressure, muscle cells may undergo rhabdomyolysis (i.e., an elevated levels of serum creatinine kinase will result).

This patient has a tibial fracture that was set and casted. Fractures usually result in some degree of hemorrhage, either from the fracture itself or from nearby torn muscles. Likewise, trauma also typically results in edema. Any fluid accumulation in a closed and restricted compartment, such as the anterior leg compartment, will lead to increased tissue pressure. A cast will exacerbate the condition.

Clinical Notes

- Acute anterior compartment syndrome is a medical emergency and requires immediate fasciotomy to relieve the increased tissue pressure.

- Sudden reestablishment of blood supply to an area that has been compromised for a period of time often results in compartment syndrome development.

- Sensory loss usually precedes motor loss: nerve damage typically occurs within 6 hours of increased pressure in the anterior compartment.

- It is important to distinguish acute from chronic anterior compartment syndrome. Chronic anterior compartment syndrome is rarely a medical emergency and is caused most often by muscle overuse. The symptoms (mostly pain) tend to disappear when the activity is decreased or ended. The chronic form of the syndrome is frequently preventable through modifications in footwear, running surface, and intensity of training.

Rhabdomyolysis Potentially fatal disease in which there is rapid destruction of skeletal muscle cells. Elevated serum creatine kinase is indicative of muscle cell breakdown. Myoglobin is in the blood and urine and may lead to organ failure

Deep Vein Thrombosis

In this condition, blood clots form in veins of the extremities, most often the lower limb. The veins are divided into superficial (subcutaneous) and deep groups. The two groups communicate via perforating veins that have valves to direct blood from superficial to deep. Either set of veins can develop thrombosis and the valves of the perforating veins typically prevent clots in the superficial veins from reaching the deep veins.

Blood in deep veins of the lower limb flows primarily due the contraction of adjacent skeletal muscles, which "milks" venous blood along the vessels. Extended periods of lower limb inactivity cause blood in the deep veins to stagnate and increase the potential for thrombus formation. The presence of thromboses is not, of itself, a medical emergency.

Clinical Note

The danger with a deep vein thrombosis is that it will become an embolus. The embolus may enter the right side of the heart and, from there, travel to the lungs where it will become lodged in branches of the pulmonary artery (pulmonary embolism). For this reason, deep vein thromboses are treated aggressively.

Signs and Symptoms

Sensory Deficits
- Focal areas of pain in leg and/or foot

Other Deficits
- Erythema
- Skin warmer than normal
- Edema

Predisposing Factors

- Long periods of inactivity (e.g., hospitalization, immobility following surgery, and long-distance travel)

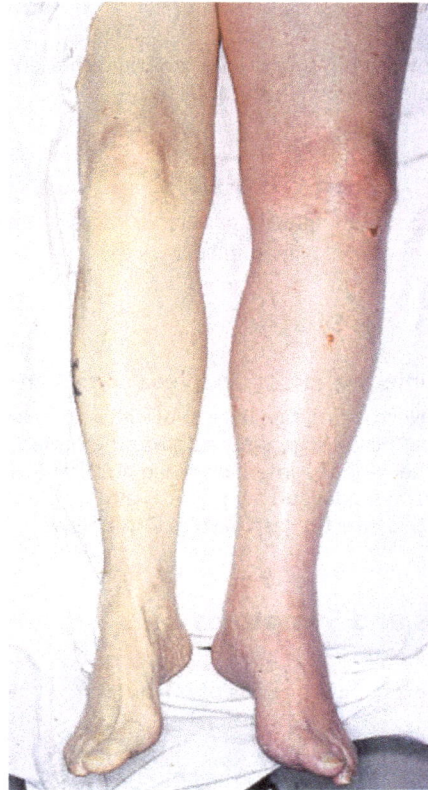

FIGURE 10.5.2 Patient with deep vein thrombosis in left lower limb, showing edema and erythema. *Source:* Fig. 12-34 in *The Atlas of Emergency Medicine,* 3e.

- Trauma (with or without surgery; with or without casting)
- Fracture

In deep vein thrombosis, a clot reduces blood drainage from a region, in this case the anterior compartment of the leg. The accumulated blood will cause swelling of the affected region and redness of the skin over the area (**Fig. 10.5.2**). Since veins interconnect extensively and randomly, there is usually sufficient collateral flow to permit some drainage from the affected area.

Thrombus A fixed mass of platelets and/or fibrin (clot) that partially or totally occludes a blood vessel or heart chamber. An embolism is a mobile clot in the cardiovascular system

Embolus A mobile clot in the cardiovascular system, often derived from a thrombus, that is obstructive

Erythematous Reddened skin
Edema Swelling of skin due to abnormal accumulation of fluid in subcutaneous tissue

Clinical Note

Following hip surgery, up to 50% of patients develop deep vein thrombosis.

Necrotizing Fasciitis

This is a rapidly spreading bacterial infection that tends to follow fascial planes and causes tissue necrosis.

Clinical Notes

- Infective agents can spread up to 3 cm/hr
- Necrotizing fasciitis is sometimes referred to a "hospital gangrene," since the patient may develop the infection while hospitalized. The media and lay-public typically refer to this infection as "flesh-eating disease."

Signs and Symptoms

Sensory Deficit

- Severe pain in areas of infection

Other Deficits

- Erythema
- Bullae
- Skin ulcerations
- Fever and chills
- Positive Gram stains

Predisposing Factors

- Trauma
- Surgery
- Diabetes
- Immunosuppression (diabetes, liver disease, chemotherapy)
- Fracture
- Bacterial contamination of skin cuts or abrasions

Necrosis Pathologic death of cells, tissues, or organs
Bulla Blister (e.g., on skin or lung surface); >1 cm
Ulcer A lesion on skin or mucous membrane

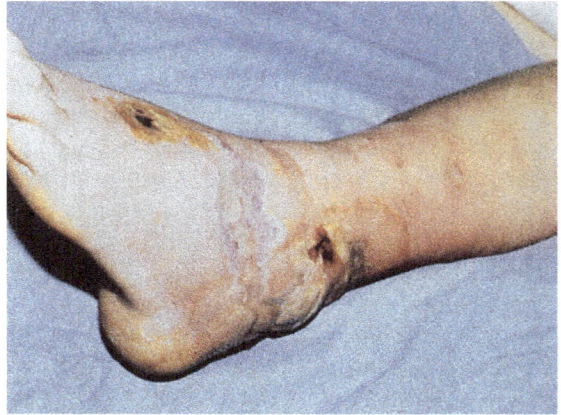

FIGURE 10.5.3 Patient with necrotizing fasciitis in left lower leg and foot. *Source:* Fig. 55-3 in *Principles of Critical Care*, 3e.

This condition is often caused by several bacteria (polymicrobial) that act synergistically. The initial infection frequently is caused when a traumatic or surgical wound is exposed to nonsterile conditions (e.g., dirt, failure to follow universal precautions, and poor personal hygiene). Immunosuppressed individuals may contract the disease through ingesting contaminated food, especially uncooked shellfish.

The necrosis is caused by toxins produced by bacteria. The toxins reduce or prevent an immune response, cause tissue hypoxia, dissolve specific tissues, or some combination of these (**Fig. 10.5.3**).

Clinical Note

This disease is often difficult to treat since progression is rapid and the causative agent(s) may be antibiotic resistant. In some cases, the spread is so rapid that amputation is the treatment of choice.

DIAGNOSIS

The patient presentation, medical history, physical examination, and laboratory tests support a diagnosis of **acute anterior compartment syndrome** secondary to a traumatic tibial fracture.

Acute Anterior Compartment Syndrome

The contents of the anterior compartment of the leg are tightly bound in place by the shaft of the tibia, crural fascia, and the interosseous membrane. Accumulation of fluid (edema, hemorrhage) in the closed anterior compartment can quickly lead to a condition in which compartment pressure exceeds arterial pressure, limiting vascular perfusion of the tissues. Acute anterior compartment syndrome is a medical emergency. Progressive increase in anterior compartment pressure due to muscle swelling, edema, and/or hemorrhage will compromise vascular supply, leading to ischemia and, potentially, necrosis.

This patient presents several signs and symptoms that indicate compromise to the tissues of the anterior leg compartment:

- Skin pallor on anterior leg
- Severe pain with foot movement
- Specific area of sensory loss
- Increased intracompartment pressure
- Elevated serum creatinine kinase

Deep Vein Thrombosis

Blood clots (thrombi) may form in either superficial or deep veins. Thrombi are typically associated with the deep veins of the lower limb. They usually form due to periods of inactivity since the primary mechanism for blood movement in these vessels is from pressure applied by adjacent muscles as they contract. By themselves, these clots are not a medical emergency. However, a thrombus may be dislodged and travel as an embolism through the heart and become lodged in the pulmonary vasculature.

Deep vein thrombosis can be ruled out because:

- The patient's pain is not related to any specific movement.
- There was no specific sensory loss.
- The limb is characteristically erythematous.

Necrotizing Fasciitis

This bacterial infection follows along fascial planes. It often involves multiple bacteria working synergistically. Tissue necrosis results from toxins released by the bacteria.

This patient does not have necrotizing fasciitis due to:

- Absence of systemic symptoms (fever and chills) due to bacterial infection
- Lack of bullae
- Lack of ulcerations

Erythematous Reddened skin

Hemorrhage Escape of blood from vessels:
- Petechia: <2 mm diameter
- Ecchymosis (bruise): >2 mm
- Purpura: a group of petechiae or ecchymoses
- Hematoma: hemorrhage resulting in elevation of skin or mucosa

10.6

Ankle Fracture

Patient Presentation

A 17-year-old female high school volleyball player presents in the emergency department with severe ankle pain and swelling. She arrives with her foot and ankle wrapped in ice.

Relevant Clinical Findings

History

The patient relates that during a volleyball match, she jumped to block an opponent's hit and landed on the opponent's foot, "rolling" her ankle. In the 2 hours since the injury, her ankle has become progressively swollen and dark.

Physical Examination

The following findings were noted on physical examination:

- Considerable edema of entire ankle region
- Large hematoma on lateral aspect of the ankle
- Severe ankle pain, regardless of foot position
- Pain exacerbated with pressure over lateral malleolus
- Positive anterior drawer test of ankle

Imaging Studies

- Anterior and lateral radiographs revealed an ankle joint fracture.

Clinical Note

The **anterior drawer test for the ankle** is used to assess the anterior talofibular ligament. One of the examiner's hands supports the patient's leg, while the other hand cradles the heel. If the ligament is not intact, pulling on the heel in the long axis of the foot during slight plantar flexion will cause the talus to translate on the tibia, yielding a positive test.

Clinical Problems to Consider

- Ankle sprain
- Avulsion fracture of the lateral malleolus
- Rupture of calcaneal (Achilles) tendon

LEARNING OBJECTIVES

1. Describe the structure and relationships of the ankle joint.
2. Describe the movements that occur at the ankle joint and the muscles involved.
3. Explain the anatomical basis for the signs and symptoms associated with this case.

Edema Swelling of skin due to abnormal accumulation of fluid in subcutaneous tissue

Hematoma Localized extravasation of blood, usually clotted

RELEVANT ANATOMY

Ankle Joint

The ankle, or talocrural joint, is formed by the articulation of the talus with the distal tibia and fibula. It is a synovial, uniaxial, hinge-type joint. As such, dorsiflexion and plantar flexion are the two movements possible. In contrast, inversion and eversion do not involve the ankle joint. These movements of the foot occur in the intertarsal joints, mostly in the subtalar (talocalcaneal) joints.

Avulsion Forcible detachment of a part

TABLE 10.6.1	Relationship of tendons, arteries, and nerves to the ankle.	
Aspect of ankle joint	**Structures crossing ankle joint**	**Notes**
Anterior	▪ Tendon of tibialis anterior ▪ Tendon of extensor hallucis longus ▪ Tendon of extensor digitorum longus ▪ Tendon of fibularis tertius ▪ Anterior tibial artery ▪ Deep fibular nerve	These structures enter the dorsum of foot
Posteromedial	▪ Tendon of tibialis posterior ▪ Tendon of flexor hallucis longus ▪ Tendon of flexor digitorum longus ▪ Posterior tibial artery ▪ Tibial nerve	These structures enter the plantar foot
Posterolateral	▪ Tendon of fibularis longus ▪ Tendon of brevis	These structures enter the plantar foot
Posterior	▪ Calcaneal (Achilles) tendon ▪ Tendon of plantaris	These structures insert on the calcaneus

The medial malleolus of the tibia and lateral malleolus of the fibula form a mortise, or deep socket, into which the trochlea of the talus is tightly inserted.

- The tibia articulates with the talus at two points: (1) the medial malleolus articulates with the medial surface of the talus, and (2) the inferior surface of the distal end rests on the domed superior surface of the trochlea.
- The lateral malleolus of the fibula articulates with the lateral surface of the talus.

Strong ligaments pull the distal ends of the tibia and fibula together, forming a tight grip on the trochlea of the talus. The trochlea has its greatest transverse dimension anteriorly and the least width posteriorly. Thus, the ankle is most stable in dorsiflexion when the anterior (broadest) part of the trochlea is between the malleoli. Conversely, the ankle is least stable in plantar flexion when the trochlea is more loosely fitted into the mortise.

Structures Crossing the Ankle Joint

Many muscles, vessels, and nerves of the anterior, lateral, and posterior compartments of the leg cross the ankle to enter the foot (**Table 10.6.1**). Except for the calcaneal (Achilles) and plantaris tendons, all of the structures listed in this table pass deep to retinacula as they cross the ankle.

Ligaments of Ankle Joint

A series of strong ligaments bind the bones of the ankle joint (**Fig. 10.6.1**). These can be divided into three groups (**Table 10.6.2**):

1. Those that unite the fibula and tibia.
2. Those that attach the medial malleolus to tarsal bones.
3. Those that attach the lateral malleolus to tarsal bones.

As in most major synovial joints, many of these ligaments blend with the external surface of the fibrous capsule to reinforce it.

Clinical Notes

- Most injuries to the ankle joint occur when the foot is plantar flexed, since this is the least stable ankle position.
- Forced inversion or eversion will put stress on the ligaments of the ankle joint and may cause fractures of the malleoli or talus.

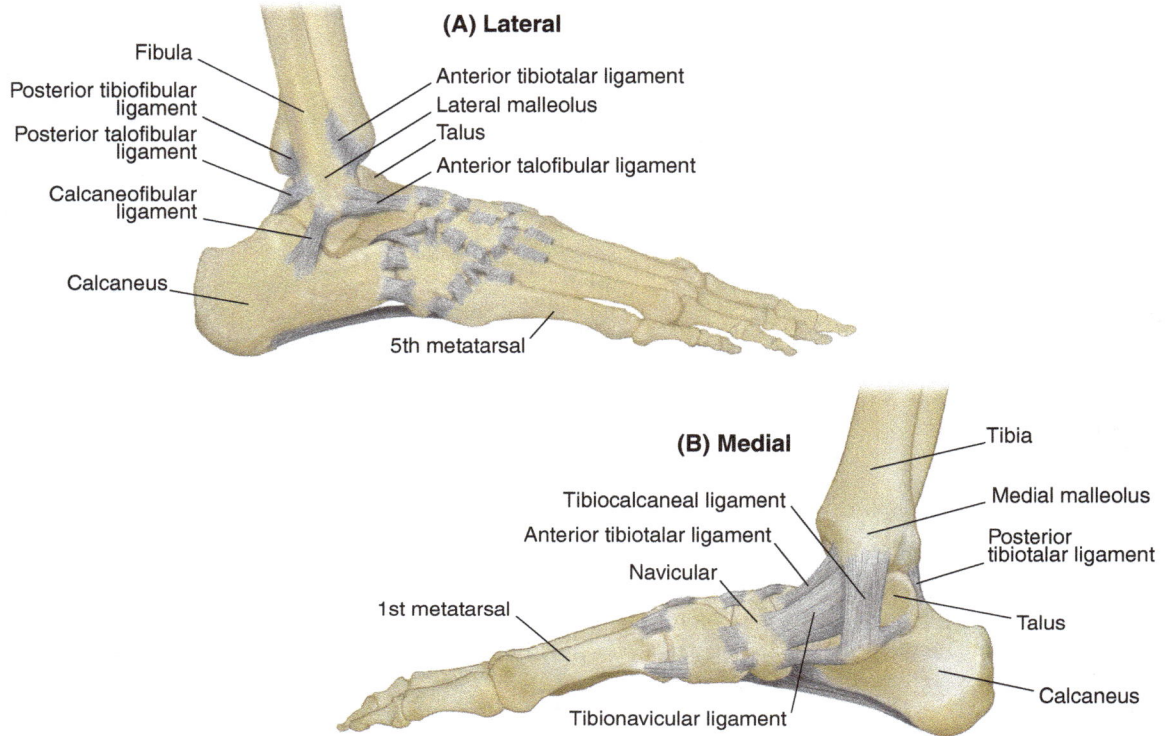

FIGURE 10.6.1 Lateral (**A**) and medial (**B**) views showing the ligaments of the ankle joint.

CLINICAL REASONING

This patient presents with signs and symptoms indicating a clinical condition of severe ankle pain following forced inversion.

Ankle Sprain

This is a situation in which the ligaments of the ankle have been overstretched but not completely ruptured. Sprains range from mild to severe. With severe sprains, the pain may be so intense that walking is not possible and the ankle often shows signs of instability. Most ankle sprains are caused by forced inversion of the foot during plantar flexion of the ankle. Stepping on uneven surfaces may cause the foot to roll inward (inversion), thereby putting increased stress on the lateral ligaments of the ankle.

Signs and Symptoms

Sensory Deficit
- Ankle pain

Other Deficits
- Swelling
- Bruising
- Ankle instability in severe cases

Predisposing Factors
- Participant in activities that require jumping (basketball, volleyball)

Distinguishing ankle sprains from fractures requires a thorough evaluation of ankle range of motion. Except in cases of mild sprain, radiographic imaging is also needed.

Sprain Excessive stretching of a ligament; may include partial tearing

TABLE 10.6.2	Ligaments of the ankle joint.	
Attachments	**Ligaments**	**Notes**
Distal ends of fibula and tibia to trochlea of talus	▪ Interosseous ligament ▪ Anterior tibiofibular ▪ Posterior tibiofibular	The interosseous ligament is the inferior part of interosseous membrane.
Medial malleolus (tibia) to tarsal bones	▪ Tibionavicular ▪ Tibiocalcaneal ▪ Anterior tibiotalar ▪ Posterior tibiotalar	Collectively, these ligaments are called the **deltoid ligament**.
Lateral malleolus (fibula) to tarsal bones	▪ Calcaneofibular ▪ Anterior talofibular ▪ Posterior talofibular	

Clinical Notes

- Ankle sprains represent 30–40% of athletic injuries. Ankle injury occurs in as many as 80% of basketball and volleyball players. As the athlete returns to the floor at the end of a jump, the ankle is plantar flexed to allow the forefoot to have initial contact. In this position, the ankle joint is least stable.

- Stepping on an uneven surface in plantar flexion will typically cause the foot to invert, placing the body weight on the malpositioned foot. This places increased stress on the lateral ankle ligaments.

Avulsion Fracture of Lateral Malleolus

The ligaments that attach the fibula to tarsal bones are very strong and the most common type of lateral malleolar fracture is the avulsion type. In these fractures, a portion of cortical bone is separated from the malleolus; the ligament remains intact with a fragment of detached cortical bone at its proximal end (**Fig. 10.6.2**). The ligament involved most frequently in this type of fracture is the **anterior talofibular**. The mechanism for this injury is the same as for an ankle sprain, that is, forceful inversion of the plantar flexed foot (see above).

Signs and Symptoms
Sensory Deficit
- Lateral ankle pain

Other Deficits
- Swelling
- Ecchymosis
- Positive anterior drawer test

Predisposing Factors
- Participant in activities that require jumping (basketball, volleyball)

Diagnosis of malleolar fracture based on the physical examination may be difficult because of swelling and patient discomfort. Radiological evidence is usually required.

Clinical Notes

- **Unimalleolar fractures** account for 70% of ankle fractures, and the majority of these involve the lateral malleolus.

- Lateral malleolar fracture decreases the stability of the talus and the forced inversion may cause the talus to be displaced medially and fracture the medial malleolus as well. These **bimalleolar fractures** are often referred to as Pott's fractures.

Rupture of Calcaneal (Achilles) Tendon

Rupture of the calcaneal tendon occurs typically in individuals involved in court or field athletic activities. Often, it occurs while initiating a jump or exerting a sudden burst of speed. These movements

---

Ecchymosis Hemorrhage into skin (bruising)

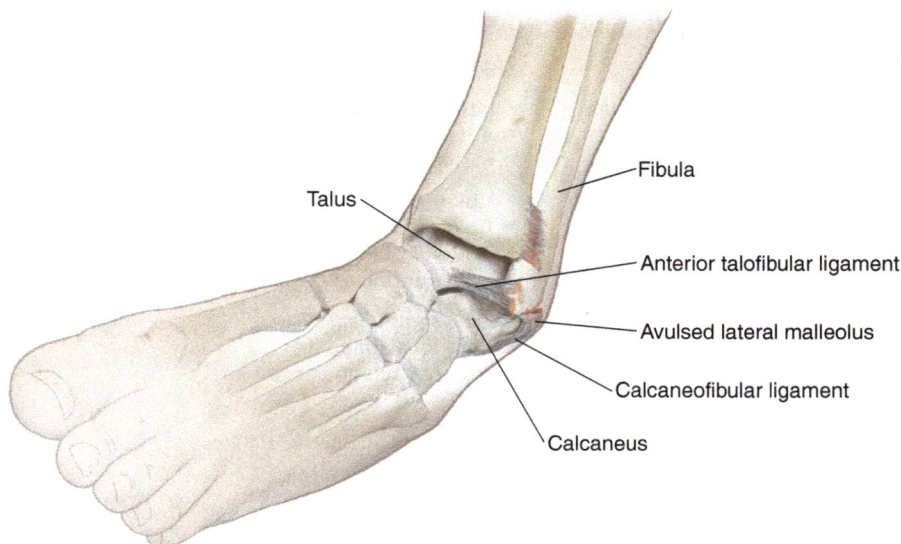

FIGURE 10.6.2 Avulsion fracture of the lateral malleolus.

require powerful plantar flexion produced by the gastrocnemius and soleus muscles using their combined, calcaneal tendon.

Signs and Symptoms

Sensory Deficit
- Pain in calf region

Other Deficits
- Swelling in calf region
- Gap in calcaneal tendon approximately 5 cm proximal to heel
- Thick ball of tissue at mid-calf representing the unattached and contracted gastrocnemius and soleus muscle bellies
- Loss of normal resting plantar flexion (**Fig. 10.6.3**)

Predisposing Factors
- Sex and age: middle-age male
- Participant in activities that require jumping and running
- Poor training ("week-end warrior")

Clinical Notes

- Patients often describe a "pop" at the time of injury that sounds like small caliber rifle fire.

- The examiner should not be misled because the patient may still be able to plantar flex the ankle and walk. Other unaffected muscles of the posterior leg (e.g., tibialis posterior) may perform these actions. A simple test for integrity of the calcaneal tendon is to grasp the calf muscle bellies and squeeze them firmly. If the tendon is intact, the ankle will plantar flex.

DIAGNOSIS

The patient presentation, medical history, physical examination, and imaging studies support a diagnosis of **avulsion fracture of the lateral malleolus**.

Avulsion Fracture of the Lateral Malleolus

Ankle injuries most often occur during "landing" activities in which the foot comes into contact with an uneven surface and the foot is forcibly inverted (a movement in the subtalar joints). This places increased stress on the lateral ankle ligaments and may cause them to be sprained (overstretched), partially ruptured, or completely ruptured. These ligaments are very strong and, in severe injuries, the ligaments may remain intact and a portion of the lateral malleolus is avulsed.

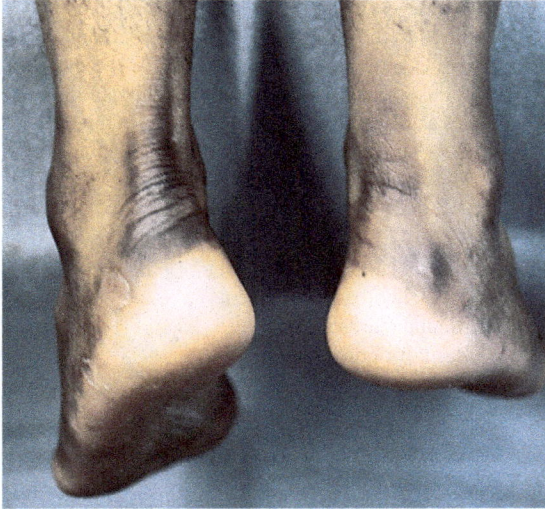

FIGURE 10.6.3 A patient with Achilles tendon rupture (note the swelling over the right tendon), showing loss of normal resting plantar flexion. *Source:* Fig. 11-76 in *The Atlas of Emergency Medicine,* 3e.

Distinction between malleolar fracture and ligament sprain may be difficult during the physical examination due to edema and pain elicited when manipulating the foot.

- In this case, as in most, the radiologic evidence distinguishes between a lateral malleolar, avulsion-type fracture, and an ankle sprain.

Rupture of the Calcaneal (Achilles) Tendon

The calcaneal tendon is formed by the union of the tendons of the soleus and gastrocnemius muscles. These powerful muscles lift the heel, via the calcaneal tendon, at the initial stages of locomotion. Calcaneal tendon rupture typically occurs during a "push-off" maneuver involving powerful plantar flexion of the ankle joint.

- Rupture of the calcaneal tendon is ruled out because the presenting clinical signs and symptoms involve the lateral ankle and not the posterior leg.

REVIEW QUESTIONS

1. A 48-year-old male dropped a kitchen knife that pierced his canvas shoe and cut the skin on the dorsum of his foot, near the third and fourth toes. He did not seek immediate medical attention, but the next day became aware of numbness in the area of the cut. A branch of what nerve is most likely injured?

   A. Deep fibular
   B. Medial plantar
   C. Saphenous
   D. Superficial fibular
   E. Sural

2. A 16-year-old soccer player "rolled" her ankle when she stepped on another player's foot. Physical examination reveals swelling and pain over the lateral aspect of the ankle joint. Imaging does not reveal a fracture. The physician assistant diagnoses an inversion sprain. Which ligament is most likely injured?

   A. Calcaneofibular
   B. Deltoid
   C. Dorsal talonavicular
   D. Long plantar
   E. Plantar calcaneonavicular ("spring")

3. A 45-year-old jockey presents in the orthopedic clinic with a painful right knee. He reports that the pain has become progressively worse. He walks with a conspicuous limp, favoring the right limb. A CT scan reveals a popliteal artery aneurysm that is compressing the adjacent nerve. Which nerve is being compressed?

   A. Common fibular
   B. Obturator
   C. Saphenous
   D. Sciatic
   E. Tibial

4. A 38-year-old female bicyclist was struck by a motorist and suffered a comminuted fracture of the proximal one-third of her left tibia. During repair, the surgeon notes that a nerve on the anterior surface of the interosseous membrane had been partially severed by bone fragments. Which movement would be compromised in this patient?

   A. Abduction of toes II–IV
   B. Dorsiflexion
   C. Eversion
   D. Flexion of great toe (toe I)
   E. Flexion of toes II–V

5. An apprentice butcher, working at the "boning table," lacerates the skin over his femoral triangle, 3 cm inferior to the right inguinal ligament. Assessment in the emergency department reveals no damage to the femoral vessels or nerve. Which is *not* a valid neurological test for the femoral nerve?

   A. Hip adduction
   B. Knee extension
   C. Patellar tendon reflex
   D. Sensation along medial margin of foot
   E. Sensation over patella

6. A 39-year-old welder is admitted to the emergency department after his left buttock came into contact with a hot piece of metal. Evaluation shows a third degree burn approximately 2 cm square near the inferior border of the buttock. Somatic pain sensations from this injury will enter the spinal cord at:

   A. L4
   B. L5
   C. S1
   D. S2
   E. S3

7. An EMS unit delivers an 87-year-old female to the emergency department following a 911 call because she fell at home and could not get up. Radiologic imaging reveals a fracture of the femoral head. Which artery does *not* provide blood to the femoral head?

   A. Acetabular
   B. Deep circumflex iliac
   C. Medial circumflex femoral
   D. Obturator
   E. Retinacular arteries

8. A 17-year-old male high school football player is admitted to the emergency department complaining of severe pain in his right knee. He is the team's center and relates that an opponent charged into his thighs as he stood up at the beginning of a play. The "drawer test" reveals a 1 cm posterior displacement of the tibia on the femoral condyles, as compared to the left knee. This sign indicates damage to the:

   A. Anterior cruciate ligament
   B. Lateral collateral ligament
   C. Medial collateral ligament
   D. Medial meniscus
   E. Posterior cruciate ligament

9. A 69-year-old sedentary male has recently begun walking in his neighborhood. He relates that frequently after walking about two blocks, he experiences excruciating pain in both calf regions. If he sits on the curb for a few minutes, the pain subsides, but often returns again when he resumes walking. This intermittent claudication (pain) is due to:

   A. Arterial vascular disease
   B. Deep vein thrombosis
   C. Disuse atrophy of his calf muscles
   D. Rheumatoid arthritis
   E. Varicose veins

10. A 19-month-old toddler examined in a pediatrics clinic has a distinctive "waddling" gait. The physician informs the parents that this is called coxa vara, which is due to:

    A. A bowed femoral shaft
    B. A gynecoid-type bony pelvis
    C. Decreased angulation of the femoral neck
    D. Malposition of the acetabulum in the hip bone
    E. Weak rotator muscles of the hip

Appendix 1
List of Clinical Terms (Referenced by Case)

Term	Definition	Case(s)
Abscess	Collection of purulent exudate (pus)	6.3
Acropachy	Clubbing of fingers and/or toes caused by edema and periosteal changes	6.2
Adenitis	Inflammation of a lymph node(s)	6.3
Adenoiditis	Inflammation of the pharyngeal tonsils (adenoids)	6.3
Adnexa	Structures accessory to an organ; used commonly with reference to the uterus to refer to uterine tube, ovaries, and uterine ligaments	3.4
Amaurosis fugax	Transient blindness (Greek, amaurosis=dark; Latin, fugax=fleeting)	6.4
Amenorrhea	Absence or abnormal cessation of menses (periodic hemorrhage related to the menstrual cycle)	6.2
Analgesia	Loss of painful sensation	8.1
Anemia	Reduced erythrocytes, hemoglobin, or blood volume	4.5
Anesthesia	Loss of sensation	7.2, 9.5
Aneurysm	Circumscribed dilation of an artery, in direct communication lumen	7.6
Anhidrosis	Absence of sweating	7.6
Anisocoria	Unequal size of pupils	7.6
Ankylosis	A disease condition that results in stiffening or joint fixation	7.3
Anosmia	Loss of smell	7.2
Arrhythmia	Irregular heart beat	2.2, 6.1
Arthritis	Joint inflammation	7.3
Asthenia	Overall weakness due to debility	2.6
Atelectasis	Reduction or absence of air in all or part of a lung (lung collapse)	2.4, 2.6
Atrophy	Wasting of tissues, organs, parts of the body, or the entire body	9.4, 9.5
Aura	Symptom that precedes partial epileptic seizure or migraine	6.4
Auscultation	A diagnostic method, usually with a stethoscope, to listen to body sounds (e.g., heart, breath, and gastrointestinal sounds)	2.1, 2.2, 2.3, 2.6

Term	Definition	Case(s)
Avulsion	Forcible detachment of a part	10.6
Bacteriuria	Bacteria in urine	3.5
Bleb	Small bulla (<1 cm diameter)	2.4
Bradycardia	Decreased heart rate: <55 bpm (normal adult heart rate: 55–100 bpm)	7.1
Bruxism	Grinding or clenching of teeth	7.3
Bulla	Blister (e.g., on skin or lung surface); >1 cm	10.5
Calculus	An abnormal concretion, usually of mineral salts	3.5
Cardiomegaly	Enlarged heart	2.2
Chemosis	Edema of bulbar conjunctiva	7.6
Cholecystitis	Gallbladder inflammation	3.3
Cholelithiasis	Gallstones	3.3
Cholestasis	Biliary obstruction	3.3
Chylothorax	Accumulation of lymph in the pleural cavity	6.1
Claudication	Muscle pain caused by vascular insufficiency	7.3
Comatose	Profound consciousness	7.1
Comminuted fracture	A fracture with three or more pieces	3.6, 7.2, 10.2
Compound fracture	A fracture in which bone fragment penetrates the skin or mucous membrane (synonym: open fracture)	7.2
Contralateral	On the opposite side	7.4
Contusion	Mechanical injury beneath skin that results in subcutaneous hemorrhage (i.e., a bruise)	5.1, 10.2
Crackle	Crackling noise heard with lung disease (also known as a rale)	2.1, 2.2
Crepitus	Grating, crackling, or popping sound in joints and subcutaneous tissues	7.3
Culdocentesis	Transvaginal aspiration of fluid from rectouterine pouch (of Douglas); also known as the cul-de-sac. The procedure involves inserting a needle through the posterior vaginal fornix to access the rectouterine pouch	4.2
Curettage	Removal of material (e.g., growths) from the wall of a body cavity or surface with a curet (spoon-shaped instrument)	4.3
Cyanosis	Bluish color of skin and mucous membranes from insufficient blood oxygen	6.3
Cyst	Abnormal, membrane-bound sac that contains gas or fluid	2.5
Cystinuria	Elevated urinary secretion of cysteine, lysine, arginine, and ornithine	3.5

Term	Definition	Case(s)
Cystitis	Inflammation of the urinary bladder	3.5, 4.4
Decubitus	Lying down	2.1
Deglutition	Swallowing	6.2
Dermopathy	A disease of the skin	6.2
Diaphoresis	Sweating	6.2
Diplopia	Double vision	7.5, 7.6
Dizziness	Various symptoms, including lightheadedness or unsteadiness (may be confused with vertigo)	7.4
Dysarthria	Disturbance of speech	6.4
Dysesthesia	Pain	10.1, 10.2
Dysgeusia	Altered sense of taste	7.4
Dysmenorrhea	Difficult or painful menstruation	4.2
Dyspareunia	Pain during sexual intercourse	3.4, 4.2, 4.3, 4.4
Dyspepsia	Indigestion	4.5
Dysphagia	Difficulty swallowing	2.6, 6.3
Dysphoria	Generalized mood depression	6.2
Dysplasia	Abnormal tissue development	4.3
Dyspnea	Difficulty breathing, shortness of breath	2.1, 2.2, 2.3, 2.4, 2.6, 3.6, 6.3, 8.2
Dystocia	Difficult birth	9.3
Dysuria	Pain during urination	3.4, 3.5, 4.2, 4.4, 5.2
Ecchymosis	Hemorrhage into skin (bruising)	7.2, 7.5, 7.6, 9.1
Edema	Swelling of skin due to abnormal accumulation of fluid in subcutaneous tissue	2.1, 2.2, 6.2, 6.3, 7.2, 7.5, 7.6, 10.4, 10.5, 10.6
Effusion	Abnormal collection of fluid (e.g., blood, lymph, synovial, pleural, or pericardial)	2.3
Embolus	A mobile clot in the cardiovascular system, often derived from a thrombus, that is obstructive	10.5
Enophthalmos	Recession (sinking in) of the eye into the orbit	7.5, 7.6
Epiphora	Overflow of tears onto the face due to inadequate drainage	7.4
Epistaxis	Bleeding from the nose	7.2

Term	Definition	Case(s)
Erythematous	Reddened skin	5.3, 5.4, 6.3, 10.5
Esotropia	Inward (median) deviation of one eye relative to the other, a form of strabismus (misalignment of the eye)	7.5, 7.6
Etiology	Underlying cause of a disease or condition	7.3
Excoriation	Scratched or abraded area of skin	5.4
Exophthalmos	Protrusion of the eye (synonym: proptosis)	6.2
Exotropia	Outward (lateral) deviation of one eye relative to the other, a form of strabismus (misalignment of the eye)	7.5, 7.6
Extravasation	Escape of body fluid into surrounding tissues	5.1
Exudate	Fluid released from tissue due to inflammation or injury	7.2
Fasciotomy	Incision through a fascia	10.4
Febrile	Elevated body temperature, i.e., a fever (normal body temperature: 36.0–37.5°C (96.5–99.5°F)	2.3, 2.4
Fecalith	Fecal stone	3.4
Fibrillation	Rapid contraction or twitching of muscle fibrils but not of the muscle as a whole	2.1
Fistula	An abnormal passage between two epithelialized surfaces (e.g., anus and skin)	5.4
Flail	A joint with impaired or absent function	9.3
Flatus	Gas or air in gastrointestinal tract that is expelled through the anus	4.4
Fluctuance	Indication of pus in a bacterial infection in which redness and induration (hardening) develops in infected skin	5.3, 5.4
Gangrene	Necrosis due to loss of blood supply	3.4, 4.5
Hematemesis	Vomiting blood	3.2
Hematochezia	Bloody stool	4.3, 4.5
Hematocrit	A blood test that measures the percentage of erythrocytes in whole blood	3.6
Hematogenous	Spread via vasculature	4.1, 4.5
Hematoma	Localized extravasation of blood, usually clotted	6.1, 6.4, 7.2, 10.3, 10.6
Hematuria	Blood in urine	3.5, 4.1, 4.4, 5.1, 5.2
Hemiparesis	Unilateral weakness	6.4, 7.4
Hemiplegia	Paralysis of one side of the body	7.4
Hemoperitoneum	Blood in peritoneal fluid within the peritoneal cavity	3.6, 4.2
Hemopneumothorax	Blood and air (gas) in pleural cavity	3.6
Hemothorax	Blood in pleural cavity	3.6, 6.1, 9.1
Hemoptysis	Blood in sputum from airway hemorrhage	2.1, 2.6

Term	Definition	Case(s)
Hemorrhage	Escape of blood from vessels: ■ Petechia: <2 mm diameter ■ Ecchymosis (bruise): >2 mm ■ Purpura: a group of petechiae or ecchymoses ■ Hematoma: hemorrhage resulting in elevation of skin or mucosa	3.6, 10.4
Hyperacusis	Increased sensitivity to sound	7.4
Hyperbilirubinemia	Elevated serum bilirubin	3.3
Hypercalciuria	Elevated urinary secretion of calcium	3.5
Hypercholesterolemia	Elevated serum cholesterol (total cholesterol level >200 mg/dL)	6.4, 7.4
Hyperhidrosis	Excessive sweating	2.3, 2.4
Hyperhomocysteinemia	Elevated serum homocysteine	6.4
Hyperlipidemia	Elevated serum cholesterol and/or triglycerides	7.4
Hyperplasia	Increase in size of a tissue or organ due to increased cell numbers (antonym: hypertrophy)	4.1, 5.2
Hyperreflexia	Exaggerated deep tendon reflex	6.2
Hypertension	Abnormal increase in arterial pressure	2.1, 2.2, 6.4, 7.1, 7.4
Hypertonicity of muscle	Abnormal increase in muscle tone	5.3, 5.4
Hypertrophy	Increase in size of a tissue or organ due to increased cell size, i.e., without increased cell number (antonym: hyperplasia)	5.4, 10.1
Hypertrophy of muscle	Increased muscle size	5.3, 5.4
Hypertropia	Upward (superior) deviation of one eye relative to the other, a form of strabismus (misalignment of the eye)	7.5, 7.6
Hypesthesia	Diminished sensation	7.5, 7.6, 9.3, 9.5
Hyphema	Blood in the anterior chamber of the eye	7.5, 7.6
Hypoageusia	Diminished sensation of taste	6.4
Hypoperfusion	Reduced blood flow relative to the metabolic requirements of an organ or tissue	7.4
Hypotension	Abnormal decrease in arterial pressure	3.6, 6.1
Hypotropia	Downward (inferior) deviation of one eye relative to the other, a form of strabismus (misalignment of the eye)	7.5, 7.6
Hypovolemic	Low or inadequate blood volume	6.1
Iatrogenic	Caused by medical or surgical treatment	3.1
Icterus	Jaundice (results from increased serum bilirubin)	3.3
Idiopathic	A condition that arises spontaneously or whose cause is obscure or unknown	7.3, 8.2
Incarcerated	Trapped	5.1

Term	Definition	Case(s)
Intramural	Within the wall of an organ or vessel	6.4
Ipsilateral	On the same side	7.4
Ischemia	Local anemia due to vascular obstruction	2.3, 10.4
–itis	Suffix used when referring to an inflammatory disease	4.5, 10.1
Jaundice	Yellowish color of the skin, mucous membranes, and/or conjunctiva. The term jaundice is synonymous with icterus	3.3
Lacerated	Torn	5.1
Lacrimation	Secretion of lacrimal (tear) fluid; sometimes refers to excess secretion	7.4
Lagophthalmos	Inability to close the eyelids completely	6.2
Libido	Sexual desire (female or male)	6.2
Lithotomy position	Supine with buttocks at edge of examination table, hips and knees flexed, feet in stirrups	4.3
Lymphadenitis	Inflammation of a lymph node or nodes	3.1
Lymphadenopathy	Disease of the lymph nodes; used synonymously to mean swollen or enlarged lymph nodes	6.2, 7.2
Lymphatogenous	Spread via lymphatic vasculature	2.6, 4.5
Malaise	Feeling of general body weakness or discomfort, often marking the onset of an illness	7.3
Menarche	Age of first menstrual cycle	2.5
Menopause	Cessation of menstrual cycles	2.5
Miosis	Excessive pupillary constriction (opposite of mydriasis)	6.4, 7.6, 9.3
Murmur	Variable vibrations produced by turbulence of blood flow	2.2
Mydriasis	Excessive pupillary dilation (opposite of miosis)	7.5, 7.6
Myositis	Inflammation of muscle	10.4
Myxedema	A dermopathy associated with hypothyroidism caused by accumulation of subcutaneous mucoid material. It may be pronounced on the face and shin	6.2
Myxoma	Benign neoplasms derived from connective tissue	2.1
Necrosis	Pathologic death of cells, tissues, or organs	2.3, 5.1, 10.2, 10.4, 10.5
Neoplasm	Abnormally increased tissue growth by cell proliferation	2.5
Neuralgia	Severe, throbbing, or stabbing pain along the course or distribution of a nerve	5.2, 7.3
Neuropathy	Damage to one or more nerves that results in impaired function	9.4, 9.5
Nociception	Nerve modality related to pain	2.3, 2.4, 6.3
Nocturia	Excessive urination at night	4.1, 5.2
Nulliparous	Having never borne children	2.5, 4.4
Obstipation	Intestinal obstruction; severe constipation	4.5

Term	Definition	Case(s)
Occlusion	Blockage (e.g., of a blood vessel, canal)	6.4
Occult	Hidden	4.5
Odynophagia	Pain on swallowing	6.3, 7.2
Oligomenorrhea	Reduced menses (periodic hemorrhage related to the menstrual cycle)	6.2
Oliguria	Decreased urine output	4.3
Ophthalmopathy	A disease of the eye	6.2
Ophthalmoplegia	Paralysis or paresis in one or more extraocular muscles	7.6
Orthosis	Correction of limbs or spine using braces or other similar devices for alignment or support	8.2
Orthopnea	Difficulty breathing, shortness of breath when lying down	2.1, 2.2
Ostium	Opening	2.2
–osis	Suffix used when referring to an abnormal or diseased condition	4.5
Otalgia	Ear pain	7.4
Pallor	Pale skin	6.4, 10.4, 10.5
Palpitation	Forcible or irregular pulsation of the heart that is perceptible to the patient	2.1, 2.2, 6.2
Palsy	Paralysis (loss of voluntary motor function; may be caused by disease or injury) or paresis (partial or incomplete paralysis)	9.3, 9.4
Paracentesis	Puncturing a cavity, usually using a needle or other hollow tube, to aspirate fluid	3.2
Paralysis	Loss of muscle function, particularly related to voluntary movement	7.4
Parenteral	A route other than the gastrointestinal tract (e.g., subcutaneous, intramuscular, and intravenous) to introduce nutrition, medication, or other substance into the body (Greek, para = around + enteron = bowel)	6.1
Paresis	Partial or incomplete paralysis	7.1, 7.6
Paresthesia	Numbness and tingling	7.4, 9.2, 9.4, 9.5, 10.1, 10.2
Parity	Having given birth	3.4
Paroxysmal	Sudden onset of a symptom or disease	2.1
Patent	Open or exposed	3.1
Pathognomonic	Characteristic symptom that points unmistakably to a specific disease	6.2, 7.2, 10.1, 10.2
Pedunculate	Having a stalk	2.1
Percuss	Diagnostic procedure in which a body part is tapped gently with a finger or instrument; used to assess organ density or to stimulate a peripheral nerve	2.4, 9.5

Term	Definition	Case(s)
Perfusion	Delivery of blood through a capillary bed to an organ or tissue	7.4
Periostitis	Inflammation of periosteum	10.4
Peristalsis	Waves of alternate contraction and relaxation along a muscular tube	3.2
Peritonitis	Inflammation of peritoneum	3.4
Plegia	Suffix meaning paralysis	7.4
Pneumoperitoneum	Air or gas in the peritoneal cavity	3.2
Pneumothorax	Air or gas in pleural cavity	3.6, 6.1, 9.1
Polyuria	Excessive excretion of urine	6.2
Postmenopausal	A period in female life following menopause	9.2
Presyncope	Feeling faint or lightheaded (syncope is actually fainting)	2.1, 6.1, 6.4
Prolapse	To fall or slip out of place	2.1
Proptosis	Protrusion of the eye (synonymous with exophthalmos)	7.6
Prostatitis	Inflammation of the prostate	3.5
Proteinuria	Elevated protein in urine	3.5
Pruritus	Itching	5.3
Ptosis	Drooping eyelid	6.4, 7.5, 7.6, 9.3
Purulent	Containing, discharging, or causing the production of pus	4.2, 5.3, 5.4, 7.2
Pus	An viscous exudate (fluid) produced during an infection at an inflammatory site; it consists of leukocytes and elements of dead cells and tissues	6.3
Pyrexia	Fever	6.3
Pyuria	Pus in urine	3.5, 4.4
Rhabdomyolysis	Potentially fatal disease in which there is rapid destruction of skeletal muscle cells. Elevated serum creatine kinase is indicative of muscle cell breakdown. Myoglobin is in the blood and urine and may lead to organ failure	10.5
Rhinoscopy	Procedure to inspect nasal cavity	7.2
Salpingitis	Inflammation of the uterine tube	4.2
Salpinx	Greek: trumpet or tube	4.2
Scotoma	Loss or absence of vision from an area of visual field	6.4
Sepsis	Presence of pathogenic organisms, or their toxins, in blood or tissues	3.5
Sibilant rhonchus	High-pitched, whistling lung sound caused by airway narrowing or obstruction	2.6
Simple fracture	A fracture in which the skin or mucous membrane is intact (synonym: closed fracture)	7.2
Sinusitis	Inflammation of one or more paranasal sinus	7.2

Term	Definition	Case(s)
Spirometry	Pulmonary function test that measures volume and rate of air flow	2.2
Splenomegaly	Enlarged spleen	3.6
Spondylolisthesis	A condition in which a vertebra "slips" out of alignment with adjacent vertebrae	8.2
Sprain	Excessive stretching of a ligament; may include partial tearing	9.1, 10.6
Stasis	Reduction or cessation (stagnation) of flow of a body fluid	7.2
Steatorrhea	Presence of excess fat in feces	3.3, 6.1
Stenosis	Narrowing a canal (e.g., blood vessel and vertebral canal)	2.1, 2.2, 6.4
Strabismus	Misalignment of the eyes	7.5
Strangulated	Constricted or twisted to prevent air or blood flow	5.1
Stridor	Noisy respiration, usually a sign of airway obstruction (especially involving the trachea or larynx)	6.3
Subluxation	Partial dislocation of a joint (*luxation* is a complete dislocation)	9.1
Sudomotor	Related to sweat glands	7.6
Suppurate	To form pus	5.3, 6.3
Syncope	Loss of consciousness (fainting)	6.4
Synovitis	Inflammation of synovial membranes	7.3
Tachycardia	Increased heart rate: >100 bpm (normal adult heart rate: 55–100 bpm)	2.2, 2.3, 2.4, 3.6, 6.2, 6.3
Tachypnea	Increased respiration rate (normal adult respiration rate: 14–18 cycles/min)	2.3, 2.4, 3.6
Tendinitis	Inflammation of a tendon	10.4
Tenesmus	Straining and painful sphincter spasm related to the sensation of incomplete evacuation of either bowel or bladder	4.5
Thrombophlebitis	Venous inflammation with thrombus formation	7.6
Thrombus	A fixed mass of platelets and/or fibrin (clot) that partially or totally occludes a blood vessel or heart chamber. An embolism is a mobile clot in the cardiovascular system	2.3, 2.4, 5.3, 6.1, 10.4, 10.5
Tic	Habitual, repeated contraction of specific muscles (e.g., facial tic); it may be suppressed voluntarily	7.3
Tinnitus	Sensation of ringing or other noises in the ears	6.4, 7.4
Tonsillitis	Inflammation of the palatine tonsils	6.3
Torticollis	Spasmodic contraction of neck muscles (often those innervated by CN XI)	6.3
Transmural	Extending through, or affecting, the entire thickness of the wall of an organ or cavity	6.1
Trismus	Jaw stiffness	6.3, 7.4

Term	Definition	Case(s)
Ulcer	A lesion on skin or mucous membrane	5.3, 10.5
Urethritis	Inflammation of the urethra	4.4
Varico-	Relating to dilated or distended vein	5.3
Vasculitis	Inflammation of blood vessels	7.3
Vertigo	Sensation that the environment is moving (spinning or "whirling") when there is not actual movement	7.4
Volar	Related to the palm of the hand or sole of the foot	9.2
Vulva	Female external genitalia	4.3, 8.1

Appendix 2
Explanations for Correct Answers

Chapter 1: Nervous System

1. **Answer B.** The **hypoglossal nerve** supplies all intrinsic and most extrinsic muscles of the tongue. Bilateral contraction of the genioglossus (an extrinsic tongue muscle) results in protrusion of the tongue in the midline. Unilateral palsy of the hypoglossal nerve will cause the tongue to deviate from the midline and toward the affected side.

2. **Answer A.** The **facial nerve** supplies the muscles of facial expression. Therefore, damage to this nerve results in weakness of facial muscles on the affected side. The orbicularis oculi (a muscle of facial expression) is located in the eyelids and around the orbit. Contraction of this muscle is responsible for closing the lids tightly.

3. **Answer A.** The anterior and posterior vagal trunks derived from the vagus nerves enter the abdomen through the esophageal hiatus of the diaphragm. These nerves carry the preganglionic parasympathetic component for abdominal organs that are supplied by the celiac trunk and the superior mesenteric artery. Postganglionic parasympathetic axons arise from terminal ganglia located near or in these organs. The **ascending colon** is supplied by the superior mesenteric artery and, therefore, would lose its parasympathetic innervation if the vagal trunks are damaged.

4. **Answer E.** The cell bodies for somatic motor neurons that supply skeletal muscle of the trunk and limbs are located in the **ventral (anterior) horn of the spinal cord**.

5. **Answer B.** Regardless of the source of the pain, all afferent (sensory) axons from the body wall and viscera enter the spinal cord via the **dorsal (posterior) roots**.

6. **Answer C.** Dorsal and ventral roots unite to form the **spinal nerve**, which is located in the intervertebral foramen.

7. **Answer A.** The sympathetic (paravertebral) chain ganglia contain **postganglionic neuronal cell bodies**. Preganglionic sympathetic cell bodies are located in the lateral horn of the T1-L2 spinal cord.

8. **Answer B.** All afferent neuronal cell bodies that supply the skin of the trunk and limbs are located in dorsal root ganglia (DRG). Those associated with the T4 dermatome would be found in the **T4 dorsal root ganglia**.

9. **Answer B.** Preganglionic sympathetic axons that originate in the T1-T5 lateral horn enter a sympathetic (paravertebral) ganglion through a white ramus communicans. These axons either synapse in the ganglion at the level of entry, or ascend in the sympathetic chain to synapse in a cervical paravertebral ganglion. Some of the postganglionic fibers that arise in these ganglia form cardiac splanchnic (visceral) nerves that supply the heart. The pathway for these postganglionic fibers does not include **gray rami communicantes**.

10. **Answer E.** Visceral pain from the appendix results with overdistension of its lumen, spasm of its smooth muscle wall, or irritation of its mucosa. Visceral afferent fibers accompany sympathetic nerves that supply the appendix. These fibers travel through the **superior mesenteric periarterial plexus** and lesser splanchnic nerve to end in neuronal cell bodies in the T10-T11 dorsal root ganglia.

Chapter 2: Thorax

1. **Answer A.** The triangle of auscultation overlies the **inferior lobe** of both lungs. Most of the thoracic wall is covered by muscle and bone associated with the pectoral girdle. The triangle of auscultation is an area of the posterior thoracic wall that is subcutaneous, providing for better auscultation. The triangle is bounded by the latissimus dorsi, trapezius, and medial border of the scapula. The 6th intercostal space lies in the floor.

2. **Answer D.** The best position to hear the sounds produced by the **pulmonary valve** is parasternal in the left 2nd intercostal space. Only the aortic valve sound can be auscultated on the right side of the thorax.

3. **Answer C.** The pericardial cavity is a **potential space between the parietal and visceral layers of serous pericardium**. This space normally contains only enough fluid to allow the opposing membranes to glide freely during heart movement. Acute increases in fluid volume may compromise cardiac function.

4. **Answer B.** The relatively short left coronary artery ends by giving rise to the anterior interventricular (anterior descending) and circumflex arteries.

5. **Answer A.** A right dominant heart implies that the **posterior interventricular artery is derived from the right coronary artery**. This is the most common pattern (67% in the general population). In a left dominant heart, the posterior interventricular artery is derived from the circumflex artery, a branch of the left coronary.

6. **Answer E.** Malignant neoplasms in the breast tend to invade the **connective tissue septae of the breast**. These septae extend between the dermis and the connective tissues overlying pectoral musculature, dividing the breast into lobes. **Invasion of the septae puts traction on them**, creating dimpling of the overlying skin. Benign breast neoplasms usually do not invade the septae.

7. **Answer C.** The **pulmonary valve** lies deep to the body of the sternum at the level of the third costal cartilage. This would be vertebral level T6-T7.

8. **Answer A.** The **ascending aorta** is within the pericardial sac. The pericardial sac and its contents are in the middle mediastinum.

9. **Answer A.** The body wall and extremities do not contain **parasympathetic** components of the autonomic nervous system.

10. **Answer B.** Two-thirds of the heart lies to the left of midline. To accommodate the heart, the **parietal pleura on the left begins its deviation from midline at the 4th intercostal space**.

11. **Answer E.** Three major structures occupy the concavity of the arch of the aorta: left main bronchus, ligamentum arteriosum, and the **left recurrent laryngeal nerve**. Compression of the nerve by a bronchial mass (or aneurysm) may manifest as hoarseness due to unilateral paralysis of intrinsic laryngeal muscles.

Chapter 3: Abdomen

1. **Answer C.** The **right gastric artery** supplies the right part of the lesser curvature of the stomach, including the superior border of the pyloric part of the stomach. It anastomoses with the left gastric artery.

2. **Answer A. Pararenal fat** lies external to the renal fascia and is most obvious posteriorly. This layer of fat is in contact with muscles of the posterior abdominal wall. In a posterior surgical approach to the kidney, this layer would be contacted first, followed by renal fascia, perirenal fat, and the renal capsule.

3. **Answer E.** The **splenic artery** is typically the largest branch of the celiac trunk. It courses along the posterior aspect of the pancreas. With chronic pancreatitis and resulting erosion of gland, this artery may be damaged.

4. **Answer B.** Indirect inguinal hernias result when the processus vaginalis remains patent after birth. Abdominal structures (most likely small bowel) may pass into the patent processes vaginalis at the deep inguinal ring, which is located **lateral to the inferior epigastric vessels**.

5. **Answer D.** The **lesser (peritoneal) sac** is a diverticulum of the greater (peritoneal) sac that lies posterior to the lesser omentum and the stomach. The lesser sac communicates with the greater sac through the omental foramen. The retroperitoneal space lies posterior to the parietal peritoneum and, therefore, is not part of the peritoneal cavity.

6. **Answer A.** The **left colic artery** is a branch of the inferior mesenteric and supplies the descending colon.

7. **Answer A.** Visceral afferents associated with pain receptors in the wall cystic duct travel along the **greater splanchnic (sympathetic) nerves**. The afferent cell bodies associated with these receptors are located in the T5-T9 dorsal root ganglia. Visceral pain from the biliary system will be felt primarily in the RUQ and epigastric region, and referred along the T5-T9 dermatomes.

8. **Answer A.** The **bile duct**, which is part of the portal triad (hepatic portal vein, proper hepatic artery, and bile duct), travels in the hepatoduodenal ligament. If a stone blocks the cystic duct, bile cannot enter or exit the gallbladder. Since the liver continues to produce bile, the bile duct becomes dilated and acts as a reservoir until bile can be released into the duodenum.

9. **Answer B.** Short gastric arteries supply the **fundus of the stomach**. These vessels arise from the splenic artery near the hilum of the spleen. During splenectomy, the splenic vessels must be ligated distal to the short gastric vessels. The left gastro-omental artery, which arises from the splenic artery, is also vulnerable in this procedure.

10. **Answer C.** The **common hepatic duct** is typically formed near the liver, by the union of the right and left hepatic ducts. It is not near the head of the pancreas.

11. **Answer C.** The perirenal abscess in this patient irritates **parietal peritoneum on the adjacent, abdominal (inferior) surface of the diaphragm**. This peritoneum transmits sensory information via somatic afferent fibers in the phrenic nerve (C3-C5). Sensation from the irritated peritoneum presents as aching or sharp pain that is referred to the shoulder and/or base of the neck (i.e., along the C3-C5 dermatomes).

12. **Answer C.** The **head of the pancreas** receives its blood from two sources: (1) the celiac trunk supplies the pancreas via the superior pancreaticoduodenal artery, a branch of the gastroduodenal, and (2) the superior mesenteric artery via its inferior pancreaticoduodenal branch. There are significant anastomoses between these vessels, so the head of the pancreas would remain well perfused. All other organs listed are supplied solely by branches of the celiac trunk.

13. **Answer C.** Functionally, the liver is divided into right and left lobes by its blood supply and biliary drainage. The left functional lobe includes the caudate, a portion of the quadrate, and the left anatomical lobes. The right functional lobe is formed by the **right anatomical lobe and a portion of the quadrate lobe**.

Chapter 4: Pelvis

1. **Answer D.** The **uterine artery** courses within the transverse cervical (cardinal) ligament along the inferior portion of the mesometrium of the broad ligament. The artery reaches the uterus near the junction of the body with the cervix. This position places the artery near the lateral part of the vaginal fornix.

2. **Answer A.** From the deep inguinal ring, the vas deferens from either side descends across the pelvic brim, crosses superior to the ureter, and lies on the posterior surface of the urinary bladder. The **ampullae of the vasa deferentia**, the dilated, terminal portions of these ducts, lie near the midline.

3. **Answer C.** The most inferior portion of the greater sac of the peritoneal cavity is the **rectouterine pouch** (Douglas), also known as the cul-de-sac. It is formed as peritoneum reflects from the uterus onto the middle portion of the rectum. This pouch lies adjacent to the posterior vaginal fornix.

4. **Answer D.** Preganglionic parasympathetic axons from cells in the **S2-S4 spinal cord** enter the inferior hypogastric plexus via pelvic splanchnic nerves. Some of these axons enter the prostatic plexus. This plexus is responsible for distributing parasympathetic innervation to the prostate and erectile tissues.

5. **Answer B.** The **mesovarium** is the part of the broad ligament that supports the ovary. Hysterectomy in a 44-year-old would not normally include removal of the ovaries (ovarectomy) because of their function in hormonal balance in the premenopausal woman.

6. **Answer D.** A situation in which the fundus of the uterus is in contact with the rectum would most likely occur when the **cervix is retroverted and the body of the uterus is retroflexed**.

7. **Answer C.** Kegel exercises increase or reestablish tone in pelvic floor muscles, and sphincters of the genitourinary system and terminal part of the digestive tract. **Obturator internus** has no direct action on the genitourinary or gastrointestinal systems. This muscle is attached to the lateral pelvic wall and is a lateral rotator of the hip.

8. **Answer D.** While there is variability in the branching pattern for the internal iliac artery, the superior vesicle artery is usually a branch of the patent portion of the **umbilical artery**.

9. **Answer B.** The diagonal conjugate is measured during pelvic examination. With the gloved index and middle fingers in the vagina, the **distance between the sacral promontory and *inferior* aspect of the pubic symphysis** is determined. The conjugate (obstetrical) dimension is determined from images and represents the distance from the sacral promontory to the *superior* aspect of the pubic symphysis.

10. **Answer E.** Lymph from the sigmoid colon first enters **lumbar nodes**.

Chapter 5: Perineum

1. **Answer C.** The **obturator nerve** does not innervate skin of the thigh involved in this reflex and does not provide motor innervation to the cremaster muscle (located in the middle layer of cremasteric fascia).

2. **Answer A.** The lateral wall of the ischioanal fossa is formed by the **muscular fascia covering obturator internus muscle**.

3. **Answer B.** The fluid of the hydrocele lies external to the testicle. In this patient, the fluid within the hydrocele "communicates" with the peritoneal cavity via the patent processus vaginalis. The **tunica albuginea** is the capsule of the testicle and lies internal to the processus vaginalis.

4. **Answer C.** Superficial inguinal nodes receive lymph from the subcutaneous tissues of the lower limb (except posterior leg), external genitalia (except testes), the anterior abdominal wall inferior to the umbilicus, the body of the uterus (along the round ligament), and the inferior portion of the anal canal. The **inferior one-third of the rectum** is drained through internal iliac nodes.

5. **Answer C.** The inferior one-fifth of the vagina and walls of the vaginal vestibule receive somatic innervation via the **pudendal nerve**. Somatic afferent impulses (e.g., temperature, pain, and touch) travel in this nerve.

6. **Answer A.** The inferior part of the anal canal receives its blood supply from the **inferior rectal vessels**. All veins along the anal canal anastomose and all are valveless. The inferior anal canal directs its venous blood preferentially into the inferior rectal vessels (and eventually the inferior vena cava), and the superior anal canal directs its venous blood into the superior rectal vein (and eventually the hepatic portal vein).

7. **Answer B.** The **dartos muscle** (smooth muscle) is in the subcutaneous layer of the scrotum. The fibers insert into the dermis and are important in regulating the size and surface texture of the scrotum. Dartos muscle tone decreases as a normal part of aging.

8. **Answer D.** The **ischiocavernosus muscle** covers the crus of the clitoris. Both the erectile body and the muscle are attached along the inferior surface of the ischiopubic ramus, at the lateral limits of the urogenital triangle of the perineum.

9. **Answer C.** The **inferior rectal nerve** courses from lateral to medial across the inchioanal fossa to innervate the external anal sphincter.

10. **Answer C.** The greater vestibular gland is a mucous-secreting gland with a duct that empties into the vestibule. It is located **deep to the bulbospongiosus muscle**. Contraction of this muscle during sexual arousal compresses the gland to help expel its secretions.

Chapter 6: Neck

1. **Answer E.** Endocrine secretion from the thyroid gland is regulated *hormonally* by thyrotropin (thyroid-stimulation hormone, TSH) released from the pituitary gland. Therefore, **none of the nerves** listed stimulate secretion from the thyroid gland.

2. **Answer A.** The inferior thyroid veins drain into the **left brachiocephalic vein**. The superior and middle thyroid veins are tributaries of the internal jugular.

3. **Answer C.** The **recurrent laryngeal nerve** innervates all intrinsic laryngeal muscles except cricothyroid. In this case, the right recurrent laryngeal nerve was damaged during surgery, leaving the right vocal fold paralyzed.

4. **Answer B.** Damage of the **phrenic nerve** is a possible complication with subclavian venous catheterization due to its course on the anterior surface of the anterior scalene muscle. The nerve passes posterior to the subclavian vein as it enters the thorax, where it may be damaged by an improperly placed catheter.

5. **Answer B.** The **lesser supraclavicular fossa** is a shallow skin depression between the clavicle and the sternal and clavicular heads of sternocleidomastoid. The internal jugular vein lies deep to this fossa, immediately lateral to the common carotid artery.

6. **Answer C.** The **palatine tonsils** are masses of lymphoid tissue that lie in the tonsillar fossa (bed). This fossa is defined by the palatoglossal arch (anteriorly) and palatopharyngeal arch (posteriorly).

7. **Answer D. Investing fascia** forms the "roof" of the posterior triangle of the neck; this fascia lies immediately deep to skin and superficial fascia. **Prevertebral fascia** surrounds cervical vertebrae and associated muscles. The prevertebral fascia that covers splenius capitis, levator scapulae, and the posterior and middle scalene muscles forms the "floor" of the posterior triangle.

8. **Answer A.** The "safe" portion of posterior cervical triangle is superior to the **accessory nerve**. This nerve courses deep to sternocleidomastoid, which it supplies. It emerges from the posterior border of this muscle, approximately half way between its attachments, to enter the posterior triangle. It courses inferiorly and posterolaterally across the triangle to reach trapezius. The "safe" area lacks major nerves and blood vessels.

Chapter 7: Head

1. **Answer D.** Sympathetic fibers supply the dilator pupillae and the tarsal muscles. Postganglionic sympathetic fibers originate in the superior cervical ganglion and join the **internal carotid (periarterial) plexus**. The ophthalmic branch of the internal carotid artery carries postganglionic sympathetic fibers into the orbit. The pupil was constricted due to interruption of the oculosympathetic pathway and the resulting unopposed action of the constrictor pupillae muscle (innervated by parasympathetic fibers). Partial ptosis was due to sympathetic interruption to the tarsal muscles.

2. **Answer C.** The **nasociliary nerve**, a branch of the ophthalmic division of the trigeminal nerve (CN V1), provides sensory innervation to the eye. These sensory axons reach the eye through two pathways: (1) directly through long ciliary nerves or (2) indirectly via the ciliary ganglion (no synapse) and short ciliary nerves.

3. **Answer E.** The **trochlear nerve** innervates the superior oblique muscle. The primary action of this muscle is *depression* of the eye, and it is most effective when the eye is adducted (i.e., looking inward). In this patient, therefore, when the right eye is adducted, it "drifts" upward due to the unopposed action of the inferior oblique (innervated by the oculomotor nerve—CN III), whose primary action is elevation.

4. **Answer C.** Secretomotor (postganglionic parasympathetic) axons that supply the lacrimal gland originate in the **pterygopalatine ganglion**. These axons join the maxillary nerve (CN V2) and then follow its zygomatic branch along the orbital surface of the zygomatic bone to the lacrimal gland.

5. **Answer B.** Sound causes vibration of the tympanic membrane, which is transferred via the ossicles (malleus, incus, stapes) to the cochlea. The stapedius muscle in the middle ear is attached to the stapes and regulates its movement. It is innervated by the nerve to stapedius, a branch of the **facial nerve (CN VII)**. Injury to facial nerve prior to the branching of nerve to stapedius may result in increased sensitivity to sound.

6. **Answer D.** The **mandibular nerve (CN V3)** is the third division of the trigeminal nerve (CN V). It is a mixed nerve with sensory and motor components. The afferent cell bodies for the mandibular nerve are located in the trigeminal ganglion, which is located in the middle cranial fossa. Damage to the sensory component of the mandibular nerve would lead to the symptoms described by this patient.

7. **Answer B.** The vagus nerve supplies the majority of muscles of the soft palate, including levator veli palatini. Bilateral contraction of the levator veli palatini muscles elevates the soft palate with the uvula in the midline. Unilateral paralysis of this muscle will cause the soft palate and uvula to deviate from the midline away the affected side (i.e., toward the remaining, functional muscle). In this patient, the uvula deviated to the right, which indicates that the **left vagus nerve** was affected.

8. **Answer C.** The facial and vestibulocochlear nerves enter the **internal acoustic meatus**. A tumor at this opening would impact both nerves. Cardinal symptoms include hearing loss (cochlear portion of CN VIII), facial muscle weakness (CN VII), and loss of taste (chorda tympani). Additional symptoms include dizziness (vestibular portion of CN VIII), hyperacusis (nerve to stapedius), dry eye and nasal cavity (greater petrosal nerve), and dry mouth (chorda tympani).

9. **Answer A.** The **auditory tube** connects the middle ear with the nasopharynx. It allows for the equalization of pressure on both sides of the tympanic membrane. The tube is lined with mucous membrane that is continuous between the pharynx and middle ear and provides a pathway for the spread of infection.

10. **Answer B.** The middle meningeal artery, a branch of the maxillary artery, is the primary blood supply to the dura mater. It enters the cranial cavity through foramen spinosum and lies between the periosteal layer of dura mater and the internal surface of the neurocranium. This patient experienced head trauma that damaged the middle meningeal artery and resulted in an epidural hematoma. Blood from the lacerated vessels collected between the periosteal dura matter and the skull. Epidural hematoma was confirmed in this patient with CT imaging that revealed a characteristic biconvex hemorrhage.

11. **Answer E.** The lingual nerve, a branch of the mandibular nerve, carries general sensory information from the anterior two-thirds of the tongue. In the infratemporal fossa, it is joined by chorda tympani, a branch of the facial nerve. Chorda tympani carries information regarding taste (special sensory) from the same region of the tongue. This patient describes a loss of general sensation, but taste is not affected. Therefore, the tumor impacted the **lingual before it was joined by (i.e., proximal to) chorda tympani.** Salivation in this patient was not affected, further supporting the conclusion that chorda tympani was not involved.

12. **Answer C.** The maxillary sinus is the largest paranasal sinus. It medial wall is directed toward the lateral nasal wall and its roof forms the floor of the orbit. The floor of the sinus is related to the roots of the maxillary premolars and molars. **Maxillary sinusitis** and maxillary dental pain are often difficult to distinguish because they share innervation (posterior and middle superior alveolar nerves, branches of the maxillary nerve).

13. **Answer B.** The mandibular nerve (CN V3), a branch of trigeminal, enters the infratemporal fossa through foramen ovale. This nerve supplies muscles derived from the first pharyngeal arch, including the muscles of mastication, anterior belly of digastric, mylohyoid, tensor tympani, and tensor veli palatini. **Lateral pterygoid** is a muscle of mastication and would be affected in this patient.

14. **Answer D.** The mucous membrane on the nasal septum is supplied by anastomosing branches of the ophthalmic, maxillary, and facial arteries. The *superior part* of the septum receives blood from the anterior and posterior ethmoidal arteries (ophthalmic). The **sphenopalatine artery** (maxillary) supplies the *posteroinferior portion* of the nasal septum. The *anteroinferior part* of the septum is supplied by the greater palatine artery and the septal branch of the superior labial artery (facial).

Chapter 8: Back

1. **Answer D.** Cerebrospinal fluid is contained in the subarachnoid space. To reach this space, the needle would pass through (in order) ligamentum flavum, dura mater, and arachnoid mater. **Pia mater** intimately invests the surface of the spinal cord and would not be penetrated during a successful lumbar puncture.

2. **Answer D.** To avoid spinal cord damage during a lumbar puncture, the needle should be introduced inferior to the medullary cone (conus medullaris). This cone most frequently lies at the L2 vertebral level and, thus, a needle introduced at the L3-L4 vertebral level is considered safe. A commonly accepted landmark for this level is the **supracristal plane**, a transverse plane connecting the superior-most aspect of the iliac crests.

3. **Answer D.** Cauda equina is located within the dural sac and is formed by **lower lumbar and sacral dorsal and ventral roots**. Roots converge in the intervertebral foramen to form the short spinal nerves. In the lumbar region, these branch almost immediately form dorsal and ventral rami that distribute branches to peripheral targets. Sacral spinal nerves divide into rami within the sacral vertebral canal and then exit through separate dorsal and ventral sacral foramina.

4. **Answer A.** The **dorsal ramus of the C2 spinal nerve**, also known as the greater occipital nerve, supplies the skin on the back of the head to the vertex.

5. **Answer A.** The spinal **epidural space** is located between the dura mater and inner aspect of the vertebral canal. It contains the internal vertebral venous plexus and variable amounts of fat. The dural sac ends at the S2 vertebral level and sacral spinal nerves (sheathed by dura) and their rami cross the epidural space. Therefore, the anesthetic injected through the sacral hiatus and into the epidural space bathes these nerves to provide anesthesia.

Chapter 9: Upper Limb

1. **Answer B.** The intervertebral disc herniation affects the C6 spinal nerve and its dermatome. This dermatome is located along the lateral aspect of the forearm, which is supplied by the **lateral cutaneous nerve of the forearm** (the terminal branch of the musculocutaneous nerve).

2. **Answer C.** Motor fibers from the C6 spinal cord enter nerves that supply muscles that control **shoulder abduction** (deltoid and rotator cuff muscles) **and elbow flexion** (biceps brachii and brachialis).

3. **Answer A.** The **long thoracic nerve** supplies the serratus anterior muscle, which is necessary for the upward rotation of the scapula when raising the abducted arm above the head.

4. **Answer B.** This injury affects the ulnar nerve. The **lateral half of the palm** is supplied by the median nerve and is, therefore, not affected in this injury.

5. **Answer C.** Nerves derived from the posterior cord supply muscles of the shoulder (e.g., axillary nerve—deltoid muscle; thoracodorsal nerve—latissimus dorsi), and the arm and forearm (radial nerve—all extensors, brachioradialis, supinator, and the snuffbox muscles). Therefore, posterior cord injury would affect extensor functions, thumb abduction (abductor pollicis longus—a snuffbox muscle), and elbow flexion (brachioradialis). The primary **pronators** (teres and quadratus) would not be affected by this injury because they are supplied by the median nerve, which receives contributions from the medial and lateral cords.

6. **Answer B.** Collateral circulation around the scapula is formed by the anastomoses between the suprascapular artery (commonly a branch of the thyrocervical trunk), the **dorsal scapular artery** (usually a branch of the subclavian), and the **circumflex scapular artery** (a branch of the subscapular, the largest branch of the axillary artery). Small, dorsal branches of posterior intercostal arteries also anastomose with the dorsal scapular and, thereby, contribute to this circulation.

7. **Answer B.** Palmar branches of the median and ulnar nerves supply the proximal portion of the palm. The palmar branches arise in the distal forearm and course in subcutaneous tissue to reach the palm. Therefore, these nerves do not pass through the carpal tunnel or ulnar canal (Guyon) and are at risk for injury with superficial laceration. In this patient, the skin of the thenar eminence (i.e., lateral palm) is involved, indicating that the **palmar branch of median nerve** is damaged.

8. **Answer D.** Skin on the posterior aspect of the hand is supplied by the three nerves, and the posterolateral aspect is supplied by the **superficial branch of radial nerve**. The posteromedial aspect is supplied by the ulnar nerve, and the distal phalanges of the lateral 3½ fingers is supplied by the median nerve.

9. **Answer A.** As the **subclavian vein** arches over the first rib, it lies posterior and inferior to the clavicle. The vein is partially protected by the subclavius muscle, which attaches to the inferior surface of the clavicle. Clavicular fracture may compress the vein between the clavicle and first rib, or cause its rupture due to the fascial connections between the vein, the subclavius muscle, and the clavicle.

10. **Answer D.** The finger movements involved in this test are abduction of the middle finger and adduction of the index finger (the reference line for abduction/adduction of fingers 2-5 passes through the middle finger). In this patient, you detect weakened actions of the **first palmar interosseous** (adduction of the index finger) **and second dorsal interosseous** (abduction of the middle finger). A mnemonic for the actions of the interosseous muscles of the hand is DAB (dorsal abducts) and PAD (palmar adducts).

11. **Answer E.** The **superficial palmar arterial arch lies more distal**. The superficial arch is formed mainly by the superficial branch of the ulnar artery. It courses immediately deep to the palmar aponeurosis and anastomoses with the superficial branch of the radial artery.

12. **Answer C.** The **scaphoid bone**, the most commonly fractured carpal bone, lies in the anatomical snuffbox. Scaphoid fractures may result from a fall on the outstretched hand (FOOSH) or from a direct blow to the palm.

Chapter 10: Lower Limb

1. **Answer D.** The **superficial fibular nerve** provides sensory innervation to most of the dorsum of the foot. A small area of skin between the 1st and 2nd metatarsals and the adjacent sides of those same toes is supplied by deep fibular nerve.

2. **Answer A.** Forced inversion of the foot places increased stress on the **calcaneofibular ligament** as the lateral malleolus and calcaneus are forced apart.

3. **Answer E.** When the popliteal artery enters the popliteal fossa, it courses parallel with, and adjacent to, the **tibial nerve**.

4. **Answer B.** The injured nerve is the deep fibular, which supplies muscles of the anterior leg compartment and on the dorsum of the foot. A primary action for the anterior compartment muscles is **dorsiflexion**.

5. **Answer A.** The obturator nerve innervates most **adductor muscles of the hip**. While pectineus is innervated by femoral nerve, detecting weakened hip adduction by this muscle would be difficult due to the comparative strength of the other adductor muscles. Pectineus may have dual innervation: obturator and femoral.

6. **Answer E.** The inferior gluteal region is within the **S3** dermatome.

7. **Answer B.** An extracapsular vascular ring is derived from the medial and lateral circumflex femoral arteries. From this vascular ring, a series of small retinacular arteries pierce the joint capsule to provide blood to the femoral neck and head. An acetabular artery, which also supplies the femoral head, is derived from either the obturator or medial circumflex femoral artery. The **deep circumflex iliac artery** does not supply the femoral head.

8. **Answer E.** A primary function of **the posterior cruciate ligament** is to prevent posterior translation (gliding) of the tibia on the femoral condyles. Conversely, if the femur is the reference bone, then the posterior cruciate ligament prevents the femoral condyles from gliding anteriorly on the tibial plateau.

9. **Answer A.** Intermittent claudication results from **peripheral arterial (vascular) disease** in which sclerotic buildup in arteries decreases perfusion to tissues. Increased perfusion is needed when muscles work and the inability to supply adequate oxygen produce ischemia and resultant pain.

10. **Answer C.** The proper angle of the neck of the femur as it connects the head to the shaft is essential for proper posture and gait. On average, the angle is 135°, but must be adjusted for age and sex. **Angulations of less than 120°** may cause coxa vara, whereas increased degrees of angulation may lead to coxa valga.

References & Suggested Readings

GENERAL REFERENCES

Cahill DR. *Lachman's Case Studies in Anatomy*. 4th ed. New York, NY: Oxford University Press; 1997.

Clemente CD. *Anatomy: A Regional Atlas of the Human Body*. 6th ed. Philadelphia, PA: Lippincott Williams & Wilkins; 2011.

Drake RL, Vogl AW, Mitchell AWM. *Gray's Anatomy for Students*. 2nd ed. Philadelphia, PA: Elsevier; 2010.

Ellis H. *Clinical Anatomy: Applied Anatomy for Students and Junior Doctors*. 11th ed. Malden, MA: Blackwell Publishing; 2006.

Federative Committee on Anatomical Terminology. *Terminologica Anatomica*. New York, NY: Thieme Medical Publishers; 1998.

Gilroy AM, MacPherson BR, Ross LR. *Atlas of Anatomy*. 2nd ed. (based on work of Schuenke M, Schulte E, Schumacher U). New York, NY: Thieme Medical Publishers; 2012.

Hansen JT. *Netter's Clinical Anatomy*. 2nd ed. Philadelphia, PA: Elsevier; 2009.

Moore KL, Dalley AF, Agur AMR. *Clinically Oriented Anatomy*. 6th ed. Philadelphia, PA: Lippincott Williams & Wilkins; 2010.

Morton D, Albertine K, Foreman B. *The Big Picture: Gross Anatomy*. Chicago, IL: McGraw-Hill; 2011.

Netter F. *Atlas of Human Anatomy*. 5th ed. Philadelphia, PA: Elsevier; 2011.

Rosse C, Gaddum-Rosse P. *Hollinshead's Textbook of Anatomy*. 5th ed. Philadelphia, PA: Lippincott–Raven; 1997.

Schuenke M, Schulte E, Schumacher U. *THIEME Atlas of Anatomy: General Anatomy and Musculoskeletal System*. Ross LR, Lamperti ED, consulting eds. Telger T, trans. New York, NY: Thieme Medical Publishers; 2006.

Schuenke M, Schulte E, Schumacher U. *THIEME Atlas of Anatomy: Neck and Internal Organs*. Ross LR, Lamperti ED, consulting eds. Telger T, trans. New York, NY: Thieme Medical Publishers; 2006.

Schuenke M, Schulte E, Schumacher U. *THIEME Atlas of Anatomy: Neck and Neuroanatomy*. Ross LR, Lamperti ED, Taub E., consulting eds. Telger T, trans. New York, NY: Thieme Medical Publishers; 2007.

Seidel HM, Ball JW, Dains JE, Benedict GW. *Mosby's Guide to Physical Examination*. Philadelphia, PA: Elsevier; 2006.

Slaby FJ, McCune SK, Summers RW. *Gross Anatomy in the Practice of Medicine*. Philadelphia, PA: Lea & Febiger; 1994.

Standring S. *Gray's Anatomy: The Anatomical Basis of Clinical Practice*. 40th ed. Philadelphia, PA: Elsevier; 2008.

Stedman's Medical Dictionary for the Health Professions and Nursing. 5th ed. Philadelphia, PA: Lippincott Williams & Wilkins; 2005.

Wilson-Pauwels L, Akesson EJ, Stewart PA. *Cranial Nerves: Anatomy and Clinical Comments*. Philadelphia, PA: B.C. Decker, Inc.; 1988.

SUGGESTED CLINICAL READINGS

Chapter 2: Thorax

Bashore TM, Granger CB, Hranitzky P, Patel MR. Acute inflammatory pericarditis. In: McPhee SJ, Papadakis MA, eds. *Current Diagnosis and Treatment*. 51st ed. New York, NY: McGraw-Hill; 2012:chap 10.

Bayés-de-Luna A, Goldwasser D, Fiol M, Bayés-Genis A. Surface electrocardiography. In: Fuster V, Walsh RA, Harrington RA, eds. *Hurst's the Heart*. 13th ed. New York, NY: McGraw-Hill; 2011.

Chesnutt M. Pulmonary disorders. In: McPhee SJ, Papadakis MA, eds. *Current Medical Diagnosis and Treatment*. 51st ed. New York, NY: McGraw-Hill; 2012:chap 9.

Chiles C, Gulla SM. Radiology of the chest. In: Chen MYM, Pope TL, Ott DJ, eds. *Basic Radiology*. 2nd ed. New York, NY: McGraw-Hill; 2011.

Eberli FR, Russi EW. Chest pain. In: Siegenthaler W, ed. *Siegenthaler's Differential Diagnosis in Internal Medicine*. 1st ed. New York, NY; Thieme; 2007:chap 6.

Euhus DM. Breast disease. In: Schorge JO, Schaffer JI, Halvorson LM, Hoffman BL, Bradshaw KD, Cunningham G, eds. *Williams Gynecology*. 2nd ed. New York, NY: McGraw-Hill; 2008:chap 12.

Giuliano AE, Hurvitz SA. Breast disorders. In: McPhee SJ, Papadakis MA, eds. *Current Diagnosis and Treatment*. 51st ed. New York, NY: McGraw-Hill; 2012:chap 17.

Goldblatt D, O'Brien KL. Pneumococcal infections. In: Longo DL, Fauci AS, Kasper DL, Hauser SL, Jameson JL, Loscalzo J, eds. *Harrison's Principles of Internal Medicine*. 18th ed. New York, NY: McGraw-Hill; 2012.

Horn L, Pao W, Johnson DH. Neoplasms of the lung. In: Longo DL, Fauci AS, Kasper DL, Hauser SL, Jameson JL, Loscalzo J, eds. *Harrison's Principles of Internal Medicine*. 18th ed. New York, NY: McGraw-Hill; 2012.

Joshi N. The third heart sound. *Southern Med J*. 1999;92:756-791.

LeBlond RF, Brown DD, DeGowin RL. The chest: chest wall, pulmonary and cardiovascular systems; the breasts. In: LeBlond RF, Brown DD, DeGowin RL, eds. *DeGowin's Diagnostic Examination*. 9th ed. New York, NY: McGraw-Hill; 2009.

McSweeney JC, Cody M, O'Sullivan P, Elberson K, Moser DK, Garvin BJ. Women's early warning symptoms of acute myocardial infarction. *Circulation*. 2003;108:2619-2623.

Poggi MM, Harney, K. The breast. In: DeCherney AH, Nathan L, eds. *Current Diagnosis and Treatment: Obstetrics & Gynecology*. 10th ed. New York, NY: McGraw-Hill; 2012:chap 63.

Raviglione MC, O'Brien RJ. Tuberculosis. In: Longo DL, Fauci AS, Kasper DL, Hauser SL, Jameson JL, Loscalzo J, eds. *Harrison's Principles of Internal Medicine*. 18th ed. New York, NY: McGraw-Hill; 2012.

Wang S. Bronchogenic Carcinoma. In: McPhee SJ, Papadakis MA, eds. *Current Diagnosis and Treatment*. 51st ed. New York, NY: McGraw-Hill; 2012:chap 39.

Yeh ETH, Bickford CL, Ewer MS. The diagnosis and management of cardiovascular disease in patients with cancer. In: Fuster V, Walsh RA, Harrington RA, eds. *Hurst's the Heart*. 13th ed. New York, NY: McGraw-Hill; 2011.

Chapter 3: Abdomen

Asplin JR, Coe FL, Favus M. Nephrolithiasis. In: Longo DL, Fauci AS, Kasper DL, Hauser SL, Jameson JL, Loscalzo J, eds. *Harrison's Principles of Internal Medicine*. 18th ed. New York, NY: McGraw-Hill; 2012.

Bacon BR. Cirrhosis and its complications. In: Longo DL, Fauci AS, Kasper DL, Hauser SL, Jameson JL, Loscalzo J, eds. *Harrison's Principles of Internal Medicine*. 18th ed. New York, NY: McGraw-Hill; 2012.

Bjerke HS, Bjerke JS. Splenic rupture. http://emedicine.medscape.com/article/432823-overview. Accessed June 21, 2012.

Bonheur JL, Ells PF. Biliary obstruction. http://emedicine.medscape.com/article/187001-overview. Accessed June 21, 2012.

Conwell DL, Wu B, Banks PA. Chronic pancreatitis. In: Greenberger NJ, Blumberg RS, Burakoff R, eds. *Current Diagnosis and Treatment: Gastroenterology, Hepatology and Endoscopy*. 2nd ed. New York, NY: McGraw-Hill; 2012.

Coomes J, Platt M. Abdominal pain. In: Tintinalli JE, Stapczynski JS, Cline DM, Ma OJ, Cydulka RK, Meckler GD, eds. *Tintinalli's Emergency Medicine: A Comprehensive Study Guide*. 7th ed. New York, NY: McGraw-Hill; 2011.

Craig S, Incescu L, Taylor CR. Appendicitis. http://emedicine.medscape.com/article/773895-overview. Accessed June 21, 2012.

DelValle J. Peptic ulcer disease and related disorders. In: Longo DL, Fauci AS, Kasper DL, Hauser SL, Jameson JL, Loscalzo J, eds. *Harrison's Principles of Internal Medicine*. 18th ed. New York, NY: McGraw-Hill; 2012.

Dempsey DT. Stomach, gastritis and stress ulcer. In: Brunicardi FC, Andersen DK, Billiar TR, Dunn DL, Hunter JG, Matthews JB, et al, eds. *Schwartz's Principles of Surgery*. 9th ed. New York, NY: McGraw-Hill; 2010.

Fulop T, Shoff WH, Green-McKinzie J, Edwards C, Behrman AJ, Shepherd SM. Acute pyelonephritis. http://emedicine.medscape.com/article/245559-overview. Accessed June 21, 2012.

Henry PH, Longo DL. Enlargement of lymph nodes and spleen. In: Longo DL, Fauci AS, Kasper DL, Hauser SL, Jameson JL, Loscalzo J, eds. *Harrison's Principles of Internal Medicine*. 18th ed. New York, NY: McGraw-Hill; 2012.

Lee SL, DuBois JJ, Shekherdimian S. Hydrocele. http://emedicine.medscape.com/article/438724-overview. Accessed June 21, 2012.

Ma OJ, Reardon RF, Sabbaj A. Emergency ultrasonography. In: Tintinalli JE, Stapczynski JS, Cline DM, Ma OJ, Cydulka RK, Meckler GD, eds. *Tintinalli's Emergency Medicine: A Comprehensive Study Guide*. 7th ed. New York, NY: McGraw-Hill; 2011.

Mahoney LK, Doty CI. Rib fracture. http://emedicine.medscape.com/article/825981-overview. Accessed June 21, 2012.

Manthey DE, Nicks BA. Urologic stone disease. In: Tintinalli JE, Stapczynski JS, Cline DM, Ma OJ, Cydulka RK, Meckler GD, eds. *Tintinalli's Emergency Medicine: A Comprehensive Study Guide*. 7th ed. New York, NY: McGraw-Hill; 2011.

Melville SC, Melville DE. Abdominal trauma. In: Tintinalli JE, Stapczynski JS, Cline DM, Ma OJ, Cydulka RK, Meckler GD, eds. *Tintinalli's Emergency Medicine: A Comprehensive Study Guide*. 7th ed. New York, NY: McGraw-Hill; 2011.

Mortele KJ. State-of-the-art imaging of the gastrointestinal system. In: Greenberger NJ, Blumberg RS, Burakoff R, eds. *Current Diagnosis & Treatment: Gastroenterology, Hepatology, & Endoscopy*. 2nd ed. New York, NY: McGraw-Hill; 2012.

Mulherjee S, Sepulveda S. Chronic gastritis. http://emedicine.medscape.com/article/176156-overview. Accessed June 21, 2012.

Schorge JO, Schaffer JI, Pietz J, et al. Pelvic mass. In: Schorge JO, Schaffer JI, Pietz J, Halvorson LM, Hoffman BL, Bradshaw KD, et al, eds. *Williams Gynecology*. New York, NY: McGraw-Hill; 2012.

Shepherd SM, Shoff WH, Behman AJ. Pelvic inflammatory disease. In: Tintinalli JE, Stapczynski JS, Cline DM, Ma OJ,

Cydulka RK, Meckler GD, eds. *Tintinalli's Emergency Medicine: A Comprehensive Study Guide*. 7th ed. New York, NY: McGraw-Hill; 2011.

Sherman V, Macho JR, Brunicardi CF. Inguinal hernias. In: Brunicardi FC, Andersen DK, Billiar TR, Dunn DL, Hunter JG, Matthews JB, et al, eds. *Schwartz's Principles of Surgery*. 9th ed. New York, NY: McGraw-Hill; 2010.

Tablang MVF, Grupka MJ, Wu GY. Viral gastroenteritis. http://emedicine.medscape.com/article/176515-overview. Accessed June 21, 2012.

Wu B, Conwell DL, Banks PA. Acute pancreatitis. In: Greenberger NJ, Blumberg RS, Burakoff R, eds. *Current Diagnosis and Treatment: Gastroenterology, Hepatology and Endoscopy*. 2nd ed. New York, NY: McGraw-Hill; 2012.

Chapter 4: Pelvis

American Cancer Society. Endometrial (uterine) cancer. www.cancer.org/cancer/endometrialcancer/detailedguide/endometrial-uterine-cancer. Accessed June 21, 2012.

Cappell MS. From colonic polyps to colon cancer: pathophysiology, clinical presentation, and diagnosis. *Clin Lab Med*. 2005;25:135.

Chang GJ, Shelton AA, Welton ML. Large Intestine. In: Doherty GM, ed. *Current Diagnosis & Treatment: Surgery*. 13th ed. New York, NY: McGraw-Hill; 2010.

Compton C, Hawk E, Grochow L, Lee F Jr, Ritter M, Nieederhuber JE. Colon cancer. In: Abeloff MD, Armitage JO, Niederhuber JE, Kastan MB, McKenna WG, eds. *Abeloff's Clinical Oncology*. 4th ed. Philadelphia, PA: Elsevier; 2008.

Cooper CS, Joudi FN, Williams RD. Urology. In: Doherty GM, ed. *Current Diagnosis & Treatment: Surgery*. 13th ed. New York, NY: McGraw-Hill; 2010.

Corn P, Logothetis. Prostate cancer. In: Kantarkian HM, Wolff RA, Koller CA, eds. *The MD Anderson Manual of Medical Oncology*. 2nd ed. New York, NY: McGraw-Hill; 2011.

Craig S, Incescu L, Taylor CR. Appendicitis. http://emedicine.medscape.com/article/773895-overview. Accessed June 21, 2012.

Deng DY. Urinary incontinence in women. *Med Clin North Am*. 2011;95:101-109.

Deters LA, Constabile RA, Leveillee RJ, Moore CR, Patel VR. Benign prostatic hypertrophy. http://emedicine.medscape.com/article/437359-overview. Accessed June 21, 2012.

Dubois RN. Neoplasms of the large and small intestine. In: Goldman L, Ausiello D, eds. *Cecil Medicine*. 24th ed. Philadelphia, PA: Elsevier; 2008.

Friedman S, Blumberg RD. Inflammatory bowel disease. In: Longo DL, Fauci AS, Kasper DL, Hauser SL, Jameson JL, Loscalzo J, eds. *Harrison's Principles of Internal Medicine*. 18th ed. New York, NY: McGraw-Hill; 2012.

Heidelbaugh JJ. Gastroenterology. In: Rakel RE, Rakel DP, eds. *Textbook of Family Medicine*. 8th ed. Philadelphia, PA: Elsevier; 2011.

Helm CW. Ovarian cysts. http://emedicine.medscape.com/article/255865-overview. Accessed June 21, 2012.

Holschneider CH. Surgical diseases and disorders in pregnancy. In: DeCherney AH, Nathan L, eds. *Current Diagnosis & Treatment Obstetrics & Gynecology*. 10th ed. New York, NY: McGraw-Hill; 2007.

Krupski TL, Theodorescu D. Prostate cancer. http://emedicine.medscape.com/article/1967731-overview. Accessed June 21, 2012.

LeBlond RF, Brown DD, DeGowin RL. *DeGowin's Diagnostic Examination*. 9th ed. New York, NY: McGraw-Hill; 2009.

Mayer RJ. Gastrointestinal tract cancer. In: Longo DL, Fauci AS, Kasper DL, Hauser SL, Jameson JL, Loscalzo J, eds. *Harrison's Principles of Internal Medicine*. 18th ed. New York, NY: McGraw-Hill; 2012.

Mayo Clinic Staff. Cervical cancer. http://www.mayoclinic.com/health/cervical-cancer/ds00167. Accessed June 21, 2012).

National Cancer Institute. Stages of cervical cancer. http://cancer.gov/cancertopics/pdq/treatment/cervical/patient/page2. Accessed June 21, 2012.

Roehrborn CG. Male lower urinary tract symptom (LUTS) and benign prostatic hyperplasia (BPH). *Med Clin North Am*. 2011;95:87.

Rogers VL, Worley KC. Obstetrics and obstetric disorders. In: McPhee SJ, Papadakis MA, Rabow MW, eds. *Current Medical Diagnosis & Treatment 2012*. 51st ed. New York, NY: McGraw-Hill; 2012.

Schorge JO, Schaffer JI, Pietz J, et al. Ectopic pregnancy. In: Schorge JO, Schaffer JI, Pietz J, Halvorson LM, Hoffman BL, Bradshaw KD, et al, eds. *Williams Gynecology*. New York, NY: McGraw-Hill; 2012.

Shepherd SM, Shoff WH, Behman AJ. Pelvic inflammatory disease. In: Tintinalli JE, Stapczynski JS, Cline DM, Ma OJ, Cydulka RK, Meckler GD, eds. *Tintinalli's Emergency Medicine: A Comprehensive Study Guide*. 7th ed. New York, NY: McGraw-Hill; 2011.

Stern JL. Uterus: Women's Cancer Information Center: Endometrial carcinoma. http://www.womenscancercenter.com/info/types/new/uterus.html. Accessed June 21, 2012.

Stern JL. Women's Cancer Information Center: Cervix. http://www.womenscancercenter.com/info/types/old/cervix.html. Accessed June 21, 2012.

Weitz J, Koch M, Debus J, Höhler T, Galle PR, Büchler MW. Colorectal cancer. *Lancet*. 2005;365:153.

Chapter 5: Perineum

Gupta K, Trautner BW. Urinary tract infections, pyelonephritis, and prostatitis. In: Longo DL, Fauci AS, Kasper DL, Hauser SL, Jameson JL, Loscalzo J, eds. *Harrison's Principles of Internal Medicine*. 18th ed. New York, NY: McGraw-Hill; 2012.

Hancock DB. Anal fissures and fistulas. *BMJ*. 1992;304:904.

Hebra A. Perianal abscess. *http://emedicine.medscape.com/article/191975-overview*. Accessed June 21, 2012.

Labat JJ, Riant T, Robert R, Amarenco G, Lefaucheur JP, Rigaund J. Diagnostic criteria for pudendal neuralgia by pudendal nerve entrapment (Nantes criteria). *Neurourol Urodyn.* 2008;27(4):306-310.

Legall I. Anal fistulas and fissures. *http://emedicine.medscape.com/article/776150-overview.* Accessed June 21, 2012.

Mansoor M, Weston LA. Perianal infections: a primer for nonsurgeons. *Curr Gastroenterol Rep.* 2010;12:270.

Martinez JM, Honsik K. Bicycle seat neuropathy. *http://emedicine.medscape.com/article/91896-overview.* Accessed June 21, 2012.

Meng MV, Stoller ML, Walsh TJ. Urologic disorders. In: McPhee SJ, Papadakis MA, Rabow MW, eds. *Current Medical Diagnosis & Treatment 2012.* 51st ed. New York, NY: McGraw-Hill; 2012.

Poritz LS. Anal fissure. *http://emedicine.medscape.com/article/196297-overview.* Accessed June 21, 2012.

Stanley JA, Hough DM, Pawlina W, Spinner RJ. Anatomical basis of chronic pelvic pain syndrome: the ischial spine and pudendal nerve entrapment. (Departments of Urology, Radiology, Anatomy, and Neurologic Surgery, Mayo Clinic, Rochester, Minnesota). *http://www.chronicprostatitis.com/pne.html.* Accessed June 21, 2012.

Tozer PJ, Burling D, Gupta A, Phillips RKS, Hart AL. Review article: medical, surgical and radiological management of perianal Crohn's fistulas. *Aliment Pharmacol Ther.* 2011; 33:5.

Chapter 6: Neck

Aithal JK, Ulrich M. Subclavian steal syndrome. *N Eng J Med.* 2010;363:e15.

Alcevedo JL, Shah RK. Pediatric retropharyngeal abscess. http://emedicine.medscape.com/article/995851-overview Accessed June 21, 2012.

Boon JM, Van Schoor AN, Abrahams PH, Meiring JH, Welch T, Shanahan D. Central venous catheterization – an anatomical review of a clinical skill, part 1: subclavian vein via the infraclavicular approach. *Clin Anat.* 2007;20:602.

Boon JM, Van Schoor AN, Abrahams PH, Meiring JH, Welch T. Central venous catheterization – an anatomical review of a clinical skill, part 2: internal jugular vein via the supraclavicular approach. *Clin Anat.* 2008;21:15.

Corbett SW, Stack LB, Knoop KJ. Chest and abdomen. In: Knoop KJ, Stack LB, Storrow AB, Thurman RJ, eds. *The Atlas of Emergency Medicine.* 3rd ed. New York, NY: McGraw-Hill; 2010.

Gomella LG, Haist SA. *Clinician's Pocket Reference.* 11th ed. New York, NY: McGraw-Hill; 2007.

Graham AS, Ozment C, Tegtmeyer K, Lai S, Braner DAV. Central venous catheterization. *NEJM.* 2007;356:e21.

Gunn JD. Stridor and drooling. In: Tintinalli JE, Stapczynski JS, Cline DM, Ma OJ, Cydulka RK, Meckler GD, eds. *Tintinalli's Emergency Medicine: A Comprehensive Study Guide.* 7th ed. New York, NY: McGraw-Hill; 2011.

Marra S, Hotaling AJ. Deep neck infections. *Am J Otolaryngol.* 1996;17:287.

McConnville JF, Kress JP. Intravascular devices. In: Hall JB, Schmidt GA, Wood LDH, eds. *Principles of Critical Care.* 3rd ed. New York, NY: McGraw-Hill; 2005.

McIntyre KE. Subclavian steal syndrome. http://emedicine.medscape.com/article/462036-overview Accessed June 6, 2012.

Pollard H, Rigby S, Moritz G, Lau C. Subclavian steal syndrome. *Australas Chiropr Osteopathy.* 1998;7:20.

Pringle E, Graham EM. Ocular disorders associated with systemic diseases. In: Riordan-Eva P, Cunningham ET, eds. *Vaughan & Asbury's General Ophthalmology.* 18th ed. New York, NY: McGraw-Hill; 2011.

Shah S, Sharieff GQ. Pediatric respiratory infections. *Emerg Med Clin N Am.* 2007;25:961.

Stephan M, Carter C, Ashfaq S. Pediatric emergencies. In: Stone CK, Humphries RL, eds. *Current Diagnosis & Treatment: Emergency Medicine.* 7th ed. New York, NY: McGraw-Hill; 2011.

Wolff K, Goldsmith LA, Katz SI, Gilchrest BA, Paller AS, Leffell DJ. *Fitzpatrick's Dermatology in General Medicine.* 7th ed. New York, NY: McGraw-Hill; 2008.

Chapter 7: Head

Albertini JG, Marks VJ, Cronin H. Temporal (giant cell) arteritis. *http://emedicine.medscape.com/article/1084911-overview.* Accessed June 21, 2012.

Amirlak B, Chim HWM, Chen EH, Stepnick DW. Malignant parotid tumors. *http://emedicine.medscape.com/article/1289616-overview.* Accessed June 21, 2012.

Beal MF, Hauser SL. Trigeminal neuralgia, Bell's palsy, and other cranial nerve disorders. In: Longo DL, Fauci AS, Kasper DL, Hauser SL, Jameson JL, Loscalzo J, eds. *Harrison's Principles of Internal Medicine.* 18th ed. New York, NY: McGraw-Hill; 2012.

Bullock JD, Warwar RE, Ballal DR, Ballal RD. Mechanisms of orbital floor fractures: a clinical, experimental, and theoretical study. *Trans Am Ophth Soc.* 1999;97:87.

Cohen AJ, Mercandetti M. Orbital floor fractures (blowout). *http://emedicine.medscape.com/article/1284026-overview.* Accessed June 21, 2012.

Dubner S. Benign parotid tumors. *http://emedicine.medscape.com/article/1289560-overview.* Accessed June 21, 2012.

Gibbons D, Pisters KM, Johnson F. Non-small cell lung cancer. In: Kantarkian HM, Wolff RA, Koller CA, eds. *The MD Anderson Manual of Medical Oncology.* 2nd ed. New York, NY: McGraw-Hill; 2011.

Goodwin J. Oculomotor nerve palsy. *http://emedicine.medscape.com/article/1198462-overview.* Accessed June 21, 2012.

Heitz CR. Face and jaw emergencies. In: Tintinalli JE, Stapczynski JS, Cline DM, Ma OJ, Cydulka RK, Meckler GD, eds. *Tintinalli's Emergency Medicine: A Comprehensive Study Guide.* 7th ed. New York, NY: McGraw-Hill; 2011.

Johnson J, Lalwani AK. Vestibular schwannoma (acoustic neuroma). In: Lalwani AK, ed. *Current Diagnosis & Treatment in Otolaryngology—Head & Neck Surgery.* 3rd ed. New York, NY: McGraw-Hill; 2012.

Kattah JG, Pula JH. Cavernous sinus syndromes. *http://emedicine.medscape.com/article/1161710-overview*. Accessed June 21, 2012.

Langford CA, Fauci AS. The vasculitis syndromes. In: Longo DL, Fauci AS, Kasper DL, Hauser SL, Jameson JL, Loscalzo J, eds. *Harrison's Principles of Internal Medicine*. 18th ed. New York, NY: McGraw-Hill; 2012.

Loh KY, Yushak AW. Virchow's node (Troisier's sign). *N Eng J Med*. 2007;357:282.

Lustig LR, Niparko JK. Disorders of the facial nerve. In: Lalwani AK, ed. *Current Diagnosis & Treatment in Otolaryngology—Head & Neck Surgery*. 3rd ed. New York, NY: McGraw-Hill; 2012.

McClay JE. Nasal polyps. *http://emedicine.medscape.com/article/994274-overview*. Accessed June 21, 2012.

Meagher R, Young WF. Subdural hematoma. *http://emedicine.medscape.com/article/1137207-overview*. Accessed June 21, 2012.

Munter DW, McGuirk TD. Head and facial trauma. In: Knoop KJ, Stack LB, Storrow AB, Thurman RJ, eds. *The Atlas of Emergency Medicine*. 3rd ed. New York, NY: McGraw-Hill; 2010.

Nguyen QA. Epistaxis. *http://emedicine.medscape.com/article/863220-overview*. Accessed June 21, 2012.

Parmar MS. Horner syndrome. *http://emedicine.medscape.com*. Accessed June 21, 2012.

Riordan-Eva P, Hoyt WF. Neuro-ophthalmology. In: Riordan-Eva P, Cunningham ET, eds. *Vaughan & Asbury's General Ophthalmology*. 18th ed. New York, NY: McGraw-Hill; 2011.

Ropper AH, Samuels MA. *Adams & Victor's Principles of Neurology*. 9th ed. New York, NY: McGraw-Hill; 2009.

Ropper AH. Concussion and other head injuries. In: Longo DL, Fauci AS, Kasper DL, Hauser SL, Jameson JL, Loscalzo J, eds. *Harrison's Principles of Internal Medicine*. 18th ed. New York, NY: McGraw-Hill; 2012.

Simon RP, Greenberg DA, Aminoff MJ. *Clinical Neurology*. 7th ed. New York, NY: McGraw-Hill; 2009.

Simon RP, Greenberg DA, Aminoff MJ. Headache and facial pain. In: Simon RP, Greenberg DA, Aminoff MJ. *Clinical Neurology*. 7th ed. New York, NY: McGraw-Hill; 2009.

Simon RP, Greenberg DA, Aminoff MJ. Stroke. In: Simon RP, Greenberg DA, Aminoff MJ. *Clinical Neurology*. 7th ed. New York, NY: McGraw-Hill; 2009.

Summers SM, Bey T. Epistaxis, nasal fractures, and rhinosinusitis. In: Tintinalli JE, Stapczynski JS, Cline DM, Ma OJ, Cydulka RK, Meckler GD, eds. *Tintinalli's Emergency Medicine: A Comprehensive Study Guide*. 7th ed. New York, NY: McGraw-Hill; 2011.

Wein RO, Chandra RK, Weber RS. Disorders of the head and neck. In: Brunicardi FC, Andersen DK, Billiar TR, Dunn DL, Hunter JG, Matthews JB, et al. eds. *Schwartz's Principles of Surgery*. 9th ed. New York, NY: McGraw-Hill; 2010.

Wright DW, Merck LH. Head trauma in adults and children. In: Tintinalli JE, Stapczynski JS, Cline DM, Ma OJ, Cydulka RK, Meckler GD, eds. *Tintinalli's Emergency Medicine: A Comprehensive Study Guide*. 7th ed. New York, NY: McGraw-Hill; 2011.

Chapter 8: Back

Chakraverty R, Pynsent P, Isaacs K. Which spinal levels are identified by palpation of the iliac crests and the posterior superior iliac spines? *J Anat*. 2007; 210:232.

Chawla J, Raghavendra M. Epidural nerve block. *http://emedicine.medscape.com/article/149646-overview*. Accessed June 21, 2012.

Hawkins JL. Epidural analgesia for labor and delivery. *N Eng J Med*. 2010;362:1503.

Hu SS, Tribus CB, Tay BK-B, Bhatia NN. Disorders, diseases, and injuries of the spine. In: Skinner HB, ed. *Current Diagnosis & Treatment in Orthopedics*. 4th ed. New York, NY: McGraw-Hill; 2006.

Mehlman CT. Idiopathic scoliosis. *http://emedicine.medscape.com/article/1265794-overview*. Accessed June 21, 2012.

PubMed Health. Kyphosis. http://www.ncbi.nlm.nih.gov/pubmedhealth/PMH0002220/. Accessed June 21, 2012.

PubMed Health. Lordosis. *http://www.ncbi.nlm.nih.gov/pubmedhealth/PMH0002220/*. Accessed June 21, 2012.

PubMed Health. Scoliosis. http://www.ncbi.nlm.nih.gov/pubmedhealth/PMH0002220/. Accessed June 21, 2012.

Robinson CM, McMaster MJ. Juvenile idiopathic scoliosis. *J Bone Joint Surg*. 1996;78A:1140.

Weiss H-R. Adolescent idiopathic scoliosis – a case report of a patient with clinical deterioration after surgery. *Patient Saf Surg*. 2007;7:1.

Chapter 9: Upper Limb

Amirlak B, Tabbal GN, Upadhyaya KP, et al. Median nerve entrapment. *http://emedicine.medscape.com/article/1242387-overview*. Accessed June 21, 2012.

Ashworth NL. Carpal tunnel syndrome. *http://emedicine.medscape.com/article/327330-overview*. Accessed June 21, 2012.

DeFranco MJ, Lawton JN. Radial nerve injuries associated with humeral fractures. *J Hand Surg*. 2006;31A:655.

Escarza R, Loeffel MF, Uehara DT. Wrist injuries. In: Tintinalli JE, Stapczynski JS, Cline DM, Ma OJ, Cydulka RK, Meckler GD, eds. *Tintinalli's Emergency Medicine: A Comprehensive Study Guide*. 7th ed. New York, NY: McGraw-Hill; 2011.

Estephan A, Gore RJ. Clavicle fracture in emergency medicine. *http://emedicine.medscape.com/article/824564-overview*. Accessed June 21, 2012.

Hak DJ. Radial nerve palsy associated with humeral shaft fractures. *Orthopedics*. 2009;32:111.

Hariri S, McAdams TR. Nerve injuries about the elbow. *Clin Sports Med*. 2010;29:655.

Mackinnin SE, Novak CB, Bartz ME. Brachial plexus injuries, obstetrical. *http://emedicine.medscape.com/article/1259437-overview*. Accessed June 21, 2012.

McMahon PJ, Kaplan LD. Sports medicine. In: Skinner HB, ed. *Current Diagnosis & Treatment in Orthopedics*. 4th ed. New York, NY: McGraw-Hill; 2006.

Mollberg M, Hagberg H, Bager B, Lilja H, Ladfors L. High birthweight and shoulder dystocia: the strongest risk factors for obstetrical brachial plexus palsy in a Swedish population-based study. *Acta Obstet Gynecol Scand.* 2005;84:654.

Palmer BA, Hughes TB. Cubital tunnel syndrome. *J Hand Surg.* 2012;35A:153.

Papp S. Carpal bone fractures. *Orthop Clin North Am.* 2007;38:251.

Pecci M, JB Kreher. Clavicle fractures. *Am Fam Physician.* 2008; 77:65–70.

Prybyla D, Owens BD, Goss TP. Acromioclavicular joint separations. *http://emedicine.medscape.com/article/1261906-overview.* Accessed June 21, 2012.

Raukar NP, Raukar GJ, Savitt DL. Extremity trauma. In: Knoop KJ, Stack LB, Storrow AB, Thurman RJ, eds. *The Atlas of Emergency Medicine.* 3rd ed. New York, NY: McGraw-Hill; 2010.

Rubino LJ, Lawless MW, Kleinhenz BP. Clavicle fractures. *http://emedicine.medscape.com/article/1260953-overview.* Accessed June 21, 2012.

Stern M. Radial nerve entrapment. *http://emedicine.medscape.com/article/1244110-overview.* Accessed June 21, 2012.

Trojian TH. Brachial plexus injury in sports medicine. *http://emedicine.medscape.com/article/91988-overview.* Accessed June 21, 2012.

Verheyden JR, Palmer AK. Cubital tunnel syndrome. http://emedicine.medscape.com/article/1231663-overview. Accessed June 21, 2012.

Chapter 10: Lower Limb

Birnbaum K, Prescher A, Hessler S, Heller KD. The sensory innervation of the hip joint. *Surg Radiol Anat.* 1999;19:371-375.

Cluett J. Acetabular fractures. *http://orthopedics.about.com/od/brokenbones/a/acetabulum.htm.* Accessed June 21, 2012.

Cluett J. Piriformis syndrome treatment. *http://orthopedics.about.com/cs/sprainsstrains/a/piriformis.html.* Accessed June 21, 2012.

Conly J. Soft tissue infections. In: Hall JB, Schmidt GA, Wood LDH, eds. *Principles of Critical Care.* 3rd ed. New York, NY: McGraw-Hill; 2005.

Raukar NP, Raikar GJ, Savitt S. Extremity trauma. In: Knoop KJ, Stack LB, Storrow AB, Thurman RJ, eds. *The Atlas of Emergency Medicine.* 3rd ed. New York, NY: McGraw-Hill; 2010.

Tubbs RJ, Savitt DL, Suner S. Extremity conditions. In: Knoop KJ, Stack LB, Storrow AB, Thurman RJ, eds. *The Atlas of Emergency Medicine.* 3rd ed. New York, NY: McGraw-Hill; 2010.

Rowdon GA, Agnew S, Goodman SB, et al. Chronic exertional compartment syndrome. *ttp://emedicine.medscape.com/article/88014-overview.* Accessed June 21, 2012.

Wheeless CR. Wheeless' Textbook of Orthopaedics. www.wheelessonline.com/ortho/blood_supply_to_femoral_head_neck. Accessed June 21, 2012.

Index